AA007447

2nd Global Congress on Construction, Material and Structural Engineering (GCoMSE 2019)

Towards Sustainable Structures and Construction

IOP Conference Series: Materials Science and Engineering
Volume 713

Melaka, Malaysia
26 - 27 August 2019

ISBN: 978-1-7138-2572-2
ISSN: 1757-8981

Printed from e-media with permission by:

Curran Associates, Inc.
57 Morehouse Lane
Red Hook, NY 12571

Some format issues inherent in the e-media version may also appear in this print version.

Institute of Physics
Dirac House, Temple Back
Bristol BS1 6BE UK

Phone: 44 1 17 929 7481
Fax: 44 1 17 920 0979

techtracking@iop.org

Additional copies of this publication are available from:

Curran Associates, Inc.
57 Morehouse Lane
Red Hook, NY 12571 USA
Phone: 845-758-0400
Fax: 845-758-2633
Email: curran@proceedings.com
Web: www.proceedings.com

TABLE OF CONTENTS

CONSTRUCTION MATERIAL AND TECHNOLOGY

LONG TERM EFFECT OF STRENGTH AND DURABILITY PERFORMANCE OF CONCRETE CONTAINING SULPHATE REDUCTION BACTERIA UNDER CHLORIDE CONDITION 1
J M Irwan, T Teddy, N Othman, F S Khalid

EXPERIMENTAL STUDY ON AFTER-MIXING TEMPERATURE CONTROL OF CONCRETE IN BATCHING PLANT PRODUCTION 7
Hooman Shokravi, Taleb Moradi Shaghaghi

STRENGTH OF CONCRETE CONTAINING SYNTHETIC WIRE WASTE AS FIBER MATERIALS 18
F S Khalid, S H Saaidin, S Shahidan, N H Othman, N A A Guntor

PERFORMANCE OF MODIFIED MORTAR CONTAINING EPOXY 24
Arisya Abdul Latip, Noorwirdawati Ali, Noor Azlina Abdul Hamid, Norhafizah Salleh, Siti Radziah Abdullah, Shahiron Shahidan

EFFECT OF LIGHTWEIGHT AGGREGATES ON EARLY-AGE BEHAVIOR OF HIGH PERFORMANCE CONCRETE 32
M Higazey, M J Shannag, A Alaskar

EFFECTS OF CURING CONDITIONS ON PROPERTIES OF LIGHTWEIGHT CONCRETE BRICK CONTAINING EXPANDED POLYSTYRENE AND PALM OIL FUEL ASH 40
N I M Yassin, S H Adnan, S Shahidan, S S Ayop, N A Kamarulzaman, M H Osman, Z Jamelloddin, M Majid, N Salleh, A Alisibramulisi, M N A Wahee Anuar

MONITORING EX SITU ELECTROKINETIC REMEDIATION (EKR) USING NITRIC ACID AS CHELATING AGENT 48
S A Ayuni, A T S Azhar, A Azim, M Aziman

PREPARATION OF SINTERED GLASS-CERAMIC FROM CRT GLASS AND RED GYPSUM WASTES FOR TILING APPLICATION 57
M M Idham, M M Haniza

DEVELOPMENT OF A MODEL FOR OTTV AND RTTV BASED ON BIMVPL TO OPTIMIZE THE ENVELOPE THERMAL PERFORMANCE 65
F Abass, L H Ismail, I A Wahab, A A Elgadi

SUSTAINABLE MATERIAL: DEVELOPMENT EXPERIMENT OF BAMBOO COMPOSITE THROUGH BIOLOGICALLY BINDING MECHANISM 73
I F Ridzqo, D Susanto, T H Panjaitan, N Putra

PROPERTIES AND FLEXURAL BEHAVIOR OF SELF-COMPACTING CONCRETE WITH ADDED COIR FIBER AND POFA 82
Afif Iman, N Mohamad, Fasya Raain, Anis Sofia, Azree Mydin, A A A Samad

ASSESSING THE PERFORMANCE OF BIO COMPOST AS SOIL MEDIA IN EXTENSIVE GREEN ROOF 90
Ming Fai Chow, Ahmad Aiman Bin Ahmad Azlan

COMBINING POZZOLANIC MATERIAL AS TERNARY BINDER IN RECYCLED AGGREGATE CONCRETE (RAC) TO DEVELOP A NEW SUSTAINABLE CONCRETE 97
M A Tambichik, A A A Samad, N Mohamad, M Z M Bosro, M. A. Iman

RESTORATION OF AGED BITUMEN PROPERTIES USING MALTENES 105
Z H Hussein, H Yaacob, M K Idham, N A Hassan, L J Choy, R P Jaya

CONSOLIDATION BEHAVIOUR OF DREDGED MARINE SOILS CONSIDERING THE INCORPORATION OF GRANULAR DRAINAGE LAYER 113
S F S M Johan, C-M Chan

STUDY ON THE EFFECTS OF PLASTIC AS ADMIXTURE ON THE MECHANICAL PROPERTIES OF CEMENT-SAND BRICKS 121
Ming Fai Chow, Muhammad Amin Khalili Rosidan

INFLUENCE OF DOLOMITE AND GRANITE WASTE CONTENT ON THE PROPERTIES OF ARTIFICIAL MARBLE 127
H Abdul Samad, R Abd Rashid

COMPARATIVE EVALUATION ON THE MOE BETWEEN EN, BS AND ASTM OF CONCRETE CONTAINING PET 135
Muhammad Amzar Othman, Shahiron Shahidan, Shamrul Mar Shamsudin, Noorwirdawati Ali, Sharifah Salwa Mohd Zuki, Faisal Sheikh Khalid, Fadzli Mohamed Nazri

DURABILITY EVALUATION OF CEMENT BOARD PRODUCED FROM UNTREATED AND PRE-TREATED EMPTY FRUIT BUNCH FIBRE THROUGH ACCELERATING AGEING 143
P Peter, N M Z Nik Soh, Z A Akasah, M A Mannan

INFLUENCE OF VARYING DENSITY ON THE MECHANICAL AND PHYSICAL PROPERTIES OF MEDIUM DENSITY FIBREBOARD (MDF) CONTAINING KENAF 152
Noorli Ismail, Muhammad Azzim Mohmed Jomali, Hasniza Abu Bakar, Tong Yean Ghing, Nik Mohd Zaini Nik Soh, Raden Maizatul Aimi Mohd Azam

THE INFLUENCE OF MOISTURE CONTENT, PH AND LIQUID LIMIT ON ELECTROKINETIC-TREATED DREDGED MARINE SEDIMENTS STABILITY BY USING CALCIUM CHLORIDE AS AN ENHANCEMENT AGENT 160
Nurul Syakeera Nordin, Chee-Ming Chan

CONCRETE, STEEL AND COMPOSITE STRUCTURE

BEHAVIOUR OF POST-TENSIONED KEMPAS TIMBER BEAM WITH TWO TENDON TYPES 169
W mushina, D Y Eng Chuan, j mushina, N Abd Ghafar, K Boon

VIBRATION BEHAVIOUR OF NATURAL TIMBER AND TIMBER CONCRETE COMPOSITE DECK SYSTEM 177
Jalal Mushina, NorHayati Abd Ghafar, David Yeoh, Wissam Mushina, Koh Heng Boon

STIFFNESS AND STRENGTH DEGRADATION OF TIMBER CONCRETE COMPOSITE UNDER FATIGUE LOADING 186
Kevin J.T Yeo, David E.C Yeoh

DYNAMIC RESPONSES OF SANDWICH CEMBOARD-FOAMED CONCRETE PANELS SUBJECTED TO LOW-VELOCITY IMPACT 198
Z M Jaini, S N Mokhatar, J Hadipramana, M Abu Seman

DYNAMIC SERVICEABILITY OF LIGHTWEIGHT COMPOSITE DECK AS FLOOR SYSTEM UNDER HUMAN EXCITATION 206
A F Norhalim, Z M Jaini, N Abd Ghafar, M A Majid, A Ahwang

SHEAR PERFORMANCE OF PRETENSIONED PRESTRESSED CONCRETE BEAM WITH STEEL FIBRE 214
Ruqayyah Ismail, Lyn Dee Goh, Nik Nurul Nadia Nik Hasan, Azerai Ali Rahman, Fariz Aswan Ahmad Zakwan

STRENGTH AND STRUCTURAL PERFORMANCE OF REINFORCED CONCRETE BEAM WITH ARTIFICIAL PLASTIC FIBRE (APF) 221
S N Mokhatar, N A N A Mutalib, A M A Budiea, A F Kamarudin, M S Md Noh

SHEAR CAPACITY OF REINFORCED CONCRETE BEAMS WITH SQUARE CROSS SECTION SUBJECTED TO BIAXIAL BENDING 228
R Thamrin, S Haris, E Dedi, E Dalmantias

MECHANICAL PROPERTIES OF SINGLE SHEAR PLANE OF BOLTED STEEL CONNECTION 235
A F Kamarudin, M K Musa, S N Mokhatar, T N Tuan Chik, S S Mohd Zuki, A Abu Bakar, J Hadipramana, H H Ahmad Johari

FINITE ELEMENT ANALYSIS OF COATED (INTUMESCENT COATING PROTECTION) CELLULAR STEEL BEAM (CSB) EXPOSE TO FIRE 243
F A A Zakwan, R R Krishnamoorthy, A Ibrahim, R Ismail

VIBRATION TEST ON EXISTING STEEL SHEET PILE 252
Sakhiah Abdul Kudus, Kunitomo Sugiura, Yasuo Suzuki

PRELIMINARY STUDY ON MODIFIED FOAMED CONCRETE-FILLED HOLLOW SECTION UNDER CYCLIC LOAD 259
Siti Amirah Azra Khairuddin, Norashidah Abd Rahman, Mohd Luqman Rossam, Anis Nazirah Mustafa

CONSTRUCTION MANAGEMENT

A STUDY ON EFFECTIVENESS OF BUILDING INFORMATION MODELLING (BIM) ON THE MALAYSIAN CONSTRUCTION INDUSTRY 266
S W R Kong, L T Lau, S Y Wong, D T Phan

FACTORS AFFECTING SAFETY PERFORMANCE OF CONSTRUCTION PROJECTS: A LITERATURE REVIEW 275
N H Abas, N Yusuf, N A Suhaini, N Kariya, H Mohammad, M F Hasmori

IMPROVEMENT FOR SIGNIFICANT CLAUSES IN THE STANDARD FORM OF CONTRACT FOR INDUSTRIALIZED BUILDING SYSTEM CONSTRUCTION 284
Mohd Ashraf Mohd Fateh, Hanah Zakariah, Shaiifa Ema Ezanee

THE ON-SITE WASTE MINIMIZATION PRACTICES FOR CONSTRUCTION WASTE 296
Muhammad Fikri Hasmori, Ahmad Faizul Md Zin, Sasitharan Nagapan, Rafikullah Deraman, Norhaslinda Abas, Riduan Yunus, Mustafa Klufallah

A REVIEW ON THE CRITICAL INFLUENCING FACTOR FOR LIFE EXPECTANCY (LE) AND THE LIFE EXPECTANCY (LE) FOR MALAYSIAN HOUSING 307
K Norfarahayu, M Hairuddin, A Nor Haslinda, Y Siti Khalijah

THE INFLUENCE OF THE STAKEHOLDERS' PERSPECTIVE FOR THE PERFORMANCE MEASUREMENT IN LOCAL AUTHORITIES 317
N Rahmat, A H Nawawi

PROJECT MANAGER'S SKILLS FOR IMPROVING THE PERFORMANCE OF COMPLEX PROJECTS IN KUWAIT CONSTRUCTION INDUSTRY: A REVIEW 324
Faisal Alshammari, Khairulzan Yahya, Zaiton Binti Haron

BUILDING AND INFRASTRUCTURE

VIBRATION CRITERIA ANALYSIS ON FLOOR DUE TO HUMAN WALKING 333
T N T Chik, P C Xian, N A Yusoff

ENHANCING THE KNOWLEDGE AND PROFICIENCY FOR INTERIOR DESIGNERS IN MALAYSIA THROUGH THE IMPLEMENTATION BUILDING INFORMATION MODELLING 343
A B Abd Hamid, M R Embi, M Z Mohd Taib, A H N Abdul Razak

A REVIEW OF GREEN ROOF: DEFINITION, HISTORY, EVOLUTION AND FUNCTIONS 351
F Abass, L H Ismail, I A Wahab, A A Elgadi

TRAFFIC NOISE ASSESSMENT AMONG RESIDENTIAL ENVIRONMENT IN BATU PAHAT, JOHORE, MALAYSIA 359
V C Segaran, Y G Tong, N H Abas, B David Daniel, S Nagapan, R Kelundapyan

FINITE ELEMENT MODELING OF INNOVATIVE SHALLOW RAFT FOUNDATION WITH GRANULAR PILE ANCHOR SYSTEM FOR EXPANSIVE CLAYS 367
Ehab Hamad Sfoog, Alvin John Lim Meng Siang, Wael M. Albadri, Nahla Naji, Sim Sy Yi, Nickholas Anting Anak Guntor

BENEFITS FOR PUBLIC HEALTHCARE BUILDINGS TOWARDS NET ZERO ENERGY BUILDINGS (NZEBS): INITIAL REVIEWS 374
M Sulzakimin, M A N Masrom, R Hazli, A A Adaji, TW Seow, A R Mohd Hilmi Izwan

EXAMINING RISK AS GUIDELINE IN DESIGN STAGE FOR GREEN RETROFITS PROJECTS: A REVIEW 384
M S N Lee, M A N Masrom, S Mohamed, KC Goh, N Sarpin, N Manap

ASSESSING THE LEVEL OF COMMUNITY INVOLVEMENT IN POST-DISASTER HOUSING RECONSTRUCTION AND CURRENT GREEN DESIGN PRACTICES TOWARDS NET-ZERO ENERGY BUILDINGS 392
M Sulzakimin, A A Adaji, A Roy Hazli, N L Mohammad Syabilee, O Roshartini, M R Indera Syahrul, Z Rozlin, I Y Ebenehi

DESIGNER'S CONFIDENCE AND ATTITUDE TOWARDS DESIGNING-FOR-CONSTRUCTION-SAFETY (DFCS) IMPLEMENTATION 400
N H Abas, R Abd Rahman, H Mohammad, M H Rahmat

Author Index

IOP Conf. Series: Materials Science and Engineering **713** (2020) 011001 doi:10.1088/1757-899X/713/1/011001

PREFACE

I am grateful that we are once again able to organize the 2nd Global Congress of Construction, Material and Structural Engineering 2019 (GCoMSE 2019). GCoMSE 2019 jointly organised by UTHM Commercial and Jamilus Research Center for Sustainable Construction (JRC) as a platform for scientists, researchers and engineers to share their knowledge and experiences in creating sustainable and environmental friendly world for the present and the future. We all know that this world is borrowed from the next generation and in other for us to ensure that it is well looked-after, we have to maintain the three important pillars of sustainability i.e. economics, environment and society. These three pillars are the catalyst to ensure that the world is livable, continue to exist and at the same time is developed to enhance the socio-culture of mankind.

The GCoMSE 2019 theme, 'Towards Sustainable Structures and Construction' can be describe as the needs to prepare the world for future generation to continue the responsibility to make the world a better place to live. It is the desire of all population and generations that mankind will continue to exist and lead the world. It has been portrayed by many in science fiction movies that the apes will rule the world or the world will be destroyed because of our negligence and too engross in the development of artificial intelligence (AI). It is possible that if we continue not to care, all the fiction can become reality. The best example is the telephone evolution. From wired to wireless and from a typical phone to a smart phone. Before this, constructing buildings was considered difficult and was limited to certain people. But now, everybody can construct a house with some skills and with help of the application of industrialized building system (IBS). With research works on reducing the usage of natural materials and increasing the use of recycled materials, pollution and negative environmental impacts can be reduced. Thus, it is our duty to ensure that we conduct the necessary steps towards sustainability of the world in whatever discipline and role that we have.

"With Wisdom We Explore"

Sincerely,

ASSOCIATE PROF. IR. TS. DR. ABDUL HALIM BIN ABDUL GHANI
Chairman of GCoMSE 2019

Conference Committee

Organizing Committee

Chairman	:	Associate Prof. Ts. Ir. Dr. Abdul Halim Bin Abdul Ghani
Deputy Chairman	:	Ts. Dr. Faisal Sheikh Khalid
		Dr. Noorwirdawati Ali
Secretary	:	Dr. Sharifah Salwa Mohd Zuki
Publication	:	Ir. Dr. Zainorizuan Mohd Jaini
		Ts. Dr. Norashidah Abd Rahman
Editing	:	Ts. Dr. Nor Haslinda ABas
		Ts. Dr. Tong Yean Ghing
		Ts. Dr. Alvin John Lim Meng Siang
		Dr. Mohd Azuan Zakaria
		Dr. Goh Wan Inn
Programme	:	Dr. Nor Hazurina Othman
		Dr. Masni A. Majid
Promotions	:	Dr. Noor Azlina Abdul Hamid
		Dr. Nor Hayati Abdul Ghafar
		Dr. Emedya Murniwaty Samsudin
		Dr. Sushilawati Ismail
Certificates	:	Dr. Siti Radziah Abdullah
		Ts. Hasniza Abu Bakar
Logistics	:	Ir. Dr. Shahrul Niza Mokhatar
		Ts. Ahmad Fahmy Kamarudin
		Afandi Abu Bakar
		Amran Abd Rahman
		Osman Abdul Rahman
Visual	:	Dr. Basil David Daniel

Dr. Fetra Venny Riza (Universitas Muhammadiyah Sumatera Utara, Indonesia)
Dr. Noor Dina Md Amin (Universiti Tun Hussein Onn Malaysia, Malaysia)
Dr. Sharifah Akmam Syed Zakaria (Universiti Sains Malaysia, Malaysia)
Dr. Mohd Nasrun Bin Mohd Nawi (Universiti Utara Malaysia, Malaysia)
Dr. Mohd Azuan Bin Zakaria (Universiti Tun Hussein Onn Malaysia, Malaysia)
Dr. Mazizah Ezdiani binti Mohamad (University College of Technology Sarawak, Malaysia)
Dr. Ahmad Zaki (Muhammadiyah University of Yogyakarta, Indonesia)
Dr. Noor Nabilah Sarbini (Universiti Teknologi Malaysia, Malaysia)
Dr. Masni binti A. Majid (Universiti Tun Hussein Onn Malaysia, Malaysia)
Dr. Nor Hazurina binti Othman (Universiti Tun Hussein Onn Malaysia, Malaysia)
Dr. Nurul Nadrah Aqilah Tukimat (Universiti Malaysia Pahang, Malaysia)
Dr. Rita Irmawaty (Hasanuddin Universiti, Indonesia)
Dr. Norhaiza Nordin (Infrastructure University Kuala Lumpur, Malaysia)
Dr. Rahmita Sari Rafdinal (Japan Development Sector)
Dr. Sharifah Salwa Mohd Zuki (Universiti Tun Hussein Onn Malaysia, Malaysia)
Dr. Khairunisa Muthusamy (Universiti Malaysia Pahang, Malaysia)
Dr. Salmia Binti Beddu (Universiti Tenaga Nasional, Malaysia)
Dr. Sharifah Maszura (Universiti Malaysia Pahang, Malaysia)
Dr. Sushilawati Binti Ismail (Universiti Tun Hussein Onn Malaysia, Malaysia)
Dr. Hairuddin Bin Mohammad (Universiti Tun Hussein Onn Malaysia, Malaysia)
Dr. Ahmed Mokhtar Albshir Budiea (Universiti Tun Hussein Onn Malaysia, Malaysia)
Dr. Tan Ooi Kuan (Universiti Tunku Abdul Rahman, Malaysia)
Dr. Norliana Binti Sarpin (Universiti Tun Hussein Onn Malaysia, Malaysia)
Dr. Chong Siaw Yah (Universiti Tunku Abdul Rahman, Malaysia)
Dr. Nur Irfah Mohd Pauzi (Universiti Tenaga Nasional, Malaysia)
Dr. Mohd Arif Bin Rosli (Universiti Tun Hussein Onn Malaysia, Malaysia)
Dr. Muad (Anbar University, Iraq)
Dr. Nor Zurairahetty Mohd Yunus (Universiti Teknologi Malaysia, Malaysia)
Dr. Taghreed (Anbar University, Iraq)
Dr. Hartini kasmin (Universiti Tun Hussein Onn Malaysia, Malaysia)
Dr. Chow Ming Fai (Universiti Tenaga Nasional, Malaysia)
Dr. Lee Yee Ling (Universiti Tunku Abdul Rahman, Malaysia)
Dr. Doris Toe Hooi Chyee (Universiti Teknologi Malaysia, Malaysia)
Dr. Nurul Hidayah Mohd Kamaruddin (Universiti Tun Hussein Onn Malaysia, Malaysia)
Dr. Nur Shaylinda Mohd Zin (Universiti Tun Hussein Onn Malaysia, Malaysia)
Dr. Mohamad Anuar Bin Kamaruddin (Universiti Sains Malaysia, Malaysia)
Dr. Ethan Lusac (Rockfield Research Center of Wales, United Kingdom)
Dr. Mohd Yazmil Mohd Yatim (Universiti Kebangsaan Malaysia, Malaysia)
Dr. Norliyati Binti Mohd Amin (Universiti Teknologi MARA, Malaysia)
Dr. Mass Kelvin (Structural Mechanics Section KU Leuven, Belgium)
Dr. Noorhazlinda Abd Rahman (Universiti Sains Malaysia, Malaysia)
Dr. Eeydzah Aminudin (Universiti Teknologi Malaysia, Malaysia)
Dr. Lee Yee Yong (Universiti Malaysia Sarawak, Malaysia)
Dr. Nur Farhayu Binti Ariffin (Universiti Malaysia Pahang, Malaysia)
Dr. Noorwirdawati Ali (Universiti Tun Hussein Onn Malaysia, Malaysia)
Dr. Lee Yeong Huei (Curtin University, Malaysia Campus)
Dr. Norsuzailina binti Mohamed Sutan (Universiti Malaysia Sarawak, Malaysia)
Dr. Norshariza Mohamad Bhkari (Universiti Teknologi MARA, Malaysia)

IOP Conf. Series: Materials Science and Engineering **713** (2020) 011003 doi:10.1088/1757-899X/713/1/011003

Peer review statement

All papers published in this volume of *IOP Conference Series: Materials Science and Engineering* have been peer reviewed through processes administered by the proceedings Editors. Reviews were conducted by expert referees to the professional and scientific standards expected of a proceedings journal published by IOP Publishing.

IOP Conf. Series: Materials Science and Engineering **713** (2020) 012001 doi:10.1088/1757-899X/713/1/012001

Long term effect of strength and durability performance of concrete containing sulphate reduction bacteria under chloride condition

J M Irwan[1], T Teddy[1], N Othman[2] and F S Khalid[1]

[1] Jamilus Research Centre for Sustainable Construction (JRC), Faculty of Civil and Environmental Engineering, Universiti Tun Hussein Onn Malaysia.
[2] Micropollutant Research Centre (MPRC), Faculty of Civil and Environmental Engineering, Universiti Tun Hussein Onn Malaysia.
Corresponding author: irwan@uthm.edu.my

Abstract. The chloride ingress in concrete cause deterioration in concrete due reinforcement corrosion. Adding bacteria in concrete can improve material properties and increase durability with mechanism resist chloride ingress in concrete. Chloride ingress in concrete by bacteria is particularly suited for applications in chloride ion penetration in concrete. The research is objective to determine the effect of adding bacteria to the concrete properties. The bacteria used is locally isolated and enriched to suite concrete environment. The type of the bacteria used is identified as Sulphate Reduction Bacteria (SRB). The SRB is added into concrete mix with concentrations of 3%, 5% and 7%. The compressive and water penetration tests at 28th, 56th, 90th, 180th and the 360th day of curing period were conducted. Cubes of size 150mm × 150mm × 150mm were prepared for compressive strength and water penetration tests. The result of compressive strength shows significant strength of 66.3 MPa on the 360 of day. The water depth penetration also shows a significant impact with addition of SRB with 80% reduction of penetrated water into concrete compared to the control specimen. The overall results of bacteria showed promising results and further study on chloride condition capability is encouraging. It can be concluded that SRB with 5% addition was the optimum percentage to obtain the optimum strength and durability performances.

1. Introduction

Corrosion of reinforcement steel bar due to chloride ingress processes by chloride penetration was the most well-known ecological assaults that lead to the decay of solid structures. Environment factors consist of chemicals in ground, ground water and relative humidity are the main concern degradation agents [1]. In fact, corrosion at reinforcement steel bars are responsible to failure due to the low durability performance [2], age of concrete structure and resistance from external factor.

The bacteria grows in dissolved zero oxygen levels and sufficient carbon with sulphate concentrations in the wastewater [3-4]. Sulfur ions was released with the sulfates and oxygen in the seawater [5]. When the bacteria derive their oxygen from dissolved oxygen, the sulfate was transform to sulfide in the wastewater [6]. The pH value plays an important role influencing the separation and solubility of molecules. This is an indirectly affect study methodologies of biological impact using microorganisms incorporate the on concrete [7]. Therefore, utilization of bacteria as bio-concrete is getting high attention from researcher to solve corrosion of steel bar in chloride condition.

The use of ureolytic and sulphate reduction bacteria isolated showed at pH 10.71 and 9.35 present an optimal growth respectively. The gram stain was positive for both types with the shape of coccus [8]. The bacteria incorporating with concrete was reduced the weight, volume variation (higher age) and water absorption [9-10]. Nosouhian et al. [9] studied concrete exposure to sulfate and enhancement of concrete containing bacteria subjected to sulfate, resulting in bacteria comprising concrete at a compressive strength of 28 days 20 percent higher than control samples. Darmawan [11] mentioned that flexural, compression strength and drying shrinkage behaviour of mortar contains bacteria are evaluated. The author claimed that the bonding behavior of mortar containing bacteria was increased compared to the control mortar.

Irwan et al. [12] indicated that both bacteria were enriched to fit the concrete environment during the enrichment period under alkaline and anaerobic conditions. The ureolytic and sulphate reduction bacteria need urea and sulphate salt, respectively to ensure only the desired strain growth. Therefore, at optimum pH 10.71 and 9.35 were optimal growth condition of ureolytic bacteria and sulphate reduction bacteria, respectively. It was claimed that concrete contains bacteria which isolated strains are able ti sustain in concrete environment.

Jonker et al. [13] claimed that bacteria in concrete can be seen in microscopic after permeability test and found that complete healing of cracks occurred in bacterial concrete. This process called as crack-sealing mechanism. This process involves a metabolic conversion of calcium lactate to calcium carbonate. It found that sealing mechanism resulted in significant sealing of 0.15 mm width cracks. Further creation of this fresh form of self-healing concrete is anticipated to lead in a more sustainable concrete that is particularly suitable for implementations in wet environments where corrosion of strengthening continues to impede the durability of traditional concrete structures.

Suwito et al. [14] studied the addition of bio-chemical in Portland cement to study strength of concrete. It found that flexural strength of beam containing bio-chemical was increased 39.4 % compared to normal beam. The author also claimed that bio-chemical in Portland cement have a resistance to environmental situation such as sulphate ions and water absorption. Gandhimathi et al. [15] studied bacterial concrete that can plugging minerals at crack area than control specimens. The author claimed that carbonate precipitation was decreases compared to control. Muynck et al. [16] studied bacteria in concrete resulted in significant decrease of water intake compared to concrete without the presence of bacteria. The author claimed that process of calcium carbonate crystals at the surface presents decreased of water absorption by 85%, depending on the porosity of specimens.

Therefore, this research investigates the influence of Sulphate Reduction Bacteria (SRB) on concrete under chloride condition in terms of strength and durability with three different of SRB concentrations.

2. Experimental program

2.1. Isolation of Sulphate Reduction Bacteria

Sulfate reduction bacteria (SRB) were collected from domestic acidic at Sungai Pelepah Kota Tinggi, Johor, Malaysia. Sodium chloride (NaCl) was added to imitate the chloride environment [17-18]. The penetration of 3% sodium chloride was used as reduction of water permeation. The compound of the media in the culture SRB process is prepared as Table 1.

Table 1. Composition of the media in the culture SRB

Sample	Nutrient Broth, ml	NaCl, ml	SRB	Total composition, ml
Control	25	10 (3%)	-	35
SRB	25	10 (3%)	1 bead	35

2.2. Materials

Concrete mix proportion was using a different percentage of SRB concentration which are 3%, 5% and 7% as shown in Table 2.

Table 2. Mix proportion of bio-concrete with different percentages of SRB prepared for fabrication

Percentage (%) of bacteria	Cement (kg/m^3)	Water (kg/m^3)	Fine aggregate (kg/m^3)	Coarse aggregate (kg/m^3)
0%	420	210.00	685	1115
3%	420	207.90	685	1115
5%	420	203.70	685	1115
7%	420	199.50	685	1115

2.3. Test procedure

Compressive strength and water absorption tests were conducted in this study using 150mm x 150mm x150mm mould. In order to simulate chloride condition, concrete specimens were cured in 3% NaCl for 28, 56, 90, 180 and 360 days of age curing. The compressive strength test was performed using a universal testing machine. Meanwhile, water permeability test was conducted to determine water depth penetration into concrete. Mechanism of cyclic wetting and drying cycles for deeper penetration of chloride condition cause at a rate 20 times by continuous immersion [19]. Specimen was transferred to sodium chloride solution tank with 15 hours of wetting and 9 hours of drying after 28 days of age curing [19].

3. Results and discussion

3.1 Compressive strength test

The compressive strength of concrete containing SRB at 3%, 5% and 7% are presented in Figure 1. The specimens had been tested after achieving the maturity of concrete on the day 28, 56, 90, 180 and 360 days. Specimens containing SRB shows slightly increase in compressive strength compared to the control specimen for 28 and 56 days. Meanwhile, the highest compressive strength occurred at 180 days was specimen containing 3% of SRB which is 63.9 MPa, followed by 5% of SRB which is 58.6 MPa and 7% of SRB which is 58.7 MPa. The specimen of concrete containing 5% and 7% SRB were increasing gradually after 180 days of curing at 63.9 MPa at 360 days.

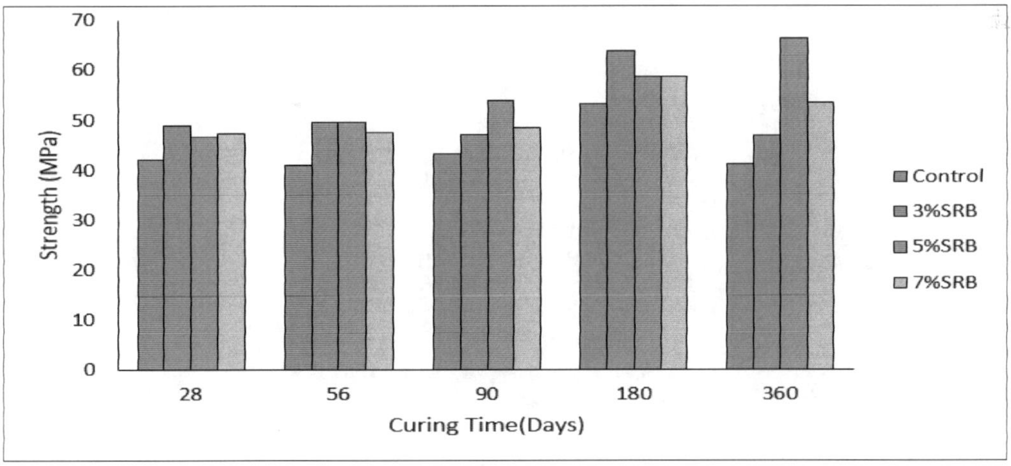

Figure 1. Compressive strength with different percentage of SRB at different curing times

This study presented concrete containing SRB increased up to 15.7%, 10.4% and 11.8% for 3%, 5% and 7% of SRB, respectively compared to control in terms of strength (28 days) as shown in Figure 2. Meanwhile, the compressive strength of SRB concrete on the day of 360 was increased up to 11.7%,

61% and 30% for 3%, 5% and 7% of SRB, respectively. It was a clear trend of increase in compressive strength of 5% of SRB specimens increased up to 61% compared to the control specimens.

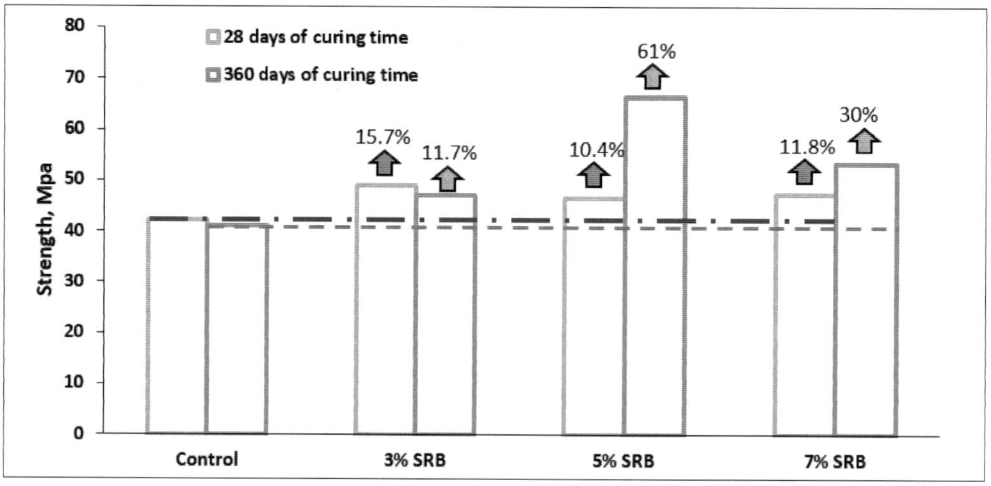

Figure 2. The comparison of concrete strength at the age of 28 and 360 days with different of SRB percentages

The density and strength of concrete was incresed due to pores in concrete is partially filled up by material growth of bacteria. It similar trend result by Kalhori and Bagherpour [17] which is applied *Bacillus subtilis* into concrete and indicates 30% of incrmenet strength comapred to control. This situation occurred because of the bacteria precipitation of calcium carbonate inside the pores of the concrete [20].

3.2. Water penetration

This study conducted 24 hours water absorption test for 28, 56, 90, 180 and 360 days age of curing a shows in Figure 3.

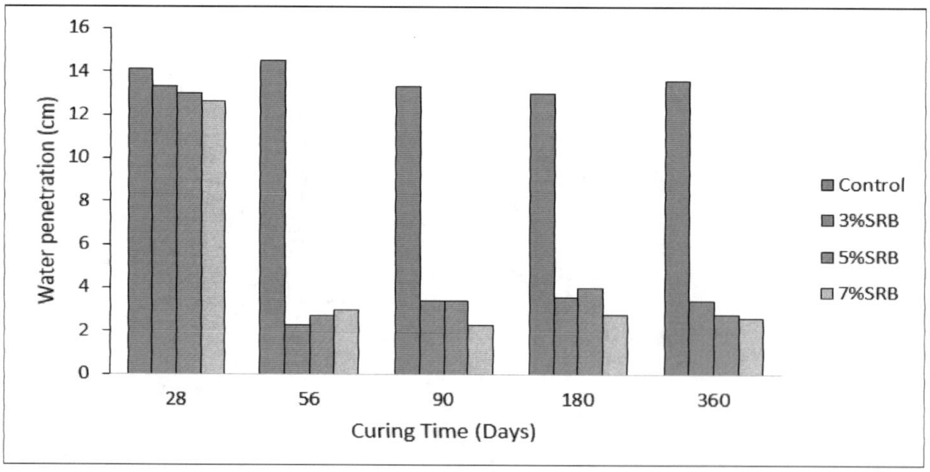

Figure 3. Water penetration with different percentage of SRB at different curing times

The specimen containing SRB shows a reduction of water depth penetration as compared to the control specimen. The control specimen shows a slight decrease from 28 days curing until 360 days curing at ranges of 6% to 10%. On the day of 180, the optimum reduction of water penetration occurred in 7% of SRB (2.8 cm) followed by 3% of SRB (3.3 cm) and 5% of SRB (4 cm). The pattern of reduction of concrete containing SRB was 75% to 80% compared to control specimen. Lastly, for the 365 of day curing shows concrete containing 7% SRB was the optimum reduction of water penetration with a reduction of 80.1% compared to control specimen as shown in Figure 4.

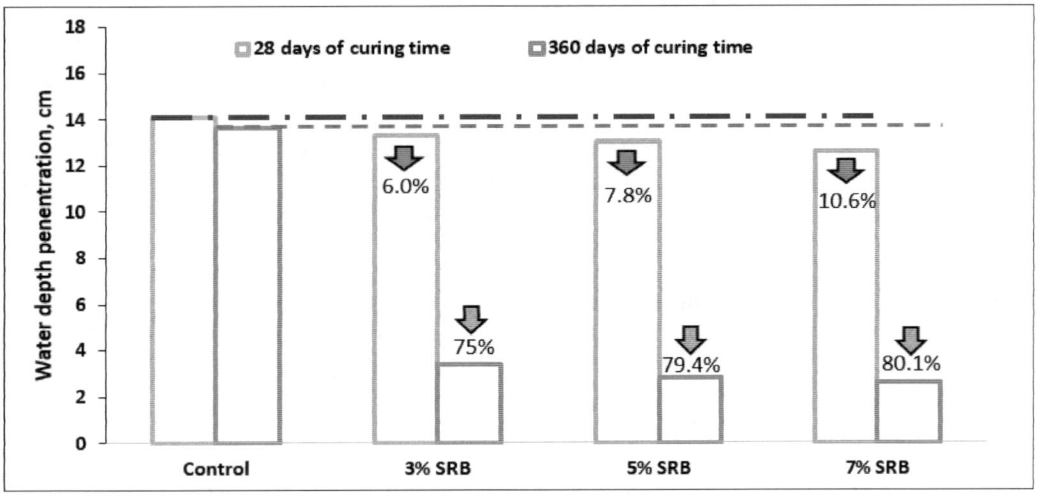

Figure 4. The comparison of water depth penetration concrete at the age of 28 and 360 days with different of SRB percentages

4. Conclusion

The study was designed to investigate the effect of sulphate reducing bacteria (SRB) on compressive strength and water depth penetrations influenced by the percentage of SRB under chloride condition. These findings were obtained according to the highest strength and water depth penetration observed based the curing period. It can be concluded that SRB with 5% addition with concrete was the optimum percentage to obtain the optimum strength and durability performances. Further study should be implemented in rapid chloride penetration test (RCPT) to determine behaviour of concrete containing SRB under chloride conditions such as chloride attack.

5.0 References

[1] Faisal S K, Mazenan P N, Shahidan S and Irwan J M 2018 Review of coal bottom ash and coconut shell in the production of concrete *IOP Conf. Series: Materials Science and Engineering* vol 342 (United Kingdom: IOP Publishing) 012032

[2] Azmi N B, Khalid F S, Irwan J M, Mazenan P N, Zahir Z and Shahidan S 2018 Performance of composite sand cement brick containing recycle concrete aggregate and waste polyethylene terephthalate with different mix design ratio. *IOP Conf. Series: Earth and Environmental Science* vol 140 (United Kingdom: IOP Publishing) 012129

[3] Jonkers H M, Thijssen,A, Muyzer G, Copuroglu O and Schlangen E 2010 Application of bacteria as self-healing agent for the development of sustainable concrete *Ecological Engineering* **36**(2) 230-235

[4] Faisal S K, Mazenan P N, Shahidan S and Irwan J M 2018 Review of coal bottom ash and coconut shell in the production of concrete *IOP Conf. Series: Materials Science and Engineering* vol 342 (United Kingdom: IOP Publishing) 012032

[5] Zivica V and Bajza A 2001 Acidic attack of cement based materials–a review Part 1 Principle of acid attack *Construction and Building Materials* **15**(8) 331–340

[6] Verma S K, Bhadauria S S and Akhtar S. 2013 Evaluating effect of chloride attack and concrete cover on the probability of corrosion *Frontiers of Structural and Civil Engineering* **7**(4) 379–390

[7] Zamer M M, Irwan J M, Othman N, Faisal S K, Anneza L H, Alshalif A F and Teddy T 2017 Bacterial carbonate precipitation improves water absorption of interlocking compressed earth block (ICEB) *IOP Conf. Series: Materials Science and Engineering* vol 271 (United Kingdom: IOP Publishing) 012061

[8] Irwan J M, Anneza L H, Othman N and Alshalif A F 2016 Compressive strength and water penetration of Concrete with Enterococcus Faecalis and calcium lactate *Key Engineering Materials* **705** 345–349

[9] Nosouhian F, Mostofinejad D and Hasheminejad H. 2015 Concrete Durability Improvement in a Sulfate Environment Using Bacteria *The Journal of Materials in Civil Engineering* **28** (1)

[10] Zamer M M, Irwan J M, Othman N, Faisal S K, Anneza L H, Alshalif AF and Teddy T 2018 Biocalcification using Ureolytic Bacteria (UB) for strengthening Interlocking Compressed Earth Blocks (ICEB) *IOP Conference Series: Materials Science and Engineering* vol 311(1) (United Kingdom: IOP Publishing) 012019

[11] Darmawan M S 2010 Pitting corrosion model for reinforced concrete structures in a chloride environment *Magazine of Concrete Research* **62**(2) 91–101

[12] Irwan J M, Anneza L H, Othman N, Alshalif A F, Zamer M M and Teddy T 2017 Mechanical Properties of Concrete with Enterococcus Faecalis and Calcium Lactate *Procedia Engineering* **171** 592–597

[13] Jonker J L, von Byern J, Flammang P, Klepal W and Power A M 2012 Unusual adhesive production system in the barnacle Lepas anatifera: An ultrastructural and histochemical investigation *Journal of Morphology* ***273***(12) 1377–1391

[14] Suwito C and Xi Y 2008 The effect of chloride-induced steel corrosion on service life of reinforced concrete structures *Structure and Infrastructure Engineering* **4**(3) 177–192

[15] Gandhimathi A and Suji D 2015 Studies on the Development of Eco-friendly Self-healing Concrete - A Green Building Concept *Nature Environment and Pollution Technology* **14**(3) 639–644

[16] Muynck W D, Beliea N D and Verstraete W 2010 Microbial carbonate precipitation in construction materials *A review, Ecological Engineering* **36** 118–136

[17] Kalhori H and Bagherpour R 2017 Application of carbonate precipitating bacteria for improving properties and repairing cracks of shotcrete *Construction and Building Materials* **148** 249–260

[18] Alshalif A F, Irwan J M, Othman N and Anneza L H. 2016 Isolation of sulphate reduction bacteria (SRB) to improve compress strength and water penetration of bio-concrete. *MATEC Web of Conferences* vol 47 (Paris: EDP Science) pp 1–6

[19] McCarter W J and Watson D 1997 Wetting and dryingof cover-zone concrete. Proceedings of the Institution of Civil Engineers *Structures and buildings* **122**(2) 227-236

[20] Faiz U A, Shaikh and Steve W M Supit 2014 Mechanical and durability properties of high volume fly ash (HVFA) concrete containing calcium carbonate (CaCO3) nanoparticles *Construction and building materials* **70** 309-321

Acknowledgements
This study has been granted by the Ministry of Education Malaysia and supported by Universiti Tun Hussien Onn Malaysia (GPPS Vot. U445).

Experimental Study on After-Mixing Temperature Control of Concrete in Batching Plant Production

Hooman Shokravi[1], Taleb Moradi Shaghaghi[1]

[1] Department of Civil Engineering, Tabriz Branch, Islamic Azad University, Tabriz, Iran
Corresponding author: ta.moradi@iaut.ac.ir

Abstract. Calculation of the concrete temperature in production phase is an important parameter to control the concrete quality, confining inconsistencies, and costly corrections. In order to obtain after mixing temperature of concrete, the before mixing temperature and weight of the constituent material as well as the moisture contents of the aggregates are collected. Using heat balance equilibrium of the concrete material is an easy and widely used method to calculate after-mixing temperature of the concrete. ACI 306R-10 has presented an equation to calculate the concrete temperature based on thermal equilibrium of the constituents. To evaluate the performance of the equation in estimation of the concrete temperature an experiment is conducted. In the experimental site test results of the output from a batching plant is recorded and graphed. The results of the after induction temperature were calculated using the formula presented in ACI. However, the proposed equation predicts smaller temperature values compared to the actual ones. In the final step the reasons for deviation in the results was discussed and an equation is derived to calculate the concrete temperature for post-mixing stages. The obtained equation was tested on other data and the results show that the method could predict the results reliably.

1. Introduction

Controlling of the fresh concrete temperature to remain within the specified range stated in standards and codes is an important factor for obtaining quality concrete in hot and cold weather concreting [1]. For large-scale concrete production, the cementitious paste is mixed together with fine and coarse aggregate in batching plant. The weight and the thermal data of aggregate, concrete, water and other ingredients are measured and recorded to obtain an estimation of the fresh concrete temperature at the batching plant output. Batching plant is a set of equipment which is used for proportioning of the materials [2] . Various methods and techniques are available for temperature control of the produced concrete. Concrete thermal conditions were controlled in accordance with the project requirements and the recommendation of the regulations [3]. Obviously, correct estimation of the concrete temperature will lead to more accurate results as well as a reduced correction costs.

Heat evolution in concrete is a very complex process involving some material and environmental variables that is controlled by interactions among other factors. The heat evolution over time can be separated into four distinctive stages including pre-induction, induction, strength gain, and steady state steps. Fresh concrete temperature can be inspected while discharging from batching plant, before delivery, and after placement. However, the concrete temperature can be regulated just by controlling the temperature of the constituent in before-mixing condition in batching plant and through controlling the temperature of its components [4]. For producing concrete in research and execution phase, several tests are conducted on the ingredients to evaluate their interaction in the final mix. However, the temperature of the fresh concrete is a parameter that has not been studied as broadly as other aspects.

It can be due to the large dependency of the fresh concrete temperature to the ambient condition and high variability due to environmental factors. The heat evolution steps of concrete are presented in Table 1.

Table 1. The heat evolution stages in concrete [4-5]

Heat evolution stage	Period *	Source of thermalization	Effect	Stage in production phase
Pre-induction	From initiation time - ~10 min	Reaction in aluminate, ferrite and silicate phases	Sharp increase in temperature	At concrete mixer
Induction	~10 min-~4 hrs. during microstructure formation	Not much hydration takes place	Lower trend compared to pre-induction	Transportation and placement
Strength gain	~4hrs-~24hrs	Reaction of the calcium silicates	Increase in temperature	In the formworks
Steady state	~24hrs-∞ and continues for years or even decades	Formation of monosulphate	Temperature stabilization	In the formwork and after formwork removal

* Depends on multiple factors such as cement type, admixtures, and w/c.

Cement hydration is an exothermic reaction and is release of energy in the form of heat. Hydration heat is greatly contributed to the chemical behaviour of the compounds, and could be determined by multiplication of the heat of hydration of the individual pure compounds into their mass proportions [6]. The total amount of the released dehydration heat is an indicator of the reactivity of concrete components. Moreover, it determines the setting and hardening characteristics of concrete at the early age of the cement paste [6]. The temperature variation of concrete due to hydration is largely varied by the constituent materials, mix proportioning and ambient parameters [7]. Calcium is the major constituent of Portland cement and development of total heat is dominated by the calcium contents in the mix [3]. Higher dehydration heat as a result of higher chemical reactivity of the cement contents may result in undesirable durability problems like cracks and shrinkage in the concrete [8,9]. The reaction between cementitious materials and water is the main source of dehydration heat. Cementitious products form the glue to hold concrete ingredients together. These materials include traditional Portland cement and by-product of industrial processes such as silica fume, fly ash, ground granulated blast furnace slag (GGBS), and limestone fines. The industry by-products are mixed either in cement production phase or combined at concrete mixer. Controlling of the type and amount of cementitious materials is an effective temperature control solution for limiting temperature rise in concrete that are sensitive to temperature variations [10]. Several researches have been conducted on the effects of temperature on the fresh concrete properties with a variety of results [11].

Several standards and specifications are contributing to the influence of cold and hot weather in properties of early age concrete and to prevent the damages caused by variation in temperature due to thermal factors. As stated in ACI 305.1-06, the concrete production temperature shall be controlled in a way that the maximum temperature at the discharge stage will not exceed maximum allowable concrete temperature specified in the standard. Hence, it is of great importance to provide cost-effective solutions for on-site concrete production. Several methods are introduced for controlling of the concrete temperature in practical applications such as artificial heating or cooling of raw materials – pre-heating or pre-cooling – and providing suitable curing condition – post heating or post cooling – [12-13]. One should consider the cost-effectiveness of the adopted additional precautionary measures.

The applied precaution actions are aimed to be served with minimum cost and maximum benefit as dual objectives. The great share of these precaution actions are applied in on site concrete production system [14-15]. Some recommendations on fresh concrete temperature are shown in Table 2.

Table 2. The temperature inspection and testing recommendation in standards and codes

Standard	Number	Fresh concrete delivery temperature		Inspection/ test	Minimum frequency
		Min (˚C)	Max (˚C)		
ACI	305	-	35	ASTM C1064 /C1064 M	ASTM C94 /C94M
BS EN	206	5	-	Measure temperature	Periodical and when new batch or load arrival
ACI	306	13	32	ASTM C1064 /C1064 M	ASTM C94 /C94M

Although, up to the present time, a great number of researches have been carried out by numerous investigators with respect to the chemical and mechanical properties of concrete and effect of the ingredients on the behaviour of concrete, thorough literature survey shows that gaps of knowledge exist on fresh concrete temperature, especially in on site concrete production system. In this paper, the thermal parameters of the concrete ingredients are investigated separately and the effect of each component and interaction with the ambient temperature is detailed. Batching plant is the only stage in which control can be applied on the produced concrete. Hence the focus in this study is on the effect of batching plant system on the concrete component temperature, and the influence of the components of the batching system on the concrete temperature. The results of the measurement of the temperature of concrete and concrete constituents are provided in a batching system. Finally, an example of a real case batching plant is brought up. The research suggests a more accurate estimation of the concrete temperature for manual medium-sized batching plant systems. By collecting and recording the weight and the thermal data of the aggregate, concrete and water, the fresh concrete temperature is reached at the batching plant output.

2. Material and methodology

In order to evaluate the quality of the material used in this experiment, several tests were carried out. Table 3 shows the summary of the conducted tests for quality control of the materials. In the first steps, several chemical and physical analyses were conducted on concrete constituents. The particle size gradation was in accordance to ASTM/C136-01, which represents the sieve analysis of the fine and coarse aggregates [16]. Degradation of the resistance was determined using ASTM/C131 for large size coarse aggregate [17]. The soluble sulphate content of the fine and coarse aggregates was determined according to ASTM/C1524 [18]. Accordingly, an investigation was made to study the potential harm of the organic impurities included in the aggregate according to ASTM C40 [19]. The potential alkali reactivity test according to ASTM C1260 was made on aggregates [20]. The test was made using the type I Pozzolan Portland cement of Oromieh factory (Iran) with maximum pozzolan content of 15%.

The concrete samples were made using the mix shown in Figure 1. The specific gravity and fineness modulus of sand are 2.51 and 3.12, respectively. The maximum aggregate size was 25.4 mm. To provide the required water-to-cement ratio (W/C) of 0.35, super-plastizer was used. The data are collected from a half-cube meter batching plant during a six-month period. The collected variables include temperature, weight and moisture content of fine and coarse aggregates, cement's temperature and weight, mixing water temperature and weight, temperature and weight of micro silica and added water to micro silica, specific gravity under a saturated surface-dry condition (SSD) of aggregates. The view of the batching plant is shown in Figure 2.

IOP Conf. Series: Materials Science and Engineering **713** (2020) 012002 doi:10.1088/1757-899X/713/1/012002

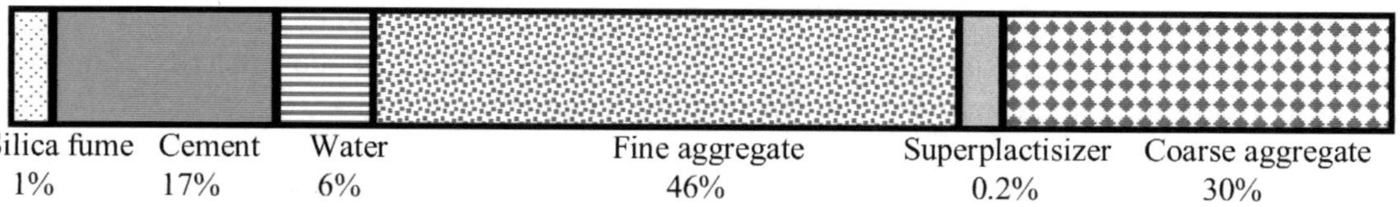

Figure 1. Mixture of the C55 concrete used for the experiment

Figure 2. Front view of the batching plant

ACI 306R-10 has proposed a formula based on heat balance equilibrium of the concrete material to calculate pre- and post-mixing temperature of the concrete. The weights, temperature of the constituents as well as the moisture content of the aggregates are used in this equation.

$$T = \frac{\left[0.22\left(T_s.W_s + T_a.W_a + T_c W_c\right) + T_w.W_w + T_s.W_{ws} + T_a.W_{wa}\right]}{0.22\left(W_s + W_a + W_c\right) + W_w + W_{wa} + W_{ws}} \quad (1)$$

Where T, T_a, T_c, T_s, T_w are temperature of concrete, coarse aggregate, cement, fine aggregate, mixing water, respectively. $W_a, W_c, W_s, W_w, W_{wa}, W_{ws}$ are weight of SSD of coarse aggregate, cement, SSD of fine aggregate, mixing water, free water on fine aggregate, and free water on coarse aggregate, respectively. The specific heat of the cement and aggregates are assumed 0.22 kcal/(kg°C).

3. Experimental results and discussion

In the following the results of the experimental case study for evaluation of the concrete constituents as well as the test carried out on the concrete are presented. Furthermore the temperature variation of the fresh concrete is discussed as possible reasons for variation of the temperature in the after mix concrete is discussed.

3.1. Material tests

Figures 3-4 show the grading results of the aggregate obtained from sieve analysis. The maximum expansion of the aggregates in the concrete samples at the age of 4, 7 and 16 days was 0.016, 0.04 and 0.112, respectively, which means the potential alkali reactivity of the aggregates was not significant. The fineness modulus of the fine aggregate is 3.79 with a specific gravity of 2.602 and silt and clay content of 0%. The results show that using silica fume has effectively improved the concrete mixture to recover the loss of required finer particles in sand.

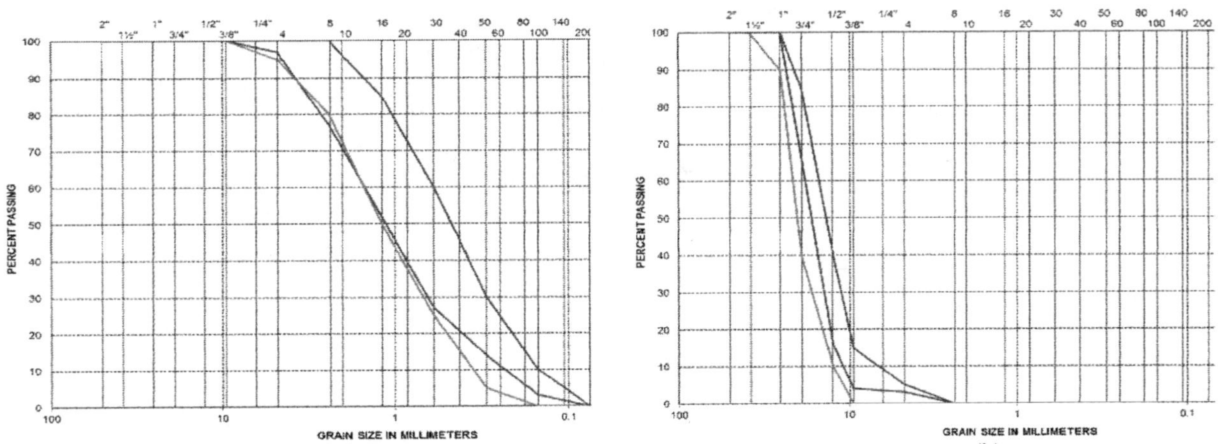

(a) (b)

Figure 3. Sieve analysis of the aggregates (a) coarse aggregates (b) fine aggregate

In order to analyse the quality of the constituent material in the concrete, several tests have been conducted that are shown in Table 3.

Table 3. Materials and their conformance to the specifications and codes

Material	Code and specification	Reference	Subject	Test results	Frequency
Cement	ASTM C595	[21]	Cement standard tests	Fineness: 3588 (cm2/gr) Setting time: 107 and 220 for initial and final, respectively Autoclave expansion: 0.05	Each shipment/ every 1000 ton
	ASTM C204	[22]	Cement test	Checking of the product data sheet	
Micro silica	ASTM C1240	[23]	Micro silica test	Checking of the product data sheet	Every shipment
Aggregates	ASTM C33	[24]	Concrete technical specification	Fulfilled	Every source
	ASTM/C136	[25]	Sieve analysis	Figure 3 and 4	Every two week
	ASTM/C117	[26]	Material passing the 0.075mm	-	Every month
	ASTM/C1524	[18]	Chloride content	0.003%	Every month
	ASTM/C131	[18]	Abrasion	19.95%	Change of material source
	ASTM/C127	[27]	Specific gravity	2.602 gr/cm^3	Every month
	ASTM/C123	[28]	Lightweight particles in aggregate		Every month
	ASTM/C142	[29]	Clay lumps and friable particles	0%	Every month
	ASTM/C127	[30]	Water absorption	0.855%	Every two weeks
	ASTM/C40	[19]	Organic Impurities	0%	Every month
	ASTM/C88	[31]	Weight loss during sulphate	0%	Every change in source
Water	ASTM D516 and AASHTO T26	[32,33]	Test for quality of water	SO4, ppm 1000; Cl, ppm 500; Compressive strength; Time of set	Every new source
Super plasticizer (SF40)	ASTM C 494/C	[34]	Chemical Admixtures for concrete	As per data sheet.	Each shipment
Concrete	ASTM C1260	[20]	Expansion	Expansion of 0.016, 0.04, and 0.112 in 4th, 7th and 16th, respectively.	Every new source

The chemical composition of the cement and water are important in hydration heat release and the heat convection parameters of the concrete. Table 4 shows the chemical composition of the used cement and water in the experiment.

Table 4. The chemical composition of the cement and water used in the experiment

Cement		Water	
Compound	**Amount**	**Compound**	**Amount**
SiO_2	26.83	Total Dissolved Solids (TDS)	267 mg/lit
Al_2O_3	5.25	Turbidity	0.6 NTU
Fe_2O_3	2.80	PH	7.9
CaO	58.97	Salinity	0.3%
MgO	1.88	Conductivity	558 μSiemens
NaO_2	0.59	Hardness	320 mg/lit
K_2O	1.02	Alkalinity test with phenolphthalein (P-ALK)	0
SO_3	1.75	Alkalinity test with methyl orange (M-ALK)	280

3.2. Analysis of the temperature data

Figure 4 shows the variation of the ambient temperature with the measured concrete temperature obtained from the output of the batching plant. The obtained data are collected within the period of six month in the temperature range of -3°C to 15°C. The plot of the temperature variation versus the ambient temperature is following a similar pattern as shown in Figure 4. However, the concrete temperature of the fresh concrete in batching plant output is higher than the ambient temperature and it was kept within 11°C to 23°C range.

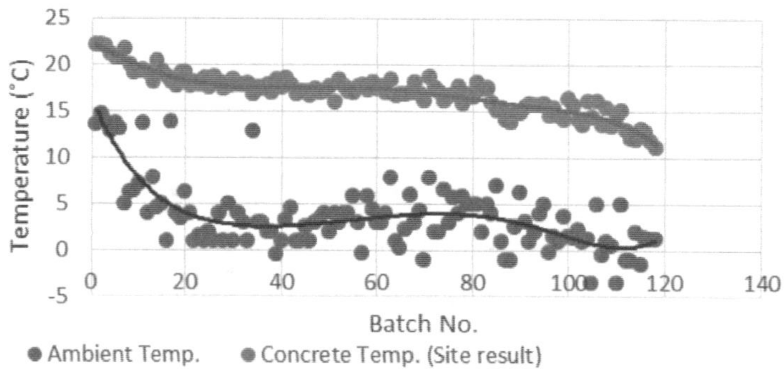

Figure 4. The concrete temperature versus the ambient temperature

The data for the materials before mixing includes the weights, temperature of the constituents and the moisture content of the aggregates as well as after mixing data including the final temperature of the produced concrete are recorded for this experiment. The after mixing temperature of the produced concrete was calculated for all available samples. The recorded temperature obtained from the site measurement of the batching plant output and the value calculated from the recorded weight, temperature and moisture of the constituent using the Eq. (1) are plotted for the comparison purpose. Figure 5 shows the results for calculated and actual results of the concrete.

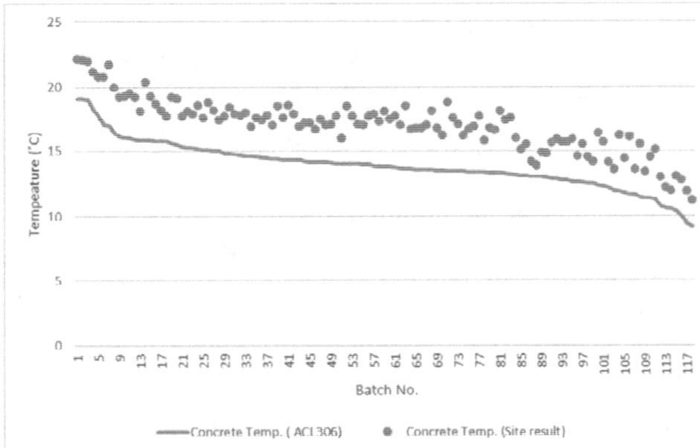

Figure 5. The concrete temperature values from the Eq. (1) and actual concrete temperature

As shown in Figure 5 there is discordance between the results obtained from the actual concrete temperature and the values obtained from Eq. (1). For both plots, the pattern of the trend are similar but the values obtained using ACI 306R-10 tend to be almost 3˚C lower in the concrete samples than their actual temperature. A concrete sample C35 with zero silica fume was also used to confirm the results obtained by C55 concrete. It is shown that there is not a significant change in the results obtained from both concrete mixtures in the final temperature using Eq. (1). In the following subsection an investigation is conducted to illuminate the discrepancy obtained from the results using two methods.

3.3. Discussion on the results

As shown in Figure 5, the actual temperature of the concrete is higher than the calculated value using Eq. (1). It can be observed from the figure that the variation is independent of any of the variables presented in Eq. (1) including weight, moisture and temperature of the constituent material. The possibility of such a variation in temperature was also mentioned in other references such as [3,35,36]. The offset of almost 3˚C in the trends of the plots supports involvement of other thermal parameters in the system. The discordance in the measurement and the calculated values of the concrete temperature could be as a result of the time required for heat transfer from the aggregates, specifically coarse aggregate, to the surrounding constituents to reach a thermal equilibrium. To investigate the effect of the required time to reach thermal equilibrium, a study was conducted. Coarse aggregate are assumed as spherical shape with diameter of 15 mm. The diameter is obtained by the grading distribution of the coarse aggregates in the sieve analysis shown in Figures. 3-4. Using the lumped capacitance method for transient conduction, the time for reaching isothermal equilibrium after imitation of exothermic reaction can be calculated. In the first step, the Biot number is calculated for the thermal system using Eq. (2). The Biot number is defined as the ratio of conduction resistance to convection resistance and imply that the external resistance associated with convectional heat transfer is dominant [37].

$$B_i = \frac{hL}{k} \tag{2}$$

Since the calculated Biot number is less than 0.1 the lumped capacitance method is valid for the present setup [38]. The required time to reach bulk fluid temperature can be calculated using Eq. (3),

$$t = \frac{\rho V_c}{hA_s} \ln \frac{\theta_i}{\theta} \tag{3}$$

Where ρ and V_c are density and volume of the assumed coarse aggregate, respectively. h and A_s are the convection heat coefficient of the concrete and outter surface area of the assumed coarse aggregate, respectively [39]. θ_i and θ are the difference between initial object temperature T_i and bulk fluid temperature T_∞ and difference between object temperature T and bulk fluid temperature T_∞

at time t. The temperature of the coarse aggregates gradually increases to reach the bulk concrete temperature. Using Eq. (3) the required time for the constituent materials to reach the isothermal temperature is calculated. In 705 seconds all constituent materials reach after mix isothermal temperature. On the other hand, the required operation time for charge, mixing, sampling and temperature measurement in a batching plant specification is shown in Table 5. As shown in Table 5, the initiation time of the process till the temperature measurement of the produced concrete is almost 310 seconds. The obtained data shows that the operation time is lower than the time required to isothermal temperature. As a result the recorded temperature is accordingly lower than the one obtained from the temperature of the constituent in thermal equilibrium. The result can demonstrate the variation of the before mix and after mix temperatures.

Table 5. The required time for charge, mixing, sampling and temperature measurement based on the batching plant specification

No.	Task name	Duration (second)
1	Charging and mixing of the concrete constituents	90
2	Discharge the mixer	10
3	Sampling and transfer for testing.	20
4	Concrete temperature measurement (2- 5 min based on ASTM 1064)	190
	Total	310

On the other hand, the obtained thermal offset might be due to other reasons amongst which probably the most important is development of hydration heat. Hydration heat is greatly contributed to the chemical compounds of the concrete constituents. Although, in this experiment, using pozzolanic cement in concrete mixtures can slightly decrease the peak amplitude of temperature and delay the temperature rising profiles, the variation is insignificant in the temperature of Portland pozzolanic cement [36]. Furthermore, the researches show that the influence of the silica fume in hydration heat is negligible [36]. Eq. (1) is based on the thermal balance among constituents in the whole medium with the assumption that all phases of the material are in thermal equilibrium. In the above assumption the effect of parameters such as hydration heat, viscous dissipation and mechanical work is neglected [3,40]. Even though the equation is obviously not sufficiently accurate to predict the after mixing concrete temperature with a high degree of confidence, it does provide a substantial insight into the output temperature. In order to obtain a more accurate estimation of the temperature it is necessary to include the effect of those missing factors in the equation. Accordingly, a nonlinear regression was performed to derive constants of the equation from the experimental data. Figure 6 shows the plot of the nonlinear regression of the data.

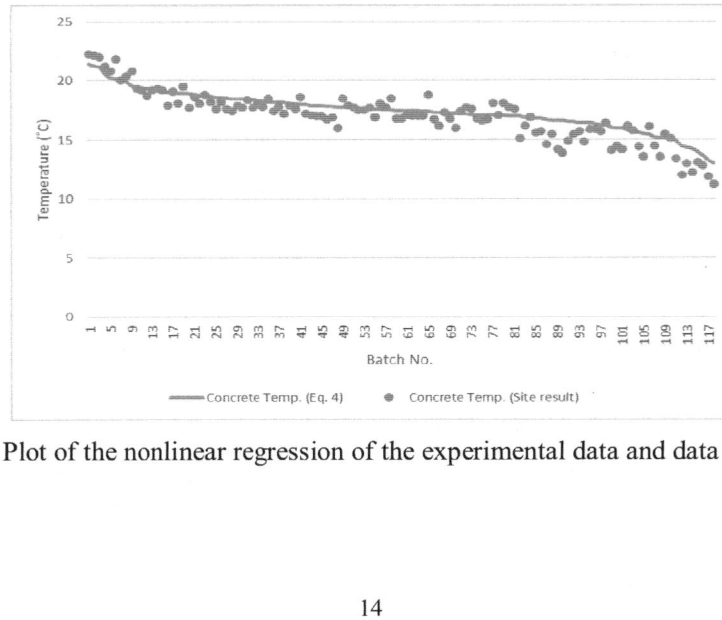

Figure 6. Plot of the nonlinear regression of the experimental data and data from Eq (4)

Based on the findings of the nonlinear regression a modified equation is derived by combining the Eq. (1) and the extracted constants as per below.

$$T=\frac{0.22\left(T_s.W_s+0.45T_a.W_a+T_cW_c\right)+T_w.W_w+T_s.W_{ws}+T_a.W_{wa}}{0.22\left(W_s+W_a+W_c\right)+W_w+W_{wa}+W_{ws}}+\frac{W_c}{100} \quad (4)$$

The results indicate that the proposed equation is capable of estimating the actual temperature with a higher accuracy than the original one. This study deals with a specific mix design of concrete [41-42]; however future study need to expand the scope to cover more other mix designs [43].

4. Conclusions

Controlling of the before-mix temperature of the concrete constituents is a simple solution to the quality problems raised by temperature inconsistencies in concrete. Several methods are introduced for controlling of the concrete temperature to the pre-determined profile in practical applications that include heating or cooling of raw materials and providing suitable curing condition. The after mixing temperature of concrete can be estimated by analysis of the before mixing temperature and weight of the constituent material as well as the moisture contents of the aggregates. ACI 306R-10 has presented an equation to calculate the concrete temperature based on thermal equilibrium of the constituents. In a case study, the data of a batching plant was collected for 120 different batches within six-month period. The required quality control tests are carried out for the concrete constituents and the output concrete of the batching plant. The result was checked by the acceptance criteria introduced by the presented standards and specification. In the next step the recorded data for the temperature, moisture and weight of the materials was put into the equation introduced by ACI to check the after mix concrete temperature. It was seen that there is a gap between the measurement of the actual temperature of the batching plant output and the calculation of the value obtained by the proposed equation. To study the reason for such a variation in the results, the heat parameters was investigated and potential reasons for the discordance were discussed. Finally, the relation between the measurement and calculation results is given by a nonlinear regression. An equation is introduced to calculate the after mix temperature of concrete with a better accuracy and precision. The obtained results show that the obtained equation can estimate the fresh concrete temperature with higher accuracy.

5. References

[1] ACI 305.1-06 2007 *Specification for hot weather concreting* (Detroit: American Concrete Institute)

[2] Ravanshadnia M and Ghanbari M. 2014 Minimizing Material Inventory in Ready Mixed Concrete Plants by Applying a Fuzzy Neural Network Approach Management *Int. J. Innovative Science Eng. & Technology* **1**(8) 134–142

[3] Neville A M 2011 *Properties of concrete 5th Edition* (Essex, England: Pearson Education Limited)

[4] Ge Z, Wang K and Gao Z 2011 Prediction of pavement concrete strength development, joint sawing, and opening time using FEMLAB *GeoHunan International Conference 2011(Hunan, China)* vol 408 (Reston:American Society of Civil Engineers)

[5] Esping O and Löfgren I 2006 Investigation of early age deformation in self-compacting concrete *2nd International Symposium on Advances in Concrete through Science and Engineering* (Quebec City, Canada)

[6] Awal ASMA and Shehu I A 2013 Evaluation of heat of hydration of concrete containing high volume palm oil fuel ash *Fuel* **105** 728–731

[7] Yang K-H, Moon G-D and Jeon Y-S 2016 Implementing ternary supplementary cementing binder for reduction of the heat of hydration of concrete *J Cleaner Production* **112**(1) 845–852

[8] Orchard D F, Curran A and Hearne R 1979 *Concrete Technology-Volume 1-Properties of*

Materials (Essex, England)

[9] Ramachandran V S and Beaudoin J J 1986 *Handbook of Analytical technique in concrete science and technology* (Norwich, New York, USA : Noyes Publication, William Andrew Publishing) pp 1–55

[10] Yu X, Chen J, Xu Q and Zhou Z 2018 Research on the influence factors of thermal cracking in mass concrete by model experiments *KSCE J. of Civil Eng.* **1** 1–10

[11] Schmidt W, Brouwers H J H, Kühne H-C and Meng B 2014 Influences of superplasticizer modification and mixture composition on the performance of self-compacting concrete at varied ambient temperatures *Cement Concrete Composites* **49** 111–126

[12] Han C-G, Son M-S and Choi H-K 2012 Development of Efficient Curing Sheet for Thermal Insulation Curing of Concrete in Cold Weather *J. Korea Institute of Building Construction* **12**(3) 291–298

[13] Kim H-C, Min T-B, Mun Y-B, Kim J-Y, Choi H-K and Lee H 2016 Effect of high early strength cement and accelerator concentrations on the low-temperature compressive strength of concrete *J. Ceramic Processing Research* **17**(6) 641–647

[14] Raghavan N and Niranjana P 2005 Construction aspects of containment structures of a typical large atomic power project *Proc. 18th Int. Conf. on Structuctural Mechanics in Reactor Technology(Beijing, China)* pp 2340-2351

[15] Karagol F, Demirboga R and Khushefati W H 2015 Behavior of fresh and hardened concretes with antifreeze admixtures in deep-freeze low temperatures and exterior winter conditions *Construction Building Material* **76** 388–95

[16] ACI 136-01 2001 *Standard test method for sieve analysis of fine and coarse aggregate* (American Society Testing Material)

[17] ASTM C131 2003 *Standard Test Method for Resistance to Degradation of Small-Size Coarse Aggregate by Abrasion and Impact in the Los Angeles Machine* American Society Testing Material

[18] ASTM C1524 A 2002 *Standard test method for water-extractable chloride form aggregate* (American Society Testing Material)

[19] ASTM C40 A 1999 *Standard Test Method for Organic Impurities in Fine Aggregates for Concrete* (American Society Testing Material)

[20] ASTM C1260 A 2007 *Standard Test Method for Potential Alkali Reactivity of Aggregates (Mortar-Bar Method)* (American Society Testing Material)

[21] ASTM C595 A 2013 *Standard Specification for Blended Hydraulic Cements* (American Society Testing Material)

[22] ASTM C204 A. 2007 *Standard Test Method for Fineness of Hydraulic Cement by Air Permeability Apparatus* (American Society Testing Material)

[23] ACI 1240-01. 2001 *Standard specification for use of silica fume as amineral admixture in hydraulic-cement concrete, mortar and grout* (American Society Testing Material)

[24] ASTM C33M AC/. Standard Specification for Concrete Aggregates. Am Soc Test Mater 2004.

[25] C136M AC/. 2001 *Standard test method for sieve analysis of fine and coarse aggregate* (American Society Testing Material)

[26] ASTM C117 A 1995 *Standard Test Method for Materials Finer than 75-μm (No. 200) Sieve in Mineral Aggregates by Washing* (American Society Testing Material)

[27] ASTM 127 128 2001 *Standard Test Method for Density, Relative Density (Specific Gravity), and Absorption of Coarse Aggregate* (American Society Testing Material)

[28] ASTM C123M AC/ 1998 *Standard Test Method for Lightweight Particles in Aggregate* (American Society Testing Material)

[29] ASTM C142 A 1997 *Standard Test Method for Clay Lumps and Friable Particles in Aggregates* (American Society Testing Material)

[30] ASTM C127 A 2001 *Standard Test Method for Relative Density (Specific Gravity) and Absorption of Coarse Aggregate* (American Society Testing Material)

[31] ASTM C88 A 2005 *Standard Test Method for Soundness of Aggregates by Use of Sodium Sulfate or Magnesium Sulfate* (American Society Testing Material)

[32] ASTM D516 A 2011 *Standard Test Method for Sulfate Ion in Water* (American Society Testing Material)
[33] AASHTO T26 A 2008 Standard method of test for quality of water to be used in concrete (American Association State Highway Transportation Officials)
[34] ASTM C494 1999 *Standard Specification for Chemical Admixtures for Concrete* (American Society Testing Material)
[35] Zain M F M, Safiuddin M and Yusof K M 1999 A study on the properties of freshly mixed high performance concrete *Cement Concrete Research* **29**((9) 1427–1432
[36] Nili M and Salehi A M 2010 Assessing the effectiveness of pozzolans in massive high-strength concrete *Construction Building Material* **24**(11) 2108–2116.
[37] Van de Velden M, Baeyens J, Brems A, Janssens B and Dewil R 2010 Fundamentals, kinetics and endothermicity of the biomass pyrolysis reaction *Renewable Energy* **35**(1) 232–242
[38] Xu B, Li P-W and Chan C L 2012 Extending the validity of lumped capacitance method for large Biot number in thermal storage application *Solar Energy* **86**(6) 1709–24
[39] Lee Y, Choi M-S, Yi S-T and Kim J-K 2009 Experimental study on the convective heat transfercoefficient of early-age concrete *Cement Concrete Composites* **31**(1) 60–71
[40] Gawin D, Pesavento F and Schrefler B A 2006 Hygro-thermo-chemo-mechanical modelling of concrete at early ages and beyond. Part I: hydration and hygro-thermal phenomena. *Int. J. Numerical Methods Eng.* **67**(3) 299–331
[41] Shokravi H 2014a Corrosion assessment of the exposed friction piles. In *International Graduate Conference on Engineering, Science and Humanities (IGCESH 2014)* vol 1 (Johor, Malaysia: Universiti Teknologi Malaysia) pp 55–68
[42] Shokravi H 2014b Mix design of the high performance concrete for highly corrosive environment. In *International Graduate Conference on Engineering, Science and Humanities (IGCESH 2014)* vol 1 (Johor , Malaysia: Universiti Teknologi Malaysia) pp 95–112
[43] Shokravi H 2011 *Effect of circular hole on compressive strength and fracture mechanism of 2d concrete* Master Thesis (Johor: Universiti Teknologi Malaysia)

Strength of Concrete Containing Synthetic Wire Waste as Fiber Materials

F S Khalid[1], S H Saaidin[1], S Shahidan[1], N H Othman[1] and N A A Guntor[1]

[1] Faculty of Civil and Environmental Engineering, Universiti Tun Hussein Onn Malaysia, 86400 Parit Raja, Johor, Malaysia
Corresponding author: faisalsh@uthm.edu.my

Abstract. The demand of concrete in construction field is growing as linear with the growing of Malaysia economy. This tendency is expected to continue in the year ahead because of improvement in technology. This study tends to determine the performance of concrete containing synthetic wire waste as fiber materials. In addition, the use of wire waste as fiber materials can reduce the amount of waste dumped into landfills. The increasing amount of wire waste which is uncontrollable may lead to environment issue. Until today there is no researchers have done the studies on performance of concrete containing synthetic wire waste as fiber materials. Therefore, an action is made to overcome the environmental problems by using wire waste as fiber materials in concrete mixture. The amount of wire waste added into concrete mixture are limited to 0%, 0.5%, 1.0%, and 1.5% respectively. Other than that, water cement ratio that used is 0.55. The test that involve in this study is workability, density, compressive strength and tensile strength tests. The compressive strength test carried at the ages of 7 and 28 days and for tensile strength test were carried at ages of 28 days. The result was compared with control mix and the optimum percentage value of using wire waste as fiber materials in concrete was 0.5% for both compressive and tensile strength test. Concrete contains 0.5% of wire waste. These mix shows significant increases about 24.4% and 42.7% increases for compressive and tensile strength respectively, compared to the normal concrete.

1. Introduction

Concrete and synthetic fiber have become more popular and attractive in recent years and frequently used for construction material. This is due to the behaviour of concrete which is low in tensile strength, energy absorption and ductility. The poor tensile behavior of concrete is cause from its low toughness and the presence of defects. Therefore, the concrete toughness needs to be improved and strengthen by add fiber material to increase interlocking tensile in concrete matrix. Therefore, a lot of studies on fiber reinforced concrete have been performed to determined optimum portion and suitable materials of fiber [1-2]. Other study investigates behaviour of pattern, strength and magnitude of fiber to increase the toughness of concrete over plain concrete are commonly observed. Many studies have been extended to investigate in shape, type, diameter and length of fibers [3].

In the 1990s asbestos were used as fiber material in concrete and after the health risks associated with asbestos were discovered, there was a need to find another replacement for the substance in concrete and other building materials. Polyvinyl alcohol (PVA), Polyethylene (PE), Polypropylene (PP), Polyvinyl chloride (PVC) and polyesters are commonly used as a fiber material in the market. One of the popular synthetic fiber is PP fiber which is widely used for construction applications such as in ground slab, pavement, shotcrete tunnel linings and blast resistant concrete [4]. The mechanism of applying synthetic fiber in concrete to improve fracture behaviour and toughness concrete performance. Synthetic fiber in concrete was acted as fiber bridging and interlocking tensile stress during tensile force was applied to the concrete. Fiber bridging tensile is considered the governing factor in improving the mechanical properties of fiber reinforced concrete. This factor is determining by bond component between fiber and concrete surface such as fiber cross section, adhesion, fibrillation and embossed surfaces to improve the fiber-concrete bond. These factors are significantly improving engineering properties of concrete namely in fracture and toughness strength. The enhanced performance of FRC compared to the plain concrete comes from its improved capacity to absorb energy during fracture. This due to the plain concrete fails in a brittle manner at the occurrence of cracking stresses which is at first crack load.

In fact, synthetic fiber was popular in the market to reduce the thermal shrinkage and shrinkage cracking of concrete. Plastic shrinkage will occur if the rate of evaporation at the concrete surface is higher compared to rate of water migration to the surface [5]. This will allow shrinkage forces cracks to appear on the surface. A small amount of fiber is significantly effectively to reduce the plastic shrinkage [5-6]. The goals of this study are to investigate the performance of concrete containing wire waste as fiber materials. A few objectives that use to achieve the goals of the study is to determine the workability, density, compressive strength, and tensile strength of the concrete containing wire waste as fiber materials.

2. Experimental

The wire waste taken from the recycle site has been used to act as fiber materials in a concrete, the wire waste percent used as a fiber material is 0.5%, 1.0%, and 1.5%. The length of wire waste was 40mm diameter with length over diameter (l/d_f) is 34.25.

Figure 1. Synthetic wire waste without copper inside.

A total of 24 cube specimens were prepared for the workability, density and compressive strength test. All cube specimens were stored in a water tank to cure the specimens for 7 and 28 days. A total of 12 cylinders with a diameter of 100 mm and a height of 150 mm were prepared for the splitting tensile strength test. The moulds were filled to the top with compaction. The concrete in the cylinder moulds are screened to ensure that the top surface is even. Table 1 shows a mix design fiber reinforced concrete for 1 m^3.

All the cylinders were stored in a water tank to cure the specimens for 28 days. This study used 0.55 as water cement ratio. Slump test was conducted immediately after the mixing process and slump value was taken to be compared to the control concrete. The cube size used was 100 mm x 100 mm, the compaction was conducted using the tamping rod manually and the concrete sample was curing for 7 and 28 days. Size of 100 mm x 150 mm was used for cylindrical mould. The compaction is conducted manually using tamping rod with 3 layers of compactions. The sample was weight before the concrete testing in order to get the density of the concrete as the physical properties. All samples were curing in water curing tank. Sample was tested using ordinary compression machine for the concrete compression cube and also for the tensile test of concrete. The compressive and tensile strength of hardened-state concrete are determined according to EN12390-3:2009 and BS EN 12390-6:2009, respectively. The result of the testing was analysed and graph was plotted to determine the behavior compressive and tensile strength.

Table 1. Mix Proportion for 1 m³ of concrete.

Cement (kg/m³)	Coarse Aggregate (kg/m³)	Fine Aggregate (kg/m³)	Water (kg/m³)	Synthetic wire waste proportions (%)	Synthetic wire waste (kg/m³)
345	805	980	189.75	0	0
345	805	980	189.75	0.5	10.65
345	805	980	189.75	1.0	21.3
345	805	980	189.75	1.5	31.95

3. Result and discussion

Test was carried out to determine the workability and performance of concrete containing synthetic wire waste as fiber materials. To achieve the objectives, a few tests have been conducted at the Universiti Tun Hussein Onn Malaysia (UTHM) Structure and Material Laboratory such as slump, density, compressive strength and tensile strength tests.

3.1. Slump

Slump test have been used to test the fresh state slump of the concrete to find the suitable workability value for concrete that contain wire waste as fiber materials. Figure 2 shows the slump value for the control and concrete containing wire waste.

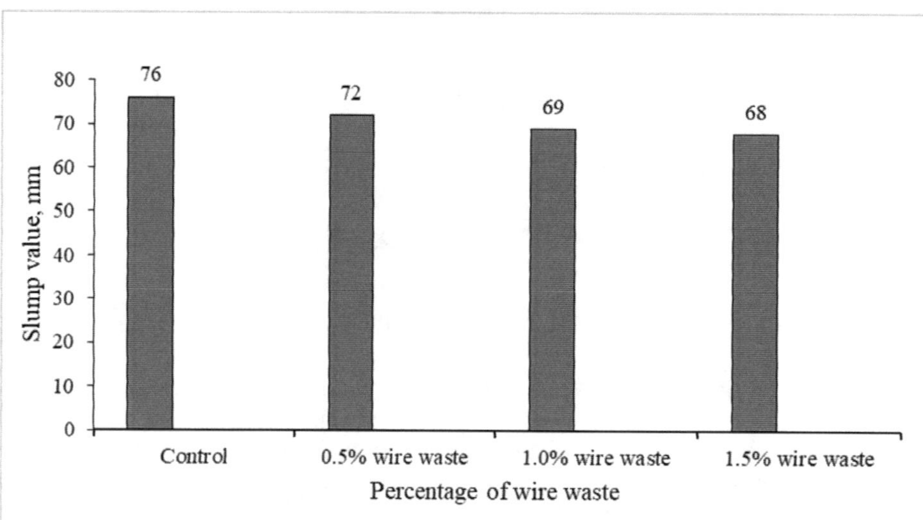

Figure 2. Slump value of concrete containing wire waste as fiber materials.

It shows that slump value for the control concrete is 76 mm. Concrete containing wire waste shows a decrease on slump value compared to control concrete. Concrete contain 0.5% wire waste has a slump value 72 mm decrease 4 mm from the control slump value. Meanwhile concrete containing 1.0% has a slump value 69 mm decrease 7 mm from the control slump. Concrete containing 1.5% wire waste has a slump value 68 mm which is decrease 8 mm from control slump

3.2. Density

Table 2 shows that the concrete containing wire waste as fiber have insignificant result compared to the control specimen. The differences between control and concrete wire waste in range of 0.55% to 1.03% for both cube concrete at 7 and 28 days. This trend is similar as steel fiber concrete [7] which is steel fiber in concrete has insignificant effect to concrete density with addition of less than 2% of fiber.

Table 2. Average density of all specimens and percentage different compared to control.

Type of samples	7 days (cube)		28 days (cube)	
	Average Density (kg/m^3)	Percentage different compared to control sample (%)	Average Density (kg/m^3)	Percentage different compared to control sample (%)
0% (control)	2353	-	2302	-
0.5%	2340	-0.55	2314	0.52
1.0%	2351	0.47	2333	0.82
1.5%	2369	0.77	2309	-1.03

3.3. Compressive strength

The percentage difference of compressive strength for 7 days of 0.5% wire waste was increased up to 17.4% compared to control concrete as shown in Table 3. However, concrete containing 1.0% waste and 1.5% wire waste fiber was decreased down to 0.4% and 1.3% compared to control specimen. The percentage difference of compressive strength for 28 days shows the same pattern. Concrete cubes containing 0.5% of wire waste was increased up to 24.4% compared to control specimen. Concrete cubes containing 1.0% and 1.5% wire waste were decreased with 18.7% and 25.6% respectively.

Table 3. Average compressive strength of concrete cube with wire waste as fiber material.

Type of samples	7 days		28 days	
	Average compressive strength (MPa)	Percentage different compared to control sample (%)	Average compressive strength (MPa)	Percentage different compared to control sample (%)
Control	22.4	-	33.2	-
Wire waste 0.5%	26.3	17.4	41.3	24.4
Wire waste 1.0%	22.3	-0.4	27	-18.7
Wire waste 1.5%	22.1	-1.3	24.7	-25.6

Based on the result compressive strength at ages of 7 and 28 days, the usage of wire waste as fiber materials in concrete was found to have significant impact on the strength of concrete. The result shows that concrete cubes that containing 0.5% of wire waste has highest strength, 26.3 MPa compared to control specimen which is, 22.4 MPa. Thus, compressive strength of concrete containing wire waste showed an increase in the concrete strength when 0.5% of wire waste were added. However, the strength decreased when the percentage of wire waste were increase up to 1.0% and 1.5%. The compressive strength at 28 days for concrete containing wire waste, 1.0% and 1.5% were decreased at 18% and 25%, respectively compared to control specimen.

IOP Conf. Series: Materials Science and Engineering **713** (2020) 012003 doi:10.1088/1757-899X/713/1/012003

It was clear that concrete specimen with 0.5% of wire waste has given better results compared to control specimen. The right amount of wire waste has filled the void in concrete thus makes the cube more solid which helped concrete strengthening [8, 9]. While the decreasing of strength started occurs at 1.0% of wire waste. The decreasing in strength happen due to larger doses of fiber that lead to improper distribution resulting in balling of fibres and air pockets which adversely affect the compressive strength of concrete [10, 11]. The overall finding shows that the concrete that contain right amount of wire waste can increase the compressive strength of the concrete.

3.4. *Tensile strength*

The tensile strength was conducted to determine the strength of concrete containing wire waste fiber at the age of 28 days. It is clearly showing that the tensile strength of concrete containing 0.5% of wire waste is higher compared to control specimen. The result of concrete containing 0.5% wire waste shows the highest tensile strength with strength 3.74 MPa. Figure 3 shows an increasing of tensile strength with additional of 0.5% wire waste but the value started to decrease when 1.0% and 1.5% of wire waste were added. The differences of tensile strength at age 28 days was compared to the control specimen with 0.5% of wire waste was with increasing of 42.7%. However, the 1.0% wire waste and 1.5% wire waste of tensile strength were decreased for 1.9% and 16.0% compared to control specimen. In other study, the inclusion of 0.6kg/m^3 nylon and polypropylene fibers in concrete resulted in increase of up to 17.1% and 9.7% in the tensile strength, respectively, compared to the control concrete [12, 13]. This pattern of increasing strength was similar pattern of this present study.

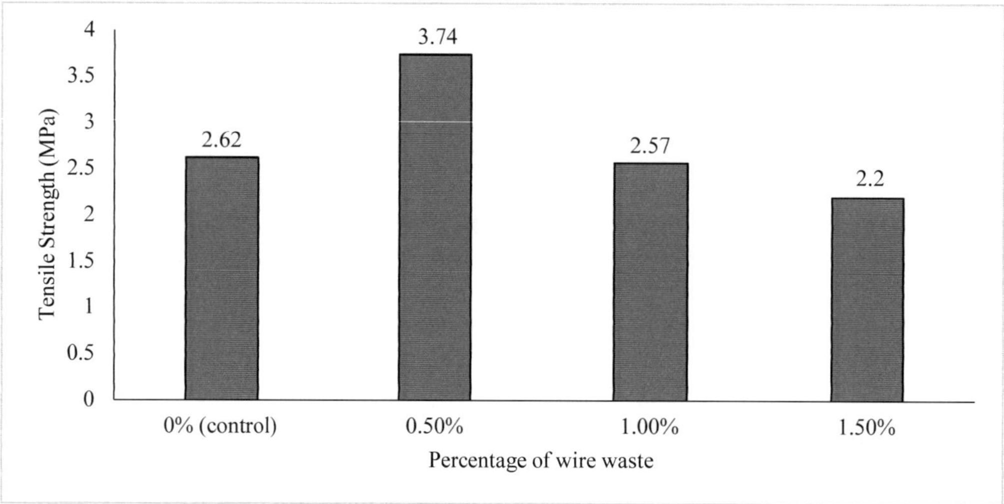

Figure 3. Tensile strength of concrete at the age 28 days.

4. Conclusion

Based on the compression and tensile results, it can be concluded that the most suitable proportion for wire waste fiber in concrete was 0.5% wire waste. These mix shows significant increases about 24.4% and 42.7% increases for compressive and tensile strength respectively, compared to the control concrete. The highest difference percentage of tensile strength of concrete containing 0.5% of wire waste with the differences 42.7 % compared to the normal concrete. Meanwhile, the smallest difference percentage and the nearest value to the normal concrete was concrete containing 1.0% of wire waste replacement with the differences about -1.9%. In fact, additional of wire waste exceed 0.5% is found to be not effective which can be seen from the reduction in performance of 1.0% and 1.5% of wire waste. It shows that concrete containing right percentage of wire waste shows a good performance for both compressive and tensile strength.

5. References

[1] Irwan Juki M, Muhamad K, Annas Mahamad M K, Boon K H, Othman N, Kadir A A, Asyraf Roslan M and Khalid F S 2013 Development of concrete mix design nomograph containing polyethylene terephtalate (PET) as fine aggregate *Advanced Materials Research* **701** 12-16

[2] Azmi N B, Khalid F S, Irwan J M, Mazenan P N, Zahir Z and Shahidan S 2018 Performance of composite sand cement brick containing recycle concrete aggregate and waste polyethylene terephthalate with different mix design ratio *IOP Conf. Series: Earth and Environmental Science* **140** 012129

[3] Ali N, Din N, Khalid F S, Shahidan S, Abdullah S R, Samad A A A and Mohamad N 2017 Compressive strength and initial water absorption rate for cement brick containing high-density polyethylene (HDPE) as a substitutional material for sand *IOP Conf. Series: Materials Science and Engineering* **271** 012083

[4] Kim S B, Yi N H, Kim H Y, Kim J H J and Song Y C 2010 Material and structural performance evaluation of recycled PET fiber reinforced concrete *Cement and Concrete Composites* **32(3)** 232-240

[5] Ali N, Shekarchi M, Mahoutian M and Soroushian P 2011 Mechanical properties of hybrid fiber reinforced lightweight aggregate concrete made with natural pumice *Construction and Building Materials* **25(5)** 2458-2464

[6] Irwan J M, Othman N, Koh H B, Asyraf R M, Khalid F S, Annas M M K and Shahrizan A M 2013 Development of mix design nomograph for polyethylene terephthalate fiber concrete *Applied Mechanics and Materials* **253-255(1)** 408-416

[7] Hassanpour M, Shafigh P and Mahmud H B 2012 Lightweight aggregate concrete fiber reinforcement - A review *Construction and Building Materials* **37** 452-461

[8] Afroughsabet V, Biolzi L and Ozbakkaloglu T 2016 High-performance fiber-reinforced concrete: A review *J. of Materials Science* **51(14)** 6517-6551

[9] Irwan J M, Muhamad K, Annas Mahamad M K, Boon K H, Othman N, Kadir A A, Asyraf R M and Khalid F S 2013 Development of concrete mix design nomograph containing polyethylene terephtalate (PET) as fine aggregate *Advanced Materials Research* **701** 12-16

[10] Khalid F S, Irwan J M, Wan Ibrahim M H, Othman N and Shahidan S 2018 Splitting tensile and pullout behavior of synthetic wastes as fiber-reinforced concrete *Construction and Building Materials* **171** 54-64

[11] Mazenan P N, Khalid F S, Azmi N B, Ayop S S, Abdul Ghani A H and Irwan J M 2018 Review of recycles concrete aggregate and polyethylene terephthalate in the manufacturing of brick *IOP Conf. Series: Materials Science and Engineering* **401** 012033

[12] Irwan J M, Annas M M K, Aeslina A K, Asyraf R M and Khalid F S 2014 Deflection behaviour of irregular-shaped polyethylene terephthalate fibre reinforced concrete beam *Advanced Materials Research* **911** 438-442

[13] Irwan J M, Annas M M K, Aeslina A K, Othman N, Koh H B, Asyraf R M and Faisal S K 2014 Cracking propagation of reinforced concrete using polyethylene terephtalate (PET) bottles as fine aggregate *Advanced Materials Research* **911** 474-478

Performance of Modified Mortar Containing Epoxy

Arisya Abdul Latip[1], Noorwirdawati Ali[1], Noor Azlina Abdul Hamid[1], Norhafizah Salleh[1], Siti Radziah Abdullah[1] and Shahiron Shahidan[1]

[1] Jamilus Research Centre, Faculty of Civil and Environmental Engineering, Universiti Tun Hussein Onn Malaysia, 86400 Batu Pahat, Johor, MALAYSIA.
Corresponding E-mail: wirda@uthm.edu.my

Abstract. Performance of FRP are highly depending on the adhesive used. FRP known to have high specific strength and specific stiffness, high durability against corrosion, light in weight, easy and quick installation therefore reduced manufacture time, great versatility and reduced maintenance. Despite all the the advantages possessed by FRP, there are actually several drawbacks which mainly causes by the epoxy resin. The organic binder are harmful towards humans, expensive and not applicable on wet surface. This study modified and elevated the binder materials by combining different percentage of mortar and epoxy. Compressive strength, density and water absorption of the specimen were tested and the optimum percentage were determined. Epoxy without harderner that were tested in this study is ranging from 0-5 % and composite cement consisted of 50% of Ground Granulated Blast Slag (GGBS) and 50% of Ordinary Portland Cement (OPC). From the experimental results, 3% of epoxy gives promising results as partial replacement of mortar.

1. Introduction

Deterioration of a concrete structure are mainly due to exposure to a harsh environment over a long period of time. Therefore the needs in finding alternative in strengthening or rehabilitate the structure is highly crucial. Strengthening of an element comes in many forms either as a composite materials or as an accessories. Fibre reinforced polymer (FRP) has been widely used in strengthening and rehabilitation of existing concrete structure. Jacketing concrete columns with steel or fibre-reinforced polymer (FRP) has been identified as an effective way of restoring and enhancing the strength and ductility of columns damaged by catastrophic events such as earthquakes, blasts and vehicle impact according [1].

FRP comprises of continuous fibres which can be either carbon, glass or aramid immerse in resin which varies from epoxy, vinylester or polyester. The resin will holds together the fibers and the concrete and exchanges the load between them. The performance of an element reinforced by FRP is depending on the adhesive or the binder materials used. Epoxy resin is a common adhesive in FRP application. Epoxy resins has been used widely mainly due to many advantages it possesses. FRP has high specific strength and specific stiffness, high durability against corrosion, light in weight, easy and quick installation therefore reduced manufacture time, great versatility and reduced maintenance. Different structures can be installed by using FRP. However, the long term durability have not been proven [2].

Polymers also known as epoxy have played an important role in repairing works mainly in constructions. Epoxy can either be used by itself as a single binder or can be incorporated in a cement mix or also known as modified epoxy [3]. According to El-Hawary and Abdul-Jaleel [4], addition of epoxy enhances the durability of concrete as it reduces the permeability of the concrete. The increase of the epoxy content able to increase the corrosion resistance of the modified concrete. In this study, modified epoxy containing different percentage of epoxy and composite cement mortar was further studied to determine the best percentage which are needed to achieve high compressive strength and density and high tensile strength.

Despite the advantages possessed by FRP, there are actually several drawbacks that can be observed. The flaws form FRP are mainly from the use of epoxy resins as adhesive. Triantafillou and Papanicolaou [5] stated that the drawbacks includes poor behaviour at temperatures above the glass transition temperature, relatively expensive cost of epoxy resins, potential hazard for manual workers, not suitable on wet surfaces or at low temperature, lack of vapour permeability, incompatibility of resins and substrate materials, absorption of water due to polar groups reduces stiffness and vulnerable towards fire since resins are organic materials.

Therefore, further studies has been conducted and an alternative has been identified. It is found out that replacing organic binder (epoxy resins) with inorganic binder (cement-based mortar) can eliminates most of the flaws form epoxy resins. However, there are some issues with the inorganic binders. The cement-based mortar has poor bond condition due to the granularity of the mortar. Hence, penetration and impregnation of fibre sheet is very difficult to achieve. The geometry structure of a concrete structure will be change affected by the mortar [5]. Modifying epoxy with composite cement mortar can elevates the performance of the modified epoxy in terms of strength and durability [6]. The effectiveness will varies depending on the percentage content of both cement based-mortar and epoxy [7].

2. Experiment

2.1. Materials

Epoxy that were used for this research is Sikadur 31 without hardener. Sikadur 31 possessed a texture similar to sand which made it easier to mix. GGBS is used as replacement at percentage of 50% of OPC which were fixed on every specimen. GGBS were mix together along with cement, sand, epoxy and water. GGBS were known to have small particles similar to cement. Ordinary Portland Cement (OPC) were used in this study. Fine aggregates that were used have an appropriate particle size of 5.00 mm.

2.2. Preparation of specimens

This study produced modified mortar containing contains cement, sand, ground granulated blast furnace slag, water together with epoxy which is Sikadur 31 without hardener. The main element of this study was the content of epoxy and the composite cement mortars which were varies in different percentages. Table 1 shows the percentage of each materials. A total of 36 specimens were casted in the shape of cube with measurement of 50×50×50 mm and cured for 7 and 28 days. The specimen were covered with damped cloth as method curing The density of the specimen were obtain by weighing using an electronic weighing machine while the compression strength were obtain from the compression test. The study were also able to determine the water absorption of the specimen. After the experiment is completed and all the data is obtained the result were evaluated and compared to determine the best percentage and concluded which perform best in terms of compressive strength, density and water absorption.

Table 1: Percentage of each materials

Composite cement		Epoxy (%)
Cement (%)	GGBS (%)	
50	50	0
49.5	49.5	1
49	49	2
48.5	48.5	3
48	48	4
47.5	47.5	5

IOP Conf. Series: Materials Science and Engineering **713** (2020) 012004 doi:10.1088/1757-899X/713/1/012004

2.3. Compressive strength test

All specimens at the age of 7 and 28 days were tested. The specimen were position on a crushing machine also known as universal testing machine. Loading were applied until the specimen breaks). The final load applied will be taken as results. The average result will be considered as the compressive strength of the specimen. The procedure of handling compressive strength test according to BS EN 12350-3:2010.

2.4. *Density and Water Absorption*

The test is simple and the procedure is direct. Before proceeding to the test, all specimens were weigh using a digital weighing machine and the weight is recorded. Water absorption were conducted by weighing the specimen before and after curing for both 7 and 28 days. The increment of weight were expressed as its absorption (in percentage)

3. Results

3.1. Compressive Strength

Compression test were conducted on 36 specimens after undergoing the process of curing for 7 and 28 days. All the results obtained from the testing were recorded and tabulated as shown in Table 2. Based on Table 2, the highest compressive strength value recorded was for sample containing 3% of epoxy at 7 days which was 23.4 MPa and 34.4 MPa at 28 days. Compressive strength for modified epoxy were slightly higher and almost similar to the control samples which was 22.4 MPa and 34.1 MPa at 7 and 28 days respectively. The lowest compressive strength were recorded at 5% of epoxy which were 18.2 MPa and 19.3 MPa at 7 and 28 days respectively.

Figure 1 illustrated the trend for overall results for compression test. Based on Figure 4.1 the compressive strength increases as the epoxy content increases up to 3% of epoxy and started to decline after 3% until 5% of epoxy content for 7 and 28 days. This shows that the optimum epoxy content is at 3%. However in comparisons with past research from Ariffin et al. [7], the optimum epoxy content should be 5%. This occurrence would probably cause by the replacement of 50% GGBS which were the difference between this research and past research.

The declination of compressive strength were most probably due to the unhardened epoxy [7]. As the epoxy content increases, the process of hardening of cement will slows downs due to the unhardened epoxy thus weaken the strength of mortar. This occurrence should be due to the absent of hardener and alkalis in the epoxy which in results retards the polymerization of epoxy. The increment of compressive strength as epoxy content increased are due to the improvement of bonding between hydrated cement to epoxy itself. Epoxy also enhance workability as a result improve the compressive strength.

Table.2: Results for Compressive Strength

Epoxy (%)	Compressive Strength (MPa)	
	7 days	28 days
0	22.4	34.1
1	18.9	27.3
2	19.3	29.9
3	23.4	34.1
4	19.4	27.2
5	18.4	19.3

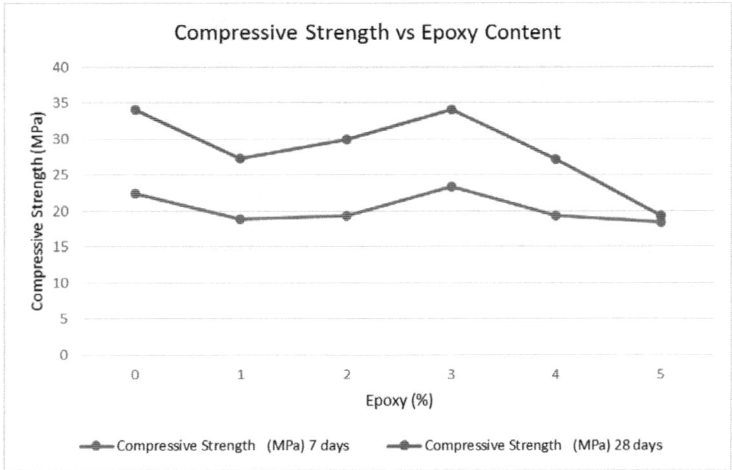

Figure 1: Graph of Compressive Strength versus Epoxy Content

3.2. Density

Before compression test were conducted, all specimen were weighed to obtain their masses before and after curing. Table 3 shows the results for density test. Based on Table 3, the highest density were at 3% of epoxy for both 7 and 28 days with 2325 and 2336 kg/m^3 of density. The density were higher than control which were 2189 kg/m^3. However after reaches the highest peak, the density decreases abruptly at 5% of epoxy for both 7 and 28 days with 2168 and 2245 kg/m^3.

Table.3: Results for density

Epoxy (%)	**Density (kg/m^3)**	
	7 days	**28 days**
0	2189	2243
1	2184	2200
2	2267	2280
3	2325	2336
4	2293	2299
5	2168	2245

The overall pattern of density for 7 and 28 days were presented in Figure 2. The optimum density for both 7 and 28 days were at 3% of epoxy which were higher than control. The increase in density were due to the ability of the epoxy to absorb water and the polymerization of epoxy [8]. The polymerization of epoxy fills the pores between cement particles thus producing denser mortar. The declination of the density are most probably due the unhardened epoxy as the epoxy content increases that were unable to fills the voids in the mortar. Other than that, the mixture of sand and epoxy were probably not uniform thus affecting the density of mortar.

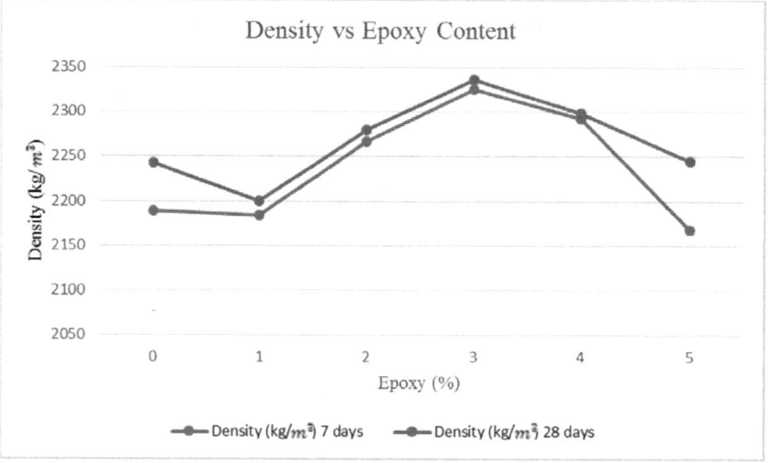

Figure.2: Graph of Density versus Epoxy Content

3.3. Water Absorption

Water absorption results were very simple to obtain which were by using simple calculation. Table 4 shows all the data of water absorption for 7 and 28 days. Based on the data recorded, the water absorption for 2%, 3% and 4% were almost the same and recorded as the highest for water absorption with 3.686%, 3.512% and 3.614% at 7 days while 3.68%, 3.515% and 3.482% at 28 days respectively. However the values of water absorption with epoxy were lower than normal mortar which was 3.924% and 3.827% at 7 and 28 days.

The overall pattern for water absorption is shown in Figure 3. The trend was inconsistent. As the epoxy content increases, the water absorption were also increases up to 4% of epoxy then decreases at 5%. The same trend were also can be shown in a study conducted by Borhan and Sutan [9]. The addition of polymer in mortar increase the water absorption but did not affect the compressive strength [9]. Based on observation during mixing process, the absorption of water can clearly be seen during 2%, 3% and 4%. The absorption of water causes the workability to improve. However, as the epoxy content gets higher, the absorption of water slows down thus makes it harder to mix and distribute during mixing. This is because epoxy possesses its own polymer membranes which prevent water from intruding [6].

Table.4: Results of Water Absorption

Epoxy (%)	**Water Absorption (%)**	
	7 days	28 days
0	3.924	3.827
1	2.244	2.814
2	3.686	3.68
3	3.512	3.515
4	3.614	3.482
5	2.31	2.494

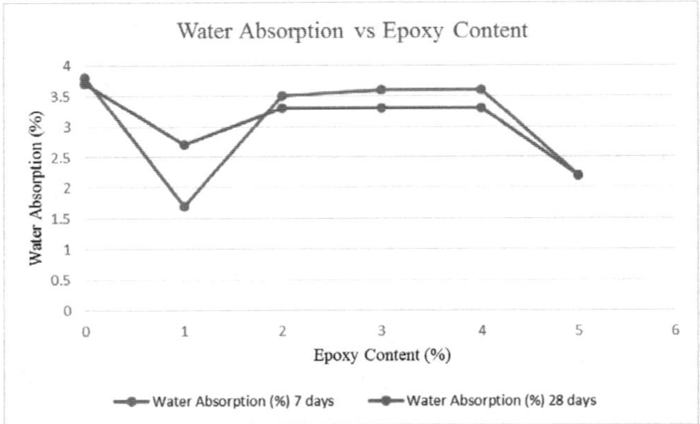

Figure.3: Graph of Water Absorption versus Epoxy Content

3.4. Relationship between Compressive Strength and Density

The relationship between compressive strength and density at 7 and 28 days were presented in Figure 4 and Figure 5. The trend of the graph showed that as the density increases the compressive strength also increases up to 3% of epoxy then decreases as the compressive strength decreases. It is found out that the optimum compressive strength and density of modified epoxy is when epoxy content is at 3% for both 7 and 28 days. The optimum density and compressive strength of modified epoxy were higher than control. This is probably due to the water absorption process were on pause at 3% of epoxy thus makes it more durable.

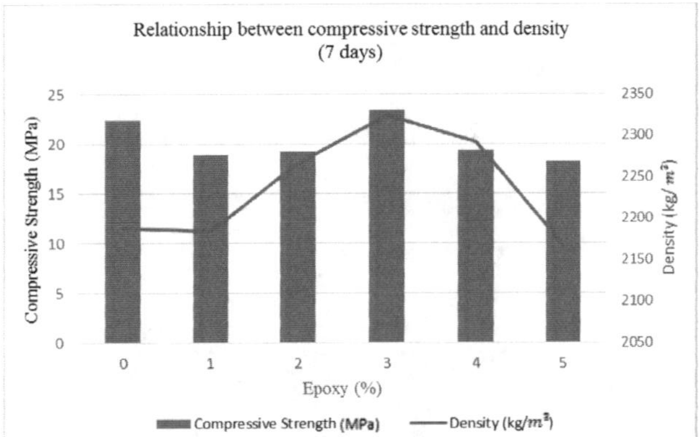

Figure. 4: Graph of Compressive Strength versus Density at 7 days

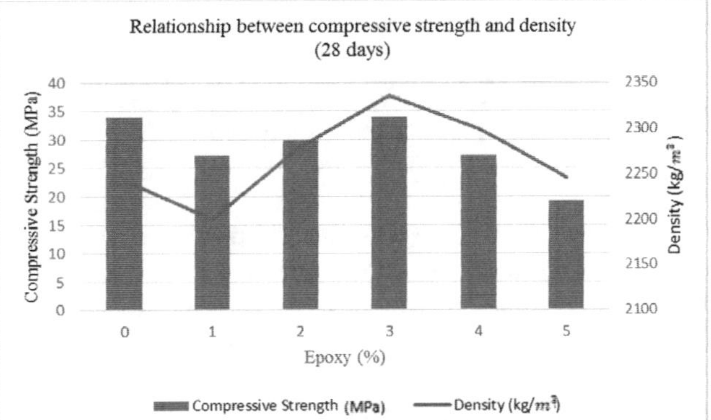

Figure.5: Graph of Compressive Strength and Density at 28 days

3.5. Relationship between Compressive Strength and Water Absorption

Figure 6 and Figure 7 concluded the relationship between compressive strength and water absorption at 7 and 28 days. The trend of the graph were inconsistent. It is found out that the compressive strength increases as the water absorption increases up to 2% of epoxy only. The water absorption were inconsistent but remains almost the same when epoxy at 2%, 3% and 4%. Water absorption slows down then declined at 5% of epoxy. In comparisons with the control specimen, modified mortar has slightly lower water absorption compared to the control specimen. However, the addition of epoxy did not help in reducing water absorption.

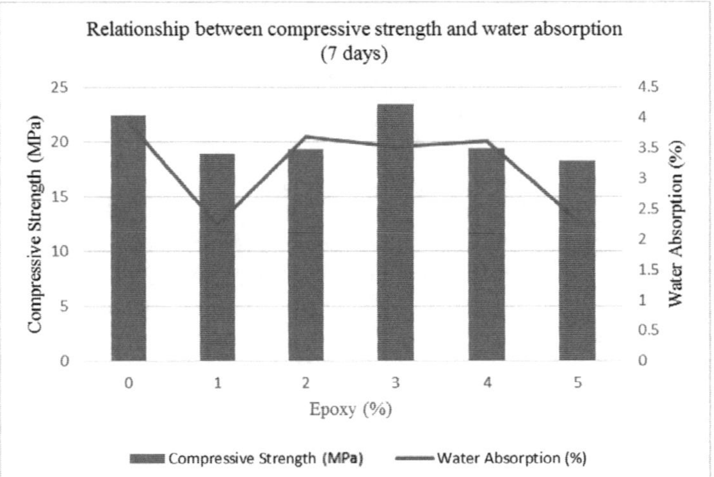

Figure 6: Graph of Compressive Strength and Water Absorption at 7 days

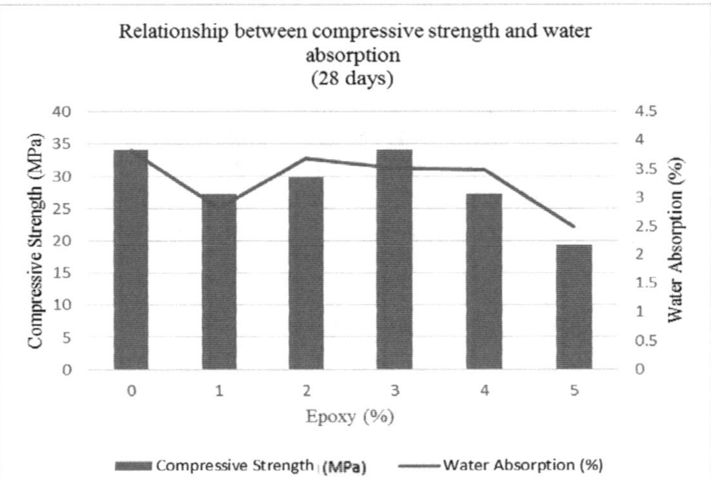

Figure 7: Graph of Compressive Strength and Water Absorption at 28 days

4. Conclusion

Epoxy plays an important role in strengthening however the production of it is costly and causes health deficit to whoever exposed to it. This study combines epoxy with mortar to reduce the usage of epoxy and elevates the performance of mortar and the results were analyzed and can be concluded as follows:

1) In terms of compressive strength, the addition of epoxy enhances the compressive strength of mortar. The compressive strength was the highest at 3% of epoxy but declined at 4% and 5%.

2) In terms of density, the addition of epoxy increases the density of mortar. The optimum density is at 3% of epoxy with 2325 kg/m3 and 2336 kg/m3 for 7 and 28 days. However at 5% of epoxy the density declined abruptly.
3) In terms of water absorption, as the epoxy increases the water absorption also increases up to 2%. However water absorption remains almost constant at 2%, 3% and 4% and started to drop at 5% of epoxy. However when compared to the control specimen the water absorption of modified mortar were lower than normal mortar.
4) The optimum percentage of epoxy for this research was 3%, 48.5% for ground granulated blast slag (GGBS) and 48.5% of OPC as it recorded the highest compressive strength of 23.4 MPa and 34.1 MPa for both 7 and 28 days respectively. The compressive strength of modified motar were slightly higher and almost the same as normal mortar which were 22.4 MPa and 34.1 MPa for 7 and 28 days respectively.
5) Overall, it can be concluded that the addition of epoxy can elevates the performance of modified mortar by enhances the compressive strength and density but not water absorption even though compressive strength should increase as the water absorption decrease.

5. References

[1] Tetta Z C and Bournas D A 2016 TRM vs FRP jacketing in shear strengthening of concrete members subjected to high temperatures *Composites Part B: Engineering* **106** 190-205
[2] Frigione, M and Lettieri M 2018 Durability issues and challenges for material advancements in FRP employed in the construction industry *Polymers* **10**(247) 1-15
[3] Aggarwal L K, Thapliyal P C and Karade S R 2007 Properties of polymer-modified mortars using epoxy and acrylic emulsions *Construction and Building Materials* **21**(2) 379–383
[4] El-Hawary M M and Abdul-Jaleel A 2010 Durability assessment of epoxy modified concrete *Construction and Building Materials* **24**(8) 1523–1528
[5] Triantafillou T C and Papanicolaou C G 2006 Shear strengthening of reinforced concrete members with textile reinforced mortar (TRM) jackets *Materials and structures* **39**(1) 93-103
[6] Anagnostopoulos C A, Sapidis G and Papastergiadis, E 2016 Fundamental properties of epoxy resin-modified cement grouts *Construction and Building Materials* **125** 184–195
[7] Ariffin N F, Hussin M W, Mohd Sam A R, Bhutta M A R, Nur N H and Mirza J 2015 Strength properties and molecular composition of epoxy-modified mortars *Construction and Building Materials* **94** 315–322
[8] Abdulla A I 2016 Thermal properties of sand modified resins used for bonding CFRP to concrete substrates *International Journal of Sustainable Built Environment* **5**(1) 176–182
[9] Borhan M and Sutan M 2011 Laboratory study of water absorption modified mortar *UNIMAS e-Journal of Civil Engineering* **2**(1) 25–30

Acknowledgments
This research is funded by Tier 1 Grant H224, Universiti Tun Hussein Onn Malaysia.

IOP Conf. Series: Materials Science and Engineering **713** (2020) 012005 doi:10.1088/1757-899X/713/1/012005

Effect of Lightweight Aggregates on early-age behavior of High Performance Concrete

M Higazey[1], M J Shannag[1] and A Alaskar[1]

[1] Department of Civil Engineering, King Saud University, Riyadh, 11421, Saudi Arabia
Corresponding author: mhigazey@gmail.com

Abstract. Laboratory experiments were conducted to investigate the effects of locally available natural lightweight aggregates (LWA's) on volume changes and mechanical properties of high-performance concrete (HPC). The normal aggregates fraction of high-performance concrete was replaced with fine and coarse LWA's at three different replacement levels, 5%, 10% and 20% by volume. The test results indicated a significant reduction in the autogenous shrinkage of high-performance concrete at 56 days without a corresponding considerable reduction in compressive strength of the same concretes. The test results have also shown that up to 20% volume fraction of coarse LWA's could be used for mitigating the volume changes of HPC with a slight reduction in compressive and splitting tensile strengths, and elastic modulus. Therefore, this investigation recommends the use of locally available natural LWA's as a viable curing technique for minimizing the volume changes and risk of cracking of HPC's without compromising their mechanical properties.

1. Introduction

Recent construction depends more and more on the utilization of HPC due to its high strength and reduced permeability. Besides cost reduction, HPC offers the engineer an outstanding combination of properties including enhanced durability and improved mechanical performance. Recently, the use of high-performance concrete has become increasingly important in high rise buildings, bridge decks, marine structures and airport pavements. As HPC is created by a high cement content and low water to cement ratio, it may have significant autogenous deformation at early ages as well as low shrinkage cracking resistance at later ages [1,2]. Proper curing of concrete structures is essential to ensure that they meet the requirements of their expected efficiency and durability. This is often achieved through external curing in ordinary building. External curing procedures are not effective in eliminating the autogenous deformation that limits the strength of HPC [3]. Many techniques were proposed to reduce the influences of autogenous shrinkage [4]. They include using shrinkage reducing admixtures [5] or expansive additives [6,7], addition of fibers [8,9], and by using different LWA [10]. Porous aggregates have been successfully used in the last few decades to mitigate shrinkage between established methods of shrinkage mitigation. Lightweight aggregates are defined as materials lighter than water and distinctly more porous than sand, gravel and ground rock, which are commonly referred to normal weight aggregates [11]. The aggregate is classified as LWA when it satisfies a dry density not higher than 2000 kg/m^3 and a bulk unit weight not higher than 1200 kg/m^3 [12]. Lightweight aggregate is capable of providing sufficient volume of water, and has a structure that enables the water to be discharged to the paste as needed, and is little enough to be spaced within the matrix [3,13]. The distribution, size, absorption and volume of LWAs are important factors that determine its effectiveness [14]. Four different mixes of HPC were designed at 0%, 5%, 10% and 20% with locally available natural coarse

LWA replacement of total volume of aggregates. Similar mixes were also designed with fine LWA to determine the effectiveness of the size of aggregates. Volume change properties including autogenous shrinkage and some mechanical properties were determined for the HPC mixtures.

2. Experimental Program

2.1. Materials

Type I cement white sand, crushed sand, coarse aggregates, water, silica fume, superplasticizer and natural lightweight aggregates are the materials used. The cement had a specific gravity of 3.15 and followed ASTM C150-16 specifications [15]. The fine aggregates used consist of 20% crushed sand and 80% white sand on weight basis. The physical properties of the normal aggregates were determined by ASTM C33-18 [16] in the laboratory and are shown in table 1. The concrete mixes are filled with drinking water. The powdered silica fume has a specific gravity of 2.2. The superplasticizer used is high range water reducer admixture (HRWRA).

Table 1. Normal aggregates' physical properties

Material	Specific gravity	Water Absorption (%)	Size (mm)	Fineness modulus
White sand	2.6	0.3	0.1- 2	1.88
Crushed sand	2.59	1.67	2 - 4	4.66
Coarse aggregates	2.62	1.45	4 - 10	7

The lightweight aggregates used in this study were coarse and fine volcanic tuffs of scoria origin as shown in figures 1 and 2, respectively. The physical properties of LWA were determined in the laboratory according to ASTM 330-04 [17] and are presented in Table 2 [18,19].

Table 2. Lightweight aggregates' physical properties

Material	Specific gravity			Water Absorption (%)	Unit Weight kg/m^3		Moisture content
	SG_{OD}	SG_{SSD}	SG_{App}		Loose	Dense	
Fine LWA	1.78	2.0	2.40	9.7	995	1120	5.23
Coarse LWA	1.54	1.72	1.87	10.6	762	860	2.88

Figure 1. Natural coarse lightweight aggregates from Saudi Arabia

Figure 2. Natural fine lightweight aggregates from Saudi Arabia

2.2. Concrete Mixtures
A series of concrete mixes were designed at a water to cement ratio of 0.3 and a cementitious materials content of 550 kg/m^3. The mixtures are designated as N, C05, C10, C20, F05, F10 and F20 as shown in Table 3. Mixture N is the reference concrete with no LWA, and Mixtures C05, C10, C20, F05, F10 and F20 are with pre-saturated natural LWA; C stand for coarse LWA and F stands for fine LWA followed by the replacement percentage 5%, 10% and 20% of the total volume of normal weight aggregates. The fine LWA replacement amount was determined by replacing the 5%, 10% and 20% of both crushed sand and white sand amounts which combine the fine aggregates by 20% for crushed sand and 80% for white sand on weight basis. The silica fume used in mixes as 10% of cement volume to get high strength concrete. The superplastizer was added as 7% of water amount. The cementitious materials to coarse aggregate ratio and cementitious materials to fine aggregate ratio are 0.59 and 0.75, respectively.

Table 3. Concrete Mix Proportions

Mix	Cement (kg/m^3)	Silica Fume (kg/m^3)	Coarse Aggregates (kg/m^3)	Crushed Sand (kg/m^3)	White Sand (kg/m^3)	Coarse LWA (kg/m^3)	Fine LWA (kg/m^3)	Water (kg/m^3)	HRWRA (Liter/m^3)
N	500	50	930	150	582	0	0	165	11.5
C05	500	50	883.5	150	582	30.5	0	165	11.5
C10	500	50	837	150	582	61.1	0	165	11.5
C20	500	50	744	150	582	122.1	0	165	11.5
F05	500	50	930	142.5	552.9	0	28.2	165	11.5
F10	500	50	930	135	523.8	0	56.4	165	11.5
F20	500	50	930	120	465.6	0	112.7	165	11.5

2.3. Vacuum Saturation
The LWAs were prewetted at a vacuum of 0.04 MPa for 3 hours to provide water for internal curing and full saturation. First, the saturated LWA was placed in the glass barrel. The bucket was then filled with water and the valve which connected the water to the barrel was closed. The valve connected to the vacuum has been opened and the vacuum is turned on for the first hour to eliminate the air from the aggregates as shown in figure 3. After one hour, the water valve was opened and the water was added until it submerged the total amount of LWA then the valve was closed. The process was maintained under constant pressure to saturate the LWA for two hours. After vacuum saturation, the coarse LWA was dried with a towel to achieve saturated surface dry (SSD) condition [14]. The SSD condition for fine LWA was determined using the cone method presented in ASTM C33-18 [16].

Figure 3. Apparatus used for Vacuum Saturation of LWA

Figure 4. Preparation of the Specimens used for Autogenous shrinkage test

2.4. *Mixing*

The concrete mixtures were prepared using a pan type mixer of 500 Liters capacity in accordance to ASTM C192-18 [20]. Firstly, to prevent absorption, the inside of the mixer was cleaned with water. The coarse aggregates were mixed first, followed by fine aggregates. Then, the quantity of water absorbed by the aggregates was added; then cement, silica fume were added. Based on the performance of trial mixes; to extend the activity time of HRWRA, it was preferred to add firstly about 75% of the water containing about 50% of the HRWRA. After that, the remaining amount of the water containing the remaining amount of the superplasticizer was added over the last 3 minutes. of the time for mixing. After casting, the samples were covered with plastic sheet and stored at 23o C and 65% relative humidity in the laboratory for 24 hours, then demolded and kept under water. Each sample was labeled as to the date of casting, mixture used, sample number, and the direction of studs for the prisms.

2.5. *Autogenous Shrinkage Test*

Autogenous shrinkage was measured on 75×75×350 mm prism specimens every week until 56 days. Three samples were measures for each mix. Stainless steel studs were embedded at the ends of the samples. The samples were removed from the molds at an age of 24 h. Upon removal of the specimens from the molds, they were placed in lime-saturated water and kept at 23 °C for 45 min before being measured for length. The samples were kept in water at 23 ° C after the initial comparator reading until they reached 7 days of age, including the period in the molds. At the end of the curing time, the prism samples were sealed with two layers of aluminum foil using adhesive tape as shown in Figure 4 [2], and a second comparator reading was taken. The samples were placed in the drying room after measurement at the end of the curing time. The readings were taken for each specimen at 7, 14, 21, 28, 35, 42, 48 and 56 days. The difference between the two readings was used for calculating the length change of the specimen at the required age [21].

2.6. *Mechanical Properties Tests*

Compressive strength test was performed using 150×150×150 mm cube samples following ASTM C39-18 test procedure [22]. Three specimens were tested at 28 days for each mixture. Molds were lubricated and compacted using a vibrating table. One day after casting, the concrete samples for the aforementioned tests were stored in the laboratory at 23◦ C for 24 hours, labeled and then cured in water till the testing day. The splitting tensile strength test was performed using three 100×200 mm cylinder specimens at 28 days following the test procedure outlined in ASTM C496-14 [23]. The Static elastic modulus test was performed using 100×200 mm cylinder specimens. Three specimens were tested at 28 days for each mixture. The sample was positioned with the strain measuring device connected to the test machine to ensure that the load transfer system is balanced properly. The elastic modulus was determined following the guidelines outlined in ASTM C469-14 [24].

3. Test Results and Discussion

3.1. *Autogenous Shrinkage*

The autogenous shrinkage can be considered as the result of self-desiccation. The results of autogenous shrinkage of the mixtures with coarse and fine LWA are shown in figures 5 and 6, respectively. It can be observed that the partial replacement of normal aggregates of HPC mixtures with LWAs lead to a decrease of autogenous shrinkage. The test results indicated that by increasing the content of LWAs in HPC, the autogenous shrinkage will decrease. The decrease of autogenous shrinkage is due to the presence of LWAs particles that served as water reservoirs for providing additional moisture to concrete. At 20% replacement level, the fine or coarse LWA lead to mitigating the autogenous shrinkage at 28 days. The fine LWA at 5% and 10% replacement levels was more effective than coarse LWA in decreasing the autogenous shrinkage. At 5% and 10% levels, fine LWA reduced the autogenous shrinkage by 38% and 93%, and the coarse LWA reduced it by 14% and 64%, respectively. The use of LWA with smaller particle size fraction filled in more spaces within the concrete mix and resulted in the distribution of the water reservoirs in the proximity of the cement paste that lead to better internal curing compared to coarse LWA [4,13].

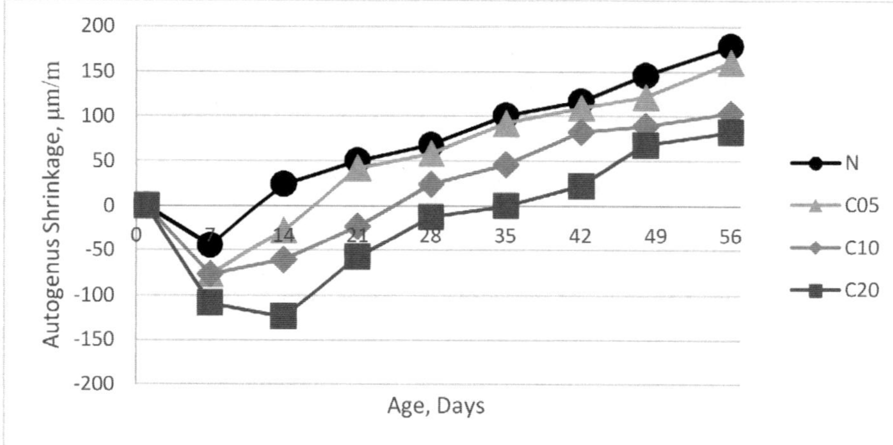

Figure 5. Autogenous Shrinkage of HPC mixes containing Coarse LWAs

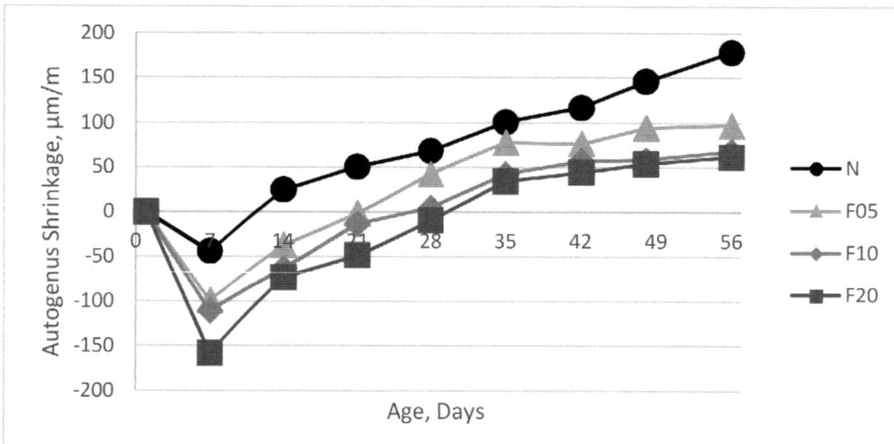

Figure 6. Autogenous Shrinkage of HPC mixes containing Fine LWAs

3.2. *Mechanical Properties*

Figures 7 through 9 show the 28-days results of compressive strength, splitting tensile strength and elastic modulus, respectively for the concrete mixtures cast in this investigation. It can be observed from figure 7 that, replacing the normal aggregates fraction of HPC with LWAs caused a slight reduction in compressive strength. The low strength of LWA lead to the compressive strength reduction [1,13]. Infact, using a 20% replacement level of LWAs resulted in about 20% loss in compressive strength, where the 86 MPa compressive strength of the control concrete decreased to as low as 70 MPa for both coarse and fine LWA mixtures. Furthermore, it can be seen that using coarse LWA at 5% and 10% replacement level resulted in 8% to 11% loss of the compressive strength, whereas, using fine LWAs mixtures at the same replacement levels caused up to 8% strength loss.

The test results shown in figure 8 indicated that using up to 20% of coarse LWA did not cause a noticeable loss in the splitting tensile strength, whereas, using fine LWAs caused up to 30% loss at 20% replacement level. The results of LWA mixtures shows a higher standard deviation than the control mixture which may explain the reduction in C05 result. The test results shown in figure 9 indicated that both types of LWAs, coarse and fine caused up to 8% and 20% loss in elastic modulus of HPC respectively at a replacement level of up to 20% on volume basis.

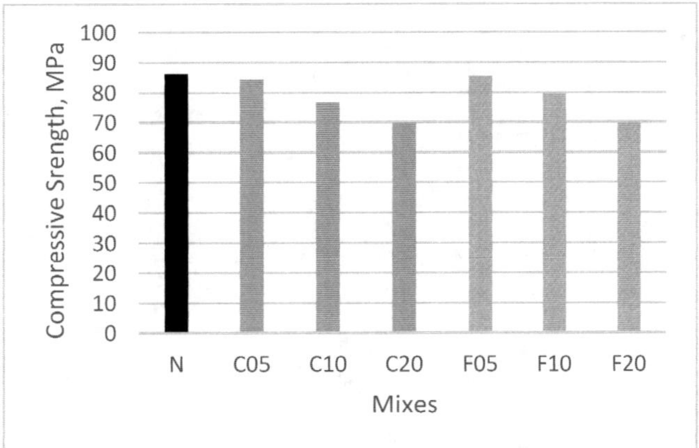

Figure 7. 28-days Compressive Strength of HPC Mixtures

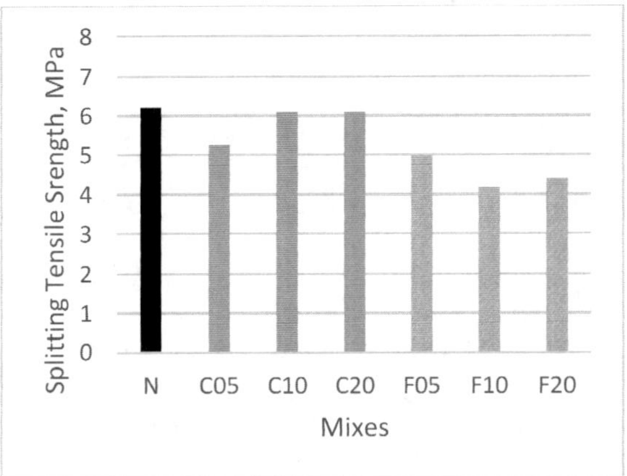

Figure 8. Splitting Tensile Strength of HPC Mixtures at 28 days

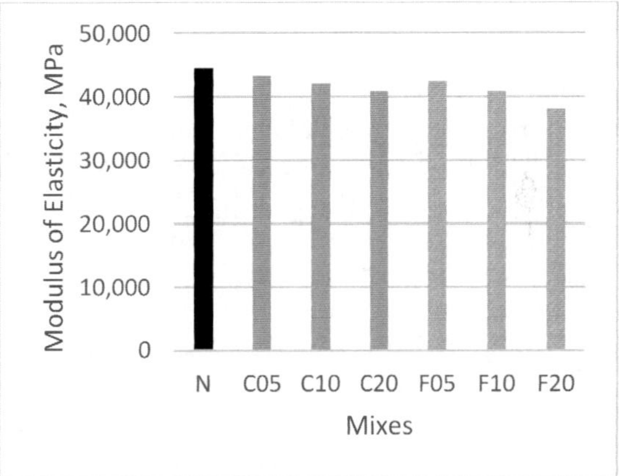

Figure 9. Elastic Modulus of HPC Mixtures at 28 days

3.3. Discussion

The test results presented in figure 10 showed a significant decrease in the autogenous shrinkage of HPC due to the use of both types of natural lightweight aggregates, coarse and fine. This could be due to the porous microstructure of the natural lightweight aggregates which served as a moisture reservoir inside the concrete and thus helped in mitigating the self-desiccation of HPC with time. On the contrary, the interconnected pores coupled with the high absorption capacity of LWAs would definitely cause some losses in its mechanical properties such as compressive, and splitting tensile strengths, and elastic modulus. However, the preliminary test results of this investigation indicated that the percentage reduction in the mechanical properties, compressive strength in particular was within the range of acceptable strength limits of HPC investigated in the literature. Further investigation is going on to explore the potential application of the natural lightweight aggregates available in the kingdom for improving the performance of high strength and high performance concretes.

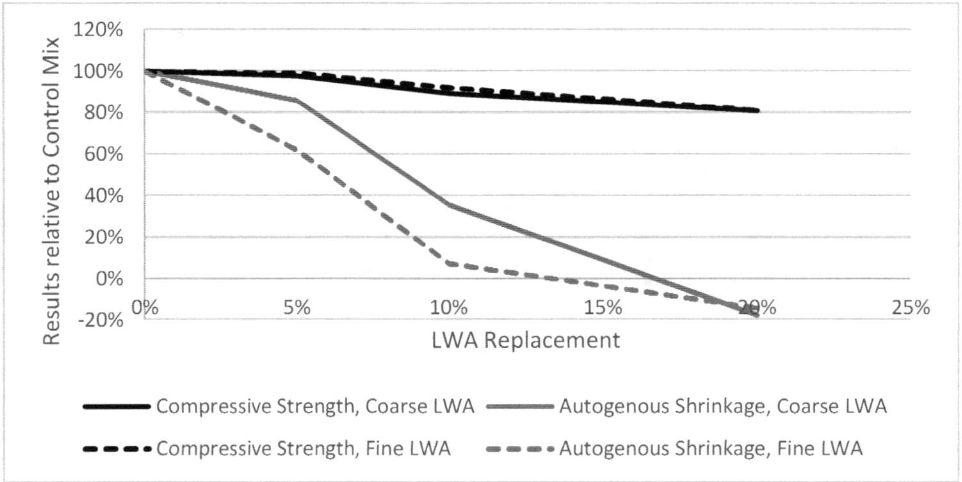

Figure 10. Effect of LWA amount on autogenous shrinkage and compressive strength at 28 days

4. Conclusions

Based on the experimental test results of this investigation the following conclusions can be drawn:

1. The autogenous shrinkage of high performance concrete (HPC) could be significantly reduced by replacing the normal aggregates fraction of HPC with natural fine or coarse LWA's at a various replacement levels of 5%, 10% and 20% on volume basis.
2. The test results reported indicated that at the highest replacement level, 20% of LWAs used, the HPC mixtures were able to maintain a residual compressive strength above 80% of the compressive strength of the control HPC mixtures without LWAs at 56 days of age.
3. This investigation recommends using locally available natural LWAs as a partial replacement of normal weight aggregates for mitigating the autogenous shrinkage and minimizing early age cracking of high performance concrete without compromising its compressive strength.

5. References

[1] Bentur A, Igarashi S and Kovler K 2001 Prevention of autogenous shrinkage in high-strength concrete by internal curing using wet LWA *Cement and Concrete Research* **31** 1587-1591

[2] Gesoğlu M, Güneyisi E, Ismael A and Öz H 2015 Internal Curing of High-Strength Concretes Using Artificial Aggregates as Water Reservoirs *ACI Materials Journal* **112** 809-819

[3] Ghourchian S, Wyrzykowski M, Lura P, Shekarchi M and Ahmadi B 2013 An investigation on the use of zeolite aggregates for internal curing of concrete *Construction and Building Materials* **40** 135-144

[4] Akcay B and Tasdemir M 2010 Effects of distribution of LWA on internal curing of concrete *Cement and Concrete Composites* **32** 611-616

[5] Bentz D 2005 Curing with Shrinkage-Reducing Admixtures: Beyond Drying Shrinkage Reduction *Concrete International* **27** 55-60

[6] Tanimura M, Mitami Y and Sato R 2005 An investigation of prediction model for autogenous shrinkage/expansion strain of low-shrinkage HSC. Self-desiccation and its importance in concrete technology *4th international research seminar* (Lund) 245–263

[7] Hori A, Marioka M, Sakai E and Daimon M 1999 Influence of expansive additives on autogenous shrinkage *Autogenous Shrinkage of Concrete* 93-104

[8] Loukili A, Khelidj A and Richard P 1999 Hydration kinetics, change of relative humidity, and autogenous shrinkage of ultra-high-strength concrete *Cement Concrete Research* **29** 577–84

[9] Farhat F, Nicolaides D, Kanellopoulos A and Karihaloo B 2007 High performance fibre-reinforced cementitious composite (CARDIFRC)-performance and application to retrofitting *Engineering Fracture Mechanics* **74** 151–67

[10] Shen D, Jiang J, Jiao Y, Shen J and Jiang G 2017 Early-age tensile creep and cracking potential of concrete internally cured with pre-wetted lightweight aggregate *Construction and Building Materials* **135** 420-429
[11] Degennaro R, Cappelletti P, Cerri G, Degennaro M, Dondi M and Langella A 2004 Zeolitic tuffs as raw materials for lightweight aggregates *Applied Clay Science* **25** 71-81
[12] Degennaro R, Cappelletti P, Cerri G, de' Gennaro M, Dondi M, Graziano S and Langella A 2007 Campanian Ignimbrite as raw material for lightweight aggregates *Applied Clay Science* **37** 115-126
[13] Yang S and Wang L 2017 Effect of Internal Curing on Characteristics of Self-Compacting Concrete by Using Fine and Coarse LWA *Journal of Materials in Civil Engineering* **29** 1-11
[14] Browning j, Darwin D, Reynolds D and Pendergrass B 2011 Lightweight Aggregate as Internal Curing Agent to Limit Concrete Shrinkage *ACI Materials Journal* **108** 638-644
[15] ASTM C150 2016 *Standard specification for portland cement* (Pennsylvania: American Society of Testing and Materials) pp 1-10
[16] ASTM C33 2018 *Standard specification for concrete aggregates* (Pennsylvania: American Society of Testing and Materials) pp 1-8
[17] ASTM C330 2004 *Standard Specification for Lightweight Aggregates for Structural Concrete* (Pennsylvania: American Society of Testing and Materials) pp 1-4
[18] Shannag M, Charif A and Dghaither S 2014 Developing Structural Lightweight Concrete Using Volcanic Scoria Available in Saudi Arabia Arabian *Journal for Science and Engineering* **39** 3525-3534
[19] ASTM C1761 2017 *Standard Specification for Lightweight Aggregate for Internal Curing of Concrete* (Pennsylvania: American Society of Testing and Materials) pp 1-8
[20] ASTM C192 2018 *Standard Practice for Making and Curing Concrete Test Specimens in the Laboratory* (Pennsylvania: American Society of Testing and Materials) pp 1-8
[21] ASTM C157 2017 *Standard Test Method for Length Change of Hardened Hydraulic-Cement Mortar and Concrete* (Pennsylvania: American Society of Testing and Materials) pp 1-7
[22] ASTM C39 2018 *Standard Test Method for Compressive Strength of Cylindrical Concrete Specimens* (Pennsylvania: American Society of Testing and Materials) pp 1-8
[23] ASTM C496 2014 *Standard Test Method for Splitting Tensile Strength of Cylindrical Concrete Specimens* (Pennsylvania: American Society of Testing and Materials) pp 1-4
[24] ASTM C469 2014 *Standard Test Method for Static Modulus of Elasticity and Poisson's Ratio of Concrete in Compression* (Pennsylvania: American Society of Testing and Materials) pp 1-5

Acknowledgments
The authors would like to thank the Deanship of scientific research for funding and supporting this research through the initiative of DSR Graduate Students Research Support (GSR).

IOP Conf. Series: Materials Science and Engineering **713** (2020) 012006 doi:10.1088/1757-899X/713/1/012006

Effects of Curing Conditions on Properties of Lightweight Concrete Brick Containing Expanded Polystyrene and Palm Oil Fuel Ash

N I M Yassin[1], S H Adnan[2], S Shahidan[1], S S Ayop[1], N A Kamarulzaman[2] M H Osman[2], Z Jamelloddin[1], M Majid[1], N Salleh[1], A Alisibramulisi[3], M N A Wahee Anuar[4]

[1]Faculty of Civil and Environmental Engineering, Universiti Tun Hussein Onn Malaysia, Batu Pahat, Johor, MALAYSIA
[2] Department of Civil Engineering Technology, Faculty of Engineering Technology, Universiti Tun Hussein Onn Malaysia, Pagoh, Johor, MALAYSIA
[3]Faculty of Civil Engineering, UiTM Shah Alam, 40450 Shah Alam, Selangor, MALAYSIA.
[4]Melayu Jati Enterprise, Batu Pahat, 83000 Johor, MALAYSIA
Corresponding author: suraya@uthm.edu.my

Abstract. This study presents the effect of curing conditions towards properties of lightweight concrete brick made up of cement, sand and some replacement materials. The replacement materials used in this study is Expanded Polystyrene (EPS) as sand replacement and Palm Oil Fuel Ash (POFA) as cement replacement. The percentage of replacement for POFA is 0%, 10% and 20% whilst 0%, 30% and 50% for EPS. The brick samples have been exposed to different types of curing conditions that is air curing and water curing until the day of test. Among the tests that have been conducted are hardened brick density, compressive strength and water absorption. Based on the findings, it has been proved that the replacement percentage of EPS and POFA has significant impacts towards the brick properties. In addition, it is found that hardened brick density and compressive strength of air curing samples are lower as compared to water curing samples meanwhile the water absorption for air curing samples are higher that water curing samples.

1. Introduction

Sustainability awareness become great concern in every construction industries in Malaysia. Due to high demand in sustainable building materials from industries, Malaysian researchers have done abundance of research on the production of sustainable building materials such as concrete and bricks.

The production of sustainable building materials can be produced by lessen the usage of natural resources such as cement and replace it with waste materials such as palm oil fuel ash (POFA), rice husk ash (RHA), silica fume and fly ash [1-4]

In addition, the production of lightweight building materials is equally important. This is because it could reduce the manufacturing and transportation cost as well as minimize the total dead load of a building [5]. One of the most preferable lightweight material use in the production of lightweight building materials is Expanded Polystyrene (EPS). Therefore, this study is focusing on the production of sustainable lightweight building materials which is lightweight concrete brick containing EPS and POFA.

According to Babu et al [6] EPS is stable low density foam of non- absorbent, hydrophobic and closed cell structure which is suitable to be used as lightweight aggregates in the production of lightweight building materials. Furthermore, based on Ferrandiz-Mas et al [7], the use of EPS will not causes any hazardous towards environment. Until now, EPS has been widely use as lightweight building material due to its characteristic which is extremely light. Previous researches show that EPS has significantly contribute to the reduction of brick density [2] [4-5] [7-21].

Meanwhile, POFA is one of the agricultural waste vastly generated by palm oil industries from the burning process of extracted palm oil fibers and shells in the palm oil mill. Since Malaysia is known as one of the largest palm oil producer in the world, hence vast amount of POFA has been generated annually. Rapid generation of POFA causes increase in the landfill areas. Besides, the waste did not give any profit to industries because it could not be used as fertilizer. Therefore, due to these problems, Malaysian researchers have conducted many of studies on the potential of POFA as the replacement of cement in the construction materials since POFA can act as the pozzolanic materials [22].

Based on previous research, an optimum amount of POFA could enhance the strength performance of concrete. For instance, 10% of cement replacement by POFA has increased the concrete strength as compared to normal concrete [23, 24].

Hence, this study has focus on properties of concrete brick containing EPS and POFA that has been exposed to different types of curing conditions. Among brick properties that has been tested are density, compressive strength and rate of water absorption.

2. Sample preparation

2.1. Materials Preparation

In this study, raw materials that have been used for the production of brick are Ordinary Portland Cement (OPC), river sand, POFA, EPS, tap water and super plasticizer.

2.1.1 Ordinary Portland cement (OPC)

The classification of OPC used in this study is type 1 based on the ASTM C150. The supplier for this cement is Tasek Corporation Berhad which certified by SIRIM. Cement need to be stored in an air tight container to prevent any moisture contact. The chemical composition of OPC has shown in Table 1.

2.1.2 Palm Oil Fuel Ash (POFA)

POFA was obtained from one of Palm Oil mill in Batu Pahat, Johor. Initially, the ash was dried in oven at 105°C for 24 hours in order to remove the moisture content. After dried, POFA was sieved by 300μm sieve passing. Table 1 shows the chemical composition of POFA obtained from the XRF analysis meanwhile Figure 1 shows the scanning electron microstructure of POFA.

Table 1. Chemical Composition of OPC and POFA

Chemical Composition	OPC	POFA
Silicon Dioxide (SiO_2)	14.6	55.20
Aluminium Oxide (Al_2O_3)	3.95	4.48
Ferric Oxide (Fe_2O_3)	3.46	5.44
Calcium Oxide (CaO)	57.1	4.12
Potassium Oxide (K_2O)	0.51	2.28
Magnesium Oxide (MgO)	1.62	2.25
Sodium Oxide (Na_2O)	-	0.1
Sulfur Trioxide (SO_3)	3.43	2.25

Figure 1. Scanning Electron Microstructure of POFA.

2.1.3 Fine Aggregates (sand)

In this study, river sand has been used as fine aggregates. River sand was sieved by sieve plate to ensure that it does not consist any impurities and in order to have a consistent size. Earlier than that, a process of oven dried for a minimum of 24 hours at the temperature of 105±5 °C was conducted. In this research, the sieve analysis of sand was as accordance to ASTM C 136.

2.1.4 Expanded Polystyrene (EPS)

The size of EPS used in this study was between 1.18mm to 2.36mm. EPS used was collected from ST Polyfoam Industries Sdn. Bhd which located in Batu Pahat, Johor.

2.2. Brick sample preparation

For this study, 9 types of concrete brick samples were prepared. The bricks were divided into 3 different group (I,II and III) according to percentage of EPS which is 0% for group I, 30% for group II and 50% for group III. Meanwhile the percentage of POFA is 0%, 10% and 20%. Water cement ratio for this study is 0.5 and super plasticizer is 8 ml for every 1 kg binder. The size of the brick for this study is 215 mm x 103 mm x 65 mm. The mix proportion of bricks as shown in Table 2.

Table 2. Mix proportions of brick (kg/m^3)

group	samples	cement	POFA	sand	EPS	Total samples
I	E0P0(c)	495.34	-	1484.72	-	15
	E0P10	445.14	50	1484.72	-	15
	E0P20	395.83	100	1484.72	-	15
II	E30P0	495.34	-	1038.19	5.98	15
	E30P10	445.14	50	1038.19	5.98	15
	E30P20	395.83	100	1038.19	5.98	15
III	E50P0	495.34	-	744.44	9.96	15
	E50P10	445.14	50	744.44	9.96	15
	E50P20	395.83	100	744.44	9.96	15

2.2.1 Mixing and casting process of bricks

Due to the tendency of EPS to float during mixing process, the sequence of mixing process should be followed in order to avoid the occurrence of segregation. Therefore, sand, POFA and cement were first added and mixed well before water and superplasticiser were added. After 5 minutes of mixing, the EPS was added and thoroughly mixed in the concrete mixture for another 3 minutes. The concrete mix was then casted into timber moulds. To minimize the segregation of EPS, the mix was compacted by hand tamping instead of vibration compaction. Once the concrete mix was fully compacted into the moulds, the samples were directly covered by plastic sheet for 24 hours before they were demoulded.

After demoulded, all the samples were exposed to two different curing conditions where the first curing condition was air curing (AC) where the samples were stored in the laboratory with room temperature. Meanwhile, the second curing condition was water curing (WC) where the samples were fully immersed into the water tank until the day of test.

3. Test Methods

3.1. Hardened brick density
In determining the density of the brick samples, the test was conducted accordance to BS EN 12390-7 in which determination of the brick mass will be as-received condition. The density of the sample is calculated by dividing the brick mass with the brick volume. The measurement of density was conducted at the age of 7 and 28 days. For this test, six samples were prepared for every mix proportion.

3.2. Compressive strength test
Compressive strength test for this study was conducted accordance to ASTM C140-11a. For this test, six brick samples were prepared for every mix proportion and all the samples tested are full-sized. Any excess moisture or any particles must be removed from the brick surface and the loading plate of the machine before conducting the test. The brick samples then exposed to air curing and water curing until the day of test. For this test, the samples were also tested at the age of 7 and 28 days.

3.3. Water Absorption Test
Water absorption test was conducted in order to determine the percentage of the water absorption by the brick. The test was accordance to the BS 1881: 122. Initially, the brick samples were dried in the oven at 100±5°C for 24 hours. The samples were then left to cool down before been immersed in the water tank for another 24 hours. The weight of samples were taken before and after the samples were immersed in the tank in order to determine the percentage of the water absorption. This test was conducted at the age of 28 days. For this test, three samples were prepared for every mix proportion.

4. Result and Discussion
The following sections discuss the outcomes from this study.

4.1.Hardened brick density
Based on Figure 2, it shows the hardened brick density for group I (a), group II (b) and group III (c). Group I, II and III show hardened brick density for brick samples with 0%, 30% and 50% of EPS respectively. Based on the result, it shows that the density of all samples are decrease as the percentage of POFA increase. The same result was found for each group. This might due to the fineness of POFA which is coarser than cement. Coarser structure of POFA has increased the formation of porosity in the brick. Therefore, this has increased the air voids which contribute in the reduction of the brick density. Similar finding has been found by Rahman et al [25] where they have figured out that, coarser size of POFA has contributed in the reduction of the masonry block density as compared to the ground POFA which are finer in structure.

In group I, it can be seen that for 7 and 28 days, all samples that exposed to air curing condition have lesser density as compared to water curing samples. For example at 7 days, E0P0 (AC) has density of 2150 kg/m^3 meanwhile E0P0 (WC) has density of 2161 kg/m^3. It shows that, density of water curing samples is 0.5% higher than air curing samples. Meanwhile, at 28 days, it has found that, density of water curing samples is 5% higher than air curing samples. This has shown that, brick ages also give impact towards brick density. This might due to the hardening process where more water were absorbed during the hydration process.

In group II, it was observed that, hardened brick density for each brick was significantly decreased. This is due to 30% of sand replacement by EPS. The same behaviour was observed in this group, where samples that exposed to air curing have lesser density as compared to water curing samples. However,

IOP Conf. Series: Materials Science and Engineering **713** (2020) 012006 doi:10.1088/1757-899X/713/1/012006

it is found that, the increment of density of brick in water curing is lesser as compared with the group I. For instance, at 7 days, E30P0 (AC) brick has density of 1778.57 kg/m^3 whilst E30P0 (WC) brick has density of 1779.83 kg/m^3. The increment of density is 0.07% and for 28 days, the increment is 2 to 3% only. This might due to the characteristic of EPS that is non-absorbent.

Next, for group III, it was observed that all samples have achieved the lightweight brick density accordance to ASTM C90 where lightweight brick density should be less than 1681.94 kg/m^3. In this group, the increment of brick density in water curing condition as compare with air curing is same as shown for group II. This is because the behaviour of brick density was affected by the existing of EPS.

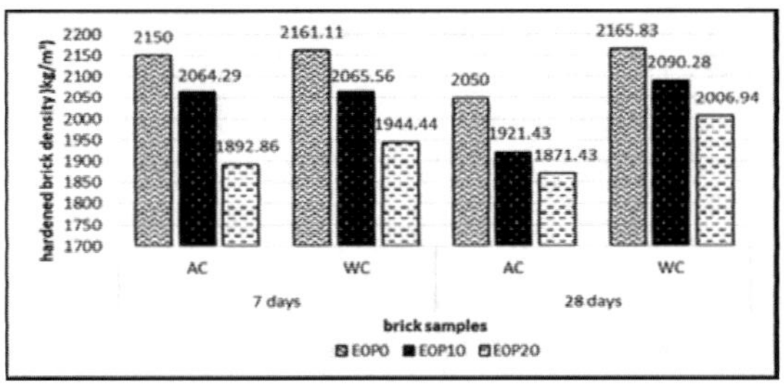

Group 1

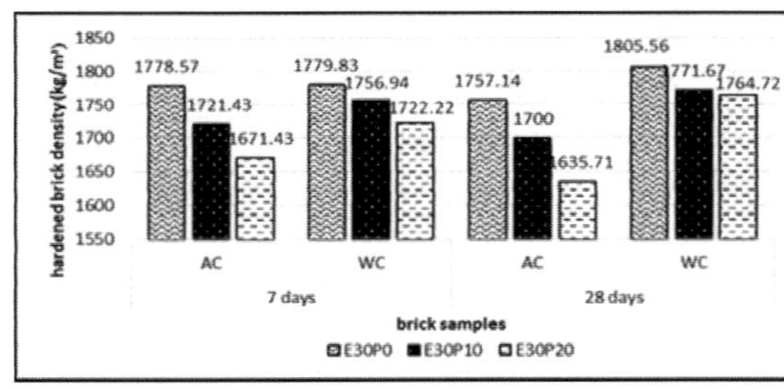

Group II

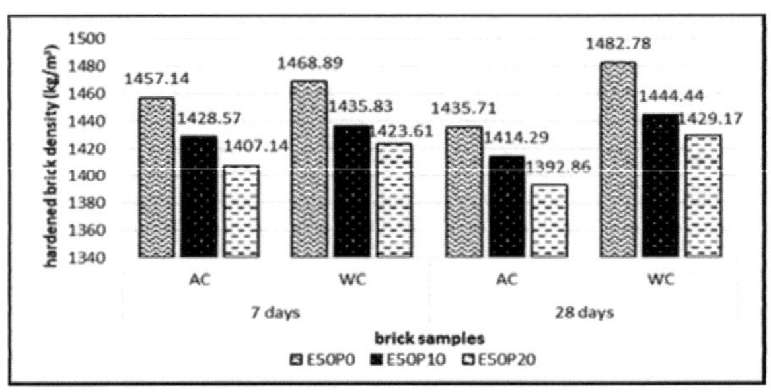

Gropu III

Figure 2. Hardened brick density of group I, II and III under different curing conditions

4.2. Compressive strength

Figure 3 summarizes the compressive strength of each group under different curing conditions. Generally, based on those figures it can be seen that, all samples in each group either with air curing or with water curing show decrement in compressive strength as the percentage of POFA increase. However, at 28 days, brick of E0P10 (AC) for group I has higher compressive strength as compared to E0P0 (AC). From this, 10% of POFA is the optimum cement replacement for air curing condition. On the other hand, for water curing, it is observed that E0P0 has the highest compressive strength compared to other samples that contains POFA. This is because, in the existing of water, the hydration process of normal brick (0% of POFA) is faster as compared to POFA brick where the pozzolanic reaction is slower at the early age.

From group I, at 7 days, the strength of E0P10 for AC and WC is same. Meanwhile, for E0P20, the compressive strength is higher at AC compared to WC. At 28days, E0P10 (WC) and E0P20 (WC) have higher strength compared AC samples. This is because, pozzolanic reaction between silicon dioxide of POFA and calcium hydroxide from cement hydration occurred due to the existing of water [2].

For group II, it can be seen that, the compressive strength of all samples is significantly decrease. This is due to the existing of EPS that is extremely low in strength [26]. However, the strength of each sample is increase in the water curing condition as compared to air curing. The same pattern is observed for brick for group III where the strength is decrease due to increase in EPS replacement and yet the strength shows an increment for water curing conditions. Although the bricks in group III are lower in strength yet the strengths are above the minimum strength for non-load bearing brick which is 3.75 Mpa as stated in ASTM C129.

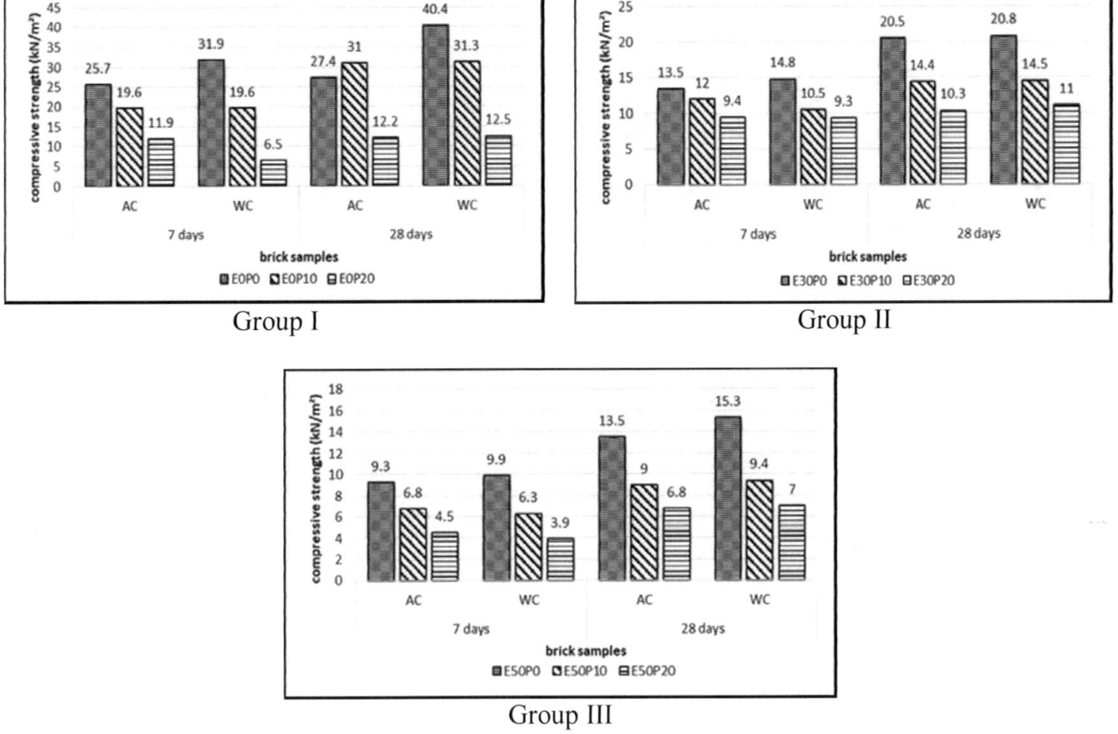

Figure 3. Compressive strength of group I, II and III under different curing conditions

4.3. Water Absorption

Figure 4 shows the summary of water absorption for 9 types of brick samples with different curing conditions. As stated earlier, water absorption test was conducted only at 28 days. Generally, from the figure, it can be seen that water absorption of air curing samples is higher compared to water curing samples. This is because, samples of air curing condition contained more air capillaries. Therefore, when all the samples immersed in the water for 24 hours, more water was absorbed into the air capillaries. Meanwhile, for water curing samples, all the capillaries were already filled with water. Therefore, water absorption of water curing samples are lower as compared to samples at air curing.

In addition, water absorption of the bricks also influenced by the percentage of replacement materials. As can be seen, the percentage of water absorption of samples increase as the percentage of POFA increase. This is due to the characteristic of POFA which is coarser in structure as compared to cement. Coarser structure of POFA has caused formation of voids in the brick samples. Additionally, POFA also has high water absorption properties. Therefore, samples with high percentage of POFA absorbed more water. On the other hand, water absorption of brick samples decrease as the percentage of EPS increase. For example, water absorption of E0P0 (WC) is 7.6% meanwhile, water absorption of E50P0 (WC) is 2.74%. The reduction of water absorption is due to the increment of EPS that is a non-absorbent materials [27].

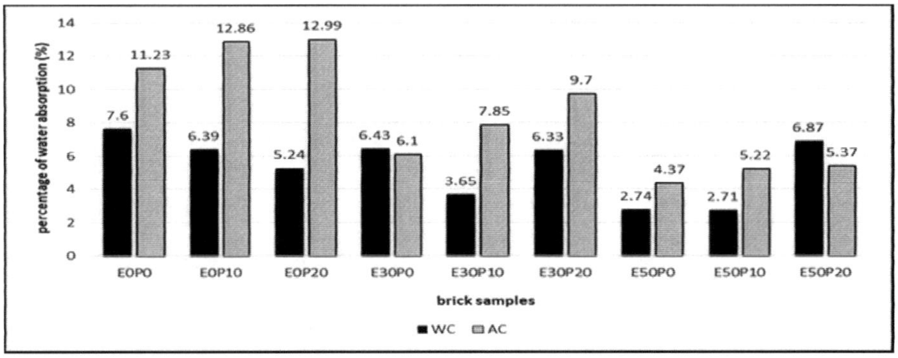

Figure 4. Water absorption of each sample under different curing conditions

5. Conclusion

In conclusion, the replacement of POFA and EPS is significantly affects the density of the brick. As the percentage of replacement materials increase, the density of the brick decrease. In this study, maximum replacement of materials is 50% of EPS and 20% of POFA has reduced the density of the brick up to 34% from normal brick. This indicates that, EPS and POFA are high potential materials for the production of lightweight. Meanwhile, the compressive strength is also highly affected by the existing of EPS and POFA in the brick. Based on the findings, generally the replacement of POFA and EPS has reduced the brick strength. However, it was found that, replacement of 10% of POFA showed an increment in brick strength. For water absorption, the findings shows that, as the percentage of POFA increase, the percentage of water absorption increase. This is due to the coarser structure of POFA which contributed to high formation of porosity in the brick.

6. References

[1] Awal A A and Hussin M W 2011 Effect of palm oil fuel ash in controlling heat of hydration of concrete *Procedia Engineering* 14 2650-2657

[2] Ling I H and Teo D C L 2011 Properties of EPS RHA lightweight concrete bricks under different curing conditions *Construction and Building Materials* **25**(8) 3648-3655

[3] Chen B, Liu J and Chen L 2011 Experimental Study of Lightweight Expanded Polystyrene Aggregates Concrete Containing Silica Fume and Polypropylene fibers *Journal of Shanghai Jiatong University (Science)* **15**(2) 129-137

[4] Herki A B, Khatib, J M and Negim E M 2013 Lightweight concrete made from waste polystyrene and fly ash. *World Applied Sciences Journal* **21**(9) 1356-1360

[5] Veiseh S and Yousefi A A 2003 The use of polystyrene in lightweight brick production *Iranian Polymer Journal* **12** 323-330

[6] Babu K G and Babu D S 2003 Behaviour of lightweight expanded polystyrene concrete containing silica fume *Cement and Concrete Research* **33**(5) 755-762

[7] Ferrándiz-Mas V, Bond T, García-Alcocel E and Cheeseman C R 2014 Lightweight mortars containing expanded polystyrene and paper sludge ash *Construction and Building Materials* **61** 285-292

[8] Hernández-Zaragoza J B, López-Lara T, Horta-Rangel J, López-Cajún, C, Rojas-González E, García-Rodríguez F J and Adue J 2013 Cellular concrete bricks with recycled expanded polystyrene aggregate *Advances in Materials Science and Engineering* **2013** 160162

[9] Bouvard D, Chaix J M, Dendievel R, Fazekas A, Létang J M, Peix G and Quenard D 2007 Characterization and simulation of microstructure and properties of EPS lightweight concrete *Cement and Concrete Research* **37**(12) 1666-1673

[10] Sayadi A A, Tapia J V, Neitzert T R and Clifton G C 2016 Effects of expanded polystyrene (EPS) particles on fire resistance, thermal conductivity and compressive strength of foamed concrete *Construction and Building Materials* **112** 716-724

[11] Tamut T, Prabhu R, Venkataramana K and Yaragal S C 2014 Partial replacement of coarse aggregates by expanded polystyrene beads in concrete *Int. J. Res. Eng. Technol.* **3**(02) 238-241

[12] Munir A 2015 Utilization of palm oil fuel ash (POFA) in producing lightweight foamed concrete for non-structural building material. *Procedia Engineering* **125** 739-746

[13] Cook D J 1973 *Expanded polystyrene beads as lightweight aggregate for concrete* (Precast Concrete No.4) pp 691-693

[14] Mulla A and Shelake A 2016 Lightweight Expanded Polystyrene Beads Concrete *International Journal of Research in Advent Technology* 17

[15] Sadrmomtazi A, Langeroudi M M, Fasihi A and Haghi A K 2009 An investigation on effect of using PP fibers and different cementitious materials on mechanical properties of EPS concrete. In *Proceeding of 3rd International Conference on Concrete & Development* (Tehran, Iran) pp 1035-1044

[16] Tang W C, Lo Y and Nadeem A B I D 2008 Mechanical and drying shrinkage properties of structural-graded polystyrene aggregate concrete *Cement and Concrete Composites* **30**(5) 403-409

[17] Xu Y, Jiang L, Xu J and Li Y 2012 Mechanical properties of expanded polystyrene lightweight aggregate concrete and brick *Construction and Building Materials* **27**(1) 32-38

[18] Nekooie M A, Goodall H and Mohamad M I 2015 March. Selection of sustainable materials for lightweight concrete. In *Proceedings of the Institution of Civil Engineers-Engineering Sustainability* vol 168(4) (Thomas Telford Ltd.) pp 159-172

[19] Laukaitis A, Žurauskas R and Kerien J 2005 The effect of foam polystyrene granules on cement composite properties *Cement and Concrete Composites* **27**(1) 41-47

[20] Babu K G and Babu D S 2004 Performance of fly ash concretes containing lightweight EPS aggregates *Cement and concrete composites* **26**(6) 605-611

[21] Babu D S, Babu K G and Wee T H 2005 Properties of lightweight expanded polystyrene aggregate concretes containing fly ash *Cement and Concrete Research* **35**(6) 1218-1223

[22] Hussin M W, Ismail M A, Budiea A, and Muthusamy K 2009 Durability of high strength concrete containing palm oil fuel ash of different fineness *Malaysian Journal of Civil Engineering,* **21**(2) 180-194

[23] Awal A S M A and Nguong S K 2010. A short-term investigation on high volume palm oil fuel ash (POFA) concrete *Proceedings of the 35th Conference on our World in Concrete and Structure* (Singapore) pp 185-192

[24] Islam M M U, Mo K H, Alengaram U J and Jumaat M Z 2016 Durability properties of sustainable concrete containing high volume palm oil waste materials *Journal of Cleaner Production* **137** 167-177

[25] Rahman M E, Boon A L, Muntohar A S, Tanim M N H and Pakrashi V 2014 Performance of masonry blocks incorporating palm oil fuel ash *Journal of Cleaner Production* **78** 195-201

[26] Miled K, Sab K and Le R, 2007. Particle size effect on EPS lightweight concrete compressive strength: Experimental investigation and modelling *Mechanics of Materials* **39** 222-240

[27] Babu D S, Babu K G and Tiong-Huan W 2006 Effect of polystyrene aggregate size on strength and moisture migration characteristics of lightweight concrete *Cement and Concrete Composites* **28**(6) 520-527

Acknowledgments
The authors gratefully acknowledge the Universiti Tun Hussein Onn Malaysia and Ministry of Higher Education for providing the financial support through the Fundamental Research Grant Scheme (FRGS), Vot 1618 and Postgraduate Research Grant (GPPS), Vot U77.

IOP Conf. Series: Materials Science and Engineering **713** (2020) 012007 doi:10.1088/1757-899X/713/1/012007

Monitoring Ex Situ Electrokinetic Remediation (EKR) using Nitric Acid as Chelating Agent

S A Ayuni[1], A T S Azhar[2], A Azim[3], and M Aziman[4]

[1] Research for Soft Soil Centre (RECESS), Universiti Tun Hussein Onn Malaysia, 86400 Parit Raja, Batu Pahat, Johor Darul Takzim
[2, 4] Faculty of Civil Engineering and Environmental, Universiti Tun Hussein Onn Malaysia, 86400 Parit Raja, Batu Pahat, Johor Darul Takzim
[3] Centre of Diploma Studies, Universiti Tun Hussein Onn Malaysia, 86400 Parit Raja, Batu Pahat, Johor Darul Takzim
Corresponding author: ayunisuied@gmail.com, saifulazhar870@gmail.com

Abstract. Heavy metal has become a serious world crisis due to natural and man-made sources. Zinc contamination in soil is caused by world urbanization and modernization. The Electrokinetic Remediation (EKR) is a green technology aiming at removing zinc contaminated soils. This paper aims to compare the electrokinetic process between Pure System and Nitric acid as chelating agents. The ex situ remediation method using soil box successfully demonstrated that nitric acid which attracted artificial contaminated zinc had been precipitated at the vicinity of cathode terminal in a form of reddish layer. In conclusion, nitric acid is highly recommended as solution for enhancing electroosmosis process.

1. Introduction

Malaysia is moving towards the status of developing country through industrialization and rapid urbanization. By referring to gross statistical value, rate of urbanization has increased from 54.3 percent to 65.4 percent between year 1991 and 2000 [1]. This is expected to increase up to 75 percent by 2020 according to NUP (full form). However, the effects compromise environmental quality as water, air and soil conditions are deteriorating. The National Policy on Environment [2] aims at continuing economic, social and cultural progresses as well as enhancement of quality living through sustainable development. The NEHAP strategies established in Malaysia Plan (MP) 9th by Ministry of Health, later on, developed and implemented in 10th of MP were to achieve objectives of making available a clean, safe, healthy, and productive environment for present and future generations. Therefore, sustainable ecosystem is the most important aspect to keep balance of the environment. Green environment is crucial when it comes to avoiding severe pollutions due to rapid development in Malaysia. Besides, heavy metal pollution in soil and groundwater in fact is a world-wide problem that each country must initiate concern to achieve a sustainable ecosystem. Heavy metals are introduced into the ecosystem by manufacturers and the use of materials containing heavy metals as well as the improper disposal of these wastes.

Heavy metals contribute toward worldwide soil contamination, thus, enhancement program must be kick started to ensure soils are clean and safe for use. In short, a literature review is widely discussed on scientific perspective and a crucial part of investigation regarding the content which may help making the future research engaging and valuable. Environment is an important element when speaking of sustaining the earth. Defined by the Environment Protection Act 1990, Part 1, Section 1(2), the environment consists of all, or any, of following media such as air, water and land; and the medium of air includes the air within buildings and the air within other natural or man-made structures above or below ground. "Pollution of the environment" means pollution due to release (into any environmental medium) from any process of substances which is capable of causing harm to man or any other living organisms, supported by the environment. Therefore, pollution can be defined as "the introduction by man into the environment of substances or energy liable to cause hazards to human health, harm to living resources and ecological systems, damage to structures or amenity, or interference with legitimate use of the environment [3-4]. Land may be considered as contaminated when it contains a sufficient quantity of toxic or otherwise harmful materials to pose a threat to the health and safety of users of the land or workers engaging in its redevelopment [2].

2. Heavy Metal Contamination in Ecosystem

Heavy metal contamination in ecosystem may include the food chain, river, sediment, soil and air. In Malaysia alone, its 14 states have different types of soil, ecosystem, marine lives and industrial activities. Heavy metal contamination in environment can be divided into two main sources which specifically anthoropogenic and natural sources. The natural sources include weathering mineral process, erosion and volcanic activities, and particles released by vegetation. The anthropogenic sources are man made from industrialization and urbanization. For example, Arsenic (As) may be produced from ore mining and smelting, and agricultural activities are also turned adversely when the emission of cadmium (Cd), and copper (Cu)[5]. Therefore, based on anthropogenic analysis conducted in Perlis, there has been found low heavy metal pollution [6]. The state of Perlis involves with anthropogenic activities such as agricultural activities, cement production, and industrialization. The activities highly contribute towards negative changes of pollution loading, decreased biodiversity, simplified structure, and emission of transportation. However, analyses on Cu, Cr, Ni, Cd and Pb concentration found that their levels were lower than allowable limit based on pollution Index (PI). The anthropogenic cycles as shown in Figure 1 [7] above highlighted that leachate from anthropogenic sources may be discharged from factories. The metal ions will aggregate and accumulate inside the soil and groundwater. The contamination in soil will be up taken by plants which are elements of food chain for animals and human beings. Besides, surface irrigations due to rain will flow into the lake and sea. Therefore marine lives are at stake of being endangered.

Marine ecosystem that plays an important role in the entire ecosystem may also involve in heavy metal pollution. A previous study by Othman et al. [8] concluded that Sungai Sedili in Johor contained a low concentration of heavy metal based on Environmental Quality Guideline for freshwater sediment. Had been highlighted heavy metals in the Sungai Sedili were Fe>Al>Zn>Cu>As>Mn>Ni>Pb>Co>Cd and had been studied metal concentration in sediment layer were found to be Fe>Al>Mn>Zn>Pb>Cu>As>Ni>Co>Cd>Ag. A survey by Shaikhah et al. [9] had been conducted at aquatic, seawater and sediment in Teluk Sengat and Pulau Setindan. The investigation proved that tissues of Strombus Canarium (*siput gonggong*) contained lower concentrations of plumbum (Pb) and Zinc (Zn) based on permissible limits recommended by Food and Agriculture Organization (FAO) and World Health Organization (WHO). Therefore, consumption should be controlled due to reason that 25 percent (%) of total Pb intake will remain in human blood and it has neurotoxic effects [10]. However, sediments of both locations recorded high concentrations of Pb and Zn which, thus, they had been declared as slightly polluted with Pb at 45.572 µg/g dry weight. Soft tissues inside clam (*kerang*) also contained some amounts of heavy metals. The data had been confirmed by Hossen et al. [11] based on Malaysia Food Regulation (1985) that levels of heavy metal concentration depended heavily on species of clam and locations of 34 sites around Malaysia. Sustainability management for soil and groundwater is in safe hands under the Environmental Quality Act 1974. However, there is no specific regulation with respect to soil and groundwater contamination according to Department of Environment (DOE), and Standard and Industrial Research Institute of Malaysia (SIRIM).

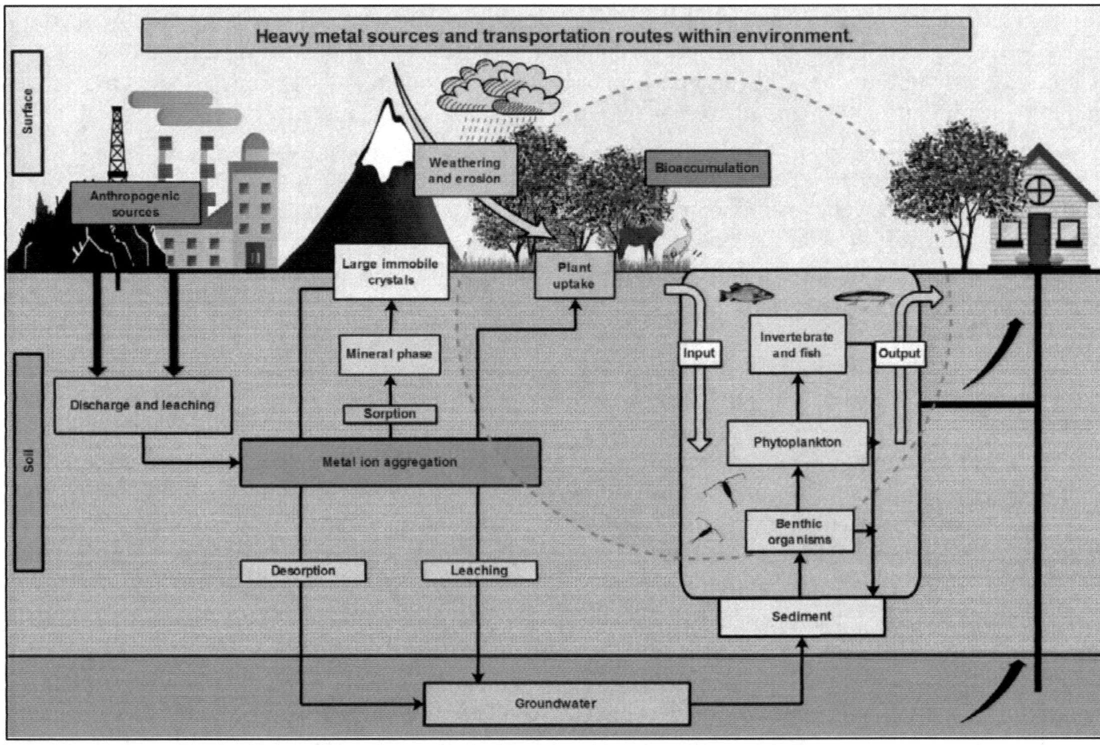

Figure 1. Anthropogenic cycle in ecosystem [7].

3. Methodology

Electrokinetic remediation (EKR) can be defined as a process of physicochemical transport of charged particles, initiated by electric potentials on formation and fluid transports in porous media. The EKR is also highly depends on types of soil, cation exchange capacity (CEC), applied current voltage, pH soil, and duration.

3.1. Artificial Zinc Contamination in Soil

Soil sampling is vital for soil classification and scientific analysis. However, in terms of laboratory studies, artificial contamination is sufficient and controllable to represent the real condition. Moreover, there are difficulties and related circumstances when it comes to extracting contaminated soil. Hence, the chemical sample should be similar to original pollutant such as arsenic, cadmium, lead, chromium, copper and zinc at the concentration of 0.5% or 5000 mg/kg [12]. The clayey soil had been mechanically mixed with dilution of zinc powder in nitric acid of close to 3269 ppm, under acceptance amount in environment 5100ppm [13] based on Equation 1 below. The sample was successfully homogeneous within at least 24 hours before the remediation had been studied.

$$g = MM_s \times M \times V \quad (1)$$

in which
g is the mass or soil (in g)
MM_s is the mole of chemical (g/mol)
M is the molarity of the solution (mol/L)
V is the volume of solution (in L)

Spiked soil is an approach for laboratory scale due to constrains to supply contaminated soils from real sites. Therefore, related heavy metals should be mechanically mixed based on environmental guideline as recorded in Table 1. Zinc nitrate-6-hydrate is a white crystal compound soluble in water molecules. The compound Zn $(NO_3)_2.6H_2O$ has molecular weight 297.49 g/moles. Consequently, the clay sample had been mechanically mixed into 0.1M zinc nitrate. The ratio

IOP Conf. Series: Materials Science and Engineering **713** (2020) 012007 doi:10.1088/1757-899X/713/1/012007

of the clay soil to 1M of zinc nitrate to distilled water (DW) followed Equation 1. Secondly, the soil sample had been mechanically mixed into dissolved zinc powder with 10 percent of nitric acid. As a result, the soil increased homogeneity after at least 24 hours mixing process. The chemical reaction for dissolution of zinc metal in 10 percent nitric acid was as shown in Equation 2.

$$\underbrace{Zn^{2+} + H^{+} + NO^{3-}}_{\text{Solution of Nitric acid}} \rightarrow Zn^{2+} + NO^{2-} + H_2O \qquad (2)$$

3.2. Electrokinetic Remediation (EKR) Laboratory Set Up

There are various heavy metals highly possible to remediate through electrokinetic process such as arsenate, cadmium, soluble copper [14], and zinc. In addition, the electrokinetic can also remove saline from greenhouse. Figure 2 shows a comparison of soil box dimension in which the electric supply is about 20 Volts across 20 cm of electrode distance. There is an acrylic tanks with three separate compartments containing a soil sample in the middle. The electrode could be in a shape of tube well or plate with good electric conductives such as titanium, platinum and graphite. However, steel plate is highly effective in terms of low cost and good conductivity. Electrolyte reservoir aims to continuously supply for electroosmosis process.

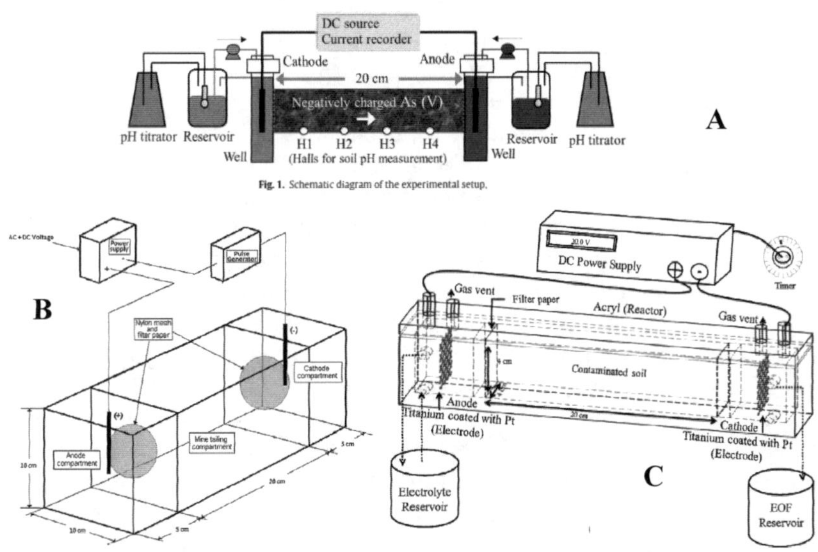

Figure 2. Comparisons of electrokinetic remediation (EKR) soil box dimensions (A= Suzuki [15]; B= Rojo [14]; C= Jo S-U [16])

Figure 3 demonstrates electrokinetic remediation (EKR) set up with regards to ex situ analysis. The soil box had been adapted from previous researches [14]-[16] with volume of 1100 cm³ (22cm x 5cm x 10 cm) for the main compartment, placed in the middle. Electrolyte will be placed at the anode and cathode compartments whilst height, width, and length were 10 cm, 10 cm, and 5 cm respectively.

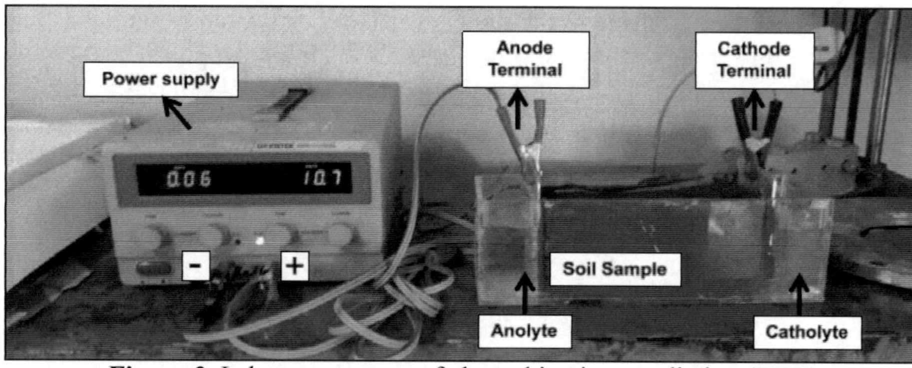

Figure 3. Laboratory set up of electrokinetic remediation (EKR)

4. Monitoring Ex Situ Electrokinetic Remediation (EKR)

Ex situ remediation had been divides into two important systems which were pure system and nitric acid as a chelating agent. The principle of chemical reaction is based on the electrolysis of water, which is reduced at the cathode terminal and oxidized at the anode terminal [17]. Therefore, it is proven that the electrolysis of water happened during Pure System for uncontaminated soil as presented in Figure 4(a) and Figure 4(b) in which there were 8 percent of soil moisture at the vicinity of cathode terminal and 19 percent increment at the vicinity of anode terminal as tabulated in Table 1. However, the electrokinetic process for uncontaminated soil using nitric acid contradicted the pure system. This was for the reason that at the vicinity of cathode (nitric acid as catholyte), bubbles appeared and they affected the soil to increase its moisture level up to 40 percent. Hence, at the vicinity of anode, soil moisture only decreased at 6 percent.

The artificial contaminant by zinc nitrate could be categorized as heavy metal. However, zinc nitrate contamination may occur from the nitrogen gas released into soils. Figure 4(c) and Figure 4(d) illustrate variations of soil moistures for Pure System by comparing electric gradients of 50V/m and 150V/m. Physical observation showed that distilled water at cathode terminal remained constant after 7 days of electrokinetic process. Next, at vicinity of anode terminal, white precipitation appeared at electric gradient 50V/cm on a second day of process while 150V/m white precipitate accumulated on third and fourth days of electrokinetic process. Therefore, the electrokinetic process using 10 Voltage of power supply equalled to 50 V/m of electric gradient, was an efficient voltage as mentioned by Bradl [17] that in fine grained soils, electroosmotic could be achieved by applying electric gradients of 100V/m.

The zinc nitrate had been neutralized by using a chelating agent aimed at determining nitric acid should be placed as catholyte or anolyte at electric gradient 50V/m. Therefore, the variations of moisture contents (mc) depicted in Figure 4(e) and Figure 4(d) resulted in accumulated white precipitate at the vicinity of cathode terminal with different soil moistures of 66 percent and 33 percent, respectively. Moreover, at the vicinity of anode terminal, it was recorded that the soil moisture became lesser than before as mentioned in Table 1. The reverse osmosis occurred through electrokinetic process placed nitric acid as catholyte, meanwhile, during third day remediation, soil moisture increased, but decreased on seventh day of soil remediation. Therefore, related electrokinetic process agreed with the fact that a period of three days was occupied to remediate 20 cm of soil sample at 50V/m of electric gradient.

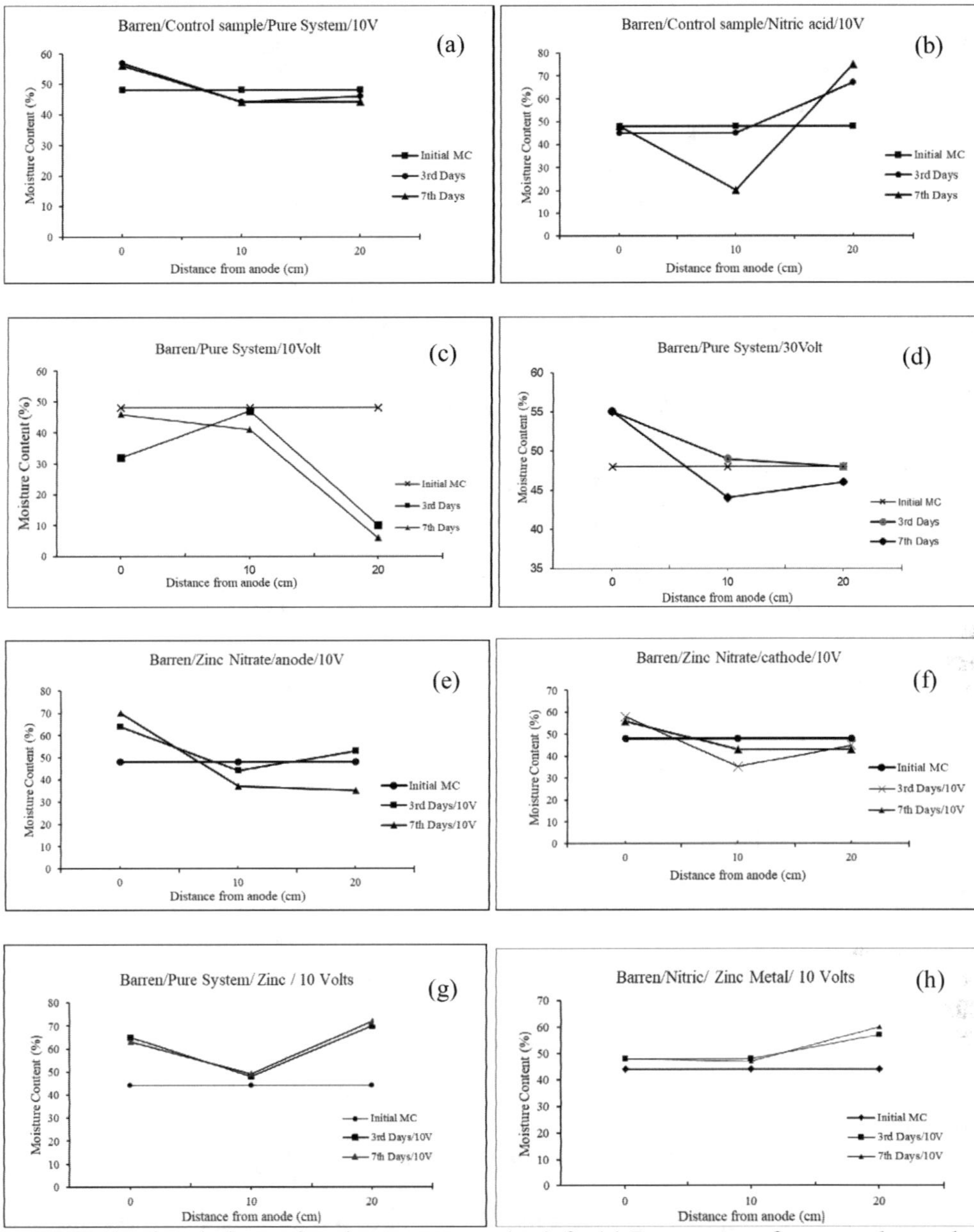

Figure 4. Graph of variations of moisture content (mc) for(a) Pure system for control sample under 10 Volt; (b) Nitric acid as a chelating agent for control sample under 10 Volt; (c) Pure system to remediate zinc nitrate-contaminated soil; (d) Nitric acid as a chelating agent to remediate zinc nitrate-contaminated soil; (e) Nitric acid as anolyte to remediate zinc nitrate-contaminated; (f) Nitric acid as catholyte to remediate zinc nitrate-contaminated; (g) Pure system to remediate artificial zinc-contaminated soil; and (h) Nitric acid as a chelating agent to remediate artificial zinc-contaminated soil.

Table 1. Observation data of ex situ analysis

Uncontaminated soil (Control Sample)				
EKR System	**Voltage (Volt)**	**Anode**	**Cathode**	**Remarks**
Pure System (DW-DW)	10	The vicinity of terminal was high moisture of about 19 percent increment	The vicinity of terminal was dry, of about 8 percent decrement.	Electrokinetic process for control sample.
Chelating agent (Nitric acid)		DW – the level had been reduced from initial level The MC also decreased about 6 percent.	Nitric acid- bubbles appeared Moisture content increased about 40 percent (%).	The nitric acid can be used as a chelating agent.
Artificial Contaminated (Zinc Nitrate) soils				
Pure System (DW-DW)	10	White precipitate (process happened on 2nd day)	At the vicinity of cathode terminal, the moisture was high and DW level remained the same.	The power supply of 10 Volts was the most efficient voltage
	30	White precipitate (process happened on 3rd and 4th days)	At the vicinity of cathode terminal, the moisture was high and DW level remained the same	
Chelating agent (Nitric acid as anolyte)	10	Dried precipitation and moisture content dropped about 16%	White precipitation with high moisture content of about 66 percent (%)	There was a significant difference in percentage value, confirming that **nitric acid** should be placed as **catholyte.**
Chelating agent (Nitric acid as catholyte)		Dried precipitation 2 percent (MC unstable) 3rd MC increased 7th MC decreased	White precipitation with high moisture content of about 33 percent (%)	
Artificial Contaminated (Zinc metal) soils				
Pure System (DW-DW)	10	Level of DW reduced/ Dried White Precipitate	The vicinity of cathode terminal was high moisture.	Based on comparison, 10 Volts was the most efficient voltage.
Chelating agent (Nitric acid)		Level of DW reduced/ Dried White Precipitate As Figure 5(a).	The vicinity of cathode terminal was high moisture and nitric acid overflow/ the **reddish layer appeared** As shown in Figure 5(b).	Nitric acid was successfully employed as enhancement solution.

The electric gradient of 50V/m became a constant parameter to analyse the electrokinetic process for zinc metal artificial contaminant soils. This study performed a comparison between Pure System and Chelating agent as enhancement solution. The electrokinetic process showed that white precipitate which was nitrate ion had been accumulated at the vicinity of anode terminal for both systems with dried region as pictured in Figure 5(a). Meanwhile, at the vicinity of cathode for both systems, there was high soil moisture. The accumulation of reddish layer proved that the anion of zinc, Zn^{2+} had been attracted towards the cathode terminal as visually presented in Figure 5(b) below.

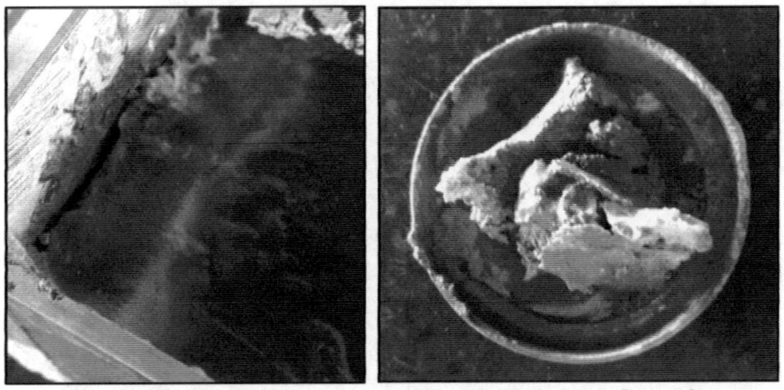

Figure 5(a). White precipitate at vicinity of anode

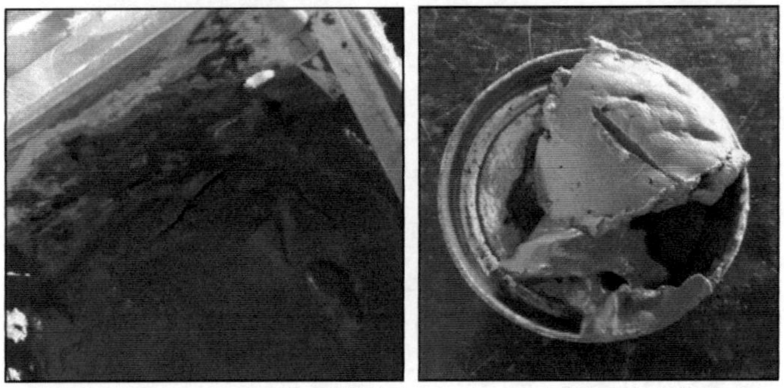

Figure 5(b). Appearance of reddish layer at vicinity of cathode terminal

5. Conclusion

Based on physical observation which had been discussed, it was verified that 10 Voltage of power supply was the most efficient compared to 30 Voltage. The right method for artificial zinc-contaminated soil was by using dilution of zinc powder with molecular weight of 64.38 based on periodic table. The white precipitation at the vicinity of anode terminal was expected to be nitrate precipitation. Therefore, spectrometry analyses such as X-Ray Diffraction (XRD), and Ion Chromatography (IC) should be conducted. In addition, analyses should be extended on electrolyte and soil particles on the electrode (steel plate). In conclusion, the method for artificial contaminated soil was succeeded in respect of ex situ electrokinetic remediation (EKR); accordingly, it should become a guideline for in situ application.

6. References

[1] NUP 2006 National Urbanization Policy Chapter 3.0; Challenges and Issues. Federal Department of Town and Country Planning Peninsular Malaysia. Ministry of Housing Local Government. Kuala Lumpur. www.townplan.gov.my/dpn

[2] DOE 2009 Contaminated Land Management and Control Guideline No. 3: Remediation Contaminated Sites. *Department of Environmental Guideline Standard* Retrieved from www.doe.gov.my pp 3-5

[3] Holdgate M W 1979 *A perspective on environmental pollution* (Cambridge:Cambridge University Press)

[4] Peter Attewell 1993 Ground Pollution: Environment, Geology, Engineering and Law (E&FN SPON) 5
[5] Dixit R, Wasiullah, Malaviya D, Pandiyan K, Singh U B, Sahu A, Shukla R, Singh B P, Rai J P, Sharma P K, Lade H and Paul D 2015 Bioremediation of Heavy Metals from Soil and Aquatic Environment: An Overview of Principles and Criteria of Fundamental Processes. *Sustainability 2015* **7** 2189-2212
[6] Norbaya S M R, Sharizal H, Lias M K and NorShahrizan M H 2014 Analysis and Pollution Assessment of Heavy Metal in Soil, Perlis *The Malaysian Journal of Analytical Sciences.* **18**(1) 155 -161
[7] Kobielska P A, Howarth A J, Farha O K and Nayak S 2018 Metal Organic frameworks for heavy metal removal from water. *Coordination Chemistry Review* **358** 92-107
[8] Othman S M, Ahmad A K, Nadzifah Y and Azmah M 2012 Metal concentration in Sungai Sedili Kecil, Johor, Peninsular Malaysia *Journal of Tropical Marine Ecosystem 1. EKOMAR* 15-23
[9] Shaikhah S, Ismid M M S and Shamila A 2014 Lead (Pb) and Zinc (Zn) Concentrations in Marine Gastropod Strombus Canarium in Johor Coastal Area *The Malaysian Journal of Analytical* Sciences **18** 37-42
[10] Hounkpatin A S Y, Edorh A P, Salifou S, Gnandi K, Koumolou L, Agbandji L Aissi, K A, Gouissi M and Boko M 2012 Assessment of Exposure Risk to Lead and Cadmium via Fish Consumption in Lacuscrian Village of Ganvie in Benin Republic *Journal of Environmental Chemistry and Ecotoxicology* **4**(1) 1-10
[11] Hossen M F, Hamdan S and Rezaur M R 2015 Review on the Risk Assessment of Heavy Metal in Malaysia Clams *The Scientific World Journal* **2015** 1-7
[12] Ahmad Tarmizi 2008 *Pelepasan Logam Daripada Tanah Tercemar yang Distabil dan Dipejalkan oleh Simen dan Abu Kulit Kayu Getah.* PhD Dissertation. Universiti Kebangsaan Malaysia.
[13] EPA 2008 Drinking Water Contaminants; United States Environmental Protection Agency (EPA): Washington DC, USA.
[14] Rojo A, Hansen H K and Cubillos M 2012 Electrokinetic remediation using pulsed sinusoidal electric field *Electrochimica Acta* **86** 124-129
[15] Suzuki T, Moribe M, Okabe Y and Ninae M 2013 A mechanistic study of arsenate removal from artificially contaminated clay soils by electrokinetic remediation *Journal of Hazardous Materials* ***254-255*** 310-317
[16] Jo S-U, Kim D-H, Yang J-S and Baek K 2012 Pulsed enhanced electrokinetic restoration of sulphate containing saline greenhouse soil *Electrochimica Acta* **86** 57-62
[17] Bradl H B 2005 Heavy Metals in the Environment: Origin, Interaction and Remediation *Interface Science and Technology* **6** 176-179

Acknowledgments
This paper is financially funded by the MyBrain15 scholarship scheme under the Ministry of Higher Education Malaysia. The author would like to thank Preston Geochem Sdn. Bhd. for the collaboration.

Preparation of Sintered Glass-Ceramic from CRT Glass and Red Gypsum Wastes for Tiling Application

M M Idham[1] and M M Haniza[1]

[1] Mineral Research Centre, Department of Mineral and Geoscience Malaysia
Jalan Sultan Azlan Shah, 31400 Ipoh, Perak.
Corresponding author: idham@jmg.gov.my

Abstract. This study aims to prepare sintered glass-ceramic (SGC) from a combination of cathode ray tube glass and red gypsum wastes specifically for tiling application. Bentonite clay which acted as a binder was also added to the prepared formulation. The starting materials were milled individually, mixed and compacted into button shape by using a uniaxial hydraulic press machine. The effects of sintering temperature and red gypsum content on the physical-mechanical properties and crystallization behaviour were investigated. The test results revealed that the glass-ceramic sintered at 850 °C with 20 wt.% of red gypsum gave the best combination of physical-mechanical properties. Modulus of rupture and linear shrinkage were measured at 33.75 MPa and 6.47 %, respectively. Water absorption was low enough and recorded at 0.4759 % while the density was measured at 2.6638 g/cm^3. The results also discovered that increasing red gypsum content by 10 wt.% and sintered at similar temperature deteriorated the quality of the glass-ceramic. Therefore, this particular sintered glass-ceramic has the potential to be used as outdoor walkway tile or brick owing to the high physical-mechanical properties, colour suitability as well as low production cost.

1. Introduction

Recycling is important in reducing wastes that cause negative impacts on the environment. Recycling conserves our natural resources and energy, saves landfill space, reduces water and air pollution, and greenhouse gas emissions that cause global warming. In the past 20 years, numerous researches have been conducted to utilize glass and industrial waste to form sintered glass-ceramic for building applications and others. Many have agreed that glass-ceramics manufacturing has been considered as a very effective method for recycling and utilizing many types of industrial waste especially by the sinter-crystallization method. Based on this method (sinter-crystallization), by optimizing the processing parameter and minimizing the powder particles size (enhanced specific surface areas) can cause the densification and crystallization process to complete in a very limited time and at relatively low temperatures. Hence, reducing the cost of waste treatment. Besides processing advantages, the sintering approach may lead to the formation of unusual crystal phases and resulted in glass-ceramic with remarkable mechanical properties. For example, E. Bernardo and his colleagues [1-5] have done a great effort for the last 10 years in demonstrating how various industrial wastes can be recycled to form glass-ceramic via the sinter-crystallization process. They have proved that glass waste (e.g. soda-lime glass and cathode ray tube glass), mining residue, low grade feldspar and lime waste from fume abatement system of the glass industry can be mixed together and recycled to form glass-ceramic with excellent physical-mechanical properties. Their studies show that some of the glass-ceramic samples exhibiting a surface hardness of over 6 GPa and modulus of rupture (MOR) of more than 100 MPa which are higher than conventional porcelain stoneware tile or the commercial Neoparies® glass-ceramic tile's properties. Similar research also was conducted by Maschio and his team [6] using paper mill sludge, glass cullet and natural red clay as main components in a body formulation to produce glass-ceramic via fast firing sintering treatment at 1040-1140 °C. The results suggested that the mixture of powder containing 42 wt.% paper mill sludge, 28 wt.% glass cullet and 30 wt.% natural red clay resulted in the glass-ceramic having good hardness and strength. They recommended that some of the produced glass-ceramics could be used in the construction industry particularly for tiling purposes.

On the other hand, Zhang et al. [7-8] had tried a different approach by adopting a reactive crystallization route to produce glass-ceramic from glass waste. They used a series of combination of kaolin, talc and chemical reagent to synthesize the crystallization promoters and incorporated them into the glass-ceramic body formulation. They managed to fabricate diopside-albite and wollastonite glass-ceramic from soda-lime glass (known for its lower tendency to crystallize) with the aid of crystallization promoter at a temperature of below 950 °C. However, the mechanical properties of the produced glass-ceramic were not as high as that being reported by E. Bernardo and his colleagues. It was also reported recently that a good quality of glass-ceramic had been produced using rice husk ash (RHA) as a silica precursor [9]. Boron oxide/sodium oxide and aluminium oxide/magnesium oxide had been added in the batch formulation to facilitate the melting process and to increase the tendency of forsterite crystal precipitation, respectively. The results obtained showed that it is possible to use RHA to produce glass-ceramic tile by a sinter-crystallization process obtaining nepheline ($Na_2O.Al_2O_3.SiO_2$) as the main crystalline phase and forsterite ($2MgO.SiO_2$) at a temperature as low as 900 °C. The sintered material also showed higher bending strength values and Mohs hardness with respect to commercial glass-ceramics such as Neoparies®. In the Malaysian context, Juoi et al. [10] managed to produce glass composite material (GCM) from incinerated scheduled waste bottom slag (BS) and soda lime silicate waste glass. The effect of BS waste loading on the GCM and the microstructural properties was studied and the results revealed that higher BS waste loading produces GCM with higher porosity, higher water absorption and lower bulk density. In contrast, the Vickers microhardness value and modulus of rupture (MOR) were decreased. In their study, MOR as high as ~70 MPa with a density around 1.9 g/cm^3 was achieved with 30 % BS waste loading.

Based on concerns for increasing trend of environmental pollution at the present time, the Mineral Research Centre (MRC) being the only research centre in the field of minerals is taking the opportunity to initiate a repurposing project involving utilization of cathode ray tube (CRT) glass and red gypsum waste by turning them into sintered glass-ceramic which can be applied in construction industry. CRT glass is a glass that primarily found in old television and monitor and it contains up to 25% of lead which make them even harder to recycle owing to the toxic nature of lead. However, the sinter-crystallization method applied in this study is a very effective method to immobilize this hazardous waste by turning it into sintered glass-ceramic which make them fairly stable physically and chemically. Meanwhile, red gypsum (RG) is a waste product during the extraction of titanium (IV) oxide from the ilmenite ores and it comes in the formed of reddish brown semi-solid mud. Therefore, the objectives of this study are: (1) to develop sintered glass-ceramic for tiling application by using recycled cathode ray tube (CRT) glass and red gypsum waste and (2) to examine the effects of sintering temperature and red gypsum content on physical-mechanical properties as well as crystallization behaviour of produced glass-ceramic.

2. Experimental Method

2.1. Materials

The cathode ray tube glass (CRT) and red gypsum (RG) wastes were used in this study (shown in Figure 1). They were sources from PUM Cullet Sdn. Bhd. (Pasir Gudang, Johor) and Tioxide (Malaysia) Sdn. Bhd. (Kemaman, Terengganu), respectively. Both materials were ground to the size of -75 µm by using mortar grinder. Bentonite clay which is act as a binder was also added to the formulation to bind the powder particles together and aiding in improving the green strength of unfired samples. The chemical composition of the CRT glass, red gypsum and bentonite clay were determined by using Shimadzu XRF-1700 X-ray fluorescence spectrometer and the results are shown in Table 1. XRD analysis for bentonite, CRT glass and red gypsum wastes also been carried out as depicted in Figure 2. As expected, CRT glass is completely amorphous while red gypsum consists of gypsum as a major mineral phase and ilmenite as a minor phase. On the other hand, bentonite clay consists of quartz, cristobalite and montmorillonite as major mineral phases. In this study, two formulations were prepared as presented in Table 2.

Figure 1. Starting materials used in this study, (a) CRT glass waste, (b) red gypsum waste and (c) bentonite clay.

Table 1. Chemical composition of CRT glass, red gypsum (RG) and bentonite clay.

Oxide	SiO_2	Na_2O	CaO	MgO	Al_2O_3	K_2O	SO_3	Fe_2O_3	F	PbO	TiO_2	Etc.
CRT Glass (%)	62.72	7.25	3.27	2.76	4.51	7.62	-	0.51	2.07	5.6	-	3.69
Red Gypsum (%)	3.12	-	32.09	1.26	0.76	-	30.01	24.71	0.69	-	6.61	1.37
Bentonite (%)	71.32	0.59	3.05	4.58	15.65	0.39	-	4.0	-	-	0.33	0.09

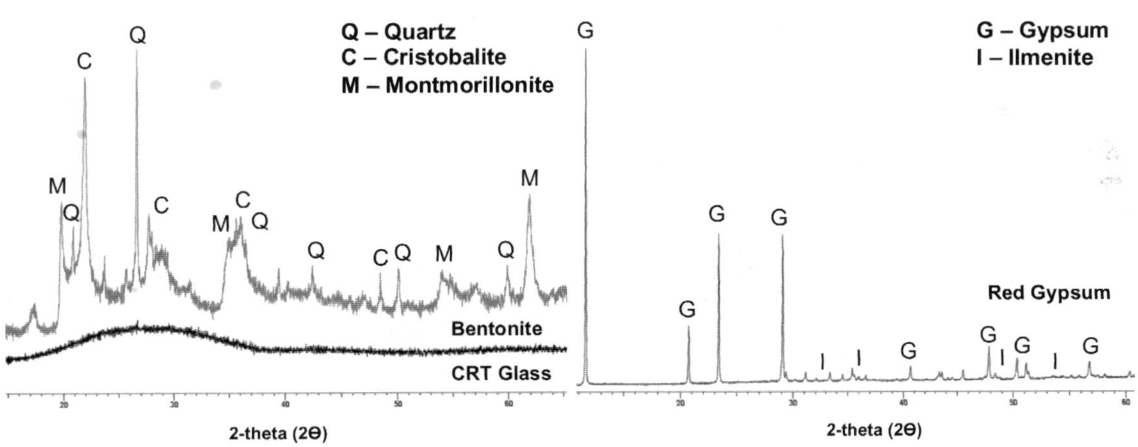

Figure 2. XRD diffractogram of starting raw materials.

2.2. Sample Preparation

All materials as shown in Table 2 were dry-mixed in the rotary mixer for at least 5 hours. Approximately 6 - 8 wt. % of water was added to the mixture to aid in the forming process. The samples in button shape were prepared by using a uniaxial hydraulic press machine at a pressing weight of 20 tons. The diameter and thickness of the buttons were approximately 35.5 mm and 10 mm, respectively. The green strength of the unfired buttons was not measured. However, it is essential to note that these buttons were strong enough to be handled, carried and moved.

IOP Conf. Series: Materials Science and Engineering **713** (2020) 012008 doi:10.1088/1757-899X/713/1/012008

Table 2. Formulation of glass powder to prepare glass-ceramic materials.

Sample	CRT Glass (wt. %)	Red Gypsum (RG) (wt. %)	Bentonite Clay (wt. %)
SGC 1	65	20	15
SGC 2	55	30	15

2.3. Sintering Process

The dry unfired buttons (SGC 1) had undergone a sintering process where the heat was gradually applied to the samples to bind the compacted powder particles. The sintering temperature was set at four different levels namely 750, 800, 850, and 900 °C. In order to investigate the effect of red gypsum content on the properties of glass-ceramic, additional samples (SGC 2) was also prepared and sintered at the same temperature. Sintering time and heating rate were set constant at 30 minutes and 5 °C/min, respectively throughout the experiment. During heating, the temperature was held for 30 minutes at 450 °C to remove the remaining water moisture and organic matters that might present in samples. After the sintering process, samples were let to cool down to room temperature naturally in the furnace. Finally, the sintered glass-ceramic samples were ready to be tested.

2.4. Sample Characterization

The sintered glass-ceramic were characterized in terms of their density, water absorption, linear shrinkage, modulus of rupture (MOR) as well as phases evolution. The density and water absorption of the samples were measured by adopting Archimedes principles and performed according to ASTM C373-88 [11]. Modulus of rupture was determined by using the universal testing machine (Instron 3367). For this purpose, the samples were cut to 30 mm (L) x 15 mm (W) in dimension. The test was carried out by using 20 mm span and 0.5 mm/min crosshead speed [12]. Linear shrinkage was determined by comparing the diameter of the samples before and after the sintering process, while the phase analysis was examined by using X-ray diffraction analyser (Bruker XRD D8 Advance) with step size and analysing range (2Θ) of 0.009° and 10° – 70°, respectively. Each quantitative data presented here is an average measurement of at least 5 different samples.

3. Results and Discussion

The prepared SGC 1 (20 wt.% RG) and SGC 2 (30 wt.% RG) are shown in Figure 3. In terms of colouration, it is observed that for both SGC 1 and SGC 2, the intensity of the reddish colour becoming slightly brighter when the sintering temperature was increased from 750 to 900 °C. It is also observed that increasing red gypsum from 20 wt.% to 30 wt.% did not affect much on the colouration of the samples as the reddish colour remains almost similar for every level of temperature. It is believed that the resulting reddish colour of the products was because of the significant presence of iron oxide in the formulation.

3.1. Physical-mechanical properties

Densification of glass-ceramic is controlled by the viscous flow of the glassy phase and it is greatly related to the amount of glassy phase in the body and its viscosity [13]. In addition, the rate of densification process in the body will slow down when the crystalline phase starts to occur because it reduces the amount of glassy phase and increases the viscosity of the glassy phase, which in turn increases the resistance of viscous flow thus, prevent further densification.

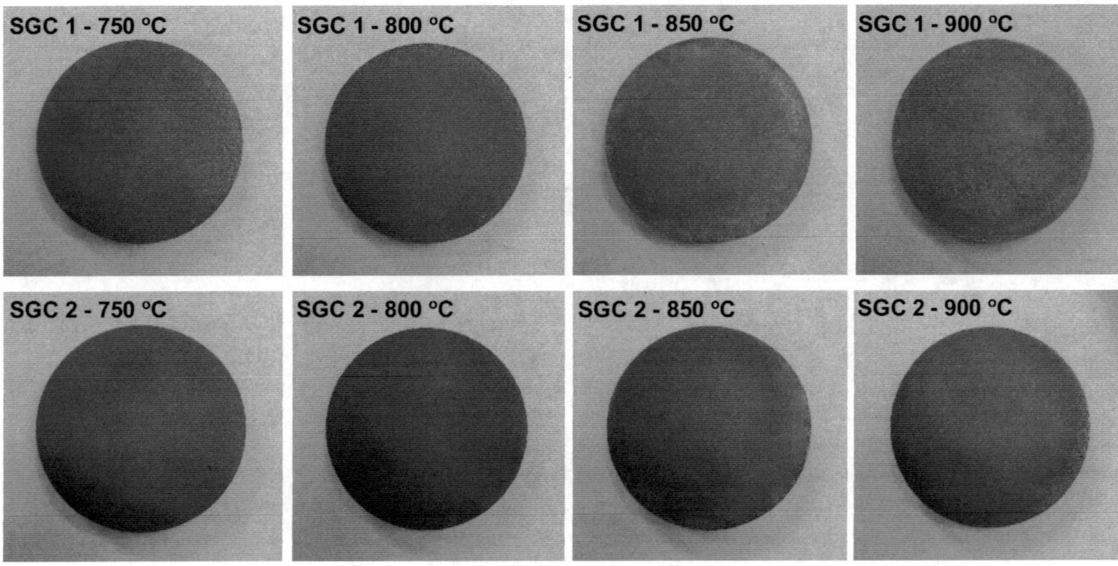

Figure 3. SGC 1 and SGC 2 sintered from 750 to 900 °C.

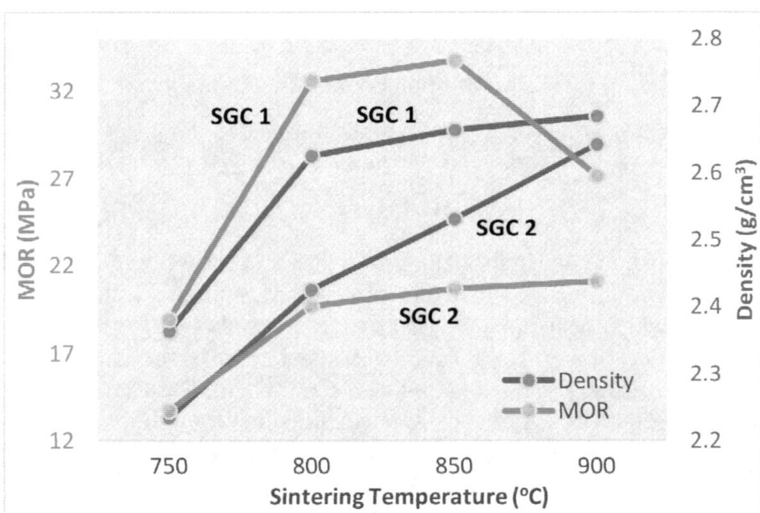

Figure 4. MOR and density in the function of sintering temperature for SGC 1 (20 wt.% RG) and SGC 2 (30 wt.% RG).

As shown in Figure 4, for SGC 1, increasing sintering temperature from 750 °C to 800 °C has witnessed a sharp increase of density from 2.3634 g/cm^3 to 2.6249 g/cm^3. A significant improvement of modulus of rupture (MOR) from 18.93 MPa to 32.60 MPa was also observed. This is mostly attributed to the lower viscous flow resistance of the glassy phase at high temperature and lead to higher densification degree of a body. In addition, the formation of crystal structure would also act as reinforcement to the glassy matrix [7]. Thus, improving the strength of glass-ceramic. Further increase the temperature to 850 °C and subsequently to 900 °C, density was slightly increased to 2.6638 g/cm^3 and 2.6837 g/cm^3, respectively. At 850 °C, the values of MOR reached its peak at 33.75 MPa before deteriorated to 27.16 MPa at 900 °C. The decline in modulus of rupture is most likely because of the growth of crystalline phases at a higher temperature which in turn enhance the viscosity of the glassy phase and inhibits further densification process as a consequence [14-15]. The microstructure also would become coarser and more porous owing to the lack of densification and enhanced crystallinity [16]. A similar trend was also observed for SGC 2 whereas the densities were improving as the temperature increased and measured at 2.2340 g/cm^3, 2.4247 g/cm^3, 2.5305 g/cm^3 and 2.6418 g/cm^3. The MOR was also increased from 13.72 MPa to 19.70 MPa and subsequently to 20.70 MPa as sintering temperature increased from 750 °C to 850 °C. Though, contrary to SGC 1, instead of slightly falling down, the MOR kept improving to 21.10 MPa at 900 °C indicating that the temperature might be not high enough to encourage crystallization that potentially prevents densification of glass-ceramic body.

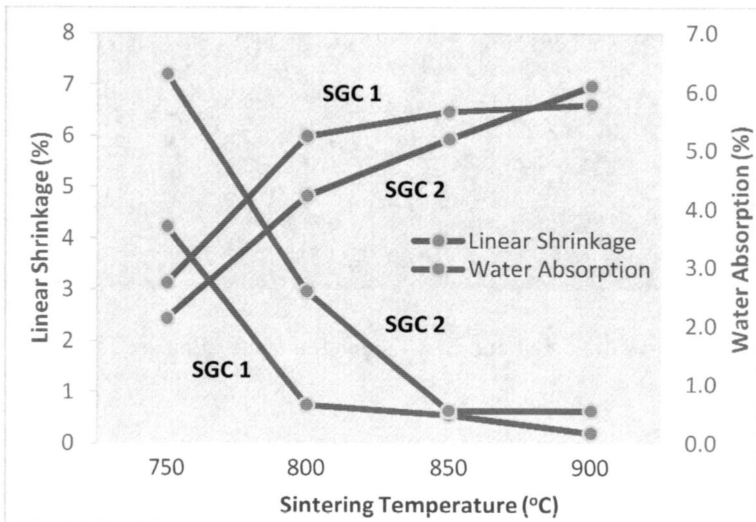

Figure 5. Linear shrinkage and water absorption in the function of sintering temperature for SGC 1 (20 wt.% RG) and SGC 2 (30 wt.% RG).

As can be seen in Figure 5, the linear shrinkages of the samples were inversely proportional to water absorption as sintering temperature increased from 750 – 900 °C. For SGC 1, the lowest water absorption was measured at 900 °C which is 0.1747 %, improving tremendously from 3.6960 % where it was measured at 750 °C. Upon heating up to 750 °C, the viscous flow of the glassy phase just began and densification was starting to take place at this point. Once the temperatures were increased to 800 °C and subsequently to 850 °C, the viscosity of glassy phase was low enough for the densification to occur at a greater rate and in turn reducing the formation of open pores and lowering its open porosity. As a result, water absorption was improved to 0.6569 % and 0.4759 %, respectively. The linear shrinkage was also enhanced from 3.13 % at 750 °C to 5.98 % at 800 °C and later to 6.47 % at 850 °C. The highest value was measured at 900 °C which is 6.61 %, thus indicated that the degree of densification is maximum at this point and contributed to the lowest value of water absorption as well as the highest linear shrinkage. On the other hand, SGC 2 which contained 30 wt.% of red gypsum also experienced similar results. The water absorptions were improving from 6.3024% at 750 °C to 2.59 % at 800 °C and later to 0.5543 % at 850 °C. The highest value was recorded at 900 °C which is 0.5531 %. The improvement of water absorption is most likely because of the lack of open pores on the surface of the samples due to more rapid densification at higher temperature. Increasing temperature from 750 – 900 °C also resulted in ascending pattern of linear shrinkage where it was measured at 2.43 %, 4.83 %, 5.93 % and 6.97 %, respectively. Comparing SGC 1 and SGC 2, in general, the densities, MOR, water absorption and linear shrinkage for SGC 1 were significantly better than SGC 2 which indicate that the sintering temperature for SGC 2 was not high enough to accelerate densification process due to the presence of the higher amount of red gypsum in SGC 2.

3.2. Crystallization Evolution

Figure 6 shows the phase diffractogram of SGC 1 and SGC 2 during thermal treatment from 750 °C to 900 °C. For both samples, upon heating up to 750 °C, no new crystalline phases were formed except quartz, cristobalite, montmorillonite and gypsum phases that originated from bentonite and red gypsum. Heating further up to 800 °C, new peaks starting to emerge which belong to crystalline phases of diopside, augite and lead oxide. When the temperature was increased to 850 °C, the peaks intensity is becoming stronger and reach its maximum at 900 °C. At 850 °C, new peaks that belong to titanian andradite was formed for both samples. However, it was observed that at 900 °C, the titanian andradite's peaks intensity for SGC 2 was more noticeable than SGC 1 possibly because of the higher content of red gypsum in SGC 2 as red gypsum contains a fairly significant amount of calcium, ferum and titanium. The formation of lead oxide in crystalline form also means that it is more stable chemically and will hardly leach out over time unlike in amorphous form. It is also believed that the formation of diopside crystal improved the mechanical strength of the glass ceramic as reported by Zhu et al and Hasheminia et al [17-18].

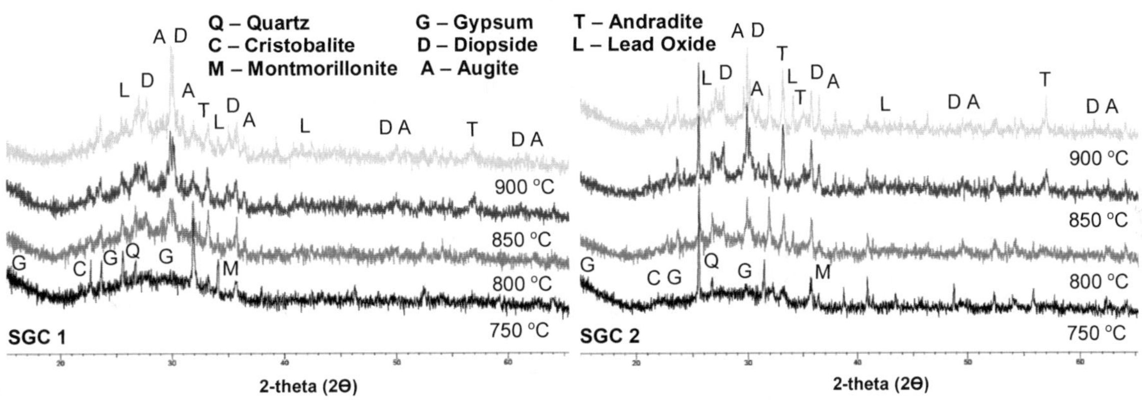

Figure 6. Phases evolution of SGC1 and SCG 2 with increasing sintering temperature from 750 °C to 900 °C

4. Conclusions

In the present research, cathode ray tube (CRT) glass and red gypsum wastes mixed with bentonite clay were converted into sintered glass-ceramic by a simple sinter-crystallization process. Two formulations with a variation of red gypsum content were prepared, pressed into a button shape and sintered at various sintering temperature. The produced glass-ceramic features obtained can be summarized as follows:

- for both SGC 1 and SGC 2, increasing the sintering temperature from 750 °C to 900 °C, the modulus of rupture, density, water absorption and linear shrinkage were improved except MOR at 900 °C for SGC 1 where a slight deterioration was observed
- in general, the modulus of rupture, density, water absorption and linear shrinkage for SGC 2 were noticeably inferior than SGC 1 which could be attributed to the higher content of red gypsum in SGC 2
- sintering a mixture of CRT glass, red gypsum and bentonite leading to the formation of new multi-crystalline phases such as diopside, augite, lead oxide and titanian andradite at a moderate temperature of 800 – 900 °C
- the developed product (particularly SGC 1) has a potential to be used as outdoor walkway tile or brick owing to the high physical-mechanical properties, colour suitability as well as low production cost (waste utilization and low sintering temperature)

5. References

[1] Bernardo E, Varrasso M, Cadamuro F and Hreglich S 2001 Vitrification of Wastes and Preparation of Chemically Stable Sintered Glass-Ceramic Products *J. Non-Crystalline Solid* **352** 4017– 4023

[2] Bernardo E, Castellan R and Hreglich S 2007 Sintered Glass–Ceramics from Mixtures of Wastes *Ceramic International* **33** 27–33

[3] Bernardo E, Doyle J. and Hampshire S 2008 Sintered Feldspar Glass-Ceramic and Glass-Ceramic Matrix Composites *Ceramics International* **34** 2037-2042

[4] Bernardo E 2008 Fast Sinter-Crystallization of a Glass from Waste Materials *J. Non-Crystalline Solid* **354** 3486-3490

[5] Bernardo E, Bonomo E and Dattoli,A 2010 Optimisation of Sintered Glass-Ceramics from an Industrial Waste Glass *Ceramic International* **36** 1675–1680

[6] Maschio S, Furlani E, Tonello G, Faraone N, Aneggi E, Minichelli D, Fedrizzi L, Bachiorrini A and Bruckner S 2009 Fast Firing of Tiles Containing Paper Mill Sludge, Glass Cullet and Clay *Waste Management* **29** 2880-2885

[7] Zhang W Y, Gao H and Xu Y 2011 Sintering and Reactive Crystal Growth of Diopside–Albite Glass–Ceramics from Waste Glass *Journal of the European Ceramic Society* **31** 1669-1675

[8] Zhang W Y and Liu H 2013 A low cost route for fabrication of wollastonite glass–ceramics directly using soda-lime waste glass by reactive crystallization–sintering *Ceramics International* **39**(2) 1943 – 1949

[9] Andreola F, Martin M I, Ferrari A M, Lancelloti L, Bondioli F, Rincon J M, Romero M and Barbieri L 2013 Technological properties of glass-ceramic tiles obtained using rice husk ash as silica precursor *Ceramics International* **39** 5427–5435

[10] Juoi J M, Arudra D, Rosli Z M, Hussain K and Jaafar A J 2013 Microstructural properties of glass composite material made from incinerated scheduled waste slag and soda lime silicate (SLS) waste glass *Journal of Non-Crystalline Solid* **367** 8–13

[11] American Society for Testing and Materials (ASTM) C 373-88 2006 *Standard Test Method for Water Absorption, Bulk Density, Apparent Porosity and Apparent Specific Gravity of Fired Whiteware Products* (West Conshohocken, United States)

[12] International Standard ISO 10545-4 2014 *Ceramic tiles - Part 4: Determination of Modulus of Rupture and Breaking Strength* (Switzerland)

[13] Prado M O, Zanotto E D and Muller R 2001 Model for sintering poly-dispersed glass particles *Journal of Non-Crystalline Solids* **279** 169–178

[14] Verné E, Bona E, Bellosi A, Vitale Brovarone C and Appendino P 2001 Na_2O-CaO-SiO_2 glass-ceramic matrix biocomposites *Journal of Material Science* **36** (11) 2801–2807

[15] Kim K D, Lee S H and Ahn H K 2004 Observation of nucleation effect on crystallization in lithium aluminosilicate glass by viscosity measurement *Journal of Non-Crystalline Solids* **336** 195–201

[16] Zhang W Y, Gao H, Li B Y and Jiao Q B 2006 A Novel Route for Fabrication of Machinable Fluoramphibole Glass–ceramics *Scripta Materialia* **55** 275–278

[17] Zhu M, Ji R, Li Z, Wang H, Liu L and Zhang Z 2016 Preparation of glass ceramic foams for thermal insulation applications from coal fly ash and waste glass *Construction and Building Materials* **112** 398–405

[18] Hasheminia S, Nemati A, Yekta B E and Alizadeh P 2012 Preparation and Characterisation of Diopside-Based Glass–Ceramic Foams *Ceramics International* **38** 2005-2010

IOP Conf. Series: Materials Science and Engineering **713** (2020) 012009 doi:10.1088/1757-899X/713/1/012009

Development of a Model for OTTV and RTTV based on BIM-VPL to Optimize the Envelope Thermal Performance

F Abass [1], L H Ismail [1], I A Wahab [1] and A A Elgadi [1]

[1] Department of Architecture, Faculty of Civil and Environmental Engineering, University Tun Hussein Onn Malaysia, Parit Raja, Batu Pahat, Johor, Malaysia
Corresponding author: fatmaelbeltagy@gmail.com

Abstract. The thermal performance of a building has a profound effect on sustaining the conditions of indoor environments, and it is critical in obtaining any energy-efficient design. Choosing an appropriate construction material is one of the most effective methods to prevent excessive energy consumption in the building, to minimize heat flows, and sustain a comfortable temperature for occupants. The majority of the existing buildings in Malaysia are mostly constructed using bricks for their outer walls; many studies nominated these materials as the producer of heat. This study is focusing on evaluating the thermal performance of construction materials in a tropical climate, with the involvement of manual calculation methods for Overall Thermal Transfer Value (OTTV) of walls and roof. Therefore, this study proposes integrating Building Information Model (BIM) and visual programming language (VPL) for supporting the calculation of OTTV based on MS1525:2014. The expected result of the proposed system will support designers on calculating OTTV coefficient comparable with the manual calculation, which is considered promising in minimizing time-consuming and human error when optimizing envelope performance during building design. This study proposed a model to access and extract construction materials properties from BIM library using VPL to estimate OTTV value using a real-time calculation tool.

1. Introduction

There is continually growing for energy, especially in the building sector, where approximately 40% of the overall global energy consumption is consumed by building [1]. The significant amount of energy requesting on buildings is to operate the mechanical cooling, which helps the occupants to achieve their level of thermal comfort. The considerable growth in CO_2 emissions alongside with the increase of energy demand is occurring in cities, where growing affluence and request for higher living standards caused by population is escalating [2]. An example, in Malaysia an enormous amount of electricity is consumed by conditioning units in commercial and residential buildings, mainly, during sunny days. Several challenges were faced by building designers when coming to maintain a comfortable environment and less energy consumption.

To increase the energy efficiency in buildings, building designers needs to become better aware of the thermal envelope performance and the selection of structural materials. Building thermal performance can be attributed to the procedure of modelling the heat transfer between the building envelopes to its surrounding environment. In simple words, thermal performance is a procedure of heat exchange through glazing and opaque wall. Modern materials like concrete, glass, and bricks are widely used in building construction in Malaysia, many studies nominated these materials as the producer of the non-comfort state for indoor spaces in a hot, humid climate. There is no doubt that the thermal transmitter and the roof construction materials are the main reasons for indoor discomfort [3]. According to Leong [4], 5% of the heat was penetrated through the roof of the single-storey dwelling, compared with 50% for double storey dwelling.

The growing concern about building thermal performance and energy efficiency led to establishing many policies. These policies have a benefit to raise awareness and interest in energy conservation, encourage effective designs for energy, and promote the development of energy-efficient products. As an example to estimate the envelope thermal performance in building in hot, humid weather, a policy identified as the Overall Thermal Transfer Value (OTTV) was established to cut heat penetration over the exterior envelope, and hence decrease the electricity requesting for air conditioning. Whereas in cold countries, thermal insulation standard and the calculation of thermal transmittance value of envelope materials are more popular.

Typically, the OTTV has required a manual calculation method which could be time requiring for project team who are inexperienced through the calculation technique, especially when human errors during the calculation of the OTTV coefficient can drive to inaccurate results. Therefore, the adaptation of Building Information Modelling (BIM) and Visual Programming Language (VPL) has been progressively applied during the design and construction stages to help the designers to evaluate their design options [5]. The new integral of analyzing building thermal performance could help to maintain more design decisions regarding the selection of building materials to prevent the building from the excessive heat flow.

The primary key of this study is to develop an integrated model based on BIM and OTTV code, which is known as BOTTVc "Building Overall Thermal Transfer Value code." .The development method could be used to evaluate the thermal performance of the building envelope compared with a manual calculation. When it comes to time-saving and minimization of human error, the proposed BOTTVc model could be a solution for analyzing building thermal performance, especially when it comes to reducing energy demand in buildings.

2. Theoretical background

2.1. Overall Thermal Transfer Value (OTTV) code

Since 2001, OTTV code has been taken into account in Malaysian Standard, especially for commercial conditioning buildings [6]. The main concern for Malaysian Standard is to consider the heat gain through windows of the building as a critical parameter. Due to energy inquiring to cool the indoor environment, the OTTV Equation generally focused on four important parameters which are U-value of structural materials, solar absorption, window-to-wall ratio, and shading coefficient; these components had a substantial impact on heat gain through the envelope [6]. In MS1525: 2014 "code of practice" [7], all the coefficients that influence the value of OTTV for walls or roof such as, solar absorptivity, orientation factors of fenestration wall, and shading coefficient were estimated based on several studies and simulations to accurate the Equation's data with the climatic characteristic of Malaysia. Equation (1) illustrates the final OTTV equation for wall and equation (2) is RTTV for roof based on MS 1525:2014.

$$OTTVi=15\,\alpha\,(1-WWR)Uw+6\,(WWR)Uf+194 \times CF \times WWR \times SC \tag{1}$$

$$RTTV=(A_r \times U_r \times T_{Deq}) + (A_s \times U_s \times \Delta T) + (A_s \times SC \times SF)/\,A0 \tag{2}$$

WWR: window to wall area ratio, α: a solar absorptivity of the wall color, U_w, U_f, U_s and U_r: the thermal transmittance (U-value) of the opaque wall/roof, window glass, and skylight, CF: the solar orientation correction factor for roof/wall, SC: the shading coefficient of the fenestration wall based on orientation, ΔT: the temperature difference, and A_s, A_r: the area of opaque wall and skylight area.

OTTV requirements is considered simple without simulation process, especially when it applies to commercial condition building. OTTV evaluation is mostly calculated using excel spreadsheets during the design process to achieve less external heat penetration from outside building to inside. While it comes to residential buildings, there is no specific equation to calculate the OTTV in Malaysia, compared with other ASEAN countries which had their Envelope Thermal Transfer Transmittance

IOP Conf. Series: Materials Science and Engineering **713** (2020) 012009 doi:10.1088/1757-899X/713/1/012009

Value (ETTV) in Singapore and Thailand. For this reason, the current OTTV code in Malaysia for the assessment of residential building is by using both techniques of traditional calculation and simulation.

2.2. The integration of Building Information Modelling (BIM) in the analysis of building performance

The incorporation of BIM theory into the green building has acquired an enormous interest from practitioners and scholars concerned with both building sustainability issues and BIM [8]. Many studies have examined the optimization of energy performance using both BIM and multi-objective optimization algorithms [9], BIM for Green material selection [10], BIM for building with vegetation systems [11], and BIM for Daylighting optimization [12]. Currently, building performance evaluation via analysis software is mostly performed after the design stage; thus, it is not integrated into the design decision-making process. To effectively evaluate the multiple performance criteria in the building design, e.g., envelope material, form, and energy efficiency, the assessment of building performance should be seamlessly integrated into the design process [13].

Several packages of authoring software that is integrated with BIM allow designers to allocate the properties of structural material. While, these tools offer their Application Program Interface (API), which help to access and modify the thermal properties of construction materials in the BIM database. The users of BIM packages (Vectorworks, Bentley, Tekla BIM, ArchiCAD, MicrosoftStation BIM, and Catia) are required to have full knowledge in programming experience in order to use their API effectively, which is not achievable for designers who were not familiar with the programming. Thus, with understanding the importance of the link of BIM and VPL software, programming companies have been developing a VPL program that gives accessibility to access BIM database and customizes building elements with a reasonable understanding of programming.

2.3. The integration of Visual Programming Language (VPL) in the analysis of building performance

VPL tool has shown promising results when implementing to estimate building performance analysis such as daylighting optimization [9], energy efficiency, acoustical analysis structural analysis [14] and more. According to Kensek [15], the VPL could aid during the early design stages of deciding the sustainable optimized design. Several studies investigated the application of Visual Programming tools to generate workflows and frameworks related to analysis of building performance. As an example, Konis [16], established a framework to optimize the passive performance of the building. Façade design and enhancing shading are one of the hottest topics associated with the implementation of a BIM and VPL tool in the area of sustainable building analysis. More so, Kensek [15] demonstrated how VPL and BIM could be adapted to design a parametric shading device for elaborating façades and energy simulation performance. A similar study by Kim [17], determined a new method to analyze energy performance for elaborate kinetic façades.

The visual programming language can offer a new workflow to optimize building performance; the workflow could provide more access to advance the demand for energy during building design. In the last years, building designers become more concerned with Visual Programming tools that allow creating, modified, powerful, and flexible form of generating algorithms without having in-depth knowledge with coding [15]. Several of these VPL based tools accessible in the market are Marionette, Grasshopper, and Dynamo. However, since the first release of Dynamo as VPL, it was connected to one of the BIM authoring tool Revit.

VPL Dynamo is a free source tool devoted to developing the design process of building and workflows in both BIM and computational design fields. According to Vandezande [18], Dynamo offers incomes for designers with the constructing programmatic relationships using a Graphical User Interface (GUI). Therefore, rather than writing 'code' from scratch; the user is capable of making a custom algorithm by connecting pre-packaged nodes with assembling custom relationships. The ability of Dynamo in controlling the BIM parameters using Revit Architecture software added a further level of created and associativity new prospects for cross-discipline and cross-platform collaboration [19]. Various sorts of data can be controlled using Dynamo, such as family geometry, parameter values, and family placement. By creating new elements in Revit using Dynamo, the designer can modify its parameters manually within Revit. Moreover, the same Dynamo script can be reused in many projects.

IOP Conf. Series: Materials Science and Engineering **713** (2020) 012009 doi:10.1088/1757-899X/713/1/012009

3. Methodology

BOTTVc is a model that is originated by Natephra *et al.* [20] to create a connection between BIM and Dynamo as a visual programming language, to enhance the calculation of the thermal performance of building envelope during the design stages based on OTTV code adapted in Thailand, Hong Kong, and Singapore. BOTTVc model is a framework used to create an approach for data extraction and management, moreover, to reveal potential mechanization workflows for further improvement of the OTTV assessment tool.

Two main steps are required in order to develop the model of OTTV. First, the assessments of the OTTV requirement involve clarifying all the related criteria, such as OTTV's coefficient data demand (input) and the specific coefficient description in MS1525:2014. In Malaysia Standard 1525:2014 "Energy efficiency and use of renewable energy for non-residential buildings - Code practice'' [7], all OTTV's requirement and the calculation methods were fully detailed in the standard.

Secondly, screening the available OTTV's coefficient request in the BIM authoring tool that is used for this study (Autodesk Revit, 2016), therefore, if requested data could be automatically extracted from the 3D BIM'S model, it is then applied to Revit functionalities to modify requested equation. In a situation where there is no functionality available for the selected coefficient, gaps will be recognized, and additional improvement is conducted using Visual Programming scripting (Dynamo).

3.1. OTTV integration framework based on BIM – VPL

In this study, Revit Architecture 2016 and Dynamo 1.3 visual programming languages are used to improve the BOTTVc based on MS1525:2014. The adaption of VPL in this study is due to the limitation of BIM authoring software (Autodesk Revit) to perform some functions of OTTV coefficients. As an example, in equation (1), there are coefficients that the code primarily depends on them (opaque wall orientation, the WWR, and SC). These parameters could not be available in BIM database and required to follow equations to calculate them, which could be easy to estimate by developing a script in VPL (Dynamo).

3.1.2. Prepare 3D Modelling. The first step to calculate the OTTV for a building based on BIM started by creating a 3D model using authoring BIM software (Revit Architecture). The 3D contained the model of building elements (walls, windows, floor, shading device, roof, and doors) for the physical and thermal properties of the construction materials used on these elements developed using scripting tool.

3.1.3. Scripting the thermal and physical properties based on the BIM database and VPL. Materials construction of building elements could be automated from the BIM database or even modified during creating the 3D model, it is available in Revit object library, and it is automatically estimated based on the selected object. In a situation where there is a variant of building material combinations, the material properties are extracted from the BIM database and modify using two types of scripting available in dynamo, which is a Python and code black node.

1- Code Black node: Before starting using this type of scripting to access and extract material properties, linking between the BIM model and Dynamo base node "SelectModelElements" is required. Figure 1 explains the process of extracting the data using "CodeBlack." Text script starts with select building element to access to material properties by "typeParameter" input; this step provides the user with access to a list of materials (IDs) in the properties list. A Parameter "ID" contains data and numerical units that scripting to obtain the value of "U" (value of wall and window glass) and the "R" (value of construction materials).

2- Python scripting is used to develop other coefficients of BOTTVc systems, by accessing to BIM material database in Revit using the API system. Some of the material coefficients could not be extracted using the Code Black node. Therefore, "ThermalAssesstClass" that symbolize the construction material properties related to energy assessment in Revit was used to access to material density, specific heat, and thermal conductivity. Figure 1 illustrates the process of extract material properties from BIM database using Code Black node and Python.

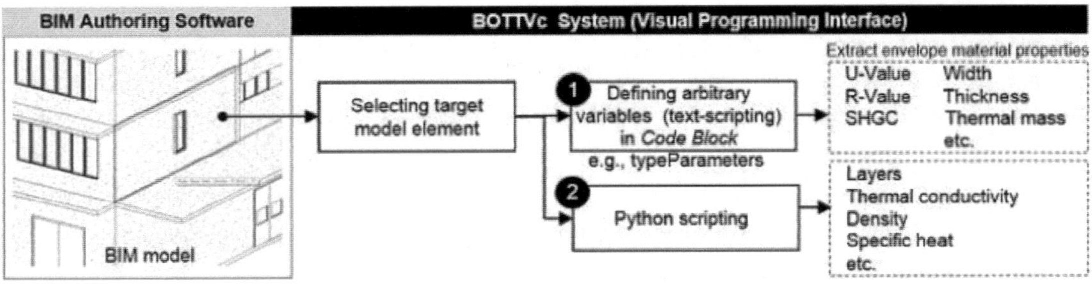

Figure 1. Process of accessing and extracting OTTV parameters from BIM database using VPL.

3.2. VPL to calculate OTTV coefficients

There are other coefficients that required estimating the value of OTTV and RTTV, which are SC, SF,CF, TDeq, A_s, A_r ΔT and WWR, besides the calculation of the area of opaque wall/roof and the area of fenestration wall/roof, and the height and width of external shading devices.

3.2.1. Obtaining WWR, A_s and A_r value based on VPL. To estimate the values of WWR, A_s and A_r using dynamo scripting, "SelectModelElements" node was applied to select the target wall, roof or skylight for every orientation. Then, "Element. Get Parameter Value by Name" node is implemented to extract the area of opaque wall/roof or fenestration wall from the 3D model using string node named "Area" after the extraction of the area a math node is adopted to calculate the value of WWR, A_s and A_r. After obtaining the surface area of the wall and roof, a solar absorption coefficient can be chosen as a constant provided in the practice code [7], in order to obtain the OTTV/ RTTV as stated in equation (1) and equation (2).

3.2.2. Obtaining SC value based on VPL According to MS1525:2014, SC could be estimated based on Equation (3):

$$SC = SC1 \times SC2 \tag{3}$$

Where: SC1 is a shading coefficient of glazing, and SC2 is the shading coefficient of the external shading device. In order to estimate SC2 for egg crate, horizontal or vertical shading device, "SelectModelElements" node is used to estimate the area width and the length of an overhang, and the sum of the area of the window and the distance between overhang and the window. After the extraction and the calculation of SC2, the output is then used to find the accurate value stated in [7]. At last, a math function code is used to estimate the final SC-based of the constant value of glazing and shading coefficient, figure 2 presents the proposed script to calculate SC based on the model of Natephra [20].

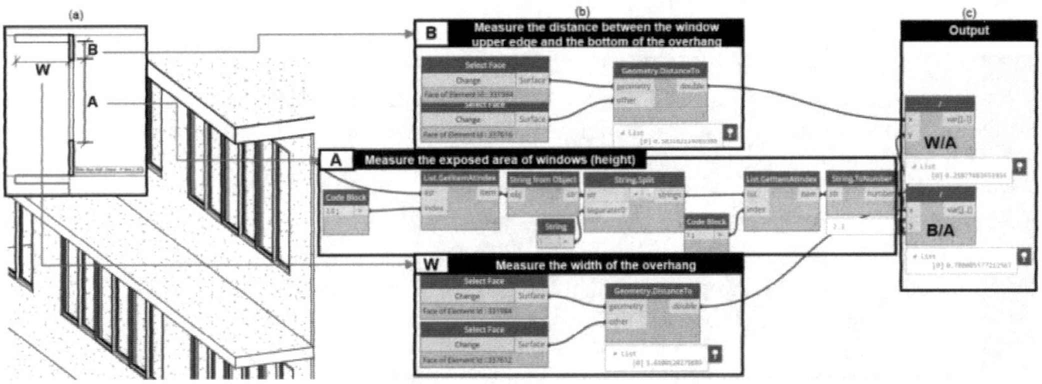

Figure 2. Process of obtaining SC coefficient using the Code Black node.

3.2.3. Obtaining SF value based on VPL

The equation of RTTV stated in section 2.1 required the calculation of A coefficient, which is the solar factor. Based on MS1525:2014, the solar factor could be estimated using equation (4)

$$SF = 323 \times OF \tag{4}$$

In this step, the math function node provided by Dynamo is used to state the numeric input value suggested by a code of practice for calculation of RTTV for Malaysia, as well as the solar orientation factor of roof (OF) coefficient. For the other coefficient provided by OTTV and RTTV calculation, including TDeq and ΔT values, were provided as constant in code of practice.

After developing the parameters of OTTV/ RTTV, the final computing is required to estimate the total value of OTTV and RTTV and then followed by comparing with the value stated in the standard. In the case that the obtained value does not comply with the standard, a modification of the design and selection of materials is required, figure 3 concludes all the process of BOTTVc from modelling stage to final output.

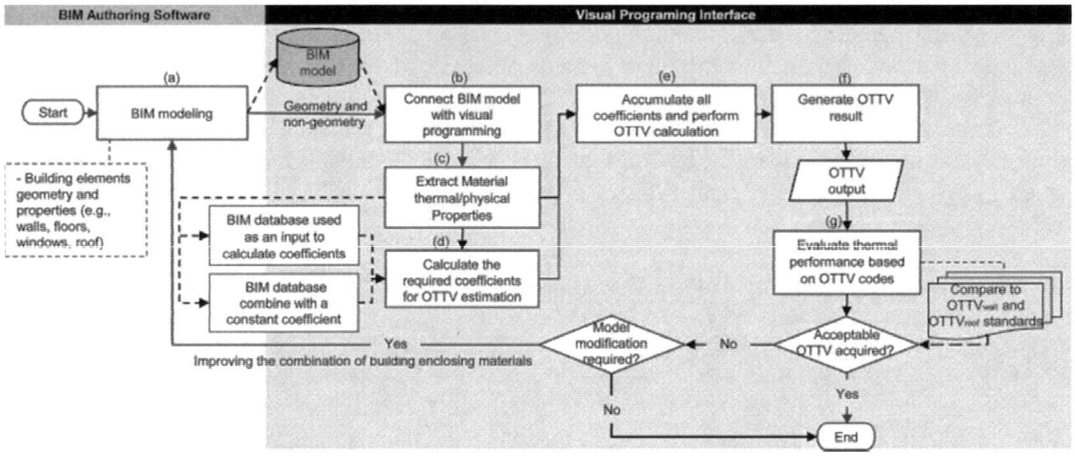

Figure 3. Process of the proposed model of BOTTVc based on BIM and VPL.

4. Discussion

This study covers part of a research that is conducted to evaluate the optimization of thermal performance of the envelope and roof of a residential building covered with green vegetation as insulation material in Malaysia. The study included two main stages; first, by detecting required coefficients for OTTV and RTTV code based on MS 1525:2014 and followed by the developing of the OTTV model based on BIM-VPL software. During the first step, prescriptive requirements of OTTV and RTTV coefficients including WWR, solar absorption factor of wall and roof, U-value of an opaque wall, fenestration wall and roof, SC, CF were reviewed and categorized. In the second phase, it involved scripting the OTTV coefficient using BIM and VPL, towards evaluating the thermal performance.

Future works are required to establish the viability of this model when compared with manual calculation, which will be computed by calculator before the input is listed in the excel sheet. Hence all the data related to opaque and fenestration walls and roof are gained by observing the 2D drawings and by calculating the properties of the elements (thermal and physical) based on the provided equations. The hand-based calculation will be evident that the manual documentation process of OTTV and RTTV is a very mind-numbing and time-consuming task. The scripting of the developed model could be applied in varying building design, with different areas and construction materials; in simple word, the developed system could be considered as a time saver for building designers. Therefore, it could provide accurate results in a short period; besides the ability of designers to easily modify their design and materials selection by running the scripting until finding the ideal design solution.

5. Summary

This study aims to enhance the growing area of research of building thermal performance and energy efficiency by proposing a model of OTTV to attain the best thermal performance, the optimization of a building envelope could achieve a good selection of construction materials, proper and accessible insulation materials, and envelope performance. The expected results of this study will show that the proposed system will lessen the time to estimate thermal performance during the design stage. Moreover, the implementation of the BOTTVc model will provide the guideline to designers when working on optimization of the thermal performance of the building envelope, especially when modifying the U-value and wall/ roof absorption factors during the design process, which are considered an essential parameters compared to the other coefficients in OTTV and RTTV formula. The next phase of this study will focus on calculating the OTTV for wall and roof of some selected case studies using both manual and a model system, before and after lowering the U value of the wall and roof by using living insulation material (green vegetation).

6. References

[1] World Energy Council 2013 *World Energy Resources Survey 2013* Retrived on Febraury 24, 2019 from https://www.worldenergy.org/publications/2013/world-energy-resources-2013-survey
[2] Wee K, Matsumoto H, Chin S and Yu F 2008 Energy Consumption and Carbon Dioxide Emission Consideration in the Urban Planning Process in Malaysia *J. Malaysian Inst. Planners* **6** 99–128
[3] Vijaykumar K, Srinivasan P and Dhandapani S 2007 A performance of hollow clay tile (HCT) laid reinforced cement concrete (RCC) roof for tropical summer climates *Energy Build.* **39** (8) 886–892.
[4] Leong A 2009 *MS1525:2007 Clause 5 : Building Envelope* (Kuala Lumpur: Department of Standards Malaysia)
[5] Fischer M 2006 *Formalizing Construction Knowledge for Concurrent Performance-Based Design* (Berlin, Heidelberg: Springer Netherlands)
[6] Djamila H, Rajin M and Rizalman A N 2018 Energy efficiency through building envelope in Malaysia and Singapore *J. Adv. Res. Fluid Mech. Therm. Sci.* **46**(1) 96–105
[7] MS1525 2014 *MS1525:Energy efficiency and use of renewable energy for non-residential buildings* (Selangor: Code of practice)
[8] Azhar S and Brown J 2009 BIM for Sustainability Analyses *Int. J. Constr. Educ. Res.* **5**(4) 276–292
[9] Asl M R, Stoupine A, Zarrinmehr S and Yan W 2011 Optimo: A BIM-based multi-objective optimization tool utilizing visual programming for high performance building design *Proc. of the 33rd Int. Conf. on Edu. and Research in Computer Aided Architectural Design in Europe* **130** (United States: CUMINCAD) pp 1-10
[10] Akanmu A, Asfari B and Olatunji O 2015 BIM-based decision support system for material selection based on supplier rating *Buildings* **5**(4) 1321–1345
[11] Lee D 2014 Building Integrated Vegetation Systems into the New Sainsbury's Building Based on BIM *Journal of KIBIM* **4** 25–32
[12] Kota S, Haberl J S, Clayton M J and Yan W 2014 Building Information Modeling (BIM)-based daylighting simulation and analysis *Energy Build.* **81** 391–403
[13] Schlueter A and Thesseling F 2009 Building information model based energy/exergy performance assessment in early design stages *Autom. Constr.* **18**(2) 153–163
[14] Makris M, Gerber D, Carlson A and Noble D 2013 Informing Design through Parametric Integrated Structural Simulation, in *Computation and Performance: Proceedings of the 31th*

conference on education and research in computer aided architectural design in Europe, eCAADe 2013 (Delft: Delft University of Technology) pp 69–78

[15] Kensek K 2015 Visual Programming for Building Information Modeling: Energy and Shading Analysis Case Studies *J. Green Build.* **10**(4) 28–43

[16] Konis K, Gamas A and Kensek K 2016 Passive performance and building form: An optimization framework for early-stage design support *Sol. Energy* **125** 161–179

[17] Kim H, Rahmani M and Yan W 2015 Parametric BIM-based Energy Simulation for Buildings with Complex Kinetic Façades, Ecaade 2015: Real Time - Extending the Reach of Computation **1** 657-664

[18] Vandezande J, Krygiel E and Dillon B 2015 *Mastering Autodesk Revit Architecture 2016: Autodesk Official Press* (Paris: Sybex) 1st ed. pp 1008

[19] Kensek K and Douglas N 2014 *Building Information Modeling: BIM in Current and Future Practice* (United States: John Wiley and Sons) pp 1689-1699

[20] Natephra W, Yabuki N and Fukuda T 2018 Optimizing the evaluation of building envelope design for thermal performance using a BIM-based overall thermal transfer value calculation *Build. Environ.* **136** 128–145

Acknowledgments

I would like to thank Assoc. Prof. Ts. Dr. Lokman Hakim bin Ismail, Dr. Izudinshah bin Abd Wahab, and Dr. Ahmed Elgadi for their discussions, guidance, and sharing of experience which enrich this study.

IOP Conf. Series: Materials Science and Engineering **713** (2020) 012010 doi:10.1088/1757-899X/713/1/012010

Sustainable Material: Development Experiment of Bamboo Composite Through Biologically Binding Mechanism

I F Ridzqo[1], D Susanto[1], T H Panjaitan[1] and N Putra[2]

[1] Department of Architecture, Faculty of Engineering, Universitas Indonesia, Depok 16424, Indonesia
[2] Department of Mechanical Engineering, Faculty of Engineering, Universitas Indonesia, Depok 16424, Indonesia
Corresponding author: dalhar3001@yahoo.com

Abstract. Bamboo as stems have been widely manufactured for composite. However, fiber as the smallest constituent component of bamboo stems supporting the strength and flexibility of the plant has not been widely employed as raw material. These strong and flexible properties, coupled with easy planting treatment and fast harvesting, apparently make bamboo highly potential developed as sustainable raw material for composite. Unfortunately, the current manufacturing process of bamboo for composite by using chemical substances would have ended bamboo up as no longer environmentally friendly. By utilizing lignocellulose content within its fiber, this research studied fabrication of a novel composite boards from bamboo fibers through biologically binding mechanism by using fungal mycelium. *Gigantochloa apus* bamboo stems are extracted into three types: long fibers, short fibers, and powder. Then, the bamboo fibers are added with water and some additional nutritions then sterilized together. These substrates are then inoculated with mycelium seed of *Ganoderma lucidum*. The fibers bound together along with the growth of mycelium. The result shows that this board is potential to be used for interior purpose in building especially high rise building with high need of light-weight insulation and partition board and expected to replace the need for building components that have been made from unsustainable raw materials and methods.

1. Introduction

Woody fiber still becomes the best option to fulfill daily material needs. Of the various plants producing woody fiber, bamboo is one of them. Bamboo stems are utilized as raw materials for various needs, such as musical instruments, furniture, and building components. Unfortunately, bamboo was once considered as "the poor timber" due to its low price and widely used by the low economy community. However, the value of bamboo as raw material begins to rise and even becomes the rich timber [1]. One of some factors that turns bamboo back in demand lies in its nature which is strong but flexile. Bamboo is easily cultivated without special treatment. Within one year, a bamboo hump can grow as many as 5 to 10 stems [2]. Furthermore, rapid growing causes bamboo stems to be harvestable at the age ranging from 3 to 5 years. With its superiorities, the needs of plant fibers as raw material which are usually supplied from wood can be gradually replaced by bamboo [3].

Bamboo *(Bambusoideae)* is a group of gigantic grass plant. It is estimated that there are about 2000 species of bamboo spreading around the world [4]. In Indonesia, the diversities of bamboo species which have been identified are as many as 145 species, which 75 species among of them are endemic. There are five bamboos which are found almost evenly across the five major islands in Indonesia, i.e

Bambusa bluemana, Bambusa vulgaris (*bambu ampel*), *Dendrocalamus asper* (*bambu petung*), *Gigantochloa atter*, and *Schizotachyum brachyladum* [5]. Due to its different morphological characters and mechanical properties, each species of bamboo is used for specific purpose, for instance *G. apus*, which has relatively small diameter stems and long fibers, is generally used to make ropes. A bamboo stem is heterogeneous. Its cells are not uniformly distributed, both horizontally and vertically along the culm [3]. The morphology of bamboo stems comprises of culm wall, fiber bundle, and elementary fiber. The highest strength and flexibility of bamboo lies in its smaller size, which is in its elementary fiber, thus is highly potential to perform as reinforcing fibers in composite materials [6]. Bamboo fiber is also referred as lignocellulose fiber, because it contains 25% lignin, 55% cellulose, and 20% hemicellulose [3].

Besides being directly used after harvested, bamboo stem is also processed in industrial scale as raw material for making composite. Generally, the manufacturing process still involves artificial adhesive chemicals to bind bamboo fibers, such as epoxy and resin [7, 8]. Unfortunately, these chemical substances would make bamboo loose its environmentally friendly nature. Although it creates very strong binding, these chemicals change the nature of bamboo as natural fibers to be unable to decompose when the material is no longer used [9].

Various methods to achieve sustainability in the process of binding composite materials from plant fibers is still continuously being developed. Some biochemical adhesives to create natural products have been made among others are resin from plants extraction [10, 11, 12] and protein-based adhesive extracted from cattle nerve [13]. Nevertheless, other binding methods which are more environmentally friendly have been developed through biologically mechanism. This method is supposed to be more effective in achieving the sustainability of a material because it passes fewer phases. Through this mechanism, microorganisms are deliberately grown on biocomposite substrates to bind fibers [14]. The ability of microorganism to process lignocellulosic fibers from agricultural waste previously has been proven by using fungal mycelium in the process of making biocomposite panels [15], acoustic panels [16], and structural components [17]. The fungal mycelium frequently used came from mushroom belonging to the Ganoderma phylum [18, 16, 17].

Fungal mycelium is a network of hyphae that part of fungi anatomy which continues to grow along its vegetative period. Mycelium continues to grow by consuming lignocellulose substance from plant fibers which become the substrate for its life [19]. The nature of hyphae which continuously becomes wider and more complex is utilized in the process of making those biocomposite materials as binder of lignocellulose fibers [18]. Beside its strength and durability performance, the mycelium as biocomposite binder has several advantages, including non-toxic, fire resistant, hydrophobic, capability for thermal insulation, easy to form, and 100% biodegradable. Compared to conventional manufacturing industries, biocomposite materials which use mycelium as binder, require much less amount of water, energy, and CO_2 emissions [18, 20].

In this experimental study, a novel biocomposite board from bamboo fiber with fungal mycelium binder have been developed. This study uses lignocellulose content from bamboo fiber which is different from previous studies using lignocellulose fibers from agricultural waste, such as wood and grain. The high strength properties of bamboo is potential to increase the value of this biocomposite board. A series of tests on this material then have been carried out to find out the physical performance that the most compatibel with function of this material in a building. This biocomposite board is expected to replace the need for building components that have been made from unsustainable materials and methods.

2. Materials

2.1. Bamboo fiber

The species of bamboo used in this study is *Gigantochloa apus* (*bambu apus*). This species is chosen because of its strong, flexile, and quite thick in diameter of its culm. The bamboo plant was cultivated in South Tangerang, Banten and harvested at the age of 5 years. Before being used in experiments, bamboo stems were first extracted to obtain its single long fibers. Bamboo stems with a length of 50 cm to 60 cm were split vertically into some thin bamboo sheets. The bamboo sheets were then flattened with hammer until its long fibers could be released one by one (Figure 1(a)). To get shorter fibers, some amount of long fibers then were cut into 1 cm to 2 cm in length (Figure 1(b)). Meanwhile, some other amount were crushed to get bamboo powder (Figure 1(c)). By this extraction process, three types of bamboo fibers were obtained; long fibers, short fibers, and powder. These three types of fiber were then used as substrate variables in the experimental process.

(a) (b) (c)

Figure 1. Three types of bamboo fibers: (a) long fibers, (b) short fibers, and (c) powders.

2.2. Fungal mycelium

The fungal mycelium used in this experiment comes from species of *Ganoderma lucidum*. This species originates from Basidiomycotina macrofungi group and Homobasidiomycetes family. This species is chosen based on previous research, its terrestrial habitat on wood trunk, so it has strong mycelium bond and capability to break down woody substances and grows relatively fast. Mycelium is in the form of F1 seeds in chopped corn medium (Figure 2). Mycelium seeds were cultured in the Microbiology division of Biology Research Center LIPI, Cibinong, Indonesia.

Figure 2. Fungal mycelium seeds in corn chopped medium.

3. Methods

3.1. Experimental method

The experiment was conducted in the Culture Laboratory at the Faculty of Engineering, Universitas Indonesia. This experimental method was carried out by following the previous studies [16, 17, 18], but with modified materials and steps' details. Three material samples, each contained different type of bamboo fiber as the main substrate were labeled as A1 for long fibers, A2 for short fibers, and A3 for powder. Into each sample, some ingredients were added to provide ready sources of C, N, and Ca which are generally needed by organisms. Then all ingredients were mixed together with water in a heat-resistant plastic molding container. After that, the samples were pasteurized to sterilize contaminants. The samples were then left until the temperature decreases. The mycelium seeds then were inoculated into the substrate. The containers were then resealed and stored in a storage box at room temperature of 30^0C to 35^0C. The mycelium process to incubate the entire substrate took 20 days.

After the mycelium incubated the entire substrate, the samples were removed from its container. The samples were then pressed using two boards placed on the top and bottom sample's surfaces so

IOP Conf. Series: Materials Science and Engineering **713** (2020) 012010 doi:10.1088/1757-899X/713/1/012010

that its thickness became 3 times thinner. Then, the samples were dried in a drying oven at temperature of 80^0C with normal pressure for 9 hours.

3.2. Testing method
Tests on samples were carried out to find out physical properties: density, swelling thickness, and internal bond. These tests aimed to determine the classification of the material function and the bonding strength of mycelium as binder. Tests were conducted at Indonesia Institute of Sciences (LIPI) Biomaterial Research Center, Cibinong by referred to the Japanese Industrial Standard (JIS) A 5905: 2003 for Fibreboards [23]. The dimension of the specimens for testing was normative size 5 cm x 5 cm. Each test was repeated 3 times. The density is calculated by the following Equation (1).

$$\rho = \frac{W}{p \times l \times t} \tag{1}$$

where ρ is the density (g/cm^3); W (g), p, l, and t (cm) are respectively the mass, length, width, and thickness of the specimen. Meanwhile, swelling thickness testing is carried out by immersing sample specimens in water for 24 hours. Calculation of swelling thickness is done by the following Equation (2).

$$St = \frac{t_2 - t_1}{t_1} \times 100 \tag{2}$$

where St is the swelling thickness (%), t_1 and t_2 (cm) are specimen's thickness before and after immersed in water. Meanwhile, internal bonding strength test is carried out with Universal Testing Machine (UTM) AG-IS Shimadzu 50 kN with a tensile speed of 2 mm/minute. The test result is obtained by the following Equation (3).

$$Ib = \frac{P'}{b \times l} \tag{3}$$

where Ib is the internal bond (N/mm^2); P' (N), b, and l (mm) are maximum load, specimen's width and length.

Observation on material's morphology to examine the distribution of mycelium to bamboo fiber is also carried out by scanning electron microscope (SEM). The process is conducted with JEOL JSM-6390 machine at Metalurgical Center for Laterite LIPI, Serpong. Because of the low density of material, the three specimens were firstly vacuumed and coated with gold to avoid charging during the scan process.

4. Results

4.1. Manufacturing Material
The growth of mycelium to incubate the entire substrate required total of 20 days (Figure 3). During the growth period, the circulation of clean fresh air into and out of the container is necessary for optimal mycelium growth. To optimize the mycelium incubation on the other sides, the substrate, which has formed a sufficiently solid volume, could be removed from molding container on the day-15. The replacement of substrate from its container is intended to let these other sides to be perfectly incubated by mycelium. Overall, there is no significant difference in the speed of mycelium growth from the three variables. Observations were also made on changes of substrate mass on day-0 and day-21 when the material is ready after being dried and pressed. Average weight loss of these three variables are about 61% to 64%.

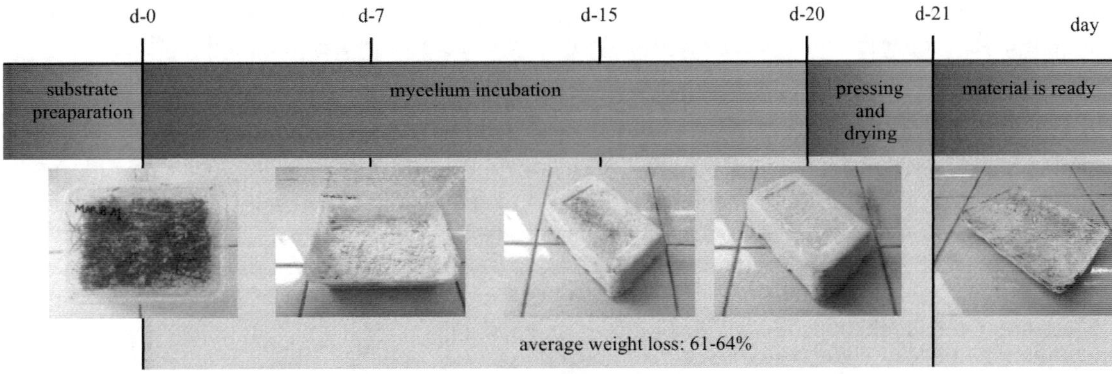

Figure 3. The material transformation day by day: from bamboo substrate, mycelium incubation throughout substrate's surfaces, and the final material.

4.2. Material's morphological characters

The final appearance of all substrate variables materials are white with brown spots in several parts. The material's surfaces are smooth and doff textured. There is no special characteristics of visual appearance to distinguish among three variables. The texture of material's surfaces actually comes from the layer of fungal mycelium that grows on the surfaces of the substrate. Initially, when the incubation period was complete, the outer layer of mycelium on the substrate's surfaces had a foam-like texture. Then, when the material has been pressed and dried, this outer layer of mycelium became dry and dense.

The long fiber (A1) sample feels most tenuous among the three variables, while the powder (A3) sample feels the most dense. Through SEM observation, there are relatively many cavities in long fiber (A1) specimens, which cause specimens to feel more tenuous. Likewise, in short fiber specimens, there are still many empty cavities found inside the material. This is different from powder (A3) specimen which relatively has fewer cavities (Table 1).

Table 1. Comparison of observations of material specimens through direct observations and SEM.

	Long Fiber Specimen (A1)	Short Fiber Specimen (A2)	Powder Specimen (A3)
Direct observation			
SEM observation at 200x magnification		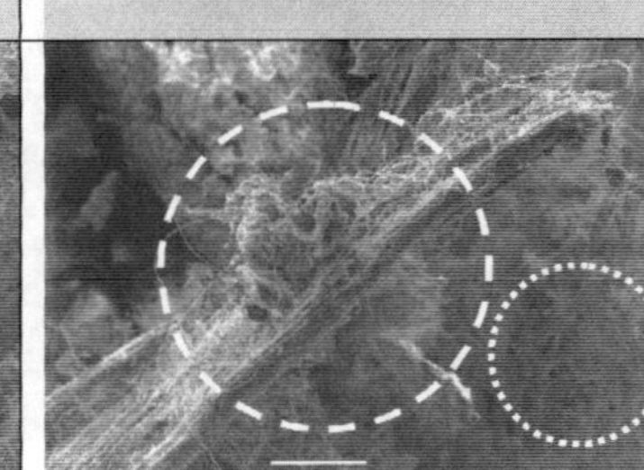	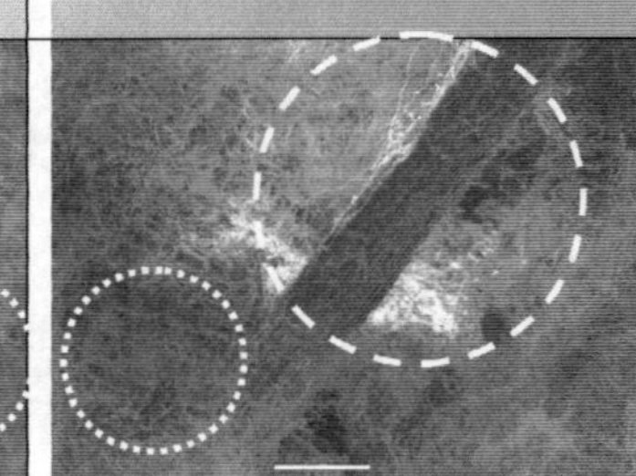

··········· indicates hyphae networks.
– – – - indicates bamboo fiber which is covered with hyphae networks (mycelium).

Through direct observation, the mycelium is identified as a white substance, while through SEM photos, the mycelium is a thin hyphae network that grew covered the bamboo fibers. Meanwhile, bamboo fiber is identified as a thicker, longer, and more straight diameter substance. The distribution of mycelium is found more widely in powder (A3) specimens. Meanwhile, the distribution of mycelium in long fiber specimens is more likely found covering some areas of bamboo fibers.

4.3. Testing Results

From three variables of materials, average range of density are obtained between 0.18 g/cm^3 to 0.23 g/cm^3. Average density of three variables from the most dense to most tenuous are powder (A3) 0.23 g/cm^3, short fiber (A2) 0.21 g/cm^3, and long fiber (A1) 0.18 g/cm^3. The density trend shows that the smaller the fiber shape, the higher the density (Figure 4(a)). Range of average materials' density are relatively low. Based on Japanese Industrial Standars (JIS) A5905 for Fibreboards, these materials are classified as insulation boards category which allows material's density under 0.35 g/cm^3.

Materials from three variations have range of average internal bonding strength between 0.0074 N/mm^2 to 0.0248 N/mm^2. Average internal bonding strength from the strongest to the weakest are powder (A3) 0.0248 N/mm^2, short fibers (A2) 0.0112 N/mm^2, and long fibers (A1) 0.0074 N/mm^2. The internal bonding strength trend shows that the smaller the fiber shape, the higher the internal bonding strength (Figure 4(b)). Range of average internal bonding strength is very low. Based on the JIS A5905 of insulation board category, the value of internal bonding strength is actually not required. But, to compare further with higher category over insulation board, namely MDF board category, this material's property of internal bonding strength is still far under the minimum of 0.2 N/mm^2 (Figure 4(c)).

All material variations have range of average swelling thickness between 0.89% to 2.48%. Average swelling thickness precentage from the smallest to the largest are powder (A3) 0.89%, long fiber (A1) 1.53%, and short fiber (A2) 1.98%. The trend shows that the smallest shape of bamboo fiber expands the least. Material's property of swelling thickness precentage is very good because of the low number. This material meets the standard values required by JIS A5905 category of insulation board which maximum number of 10%.

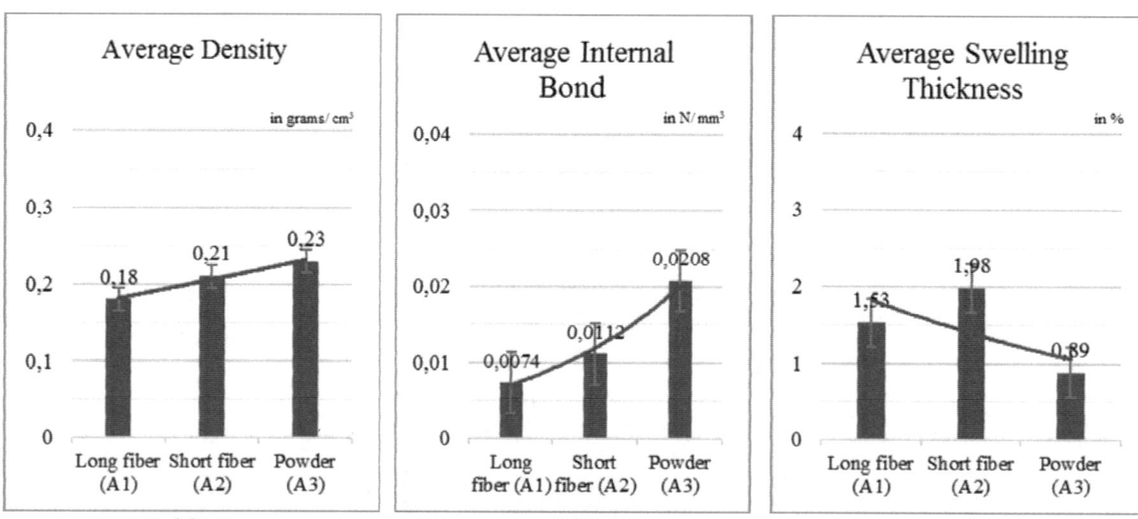

(a) (b) (c)

Figure 4. Charts of testing results: (a) average density, (b) internal bonding, and (c) swelling thickness.

5. Discussion

Based on the comparison with the standard JIS A5905 for Fibreboards, these composite boards from bamboo fiber with fungal mycelium in this study show the function of the material for non-structural purpose, that is insulation boards (Table 2). The low number of particle's density is the main indicator of the non-structural material category. In addition, the internal bonding strength also has a low number. On the other hand, the percentage of swelling thickness is considered very good, making the material very interesting and unique compared to other biocomposite boards that have been marketed.

Tabel 2. Comparison of material's testing result to JIS A5905 for Fibreboards.

Testing Type	JIS Standard A5905 for *Fibreboards*			Result of this study	Valuation
	Insulation Boards	MDF	Basis hardboard		
Density (g/cm^3)	≤ 0.35	≥ 0.35	≥ 0.80	0.18 – 0.23	Fulfilled
Internal bond (N/mm^2)	-	≥ 0.2	-	0.0074 – 0.0248	Unfulfilled
Swelling thickness (%)	≤ 10	≤ 17		0.89 – 2.48	Fulfilled

Compared to several previous studies which reported basic properties of similar biocomposite materials with fungal mycelium [15, 21, 22] (Table 3), biocomposite board from bamboo fiber produced in this study had advantage from its good swelling thickness number. Meanwhile, the board's density in this study is still within the range of similar materials that have been made in previous studies even, it is potential to be improved.

Furthermore, bamboo fiber as raw material used in this study gives a high novelty value in this study. Kenaf, hemp, flax, wood, and other agricultural wastes are some of raw materials commonly used in previous studies. These lignocellulosic fibers are the natural resources which available in their research location context. But, in this study, bamboo contributes special value to this board by aspect of study's location context, Indonesia, which has rich diversity of bamboo species. In addition, bamboo fiber has its own uniqueness by aspect of its long fiber form, giving benefit to be adjusted into different smaller size particles purposed to make different type and function of future materials.

Table 3. Comparison of material's testing resulted in this study with previous researches.

Material's function	Insulation Board	Packaging and furniture [21]	Insulation [22]	Product packaging [15]
Raw material	Bamboo Fiber	Wood	Hemp, flax, straw, wood	*Burlap*
Density (g/cm^3)	0.18 – 0.23	0.3 – 0.6	0.05977 – 0.18645	*Specific gravity: 0,10 – 0,14
Internal bond (N/mm^2)	0.0074 – 0.0208	3	-	-
Swelling thickness (%)	0.89 – 2.48	~ 43 – 44	-	-

6. Conclusions

The sustainable composite board of bamboo fiber through biologically binding mechanism with fungal mycelium is a novel biocomposite. The three variations of composite boards are highly potential to be developed for building non-structural function. Of three variations, the boards made of bamboo powder are assessed very well from the aspect of density, internal bonding, and swelling thickness. In architecture, this novel biocomposite board is expected to replace the need of non-structural composite boards which are not environmentally friendly and have been widely used in buildings. The function of this material as the insulation board is expected to be used for interior space purposes, including as thermal insulation boards, acoustic boards, space partitions, ceiling boards, and decorative boards in low-rise buildings such as residential houses to high-rise buildings with high demand of insulation components. To achieve better performance of this board, further research to determine acoustic and thermal performance needs to be done.

7. References

[1] Tuasikal R 2019 *Bambu Berpotensi Jadi Sumber Kayu Masa Depan* Retrieved on May 1, 2019 from https://www.voaindonesia.com/a/bambu-berpotensi-jadi-sumber-kayu-masa-depan/4679561.html

[2] Dransfield S and Widjaja E A 1995 *Plant Resources of South-East Asia. No. 7. Bamboos* (The Netherlands: Backhuys)

[3] Anokye R, Bakar E S, Ratnasingam J and Awang K B 2016 Bamboo properties and suitability as a replacement for wood, *Pertanika J. of Scholarly Research Reviews* **2(1)** 63-79

[4] Bystriakova N, Kapos V, Stapleton C and Lysenko I 2003 *Bamboo Biodiversity: Information for Planning Conservation And Management in The Asia-Pacific Region* (UNEP-WCMC/INBAR)

[5] Widjaja, E A 2000 *Bamboo Diversity and Its Future Prospect in Indonesia Sustainable Utilizaton of Forest Products: Socio-Economical and Ecological Management of Tropical Forests* (Kyoto: Third Wood Symposium) pp 235-240

[6] Osorio L, Trujillo E, Lens F, Ivens J, Verpoest I and Vuure A V 2018 In-depth study of the microstructure of bamboo fibres and their relation to the mechanical properties *J. of Reinforced Plastics and Composites* **37(17)** 1099-1113

[7] Krause J Q, Silva F, Ghavani K, Gomes O, and Filho R 2016 On the influence of Dendrocalamus giganteus bamboo microstructure on its mechanical behavior *Construction and Building Materials* **127** 199-209

[8] Yu Y, Ru Liu, Yuxiang Huang, Fandan Meng and Wenji Yu 2017 Preparation, physical, mechanical, and interfacial morphological properties of engineered bamboo scrimber *Construction and Building Material* **157** 1032-1039

[9] Walczyk D, Jiang L, Mcintyre G and Kuppers J 2012 *A Preliminary Investigation of Using Mycelium to Bind Natural Fibers and Core Filler Material in Sustainable Composites* Retrieved on May 1, 2019 from https://www.researchgate.net/publication/274780180

[10] Huang X and Netrabali A 2009 Biodegradable green composites made using bamboo micro/nano fibrils and chemically modified soy protein resin *Composites Science and Technology* **69** 1009-1015

[11] Corrêa A A R, Mendes L, Barbosa N, Protásio T, Campos N and Tonoli G 2015 Incorporation of bamboo particles and "synthetic termite saliva" in adobes *Construction and Building Materials* **98** 250-256.

[12] Sánchez M L, Morales L and Caicedo J 2017 Physical and mechanical properties of agglomerated panels made from bamboo fiber and vegetable resin *Construction and Building Materials* **156** 330-339.

[13] Nguyen D M, Grillet A C, Diep T and Woloszyn M 2018 Building bio-insultaion materials based on bamboo powder and bio-binders *Construction and Building Materials* **186** 686-698.

[14] Ivanov V and Stabnikov V 2018 *Construction Biotechnology: Biogeochemistry, Microbiology and Biotechnology of Construction Materials and Processes* (Singapore: Springer)

[15] Ziegler A R, Bajwa S, Holt G, Mcintyre G and Bajwa D 2016 Evaluation of physico-mechanical properties of mycelium reinforced green biocomposites made from cellulosic fibres *Applied Engineering in Agriculture* **32** 931-938

[16] Pelletier M G, Holt G, Wanjura J, Bayer E and Mcintyre G 2013 An evaluation study of mycelium based acoustic absorbers grown on agricultural by-product substrates *Industrial Crops and Products* **51** 480-485

[17] Heisel F, Schlesier K, Lee J, Rippmann M, Saedi N, Javadian A, Nugroho A R, Hebel D and Block P 2017 *Design of A Load-Bearing Mycelium Structure Through Informed Structural Engineering* Retrieved on November 5, 2018 from https://www.researchgate.net/publication/320443920.

[18] Ross P 2012 *Method for Producing Fungus Structures US 2012O135504A1* (United States of America)

[19] Stamets P and Chilton J S 1983 *The Mushroom Cultivator: A Practical Guide to Growing Mushrooms at Home* (Washington: Agarikon Press)
[20] Arifin Y H and Yusuf Y 2013 Mycelium fibers as new resource for environmental sustainability *Procedia Engineering* **53** 504-508
[21] Sun W, Tajvdi M, Hunt C, Mcintyre G and Gardner D 2019 *Fully Bio-Based Hybrid Composites Made of Wood, Fungal Mycelium and Cellulose Nanofibrils* Retrieved on May 10, 2019 from www.nature.com/scientificreports
[22] Elsacker E, Vandelook S, Brancart J, Peeters E and De Laet L 2019 *Mechanical, Physical and Chemical Characterisation of Mycelium-Based Composites with Different Types of Lignocellulosic Substrates* Retrieved on May 1, 2019 from https://www.biorxiv.org/content/early/2019/03/07/569749.full.pdf
[23] Japanese Industrial Standard A 5905 2003 *Fibreboards* (Tokyo: Japanese Standards Association)

Acknowledgments
This paper is funded by Universitas Indonesia through PITTA-B DRPM-N research grant no. NKB-0715/UN2.R3.1/HKP.05.00/2019. Gratitudes also need to be conveyed to Prof. Elizabeth Widjaja, Dr. Iwan Saskiawan, Prof. Subyakto, and Kurnia Prasetiyo, M. Si. for sharing biology and biomaterials knowledge.

Properties and Flexural Behavior of Self-Compacting Concrete with Added Coir Fiber And POFA

Afif Iman[1], N Mohamad[1], Fasya Raain[1], Anis Sofia[2], Azree Mydin[3] and A A A Samad[1]

[1] Faculty of Civil and Environmental Engineering, Universiti Tun Hussein Onn Malaysia, 86400 Parit Raja, Johor, Malaysia
[2] School of Materials Engineering, Universiti Malaysia, 02600 Arau, Perlis, Malaysia
[3] School of Housing, Building and Planning, Universiti Sains Malaysia, 11800 Gelugor, Penang, Malaysia
Correspoding author: noridah@uthm.edu.my

Abstract. This paper presents the mechanical properties and flexural behavior of self-compacting concrete (SCC) with addition of 0.2%, 0.4% and 0.6% coir fiber (CF) as filler, and 10% palm oil fuel ash (POFA) as partial cement replacement. Fresh SCC-POFA-CF mixture was tested under slump flow and J-ring tests to determine its workability. SCC-POFA-CF cubes and cylinders were tested under compressive and tensile tests to determine its compressive strength, tensile strength, and flexural strength, while prisms were tested under four point bending load to determine its flexural behavior. It is found that workability of SCC-POFA-CF decreased when CF increased. The optimum percentage of CF in SCC mixture was found to be 0.4%. SCC-POFA-CF experienced less crack compared to the control specimen. This shows that CF in SCC prisms managed to control the crack propagation.

1. Introduction

Self-compacting concrete is a non-segregated and highly flowable concrete, which is able to consolidate under its own weight. The easy flowable nature of SCC allows it to completely fill the formwork and achieve complete compaction even in the presence of congested reinforcements [1, 2]. The development of SCC has led to improvements in the working conditions which in turn, results with less energy consumption, less vibration, greater productivity, and less noise [3, 4]. SCC concrete has been proven to have improved fresh and hardened properties compared to conventional concrete [5, 6].

Meanwhile, the disposal of wastes from the activities associated with agriculture has created nuisance and environmental issues. One of the solutions to tackle this problem is by recycling these wastes as the new innovative construction materials. In this study, coir fiber (CF) and palm oil fuel ash (POFA) were used as a filler and cement replacement, respectively, in the SCC mixture. CF is a natural organic resource which is extracted from the outer shell of a coconut [7]. It has been used as reinforcement in cement paste, mortar and concrete, due to its high tensile strength [8, 9]. POFA is an agricultural waste generated by the burning of oil palm residue in the oil palm industry to produce steam for electricity generation [10]. It has been used in concrete mixture as cement or sand replacement to enhance its strength and performance [11-15].

Previous research on properties of SCC concrete added with new materials as cement replacement have shown promising results in the aspects of its fresh and hardened properties. Dinesh et al. conducted experimental study on SCC with added fly ash and silica fume at various percentages. It was found that the compressive and tensile strength decreased with the increase in the percentage of fly ash and increased with the increase in percentage of silica fume [16]. Tarun et al. [17] replaced Portland cement by fly ash in the range of 35% to 55% by the mass of cement. The results showed it obtained 28-day strengths up to 62 MPa. Kumar et al. conducted an experimental investigation on flow properties and compressive strength of SCC with ultrafine natural steatite powder (UFNSP) as cement replacement. The results showed that the addition of UFNSP influences the flow property, by reducing the flow, and increases the compressive strength up to 20% replacement [18].

This paper presents the study on the mechanical properties of SCC concrete incorporating CF as filler and POFA as cement replacement. The main aim of this work is to determine the effects of CF and POFA on the mechanical properties of SCC as compared to properties of control SCC.

2.Materials and methods

Ordinary Portland Cement (OPC) was used in every mix design of SCC concrete according to BS EN 206-1:2006 [19]. Coarse aggregate used in this study was prepared according to BS EN 812-2:1995 [20]. POFA, sieved using 75 μm, was incorporated in the SCC mixture as a cement replacement at 10% of mixture's total weight. The percentage of 10% POFA was chosen based on findings from previous research. Munir et al. replaced OPC with POFA in foamed concrete as a cement replacement at 10%, 20%, 30%, 40% and 50% by weight of cement in the foamed concrete. The results showed that 20% cement replacement by POFA in the foamed concrete is the optimum percentage [21]. From this finding, the 10% POFA was chosen because the density of SCC is much higher than foamed concrete; thus, lower POFA percentage is needed. CF was incorporated in the SCC mixture at various percentages of 0.2% to 0.6% of the total weight. Its length was varied from 12 mm to 20 mm long. These various percentage was chosen based on study by Sureshkumar et al. [22], which investigated the SCC properties with various percentages of coir fiber, 0.25, 0.5 and 1% by weight of cement. It was found that 0.5% CF is the optimum percentage to be used.

The mix design in this study was conducted according to EFNARC [23]. Table 1 shows the mix design of SCC without addition of coir fiber. Coir fiber was added in SCC in the range of 0.2%, 0.4% and 0.6%. as shown in Table 2.

Table 1. Mix Design of SCC.

Cement (kg/m^3)	Aggregates (kg/m^3)	Sand (kg/m^3)	POFA (kg/m^3)	Superplasticizer (%)	Water Cement Ratio (w/c)
360	800	800	40	2	0.6

Table 2. Mechanical properties of SCC-POFA-CF with 10% POFA and various percentages of CF.

Specimens	Cube (150 mm x 150 mm x 150 mm)		Cylinder (150 mm diameter x 300 mm height)		Prism (100 mm x 100 mm x 5 00mm)
Testing	Compressive		Tensile	MOE	4-Point Bending
Curing Days	7 Days	28 Days	28 Days		28 Days
SCC + 10% POFA + 0% CF	3	3	2	2	3
SCC + 10% POFA + 0.2% CF	3	3	2	2	3
SCC + 10% POFA + 0.4% CF	3	3	2	2	3
SCC + 10% POFA + 0.6% CF	3	3	2	2	3

Laboratory tests were divided into two phases; the first phase was to determine the fresh properties and the second phase is to determine the mechanical properties of SCC-CF-POFA mixtures. Tests for fresh properties include slump flow test and J-ring test which were conducted according to BS EN 12350: Part 8 2010 [24] and BS EN 12350: Part 12 [25], respectively. Meanwhile, tests for hardened properties included compressive test, splitting tensile test and four-point bending test which were conducted according to BS-EN 12390:2002[26], BS-EN 12390-6 [27] and BS EN 12390: Part 5 [28], respectively.

3. Results and discussions

3.1. Slump Flow

Slump flow, *S* is determined by using the formula in Equation (1). D_{max} is the maximum diameter of the flow spread while the d_{perp} is the diameter of perpendicular direction.

$$S = \frac{d_{\max} + d_{perp}}{2} \tag{1}$$

From the slump flow diameter recorded, it is noticed that the diameter decreases as the percentage content of CF increase. Meanwhile, from the T_{500} recorded, it is shown that the time recorded increased as the percentage content of CF increased. This is caused by the flow resistance in the SCC matrix developed by CF which was added in the SCC mixture. This is in good agreement with the investigation conducted by Mohamad et al. 2018[12] on fresh and hardened properties of foamed concrete consisting coir fibres where the results showed decreased slump flow.

From Figure 1, the highest value for slump flow was 600 mm with T_{500} at 2 seconds but for SCC with addition of 0.2% coir fiber, the highest slump flow was 580 mm with T_{500} at 4 seconds. This is because CF is organic material which has high water absorption; thus, increased the hydration process of SCC and generate the cohesiveness between the particles. From the results recorded, the optimum percentage of coir fibre is 0.2% where workability was highest.

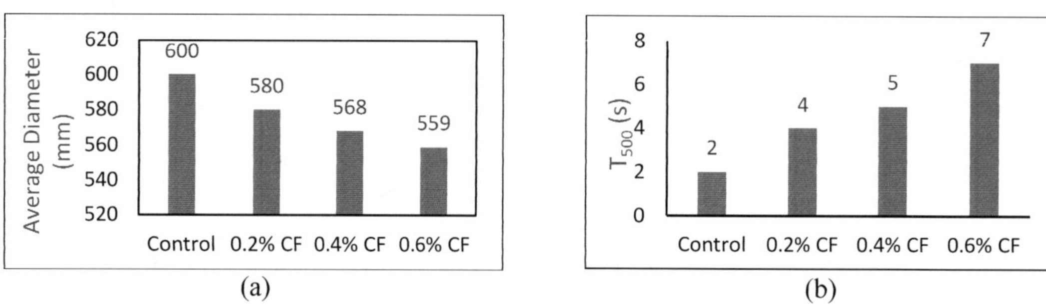

Figure 1. (a) Slump flow diameter, and (b) Time taken for 500 mm diameter.

3.2. Passing ability of SCC

J-Ring test was conducted to determine the passing ability of the SCC. The average of the differences in height at four direction will determine the height of flow, B_j, as shown below.

$$Bj = \frac{(hx1 + hx2 + hy1 + hy2) - ho}{4} \tag{2}$$

From Figure 2, the difference in the height of flow was increasing from 2.5 mm to 15 mm as the content of CF in SCC mixture increased. This is due to less water content of the SCC mixture, caused by existence of CF in the mixture, which absorb the water. The greater difference in height of flow indicated the less passing ability of the concrete.

3.3. Compressive strength

Figure 3 presents the graph of compressive strength for SCC-POFA-CF mixtures at 7 and 28 days. From the figure, it is seen that the maximum strength was achieved when the mix was added with the 0.2% of CF. The highest value of compressive strength for 7 and 28 days was 25.7 MPa and 28.5 MPa, respectively, for this SCC mixture.

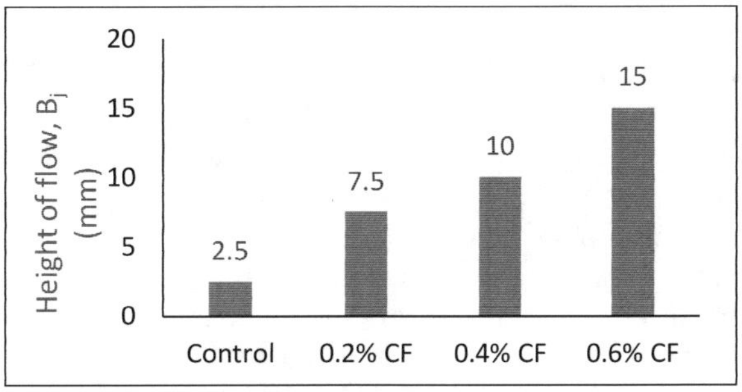

Figure 2. Height of flow versus percentage of CF from J-Ring test.

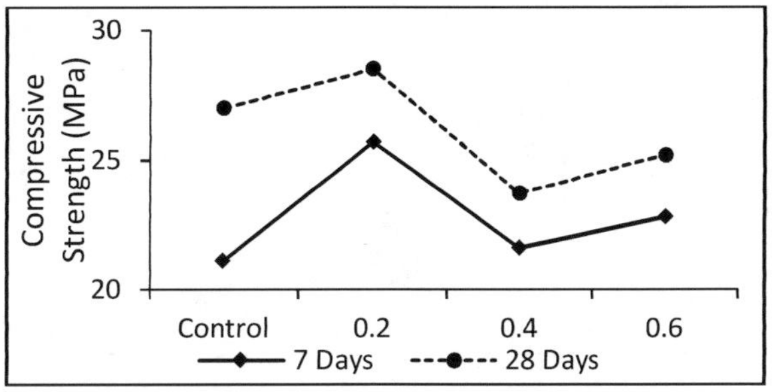

Figure 3. Compressive Strength for SCC-POFA-CF mixtures at 7 and 28 days.

The addition CF has proven to improve the compressive strength by approximately 5% with added 0.2% CF. This is in agreement with the findings from previous research, which investigated the strength of concrete containing various percentage of CF [12, 29] which examined the fresh and hardened properties of concrete when CF was utilized as filler. The results showed that CF improved the cohesive bonding between the concrete particles which tends to increase the strength in concrete.

However, the compressive strength was decreased, compared to compressive strength control specimen, when CF in the mixture was increased to 0.4% and 0.6%. This is due to CF with its high absorbance characteristic, reduced the amount of water in SCC mixture, resulted with less strength achieved.

3.4. Tensile strength

Splitting tensile strength test was carried out to determine the tensile strength of cylinder with size 150 mm diameter and 300 mm height. Figure 4 shows that the highest value of tensile strength of SCC is 2.35 MPa, achieved with addition of 0.4% of coir fiber. The lowest value of tensile strength was 2.20 MPa which is SCC with addition of 0.2% of coir fiber. Tensile strength in SCC mixture is higher when the content of coir fiber is higher because the fibrous nature of coir fiber which is higher in tensile.

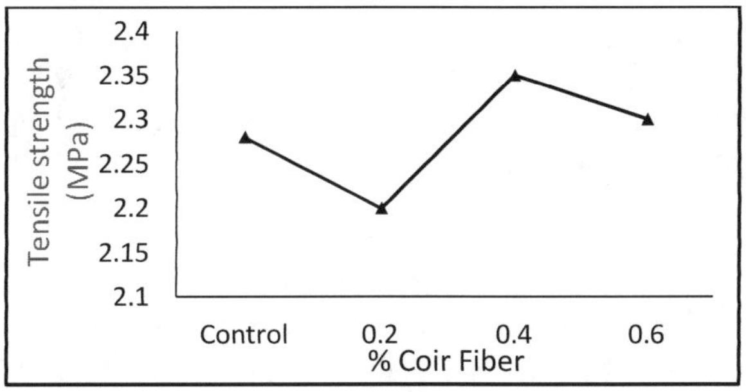

Figure 4. Tensile Strength of SCC-POFA-CF at 28 days.

3.5. Modulus of elasticity

Modulus of elasticity obtained from the experiment is shown in Figure 5. The modulus of elasticity, MOE, for control specimen recorded was 20.27 GPa and increased to 26.26 GPa which is SCC with addition of 0.2% of coir fiber. After that, the MOE value decreased to 13.49 GPa and slightly increased to 15.08 GPa. The highest value for MOE was SCC with addition of 0.2% of CF while the lowest value for MOE was SCC with addition of 0.4% of CF.

The results from Figure 5 shows that with addition of 0.2% CF, the MOE of SCC was higher compared to control specimen. However, with the increment of CF content, the MOE significantly reduced, which shows that more addition of CF reduced the elasticity of SCC. Thus, the optimum percentage concrete is 0.2%.

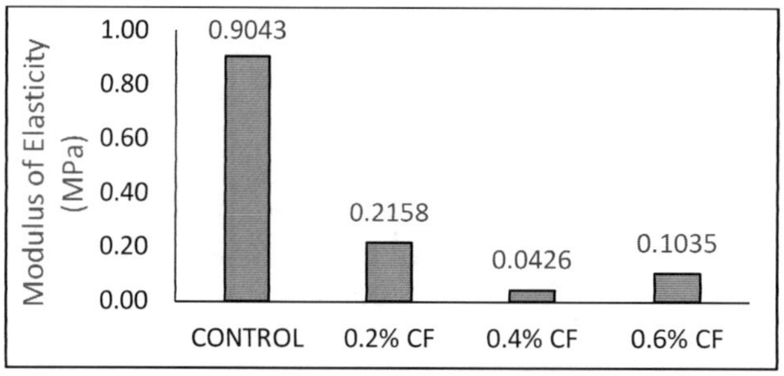

Figure 5. Modulus of Elasticity of SCC-POFA-CF.

3.6. Flexural strength, crack pattern and load-deflection profile

The flexural strength of SCC-POFA-CF mixture was obtained from the four-point bending test on prism specimens. From the test, the ultimate load, crack pattern and load-deflection curve were analyzed.

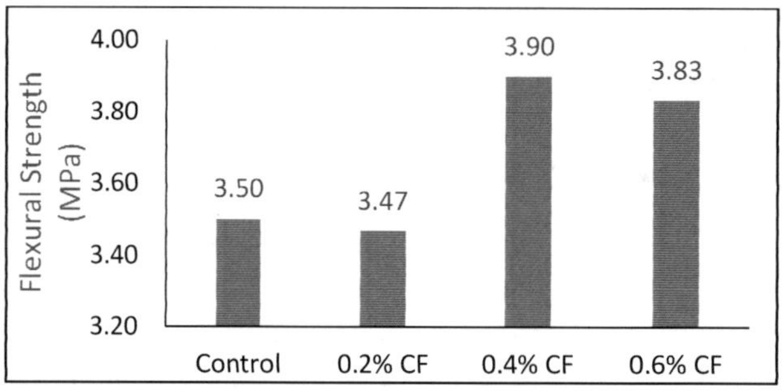

Figure 6. Flexural strength of SCC-POFA-CF.

From Figure 6, the highest value for flexural strength was 3.90 MPa, about 11% higher than control specimen, which was obtained from SCC-POFA-CF mixture incorporating 0.2% CF. This is caused by the CF, as filler in the SCC, was high in tension and elasticity. It is proven that CF as filler resulted with enhancement in the internal bonding in the mixture, which was responsible to improve the flexural capacity [12, 29]. However, for the 0.6% of coir fiber in SCC mixture, the ultimate strength decreased to 3.83 MPa. Thus, the optimum percentage of coir fiber for tensile strength of SCC mixture is found to be 0.4%.

From Figure 7, the control specimen experienced the extreme crack propagation where it finally broke into two at its ultimate load. SCC with added 0.2% and 0.6% CF developed flexural crack near its mid zone, with smaller crack width and length, compared to control specimen. Lastly, the smallest crack propagation was observed in SCC prism with 0.4% of coir fiber, which was highest in tensile strength. Based on the observation, it is concluded that in general, SCC with addition of CF experienced less crack compared to the control specimen.

IOP Conf. Series: Materials Science and Engineering **713** (2020) 012011 doi:10.1088/1757-899X/713/1/012011

Figure 7. Crack pattern.

From Figure 8, the SCC mixtures with added CF obtained lower deflection values compared to control specimens. This indicate that the CF in SCC prisms managed to create the concrete bridges in structure which in turn control the crack. The first crack load was highest in specimen with 0.4% CF, which is about 9.78 kN, continued with specimen with 0.6% CF, which is 9.56 kN. It is also seen from the figure that control specimen behaved in most ductile manner compared to the other specimens. This is caused by CF, which absorbed more water. Thus, as the SCC mixture dried up, the water content in the mixture decreased, and caused the concrete to be more brittle. Among the three specimens with different CF percentage added, it is observed that 0.4% CF resulted with more ductile behavior of SCC mixture.

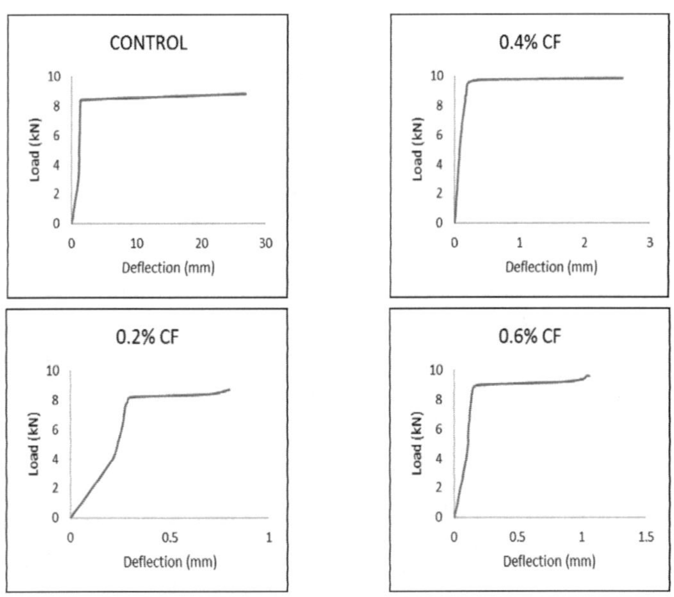

Figure 8. Load-deflection profile

4. Conclusions

From the results obtained, CF has proven to have significant effects on fresh and mechanical properties of SCC-POFA mixture. It increased the workability of SCC-POFA mixture, where the bigger content of CF resulted with higher workability. From the mechanical properties obtained, it is found that 0.2% CF content increased the compressive strength and modulus of elasticity by 11% and 30%, respectively. Meanwhile, the tensile and flexural strength achieved their highest values, 2.35 MPa and 3.9 MPa, respectively, when 0.4% CF was added to the mixture. The crack pattern recorded showed that SCC-POFA-CF prisms developed the least crack in prism with 0.4% CF added. It is also noticed that mixture with added CF attained lower deflection and less ductile behavior compared to the control specimen. This is mainly cause by the characteristic of CF, which is high water absorbance, which in turn reduce the water content in the mixture. This makes the SCC-POFA-CF mixture more brittle. Thus, considering its effects on all the properties of SCC-POFA-CF mixture, 0.2% CF was chosen as the optimum percentage of CF to be added into the mixture.

5. References

[1] Ahmad S and Umar A 2017 characterization of self-compacting concrete *Procedia Engineering* **173** 814-821
[2] Okamura H, Ozawa K and Ouchi M 2000 Self-compacting concrete *Structural Concrete* **1(1)** 3-17
[3] Schutter P G D E 2005 Self-compacting concrete *Int. Review of Chemical Engineering* **3(1)** 99-100
[4] Ahmed Z A and Mahure H 2016 High strength self-compacting concrete using fly ash *Research in Applied Science and Engineering Technology* **4(8)** 2321-9653
[5] Aijaz A Z and Khadirnaikar R B 2014 An Overview of the properties of self-compacting concrete *Int. Conf. on Advances in Engineering and Technology* pp 35-43
[6] Arulsivananantham P and Gokulan R 2017 A review on self-compacting concrete *Int. J. of ChemTech Research* **10(11)** 62-68
[7] Ali M 2011 Coconut fibre: A versatile material and its applications in engineering *Civil Engineering and Construction Technology* **2(9)** 189-197
[8] Alalwany A, Mohamad N, Samad A A A and Goh W I 2017 *Mechanical and Fresh State Properties of Medium Strength SCC Containing Polypropylene Fiber* (MATEC)
[9] Mohamad N, Samad A A A, Nadjua A, Goh W I and Anas A 2016 Fresh state and mechanical properties of SCC incorporating POFA (Jurnal Teknologi)
[10] Tangchirapat W, Saeting T, Jaturapitakkul C, Kiattikomol K and Siripanichgorn A 2007 Use of waste ash from palm oil industry in concrete *Waste Management* **27(1)** 81-88
[11] Tambichik M A, Mohamad N, Samad A A A, Bosro M Z M and Iman M A 2018 Utilization of construction and agricultural waste in Malaysia for development of green concrete: A review *IOP Conf. Series: Earth and Environmental Science* **140** 012134
[12] Mohamad N, Iman M A, Mydin M A O, Rosli J A and Noorwirdawati A 2018 Mechanical properties and flexural behaviour of lightweight foamed concrete incorporating coir fiber *IOP Conf. Series: Earth and Environmental Science* **140** 012140
[13] Munir A, Abdullah, Huzaim, Sofyan, Irfandi and Safwan 2015 Utilization of palm oil fuel ash (POFA) in producing lightweight foamed concrete for non-structural building material *Procedia Engineering* **125** 739-746
[14] Mohamad N, Samad A A A, Nadjua A, Goh W I and Anas A 2016 Fresh state and mechanical properties of self-compacting concrete incorporating POFA *Jurnal Teknologi* **78(5)** 437-441
[15] Ali N, Mohd Sobri M H A, Hadipramana J, Abdul Samad A A and Mohamad N 2017 *Potential Mixture of POFA and SCBA as Cement Replacement in Concrete: A Review* (MATEC)
[16] Dinesh A, Harini S, Jasmine Jeba P, Jincy J and Shagufta J 2017 Experimental study on self compacting concrete *Int. J. of Engineering Sciences and Research Technology* **6(3)** 42-50
[17] Tarun R N, Rakesh K, Bruce W R and Fethullah C 2012 Development of high-strength, economical self-consolidating concrete *Construction and Building Materials* **30** 463-469
[18] Kumar P, Sudalaimani K and Shanmugasundaram M 2017 *An Investigation on Self-Compacting Concrete Using Ultrafine Natural Steatite Powder as Replacement to Cement* (Advances in Materials Science and Engineering)
[19] BS 8500-BS EN 206-1 2006 *Guidance for Concrete Mixing-Part 1: Method of Specifying and Guidance for The Concrete Mixing* (London: British Standards Institution)
[20] BS EN 812-2 1995 *Testing Aggregates-Part 2: Methods of Determination of Density* (London: British Standards Institution)

[21] Munir A, Abdullah, Huzaim, Sofyana, Irfandia and Safwana 2015 Utilization of palm oil fuel ash (POFA) in producing lightweight foamed concrete for nonstructural building material *Procedia Engineering* **125** 739-746
[22] Sureshkumar E, Jegadeesan S, Manimuthu K, Sivanesan K and Rathinamoorthy S 2016 Experimental investigation of self-compacting concrete using coir fiber *J. of Structural Engineering and Management* **2(1)** 6-12
[23] The European Guidelines for Self-Compacting Concrete 2005 Specification, Production and Use (The Self-Compacting Concrete European Project Group)
[24] BS EN 12350: Part 8 2010 *Testing Fresh Concrete-Part 8: Self-Compacting Concrete-Slump-Flow Test* (London: British Standards Institution)
[25] BS EN 12350: Part 12 2010 *Testing Fresh Concrete-Part 12: Self-Compacting Concrete-J-Ring Test* (London: British Standards Institution)
[26] BS EN 12390: Part 3 2002 *Testing Hardened Concrete: Compressive Strength of Test Specimens* (London: British Standards Institution)
[27] BS-EN 12390: Part 6 2003 *Testing Hardened Concrete-Part 6: Tensile Splitting Strength of Test Specimens* (London: British Standards Institution)
[28] BS EN 12390 Part 5 2009 *Method for Determination of Flexural Strength* (London: British Standard Institution)
[29] Ede A N and Agbede J O 2015 Use of coconut husk fiber for improved compressive and flexural strength of concrete *Int. J. of Scientific and Engineering Research* **6(2)** 968-974

Acknowledgement

The authors would like to thank Universiti Tun Hussein Onn Malaysia for its financial support (GPPS Vot U788)

IOP Conf. Series: Materials Science and Engineering **713** (2020) 012012 doi:10.1088/1757-899X/713/1/012012

Assessing the performance of bio compost as soil media in extensive green roof

Ming Fai Chow[1*] and Ahmad Aiman Bin Ahmad Azlan[2]

[1]Institute of Sustainable Energy (ISE), Universiti Tenaga Nasional, Jalan IKRAM-Uniten, Selangor Darul Ehsan 43000, Malaysia
[2]Department of Civil Engineering, College of Engineering, Universiti Tenaga Nasional, Jalan IKRAM-Uniten, Selangor Darul Ehsan 43000, Malaysia
*Corresponding author: Chowmf@uniten.edu.my

Abstract. Composted green waste is a recycled material which can be produced locally, adding value to the environmental credentials of a green roof system. As the organic component of green roof growing media, composted green waste can contribute positively to the physical requirements as well as improve environmental performance. In addition it can contain nutrients for plant growth and survival. These will not typically be provided by an inert inorganic substrate and will help establish vegetation. This study is aimed to investigate the performance of bio compost as soil media in extensive green roof with respect to different composition. Runoff were tested for total suspended solids (TS), turbidity, chemical oxygen demand (COD), pH, and color measurements. The results showed that 50% of bio compost in soil media has the best performance in term of plant growth and runoff quality.

1. Introduction

Urbanization process has significantly increases the stormwater runoff and pollutant build-up at the catchment surface area [1-4]. Urban heat island (UHI) phenomenon also happens at the urban city when impervious surfaces absorbs and then returns back the solar radiation. Compared to the surrounding landscape, the average air temperatures recorded in urbanized areas are higher [5]. As a result, greater expenditures are needed to keep the buildings cool via air-conditioning systems. Therefore, green roof have been highly promoted for controlling the urban stormwater runoff and enhancing the environmental quality at the urban city [6 - 8]. Green roofs are generally classified into three types which are intensive green roofs, semi-intensive green roofs, and extensive green roofs [9]. These different types of green roofs vary in terms of the vegetation growing on them therefore requiring different depths of growing medium used [8,10]. Extensive green roof plants would typically fulfil the characteristics of reproducing efficiently, having a short height as well as mat-forming, having shallow and spreading roots, and having succulent leaves which are able to store water [6, 11].

Understanding regarding the effect of soil composition in the green roof growing medium remain limited and have vital gaps. For example, limited studies are available on the effect of growing medium composition when it comes to evapotranspiration mechanisms. The same can be said about research on the role of depth or composition of growing medium as well of type of vegetation grown on air pollution. In addition, there is no literature on the evolution of the growing medium in terms of microorganism presence which plays a huge role in the nitrogen cycle. It is significant as growing medium which are different from natural soils in terms of mineral composition and organic compounds may cause specific varieties and activities of microorganisms [12]. So far, it is argued that the growing medium plays a vital component in extensive green roof designs. More specifically, the available water in the growing medium needs to be controlled to an acceptable amount to meet the requirements of the selected vegetation. As such, this can be achieved using two methods; i.e., manipulating the composition of the growing medium or more typically, limiting the depth of growing medium [13]. There is limited studies on the usage of bio-compost for green-roof media in Malaysian. Therefore, this study is aimed to investigate the performance of bio compost as soil media in extensive green roof with respect to different composition with the environmental parameters being total solids, turbidity, chemical oxygen demand

(COD), pH, and color concentration of the green roof water runoffs.

2. Methodology

The performance of bio-compost as a potential soil substitute or additive for the growth of Portulaca grandiflora or known locally as "ros jepun" in extensive green roofs, in the context of Malaysian high temperature and moisture weather conditions was investigated. The monitoring period of the experiments is 3 months. The growth media was prepared by mixing bio-compost (BC) and clay soil (S) at different ratios. Clay soil was chosen considering its wide use as soil medium in extensive green roofs. Five BC: S growth media mixtures (on a volume basis) was prepared as: Model 1 (BC:S =0) ie. 0% bio-compost, Model 2 (BC:S =0.25) ie. 25% bio-compost, Model 3 (BC:S =0.5) ie. 50% bio-compost, Model 4 (BC:S =0.75) ie. 75% bio-compost, and Model 5 (BC:S =1.0) ie. 100% bio-compost. The bio-compost were produced from composting food waste from municipalities in Putrajaya. Thus the bio-compost used were fully organic. Green roof models were set up with a uniform 10 cm depth of growth medium. All five green roof models have dimensions of 30 cm (depth) x 30cm (width) x 50cm (length). Filter membrane was put between the soil medium and drainage layer to prevent the losses of soil medium along with runoff. Collecting tank were connected to the discharge point of the green roof model. The schematic diagram of green roof model is shown in Figure 1. The five models were located at the open space beside the civil engineering laboratory of Universiti Tenaga Nasional.

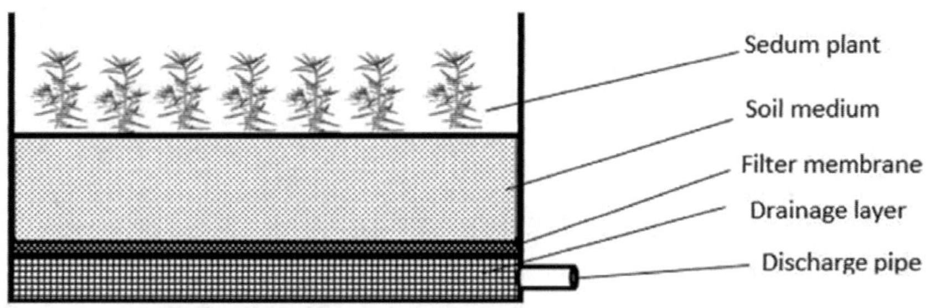

Figure 1. Schematic diagram for green roof model

The experiments were conducted throughout the months of August and September whereby the average rainfall in the region was around 200 to 250mm. The rainwater passed through the vegetation layer and substrate layer/ growing medium and then retained at the drainage layer in the green roof model. Rainwater runoff was collected from the green roof models via the collecting tanks that was installed. The runoff samples were then analyzed to determine the runoff quality from green roof models with different percentages of bio-compost. The runoff samples were tested for total solids (TS), turbidity, chemical oxygen demand (COD), pH, and color concentration.

3. Results and discussions

The runoff qualities from the green roof models with different bio-compost composition percentages were determined over the 3 months monitoring period. From the results over nine weeks, green roof model with 0% of bio-compost has the highest total solid (TS) concentration while the lowest is the green roof model with 100% bio-compost as soil medium. This observation may explain that the sample with 0% of bio-compost consists of only soil which would contain a lot of inorganic materials like salt. Throughout nine weeks of monitoring period, the readings of total solids concentration were observed to be reduced slightly over time as shown in Figure 2.

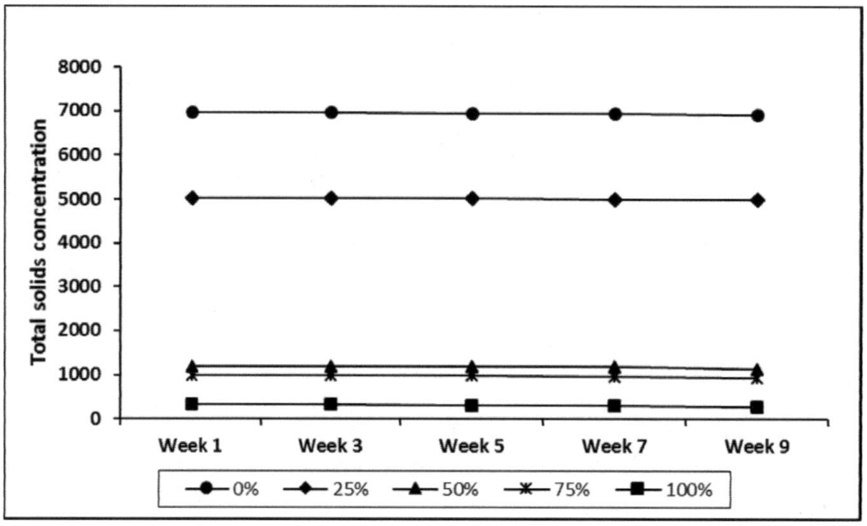

Figure 2. Total solids concentrations for different green roof models over the monitoring period

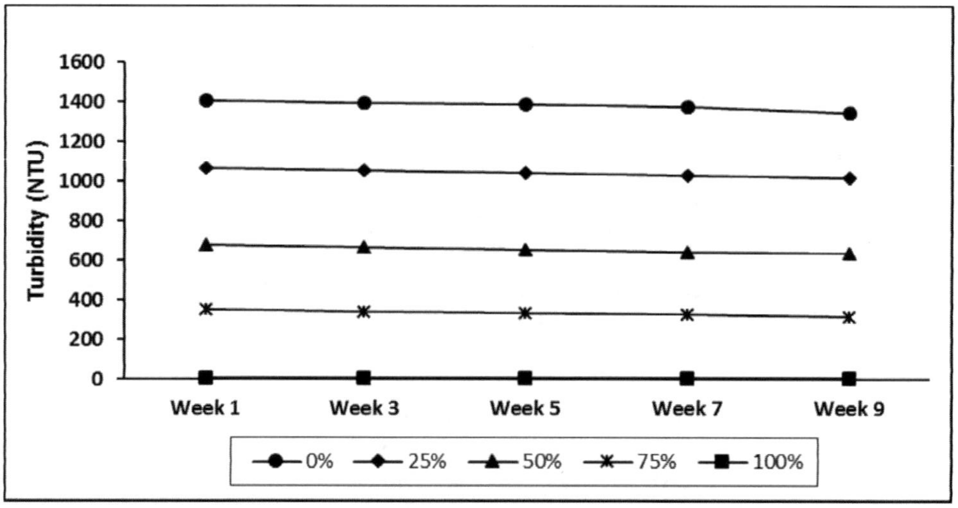

Figure 3. Turbidity measurements for different green roof models over the monitoring period

Turbidity is used as the visible indicator of runoff quality for green roof models. Molineux [14] stated that clear water is usually considered as an indicator of healthy water. From the turbidity results as shown in Figure 2, the sample that has highest turbidity concentration is sample with 0% bio-compost concentration while lowest turbidity is the runoff sample from green roof model with 100% of bio-compost. This shows that the sample with 0% of bio-compost has a high amount of suspended solids.

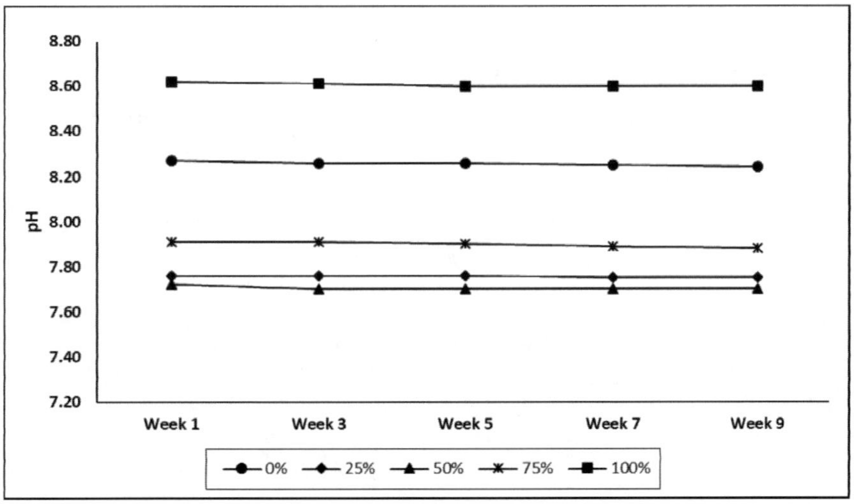

Figure 4. pH measurements for different green roof models over the monitoring period

The most commonly measured chemical attribute of water is its acidity or pH. The pH scale ranges from 1 to 14, with 1 being the most acidic and 14 being the most basic. Most streams have a neutral to slightly basic pH of 6.5 to 8.5. If streamwater has a pH less than 5.5, it may be too acidic for fish to survive in, while streamwater with a pH greater than 8.6 may be too basic. A change in streamwater pH can also affect aquatic life indirectly by altering other aspects of water chemistry. For example, low pH levels can increase the solubility of certain heavy metals. This allows the metals to be more easily absorbed by aquatic organisms as mentioned by Rostern [15]. A shift of pH in either direction from neutral may indicate the presence of a pollutant in a body of water.

Runoff samples from green roof models with 0% and 100% of bio-compost indicate high pH value as shown in Figure 3. Instead, the runoff sample from green roof model with 50% of bio-compost has the lowest pH value of 7.7 which is the nearest to neutral value. This result may suggest that 50% of bio- compost as soil medium can better control the pH value of the runoff water.

The runoff quality from the green roof models are also checked for colour, BOD and COD concentrations. The average values for these parameters are summarized in Table 1. For colour measurement, the unit is Pt/Co (platinum cobalt). From the results in Table 1, it is clearly shows that the higher the bio-compost percentage, the darker the color of the water sample. Similar observation was reported by Razzaghmanesh [16]. Biochemical oxygen demand (BOD) is used to measure the amount of oxygen used by aerobic bacteria during decomposition in the water. Lower BOD value means that the sample is less polluted. A higher BOD indicates a high content of easily degradable organic material in the water sample [17]. While, a low BOD indicates a low volume of organic materials substances which are difficult to break down. Among the five samples, the sample with 50% bio- compost composition has the lowest BOD measurement. Other than that, green roof model with 50% of bio-compost also shows the lowest COD concentration after the control green roof model with only soil medium. This shows that too much bio-compost may lead to an unwanted pollution to the environment.

Table 1. BOD, COD and colour measurements for green roof models with different percentage of bio- compost

Sample (Bio-compost %)	BOD_5 (mg/L)	COD (mg/l)	Color (PtCo)
0	98.2	78	71
25	48.8	448	1630
50	37.2	354	1650
75	47.6	714	2016
100	154.6	584	3536

In terms of physical growth of the plants, it can be observed that model with 50% of bio-compost shows the best plant growth while green roofs with 0% and 100% of bio compost show limited growth as shown in Figure 4. This is probably due to the lacking and over-abundance of nutrients for green roofs with 0% and 100% of bio compost, respectively. This observation is generally similar with previous studies done by Van Woert [18]. While bio-composting is a great way to give useful purpose to waste materials, care needs to be taken as not to pollute the aquatic environment when bio-compost is utilized as growing medium in green roof.

a) 0%

b) 25%

c) 50%

d) 75%

100%

Figure 5. Plant growth conditions for (a) 0%, (b) 25%, (c) 50%, (d) 75% and (e) 100% of bio compost in green roof models

4. Conclusion

The green roof models with bio-compost inclusion show favourable results compared to the control model with only soil as the growing medium. The results showed that higher composition of bio-compost in the growing medium cause lower total solid and turbidity values. However, it is clear from experimental result that higher bio-compost also can lead to more intense color of runoff from the green roof models. The green roof model with 50% of bio-compost composition has the lowest pH value if compared to that models with 0% and 100% of bio-compost. The 50% bio-compost of green roof model shows the lowest BOD_5 and COD measurements for the runoff quality. In addition, the plants grown on the model with 50% bio-compost appears to grow as the most healthily. Therefore, it is concluded that the ideal composition of soil to bio-compost is 1:1 in Malaysian tropical climate conditions.

5. References

[1] Chow M F and Yusop Z 2019 Contributions of dry and wet weather runoffs to annual pollutant loading in tropical urban catchments *Lecture Notes in Civil Engineering* **9** pp 1511-1521

[2] Haris H, Chow M F, Usman F, Sidek L M, Roseli Z A and Norlida M D 2016 Urban Stormwater Management Model and Tools for Designing Stormwater Management of Green Infrastructure Practices *IOP Conference Series: Earth and Environmental Science* vol 32(1) (United Kingdom:IOP Publishing) 012022

[3} Chow M F, Yusop Z and Toriman M E 2013 Level and transport pattern of faecal coliform bacteria from tropical urban catchments *Water Sci.Tech.* **67**(8) 1822-1831

[4] Chow M F, Yusop Z and Abustan I 2015 Relationship between sediment build-up characteristics and antecedent dry days on different urban road surfaces in Malaysia *Urban Water J.* **12**(3) 240-247

[5] Dunnett N P and Kingsbury N 2004 *Planting green roofs and living walls* (USA: Timber Press Oregon) pp 254

[6] Chow M F, Abu Bakar M F, Sidek L M and Basri H 2017 Effects of substrate types on runoff retention performance within the extensive green roofs *Journal of Engineering and Applied Sciences* **12**(21) 5379-5383

[7] Maclvor J S and Lundholm J 2011 Performance evaluation of native plants suited to extensive green roof conditions in a maritime climate *Ecol. Eng.* **37** 407–417

[8] Chow M F, Bakar M F A and Wong J K 2018 An overview of plant species and substrate materials or green roof system in tropical climate urban environment. *AIP Conference Proceedings* (USA: AIP Publishing) **2030** 020004

[9] Chow M F, Bakar M F A 2019 Environmental benefits of green roof to the sustainable urban development: A review *Lecture Notes in Civil Engineering* (Switzerland: Springer Nature) **9** pp. 1525-1541

[10] Banting D, Doshi H H, Li J and Missios P 2005 *Report on the environmental benefits and costs of green roof technology for the city of Toronto* City of Toronto and Ontario Centres for Excellence— Earth and Environmental Technologies Toronto

[11] Bates A J, Sadler J P, Greswell R B and Mackay R.2015 Effects of varying organic matter content on the development of green roof vegetation: a six year experiment *Ecol. Eng.* **82** 301–310

[12] Farrell C, Ang X Q and Rayner J P 2013 Water-retention additives increase plant available water in green roof substrates *Ecol. Eng.* **52** 112–118

[13] Kanechi M, Fujiwara S, Shintani N, Suzuki T and Uno Y 2014 Performance of herbaceous Evolvulus pilosus on urban green roof in relation to substrate and irrigation *Urban For. Urban Green* **13** (1) 184–191

[14] Molineux C J, Fentiman C H and Gange A C 2009 Characterising alternative recycled waste materials for use as green roof growing media in the U.K". *Ecol. Eng.* **35 (10)**, 1507–1513.

[15] Rostern T N 2017 The Effects of Some Metals in Acidified Waters on Aquatic Organisms *Fish & Ocean OPJ* **4**(4) 1-7

[16] Razzaghmanesh M, Beecham S and Kazemi F 2014 The growth and survival of plants in urban green roofs in a dry climate *Sci. Total Environ* **476** 288–297

[17] Thuring C E, Berghage R D and Beattie D J 2010 Green roof plant responses to different substrate types and depths under various drought conditions *Hort Technology* **20**(2) 395–401
[18] Woert N D, Rowe D B, Andresen J A, Rugh C L and Xiao L 2005 Watering regime and green roof substrate design affect Sedum plant growth *HortScience* **40**(3) 659–664

Acknowledgments
The authors would like to thank the Chair of Energy Economics grant (Project No. 2018003KETST) provided by Energy Commission Malaysia and Tenaga Nasional Berhad (TNB) and supports provided by Innovation & Research Management Center (iRMC) Universiti Tenaga Nasional.

IOP Conf. Series: Materials Science and Engineering **713** (2020) 012013 doi:10.1088/1757-899X/713/1/012013

Combining pozzolanic material as ternary binder in Recycled Aggregate Concrete (RAC) to develop a new sustainable concrete

M A Tambichik[1], A A A Samad[1], N Mohamad[1], M Z M Bosro[1], M. A. Iman [1]

[1] Department of Structural and Material Engineering, Faculty of Civil and Environment Engineering, Universiti Tun Hussein Onn Malaysia, Johor, Malaysia
Corresponding author: azizsamad14@gmail. com

Abstract: The demand for a sustainable concrete has increasing due to the effect of conventional concrete that led to many environmental problems such as carbon dioxide emission and depletion of natural resources. Hence, this paper discusses about the usage of ternary binder in recycled aggregate concrete (RAC). Ternary binder was developed from combination of palm oil fuel ash (POFA) Class C and rice husk ash (RHA) Class F as partial cement replacement. This study consists of two (2) phases that are; phase 1 to determine the optimum percentage of ternary binder and phase 2 to determine the compressive strength of recycled aggregate concrete with ternary binder system. Compressive test was conducted for 7 and 28 days to determine the compressive strength. The results show that existence of POFA and RHA as pozzolan for ternary binder help to improved strength of RAC. Combinations of 10% POFA + 15% RHA + 20% RA produce a compressive strength up to 32.8 MPa at 28 days. Increasing of RA content up to 20% will reduce the compressive strength. Overall it can be concluded that combination of POFA and RHA as ternary binder in RAC can produce a concrete with satisfactory strength for structural purposed.

1. Introduction

The use of sustainable concrete or green concrete in construction industry is needed due to the negative effect of conventional concrete and numerous benefits of using greener concrete. The demand of using this type of concrete has been spurred by demand for high quality concrete product, the national objectives to reduce greenhouse gas emission, the need for conservation of natural resources, and limited landfill spaces. This sustainable concrete comes in various types such as by utilizing pozzolanic material for replacing cement, geopolymer concrete, using recycled concrete aggregate, and addition of waste fiber in concrete.

In Asia, agricultural industry is one of the sources that contribute to the production of waste. Common agricultural wastes produce such as Palm Oil Fuel Ash (POFA) and Rice Husk Ash (RHA). There are about 3.1 million hectares of oil palm trees especially in Malaysia that produce over 9 million tons of crude oil. Oil palm plantation has contributed to the production of POFA. POFA is a by-product of oil mill arising from the uses of palm oil shell and palm oil bunches which are used to power oil mill plants for generate electricity [1].

Meanwhile, RHA is basically an agricultural residue obtained from rice processing mill. Only a small amount of husk produced was used as fuel in rice mill and electricity generating power plant.

Then, the by-product from burning the rice husk about 20% of its original weight is RHA [2]. Because of its huge amount of waste, the government needs to assign more hectares of landfill for disposal and spend a lot of money for transferring the waste to the landfill. The growing environmental concern, increasing scarcity of landfill, and depleting quality source of virgin aggregated are the problems that promote recycling of construction waste including concrete [3]. Recycled Aggregate Concrete (RAC) was made from using Recycled aggregate (RA) as replacement for virgin coarse aggregate in concrete. Recently, the use of more than one type of cement replacing materials in concrete, called ternary, quaternary, or multi binder concrete, has globally grown because they present some advantages over the binary binder concrete [4][14]. A ternary mixture was consist of portland cement and two other material as binder, blended either at the cement plant or at the batch plant. The materials for blended can be supplementary cementitious material (SCM) such as fly ash, metakaolin, pofa or rha.

Previous researcher has tried to utilizing binary and ternary binder in RAC to improve it mechanical properties. Faiz [5] study on the combination of silica fume with RAC and observed that 10% silica fume + 50% RA and 40% slag can improve compressive strength and indirect tensile strength. Radonjanin [6] develop a green recycled aggregate concrete (GRAC) with combination of RA with several pozzolan such as fly ash, metakaolin, milled lime stoned, and silica fume found that the strength of GRAC composite is lower than normal concrete. Hence, this study will be focusing on combining pozzolanic material as ternary binder in RAC to develop a new sustainable concrete.

2. Experimental Works

2.1. Procedure

This experimental program consists of two (2) phases that are:

Phase 1: To determine an optimum percentage for combination of POFA and RHA as ternary binder

Phase 1(a) (Optimum percentage of POFA and RHA)

In this phase, the optimum percentage for POFA and RHA in a concrete was determined by conducting a compressive test for 7 and 28 days according to BS EN 12390-3:2009[10]. POFA and RHA were replaced at 10 %, 15 %, 20 %, 25 % and 30 % by weight of cement. At 28 days, percentages for POFA and RHA that achieved the target strength of 30MPa and higher was selected for phase 1(b).

Phase 1(b) (Optimum percentage for combination POFA and RHA)

Next, the optimum percentage for combination of POFA with RHA was determined by conducting a compressive test for 7 and 28 days according to BS EN 12390-3:2009[10]. Percentage for combination was depending on compressive strength result at phase 1(a). The optimum combination percentage (POFA + RHA) was used for producing ternary binder.

Phase 2: To determine the optimum percentage for combination of ternary binder with RA

At this phase, an optimum percentage of Recycled Aggregate (RA) combined with ternary binder was determined by conducting compressive test for 7 and 28 days according to BS EN 12390-3:2009[10] The percentage of RA used was 10%, 20%, 30%, 40% and 50% by weight of natural coarse aggregate.

3. Material Preparation

3.1. Cement

Ordinary Portland cement type 1 was used for this study. The cement conforms to BS EN 197-1:2011 Cement composition, specification and conformity criteria for common cement [11].

3.2. Superplasticizer
Superplasticizer (Estop Admix AP 20L) was added to improve concrete workability. In this study, the dosage was fixed at 0.6% from weight of binder.

3.3. Coarse Aggregate
Natural aggregate and recycled aggregate with maximum size of 9.5 mm was two types of coarse aggregate used for this experiment. The natural aggregate made from crushed stone is shown clearly in Figure 1(a). The specific gravity at saturated surface dry condition (SSD) of natural aggregate was 2.75. Meanwhile, for RA, as shown in figure 1 (b), is obtained by crushing disposal concrete cubes ranging from C25 to C35 samples using a concrete crusher. Then it been sieved through a sieve with size 9.5 mm to 5mm. Lastly, Lost Angeles Abrasion (LA) machine was used for grinding the RA as to ensure the removal of excess mortar covering the RA.

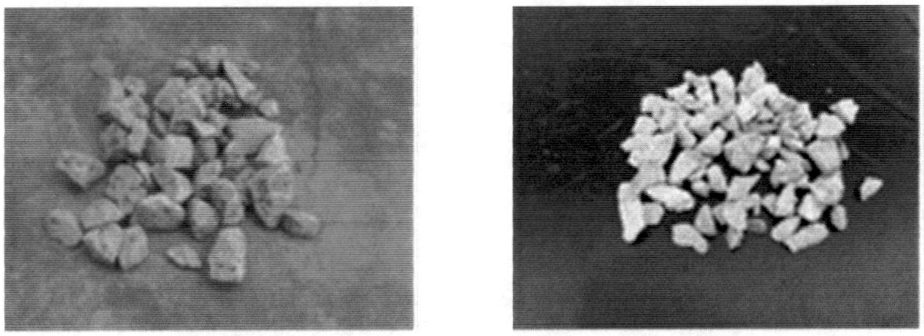

(a) (b)

Figure 1. Types of coarse aggregate (a) Natural Coarse Aggregate (b) Recycle Aggregate

3.4. Fine aggregate
For fine aggregates, river sand which passed through 5 mm sieve and had a fineness modulus and specific gravity of 2.88 and 2.65 respectively. It was stored in a storage tank to prevent moisture.

3.5. Palm Oil Fuel Ash (POFA)
Palm Oil Fuel Ash (POFA) used in this study is obtained from Bell Oil Palm Plantation located at Parit Sulong, Batu Pahat, Johor, as shown in figure 2. Initially, the ash was put in the oven at temperature of 100 ± 5 degree Celsius for 24 hours. Then it was followed by sieving using 300 micron sieves to remove impurities and also reduce carbon that affect the pozzolanic properties. After that it was put in the Los Angeles Abrasion machine with 10 steel balls for 2 hours to make it finer than its original size and to ensure it passes the 150 micron sieve.

Figure 2. Palm Oil Fuel Ash (POFA)

3.6. Rice Husk Ash (RHA)
Rice Husk Ash (RHA) used in this experiment is obtained from Kilang Beras Jelapang Selatan (M) Sdn Bhd, Muar, Johor rice as shown in figure 3. The ash has been put in the oven for 24 hours at a temperature of 100 ± 5 degree Celsius. Then it was sieved through 300 micron sieve to remove impurities and carbon. After that, it was grind using Lost Angeles Abrasion (LA) with 10 steel balls for 2 hours to make it finer than its original size and passing 150 micron sieve.

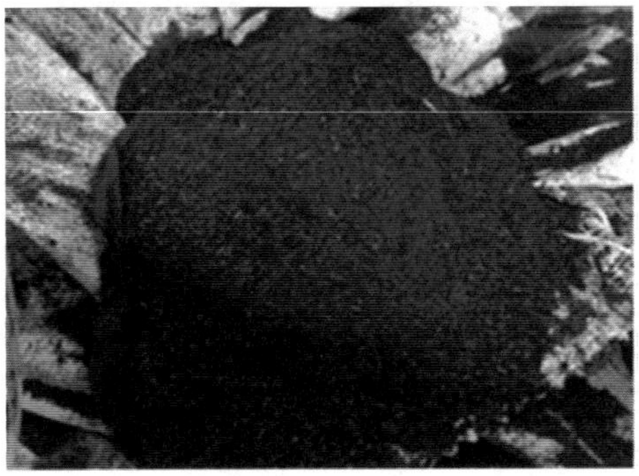

Figure 3. Rice Husk Ash (RHA)

4. Concrete mix design

American Concrete Institute (ACI) was used to design the concrete with targeted strength of 30MPa at 28 days. The water cement ratio used was 0.57 according to Table 6.3.4(a) in ACI design.

Table 1. ACI mix design for grade 30.

Mix Design				
Quantity	Cement (kg/m^3)	Water (kg/m^3)	Sand (kg/m^3)	Coarse Agg (kg/m^3)
1^{m3}	400	228	931	756

5. Result and discussion

5.1. Phase 1: Optimum percentage of ternary binder in recycled aggregate concrete (RAC)

5.1.1 Phase 1(a): Optimum percentage of POFA and RHA. In phase 1(a), compressive strength for concrete containing POFA (figure 4) and RHA (Figure 5) was determined. Percentage that achieved compressive strength of 30 MPa at 28 days was selected to be used at phase 1(b).

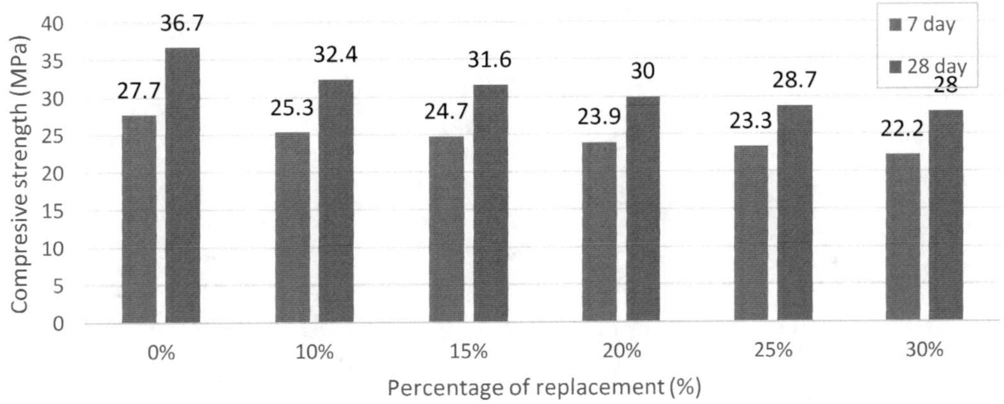

Figure 4. Compressive strength of POFA in concrete.

Based on figure 4, it shows that compressive strength of concrete with POFA decreases as the percentage of replacement increases. The highest compressive strength is recorded at 10% of replacement with 32.4 MPa but it still 13.2% lower than control specimen. Increase of POFA percentage more than 10% shows a continuous decrease in compressive strength. Reduction in compressive strength as the percentage of POFA increase is due to the excessive POFA addition which has created void inside the concrete due to problem of particle packing. However, it can be seen that 10%, 15%, and 20% replacement of POFA has achieved target strength of 30MPa at 28 days. Hence, the three percentages of replacement have been selected to be used for combination with RHA.

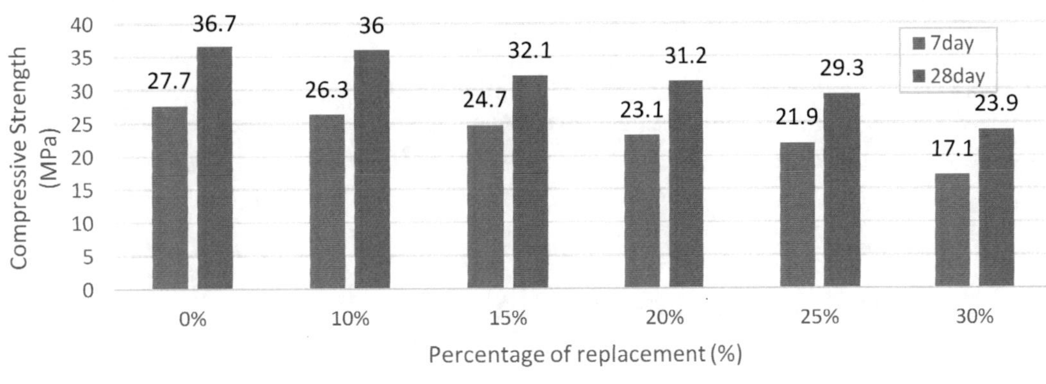

Figure 5. Compressive strength of RHA in concrete.

Figure 5 shows the compressive strength result of concrete containing RHA at 7 and 28 days. Based on the result, it also seems that increases of RHA replacement will decrease the compressive strength. 10% replacement of RHA produces a nearly similar strength with control specimen that is 36 MPa at 28 days. It can also be observed that the addition of RHA increases the compressive strength due to the filler effect. However, a low compressive strength is recorded at early age due to slower hydration process by RHA. Increase of RHA more than 10% has reduced the compressive strength and this result is similar to Kishore [12] where they also found that increase of RHA more than 10% leads to decrease in compressive strength. This result is also similar to the percentage of POFA, only 10%, 15%, and 20% of RHA replacement achieved the targeted strength of 30MPa at 28 days. Hence, this percentage is the best percentage and been selected for combination with POFA.

5.1.2 Phase 1(b): Optimum percentage for combination of POFA with RHA. In phase 1(b), the percentage for POFA and RHA that achieve target strength of 30 MPa is selected for determine the optimum combination of both materials. The compressive strength result at 7 and 28 days is shown in figure 6 below.

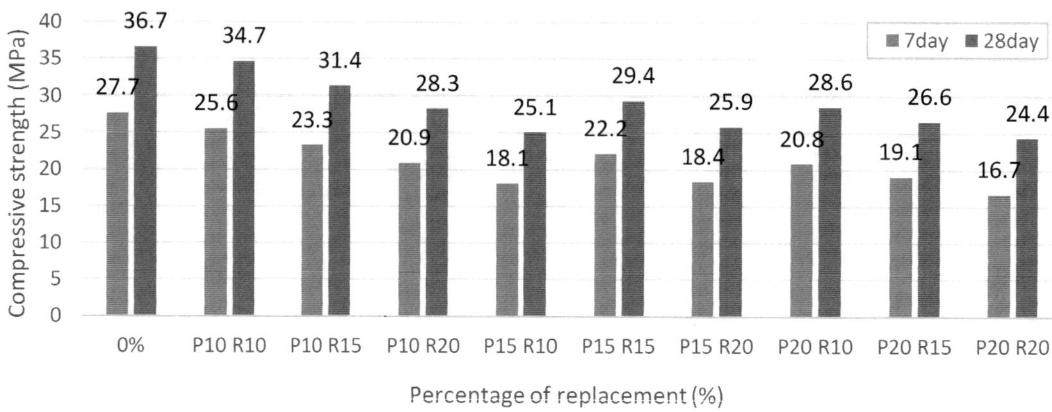

Figure 6. Compressive strength for combination of POFA and RHA.

From the result, can be seen that combination of 10% of POFA with 10% of RHA has the highest compressive strength compared to other percentage at 28 days. For P10R10, the compressive strength at 28 days is 34.7MPa and it is 5.7% slightly lower than the control specimen. Increasing percentage of RHA from 15% to 20% combined with 10% POFA has lowered the compressive strength. Reduction in compressive strength is due to the different physical and chemical properties of both materials that lead to the problem with particle packing and also lack of silica content that produces lower calcium silicate hydrate gel. Only P10 R10 and P10 R15 show a compressive strength of more than 30 MPa at 28 days. Hence, the optimum combination of POFA and RHA in concrete is at 10% POFA and 15% RHA because it has the highest replacement of waste and achieves target strength of 30MPa at 28 days. The optimum combination was selected to be used as fixed percentage for ternary binder in this study.

5.1.3 Phase 2: Optimum percentage for combination of ternary binder with RA. In phase 2, the optimum combination for POFA and RHA is combined with 10%, 20%, 30%, 40% and 50% of RA. The compressive strength result at 7 and 28 days is shown in figure 7.

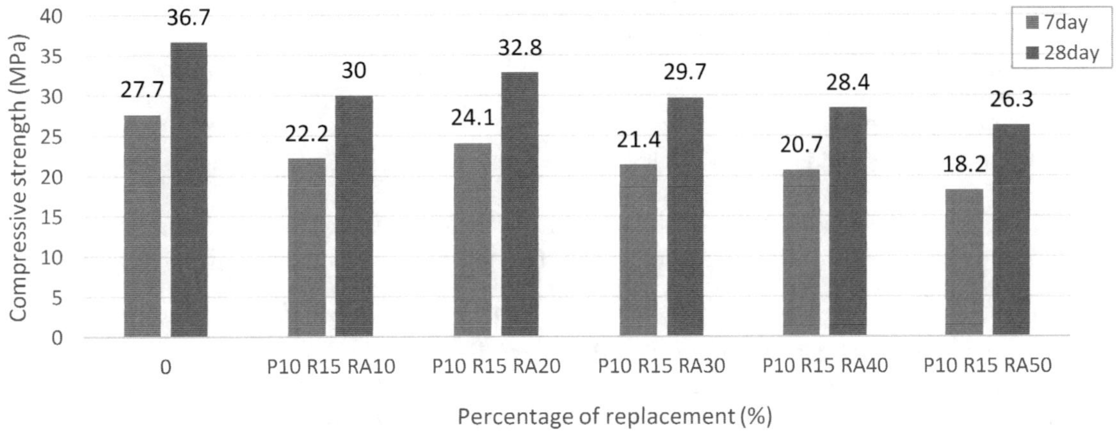

Figure 7. Compressive strength for combination of ternary binder with RA.

Based on figure 7, it shows that the combination ternary binder with 20% RA has the highest compressive strength than other percentages. At 28 days, P10 + R15+ RA 20 have a compressive strength of 32.8 MPa but it is still lower than control specimen. The result of lower compressive strength than control specimen is similar with Kwan [13] where they also observe that replacing natural aggregate with RA will decrease the compressive strength respectively. From the graph, it can also be observed that with a suitable percentage of pozzolanic material combined in a recycled aggregate concrete, it can improve RAC properties. This can be attributed to the contribution by the pozzolanic reaction of silica in the POFA and RHA with the calcium hydroxide in the matrix, which produces additional calcium silicate hydrate (C-S-H) gels and reduces the pores and densifies the ITZ of RAC with the matrix. Only P10+R15+RA10 and P10+R15+ RA20 achieve 30 MPa at 28 days compared to other percentages. Therefore, based on both percentages, P10+ R15+RA20 is the optimum percentage for combination of ternary binder with RA in concrete due to the highest waste utilized and exceed the target strength.

6. Conclusion

The following conclusion can be drawn from the study above:

1. For the phase 1, the optimum combination for ternary binder (POFA + RHA) in a concrete was at 10% POFA+15% RHA. This combination produces a concrete with a compressive strength of more than 30 MPa and higher utilization of agricultural waste. The compressive strength recorded at 28 day was 31.4 MPa but it still lower than control specimen. Meanwhile for phase 2, the combination of RAC with ternary binder system could produce a concrete with a satisfactory strength. Combinations of ternary binder (10% POFA + 15% RHA) with 20% RA produce a compressive strength of 32.8 MPa at 28 days. The strength increases up to 4.4% when ternary binder was combined with RA. However, the strength was still lower than control.
2. The optimum percentage for combination of agricultural and construction waste in concrete can be used for the construction that needed a structural strength of 30 MPa as it can reduce the harmful effects of conventional concrete to the environment. Thus, this would ultimately lead to a sustainable construction and greener environment.

7. References

[1] Awal A S M A and Hussin M W 2010 Influence of palm oil fuel ash in reducing heat of hydration of concrete *Journal of Civil Engineering (IEB)* **38 (2)** 153-157
[2] Chindaprasirt P, Homwuttiwong S and Jaturapitakkul C 2007 Strength and water permeability

of concrete containing palm oil fuel ash and rice husk–bark ash *Construction and Building Materials*, **21(7)** 1492-1499

[3] Nassar R and Soroushian P 2012 Strength and durability of recycled aggregate concrete containing milled glass as partial replacement for cement *Construction and Building Materials* **29** 368–377

[4] Tahir K E and Onder K 2008 Use of binary and ternary blends in high strength concrete *Construction and Building Materials* **22** 1477-1483

[5] Faiz S, Sachin K and Shailesh K 2015 Effect of micro silica on mechanical and durability properties of high volume fly ash recycled aggregate concretes (HVFA-RAC) *Journal of Advance Concrete Construction* **3** 317-331

[6] Radonjanin V, Malesev M, Marinkovic S and A E Saed 2013 Green recycled aggregate concrete *Construction and Building Materials* **47** 1503

[7] Raut A N and Gomez C P 2016 Thermal and mechanical performance of oil palm fiber reinforced mortar utilizing palm oil fly ash as a complementary binder *Construction and Building Materials* **126** 476–483

[8] ASTM C270-14a 2014 *Standard Specification for Mortar for Unit Masonry* (West Conshohocken PA: ASTM International)

[9] Ahmad M H and Nurazuwa M N 2008 Mix design of palm oil fiber concrete In: *International Conference on Civil Engineering (Kuantan: Pahang)*

[10] BS EN 12390-3 2009 *Testing Hardened Concrete. Compressive Strength of Test Specimens* (London: Bristish Standard)

[11] BS EN 197 2011 *Cement composition, Specifications and conformity criteria for cements* (London: British Standard)

[12] Kishore R, Bhikshma V and Prakash P J 2011 Study on Strength Characteristics of High Strength Rice Husk Ash Concrete *12 Asia-Pasific Conf. on Structural Engineering and Concstruction* vol 14 pp 2666–2672

[13] Kwan W H, Ramli M, Kam K J and Sulieman M Z 2012 Influence of the amount of recycled coarse aggregate in concrete design and durability properties *Construction and Building Materials* **26(1)** 565–573

[14] Tambichik M A, A A A Samad, M, Mohamad N, Zurisman A, Mydin A O, Bosro M Z M, and Iman M A 2018 Effect of combining Palm Oil Fuel Ash (POFA) and Rice Husk Ash (RHA) as partial cement replacement to the compressive strength of concrete *International Journal of Integrated Engineering (IJIE)* **10(8)** 61-67

Acknowledgement
The author gratefully acknowledges the Fundamental Research Grant Scheme (No. Vot 1573) funded by Ministry of Education Malaysia and Postgraduated Research Grant Scheme (GPPS) No. Vot U786 funded by Universiti Tun Hussein Onn Malaysia (UTHM).

Restoration of Aged Bitumen Properties Using Maltenes

Z H Hussein[1], H Yaacob[1*], M K Idham[1], N A Hassan[1], L J Choy[2] and R P Jaya[3]

[1] Faculty of Engineering, School of Civil Engineering, Universiti Teknologi Malaysia, Skudai, 81310, Johor, Malaysia
[2] Department of Bioprocess and Polymer Engineering, School of Chemical and Energy Engineering, Universiti Teknologi Malaysia, Skudai, 81310, Johor, Malaysia
[3] Faculty of Engineering and Earth Resources, Universiti Malaysia Pahang, 26300, Gambang, Pahang, Malaysia
Corresponding author: haryatiyaacob@utm.my

Abstract: Reclaimed asphalt pavement (RAP), a material frequently used in asphalt mixtures, is associated with several environmental and economic advantages. Many existing road construction technologies are capitalising on the availability of recycled materials. This study has looked on the usage of maltene as a rejuvenator in aged bitumen. Bitumen with Pen 60-70 was chosen as the control binder. Maltene was added into the aged bitumen at various concentrations (5%, 10%, and 15%) by weight of total binder. The characteristics of rejuvenated bitumen were examined with penetration, softening point, penetration index (PI), viscosity, and storage stability. As the maltene concentration increased, aged bitumen characteristics improved, while the level of stiffness reduced. Mixing temperature and time were significantly linked to the efficiency of blending between aged-virgin binders. Lower mixing temperatures undermined the diffusion of the virgin bitumen into the aged binder, thereby impairing blending efficiency. The results indicate that 15% of maltene can rejuvenate the properties of aged bitumen.

1. Introduction

To achieve the goal of "Zero Waste Management" by 2020, Malaysian institutions of higher education have launched a range of initiatives for managing waste. The majority of which focus on making people more aware of the necessity to cut back on the amount of waste materials disposed in landfills. A key recycling material for the production of a sustainable asphalt mixture is reclaimed asphalt pavement (RAP) [1]. Waste reduction, contribution to the resolution of issues related to discarding of materials from highway construction, counteraction of elevated preliminary expenditure, and conservation of natural resources are the major advantages of RAP [1, 2]. This material is incorporated in the majority of asphalt mixtures currently produced in developed countries. However, unforeseen untimely failure can occur due to the challenges presented by mixtures with high RAP contents to field compact, so RAP can only be added in specific proportions to new mixtures[3]. The restrictions on RAP levels primarily stem from performance problems with the aged binder, which, unlike virgin binder, display greater brittleness and vulnerability to thermal and fatigue cracking [4]. This is due to the natural short and long aging process suffered during service life that lead to increasing the viscosity[5] The chemical composition of the bituminous blend is altered by the existence of RAP binder, yielding a composite asphalt binder of greater hardness and viscosity [6], as the maltenes fraction transforms into asphaltene via oxidation due to the higher viscosity and aging of the RAP binder. However, the application specification can be met by employing rejuvenators to restore the properties of aged binder. Through supply of more maltenes and/or mediation of improved asphaltenes diffusion, rejuvenators ought to ensure that the four chemical constituents of aged binder are in equilibrium [7, 8]. This implies a reduction in viscosity and rigidity and an increase in penetration and ductility [9].

A lot of studies have been conducted to assess the effects of rejuvenators on RAP binders. The research carried out by Zaumanis et al. [10] assessed the efficacy of rejuvenating agents in producing RAP mixtures with 40-100% RAP content. A range of softening agents were used, including waste-derived oils, plant oils, refinery base oils (both non-traditional and traditional), and engineered oils. Additionally, the rejuvenating agents were waste engine oil (WEO), WEO bottoms, WEO with FT wax, distilled tall oil, naphthenic flux oil, organic blend, aromatic extract, refined tallow, and paraffinic base oil. The results demonstrated that each rejuvenating agent resulted in contrasting effects on the aged asphalt binder. The most efficacious in reducing the RAP binder's viscosity was refined tallow. However, WEO bottoms were associated with a higher viscosity level, indicating that they produced greater compaction and mixing temperatures. In the study, the researchers also took performance measurements at low temperatures, the idea being to assess indirect tensile creep compliance and indirect tensile strength. In almost all cases, rejuvenating agents enhanced the mixtures' low-temperature characteristics.

A recent study, conducted by Zhang et al. [11], employed sawdust-derived bio-oil for the rejuvenation of aged asphalt. For the base binders, PG 58-28 and PG 64-22 were used, while the bio-oil concentrations were 10%, 15%, and 20% of the overall binder weight. A range of assessments were performed to gain insight into the aged binders' properties and level of rejuvenation. The results indicated that viscosity and activation energy reduced after adding the bio-oil, whereas temperature susceptibility and viscous component content increased. Additionally, the rejuvenating agent was found to soften the aged asphalt. In terms of the rutting index, this reduced by 75.5% and 77.2% for PAV PG 58-28 and PAV PG 64-22, respectively, at 52-76 °C. Furthermore, the bio-oil improved the low-temperature crack resistance of recycled PAV PG 58-28 and PAV PG 64-22, attaining a standard greater than or equal to that of virgin asphalt. Hence, the rejuvenating agent was effective for PAV PG 58-28 and PAV PG 64-22 at a dose of 15% and 20%, respectively.

Although asphalt binder and rejuvenator recycling has been extensively studied, the rejuvenator potential of maltenes, which is extracted from virgin binder, has not been investigated. Hence, the use of maltenes from virgin asphalt to rejuvenate aged bitumen was examined in the present study. Tests were conducted on penetration, softening point, viscosity, penetration index (PI), penetration-viscosity number (PVN) and storage stability to assess physical properties of rejuvenated bitumen.

2. Materials and Experiments

2.1. Virgin bitumen

The employed control binder was PEN. 60-70 bitumen taken from KBC Malaysia Ltd, which is often used in Malaysian roadworks. Table 1 provides a number of fundamental properties the of bitumen used. Aged bitumen recycling was achieved by using a maltene as a rejuvenator. A chemical solvent and rotary evaporator enabled extraction of the maltene from virgin bitumen.

2.2. Preparation of aged bitumen

A milling process was performed for sourcing RAP from Secondlink highway with a service life of about 22 years. The bitumen from the RAP was extracted by using methylene chloride solvent as per the ASTM D2172[12]. A rotary evaporator was then used to evaporate out the methylene chloride and recover the aged bitumen as per ASTM D5404[13]. Table 1 shows the properties of aged bitumen. The term applied from this point on is aged bitumen rather than recovered bitumen, for reasons of consistency.

Table 1. Properties of virgin and aged bitumen

	Virgin Bitumen	Aged Bitumen
Penetration (dmm.)	64	7
Softening Point (°C)	50.5	83
Viscosity @ 135 °C (mPa.s)	650	7400
Viscosity @ 165 °C (mPa.s)	200	1500

2.3. Maltene extraction

A multi-stage process was conducted to extract the maltene intended as rejuvenator. The first stage involved mixing virgin bitumen with chemical solvent "petroleum ether" in a flask, which was introduced in a water bath at 50°C and stirred continuously for 120 minutes. In the second stage, the settling of the asphaltenes in the flask was achieved by keeping the dissolved virgin bitumen in a water bath for half an hour, with no stirring. Next, filter paper was employed to attain the fraction of maltene and chemical solvent from asphaltene. Maltene recovery was then achieved with a rotary evaporator at 60-90°C. Finally, complete removal of the chemical solvent was accomplished by placing the maltene in an oven at 110°C for half an hour, as shown in Figure 1. Table 2 provides basic properties of the maltene rejuvenator.

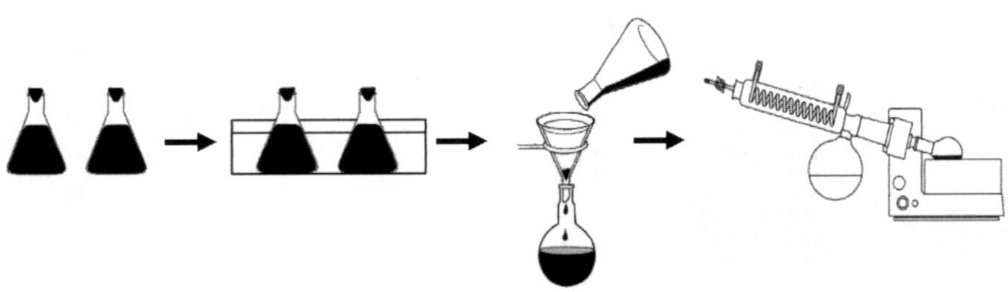

Figure 1. Maltene extraction process.

Table 2. Maltene properties

	Maltene
Density @ 20 °C (kg/cm^3)	0.955
Viscosity @ 60 °C (mPa.s)	782.7
Extracted from:	Virgin bitumen (60-70)

2.4 Samples preparation

The reclaimed bitumen (aged bitumen) was mixed with the virgin bitumen in a proportion of 30% bitumen weight at 145°C and 1500-rpm mixer rotation speed for half an hour. The blending effectiveness of RAP-virgin binders and maltene was substantially influenced by time, temperature and speed. Therefore, the selection of those parameters was based on trial and error. The preliminary study involved blending 0%, 5%, 10% and 15% maltene content by the total binder weight with virgin bitumen that incorporated reclaimed bitumen. The ideal rejuvenator concentration to be added to the mixture was established via penetration, softening point and viscosity tests.

2.5 Experiments

2.5.1. Penetration test. The distance or depth measured in tenths of millimetre permitting achievement of vertical penetration of 100 g weight for a standard needle in a sample at 25°C for 5 seconds is called

IOP Conf. Series: Materials Science and Engineering **713** (2020) 012014 doi:10.1088/1757-899X/713/1/012014

penetration. Compliance with ASTM D5/D5M was ensured when the bitumen sample consistency was assessed[14].

2.5.2. Softening point. The thermal sensitivity of the binder is reflected in the softening point, which is assessed in accordance with ASTM D36/D36M[15]. This helps to determine how the binder behaves at raised temperature. The average temperature associated with suitable binder softening and a 25-mm fall of every steel ball prior to coming into contact with the base plate was considered the softening point. The higher the softening point, the less thermally sensitive the binder is.

2.5.3. Viscosity test. In accordance with ASTM D4402/D4402M, the flow resistance and internal friction of bitumen are determined through viscosity measurement[16]. A Brookfield Thermosel viscometer was employed to assess viscosity at 135, 150 and 165°C. Expressed as units of centipoise (mPa.s), low viscosity measurement was reflected in the low friction in the binder sample.

2.5.4. Penetration index (PI). The temperature susceptibility of a bitumen is reflected in the PI [17]. Knowledge of this parameter helps to determine how it will behave in an application. The formula for PI determination is based on equation 1:

$$PI = \frac{(1951.4 - 500 \log P - 20SP)}{(50 \log P - SP - 120.14)} \quad (1)$$

In equation 1, the penetration value (dmm) and softening point value (°C) are respectively denoted by P and SP. A lower PI value indicates a higher temperature susceptibility, with extreme thermal sensitivity being indicated by a PI value of less than -2 [18]. Aged bitumen typically has an elevated PI, suggesting increased brittleness, reduced flow ability and heightened susceptibility to cracking.

2.5.5. Penetration viscosity number (PVN). McLeod N (1976) introduced the PVN as an additional option besides the PI. In accordance with standard specifications for paving asphalt, equation 2 permits PVN determination based on penetration and viscosity at 25°C and 135°C, respectively[19].

$$PVN = (-1.5)\, x \frac{(4.258 - 0.7967 \log P - \log V)}{(0.795 - 0.1858 \log P)} \quad (2)$$

In equation 2, the penetration viscosity number, penetration at 25°C and viscosity at 135°C are respectively denoted by PVN, P and V.

2.5.6. Storage stability. In line with ASTM D5892[20], the potential separation of the rejuvenator from the bitumen incorporating RAP under storage was assessed via the storage stability test. Blending was followed straight away by pouring of the rejuvenated bitumen in a 25 mm × 140 mm aluminium tube, which was introduced in the oven for 48-hours heating to 165°C. Sample solidification was afterwards achieved by putting the samples in the freezer at -10°C for four hours. The samples were then separated into three parts of identical length, with the superior and inferior parts being subjected to the softening point test. Control of the discrepancy between these two parts within 2.2°C is recommended.

3. Results and Discussion

3.1. Penetration test

The penetration test was performed on the control sample and bitumen samples rejuvenated with 5, 10 and 15% maltene by weight of total binder. As the maltene concentration increased, penetration increased as shown in Figure 2. More specifically, the penetration increased by 27%, 69%, and 110% as the maltene contents increased by 5%, 10%, and 15%, respectively. One interpretation of these results is a rise in the maltene-to-asphaltene ratio and the softening action of the maltene serving as rejuvenator.

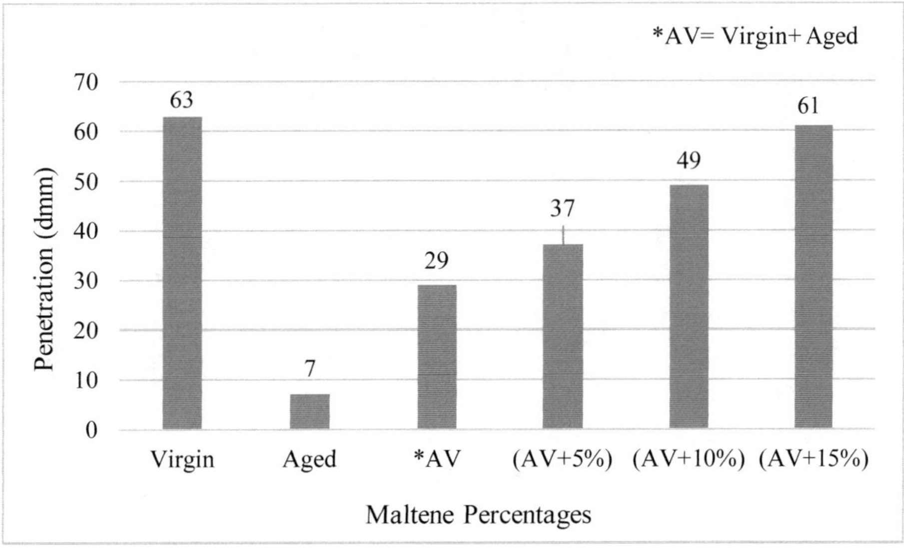

Figure 2. The correlation between penetration and maltene percentages.

3.2. Softening point

Penetration was indirectly correlated with the softening point values of the control binder and bitumen with 5, 10 and 15% maltene rejuvenation (Figure 3). As the maltene concentration increased, the softening point declined markedly. More specifically, a 15% maltene concentration lowered the softening temperature of rejuvenated bitumen to 49°C, which is close to the 50.5°C softening point of virgin bitumen. As a result of such softening point enhancement, pavement degradation owing to fatigue and low-temperature cracking was minimised.

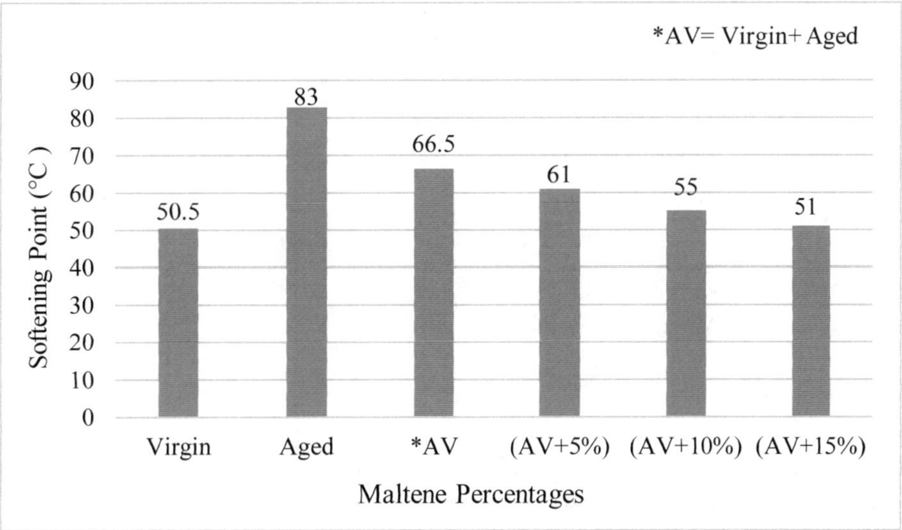

Figure 3. The correlation between softening point and maltene percentages.

3.3. Viscosity test

At 135°C, aged bitumen was more viscous than unaged bitumen, owing to the fact that the binder became markedly more rigid as ageing hardened it and led to high internal friction. Virgin and RAP aggregates must have comparable film thickness and virgin and RAP binder must be homogeneously blended. Therefore, low viscosity is required to achieve sufficient binder flow. The viscosity of virgin, aged and rejuvenated bitumen at various temperatures is indicated in Table 3. Aged bitumen kinematic viscosity was markedly reduced by 15% maltene, bringing it close to the 650 mPa.s value of virgin bitumen viscosity, as reflected by viscosity at 135, 150 and 165°C. A tight correlation exists between kinematic viscosity and the necessary mixing and compaction temperature.

Table 3. Viscosity of virgin, aged, and rejuvenated bitumen

	Viscosity (mPa.s)		
	135 °C	150 °C	165 °C
Virgin Bitumen	650	400	200
Aged Bitumen	7400	5100	1500
Aged + Virgin = (AV)	1700	800	400
(AV)+5% Maltene	1350	750	400
(AV)+10% Maltene	1050	600	300
(AV)+15% Maltene	675	400	200

3.4. Penetration index (PI)

Thermoplastic properties are possessed by all bitumen types. Whereas high-PI bitumen exhibits softness and is employed in regions with low temperatures, low-PI bitumen displays hardness and is employed in regions with high temperatures. Thus, the behaviour of different bitumen formulations to temperature alterations is quantitatively measured by the PI. Aged bitumen typically has an elevated PI, suggesting increased viscosity and brittleness, reduced flow ability and heightened susceptibility to cracking. This process should be countered by effective recovery of the properties of aged bitumen. Therefore, the quality of rejuvenation may be closely reflected in PI reduction by contrast to extracted bitumen PI. The penetration index of the control, aged bitumen and the bitumen rejuvenated with 5%, 10%, and 15% in this research are presented in Table 4. This proves that the rejuvenated bitumen improved the properties rheological, 15% rejuvenation lowering the aged bitumen PI from 0.96 to -0.99.

Table 4. The correlation between PI and maltene percentage

	PI
Virgin Bitumen	-0.525
Aged Bitumen	0.812
Aged + Virgin = (AV)	0.963
(AV)+5% Maltene	0.508
(AV)+10% Maltene	-0.076
(AV)+15% Maltene	-0.481

3.5. Penetration-viscosity number (PVN)

As it is known, high PVN indicates low thermal susceptibility. Compared to virgin bitumen, aged bitumen had higher PVN, suggesting greater cracking potential. Nevertheless, as shown in Table 5, aged bitumen PVN was markedly reduced by 15% maltene addition, making it similar to virgin bitumen PVN. As a result, cracking-related pavement degradation was minimised. The PVN and PI values were significantly discrepant, without any correlation between them.

Table 5. The correlation between PVN and maltene percentage

	PVN
Virgin Bitumen	-0.04
Aged Bitumen	0.67
Aged + Virgin = (AV)	0.39
(AV)+5% Maltene	0.36
(AV)+10% Maltene	0.34
(AV)+15% Maltene	-0.02

3.6. Storage stability
Separation of rejuvenator and aged bitumen occurs due to the lack of compatibility between binder and rejuvenator. Therefore, rejuvenated bitumen with rejuvenators is morphologically unstable in protracted storage at high temperature. This study investigated the phase separation between maltene and rejuvenated bitumen by conducting the high-temperature storage stability test. This test is indicative of how homogeneously maltene is diffused in bitumen. A bitumen sample is considered to be storage stable at high temperature and a rejuvenator is considered to have good diffusion if the inferior and superior parts have a softening point discrepancy of less than 2.2°C. Table 6 indicates how much the inferior and superior parts of the bitumen samples differed in softening point at 5% (0.5°C), 10% (0.5°C) and 15% (1.0°C) maltene addition. The softening point difference didn't exceed 2.2°C at any maltene concentration. Hence, rejuvenated bitumen constituents retained their stability and homogeneity at high temperature. In addition, the appropriateness of the chosen blending procedure was confirmed by the fact that the maltene-bitumen composite was homogeneously consistent.

Table 6. Storage stability displayed by rejuvenated bitumen

	Softening Point (°C)			
	Top	Bottom	Difference	Requirement < 2.2
(AV)+5% Maltene	63.0	63.5	0.5	Pass
(AV)+10% Maltene	56.5	57.0	0.5	Pass
(AV)+15% Maltene	51.5	52.5	1.0	Pass

4. Conclusion
The results obtained allow the formulation of several concluding remarks. The use of maltene as rejuvenator in 15% concentration enabled the successful recovery the properties of aged bitumen. Penetration was enhanced while the softening point was reduced when maltene being added to aged bitumen as rejuvenator, making aged bitumen less rigid and viscous. Thirdly, the addition of 15% maltene has also decreased PI from 0.96 to -0.48 and PVN from 0.67 to -0.20. This improvement in PI and PVN lead to minimise of cracking-related pavement deterioration. And finally, rejuvenated bitumen constituents retained their stability and homogeneity at high temperature, as confirmed by the high-temperature storage stability test.

5. References

[1] Soon J L, Serji N A, Nam W P and Kwang W K 2009 Characterization of warm mix asphalt binders containing artificially long-term aged binders *Construction and Building Materials* **23**(6)2371-2379
[2] Duraid M A, Hussain A K and Riaz A 2018 Novel Methodology to Investigate and Obtain a Complete Blend between RAP and Virgin Materials *Journal of Materials in Civil Engineering* **30**(5) 04018060

[3] Mohamed E, Christopher W and Eric C 2018 Thermal stability and evolved gas analysis of rejuvenated reclaimed asphalt pavement (RAP) bitumen using thermogravimetric analysis–Fourier transform infrared (TG–FTIR) *Journal of Thermal Analysis and Calorimetry* **131**(2) 865-871

[4] Alessandro B, Ana J D, Davide L P and Filippo G 2017 Effects of laboratory aging on properties of biorejuvenated asphalt binders *Journal of Materials in Civil Engineering* **29**(10) 04017149

[5] Cavalli M C, Martins Z, Edoardo M, Manfred N P and Lily D P 2018 Effect of ageing on the mechanical and chemical properties of binder from RAP treated with bio-based rejuvenators *Composites Part B: Engineering* **141** 174-181

[6] Noferini L, Andrea S, Cesare S and Francesco M 2017 Investigation on performances of asphalt mixtures made with Reclaimed Asphalt Pavement: Effects of interaction between virgin and RAP bitumen *International Journal of Pavement Research and Technology* **10(**4) 322-332

[7] Elseifi M A, Louay N M and Samuel B C 2011 Laboratory evaluation of asphalt mixtures containing sustainable technologies *Asphalt paving technology* **80** 227-244

[8] Shen J, Serji A and Jennifer A M 2007 Effects of rejuvenating agents on superpave mixtures containing reclaimed asphalt pavement *Journal of Materials in Civil Engineering* **19**(5) 376-384

[9] Ali A W, Yusuf A M, Aaron N, Caitlin P and Thomas B 2016 Investigation of the impacts of aging and RAP percentages on effectiveness of asphalt binder rejuvenators *Construction and Building Materials* **110** 211-217

[10] Martins Z, Rajib B M, and Robert F 2013 Evaluation of rejuvenator's effectiveness with conventional mix testing for 100% reclaimed Asphalt pavement mixtures *Transportation Research Record* **2370(**1) 17-25

[11] Zhang R, Zhanping Y, Hainian W, Mingxiao Y, Yoke K Y and Chundi Si 2019 The impact of bio-oil as rejuvenator for aged asphalt binde *Construction and Building Materials* **196** 134-143

[12] ASTM D2127 2017 Standard Test Methods for Quantitative Extraction of Asphalt Binder from Asphalt Mixtures (ASTM International, West Conshohocken, Pa)

[13] ASTM D5404/D5404M 2012 Standard practice for recovery of asphalt from solution using the rotary evaporator (ASTM International, West Conshohocken, Pa)

[14] ASTM D5/D5M 2013 Standard test method for penetration of bituminous materials, (ASTM International, West Conshohocken, Pa)

[15] ASTM D36/D36M 2014 Standard test method for softening point of bitumen (ring-and-ball apparatus), (ASTM International, West Conshohocken, Pa)

[16] ASTM D4402/4402 M 2015 Standard test method for viscosity determination of asphalt at elevated temperatures using a rotational viscometer (ASTM International, West Conshohocken, Pa)

[17] Chen J S, M C Liao, and C H Lin 2003 Determination of polymer content in modified bitumen *Materials and structures* **36**(9) 594-598

[18] Read J, David W and Robert N H 2003 The shell bitumen handbook (*Thomas Telford)*

[19] McLeod N 1976 Asphalt cements: pen-vis number and its application to moduli of stiffness *Journal of Testing and Evaluation* **4**(4) 275-282

[20] ASTM D5892-00 2015 Standard Specification for Type IV Polymer-Modified Asphalt Cement for Use in Pavement Construction (ASTM International, West Conshohocken, Pa)

Acknowledgments
The authors express their gratitude to the Ministry of Education Malaysia for funding this work through the Fundamental Research Grant Scheme (**Grant Number R.J130000.7851.5F019 and R.J130000.7846.4F827**).

IOP Conf. Series: Materials Science and Engineering **713** (2020) 012015 doi:10.1088/1757-899X/713/1/012015

Consolidation Behaviour of Dredged Marine Soils Considering the Incorporation of Granular Drainage Layer

S F S M Johan[1] and C-M Chan[2*]

[1&2] Department of Civil Engineering Technology, Faculty of Engineering Technology, Pagoh Higher Education Hub, 84600 Pagoh, Muar, Johor Darul Takzim, Malaysia
Corresponding author: chan@uthm.edu.my

Abstract. The extraction of DMS from seabed along the Malaysian coastal generate millions cubic meters every year. Dredge-transport-dump the DMS at the specified open waters are unfavourable in context of environmental and geotechnical engineering. This paper describes consolidation behaviour of DMS using a modified large- oedometer with a layer of recycled granular materials such as palm oil clinker and recycled pavement materials. DMS were obtained to have high ratio of Wc with 3.27LL and considered as silty clay. A layer of drainage materials accelerates the rate of consolidation and changes in hydraulic conductivity, *k* of DMS. At the end of consolidation, both RPM and POC layer expediate the water from DMS and it becomes stiffer and structured.

1. Introduction

The maintenance of sea bottom near the harbours is necessary for maritime activities. It is to prevent any of the vessels in or out of the ports from stranded due to the soil sedimentation. The sedimentation rapidly happens because of the current wave. Therefore, dredging works play a vital role in this case. Prevalently, Malaysia's practice by dredge and dump the sediments from the dredging site to the designated pond or open water within 10 nautical miles. Considering the disposal of unwanted sediments without any proper treatment, causes an environmental effect on the marine life either short or long term [1]. For examples, impacts on the contamination of sediments into the water, change in seabed surface and exposure of contamination to benthic organisms and fishes. Excavation near urbanized areas have high contamination with organic matter, high water content and low shear strength [2]. Wang *et al.* [3] supported that the dredging not only creates or produces huge amounts of sediment, but also brings others type of contamination to the surrounding waters. According to Bortone and Palumbo [4], mostly of the sediments that settle down at the seabed are consists of sand, silt and clay. Rocks and gravels are also considered types of dredged sediments [5]. Dredged marine soils (DMS) are generally classified as waste materials in civil engineering applications or known as *geo-waste* because of poor engineering properties such as low in shear strength, permeability and high compressibility. Hence, DMS are generally classified as waste materials (geo waste). The beneficial uses of DMS depend on the properties, dredging location and dredging method. DMS can be considered as a resource and reusable material for construction purpose unless contamination is found to be excessive. Other benefits by reusing the DMS such as mitigation control, stabilisation of coastal near the shoreline or any construction purposes [6-7]. Satoh and Kitazume [8] and Ganesalingam *et al.* [9] had successfully proven their works on land reclamation projects for Central Japan International Airport and Port Brisbane, Australia. Additionally, DMS been used as a backfilling and construction material at the Bremerhaven, Germany [10]. Another application to reuse the dredged sediment by filled-in-place the sediments into geotextile tubes for variety of functions such as dewatering, tube walls, embankment and land restoration. This practice provides a lot of benefits toward the environmental and economic near the coastal [11].

The idea of reusing waste materials in civil structures is not a new concept to be applied these days. The waste materials could be reused as backfill materials, road pavements or concrete materials. Yet, the development to advancing the applications by reusing the waste as partial reinforcement in composite materials is still on going. Using recycled materials in construction applications will reduce the negative impact the environment. It also provides an alternative to reduce the usage of aggregates that often been used in construction industry [12]. Palm oil clinker (POC) are waste by-products after the incineration of oil palm shell and fibre to produce steam for palm oil mill. POC generally have large porosity and high-water absorption. Recycled pavement materials (RPM) usually comes from the milling work during the maintenance roadworks that reach the end of design life. RPM consists of aggregates that still coated with bitumen binder. This study attempts to investigate the consolidation behaviour of DMS by using the POC and RPM as a layer in single drainage.

2. Methodology

DMS were retrieved from Kuala Perlis during the dredging works along the Peninsular of Malaysia in 2015. The backhoe dredger machine on top of barge (figure 1) was used to dredge the DMS from seabed at depth 4-6 m. The disturbed samples of DMS (figure 2a) were sealed with sampling bags and all the samples were kept inside the storage bin to avoid the loss of moisture content along way back to laboratory at Johor. The waste granular materials used in this study were palm oil clinker and recycled pavement materials. Palm oil clinker (POC) samples were collected from the palm oil mill in Kluang, Johor and RPM from the stockpile of bitumen pavement in Melaka. Both materials were kept in an open air to dry. The origin shape of POC and RPM as shown in figure 2 (b) and (c) respectively. Therefore, both need to be ground using grinder machine in low speed and sieve the particle within range of 2.00-2.36 mm. The XRD and X-ray fluorescence (XRF) were carried out for DMS, POC and RPM to determine the mineralogy and chemical composition. Preparation on DMS samples must be in homogenous state and uniformly remoulded by mixing using the conventional mixer at least 10 minutes in same speed. At the oedometer specimen mould, the outer and inner part need to be lubricated with petroleum jelly to avoid any skin friction of specimen. In this case, the modified large mould been used with dimension of 100 x 100 mm height. The configuration of specimens was conducted as shown in the illustration figure 3. Noted that 50 % of granular waste materials from the dry unit weight were used in this study. All the setup and apparatus of oedometer test was adopted from BS 1377:1990 Part 5 Clause 3 [13].

Figure 1. Backhoe Dredger machine during dredging work at Kuala Perlis

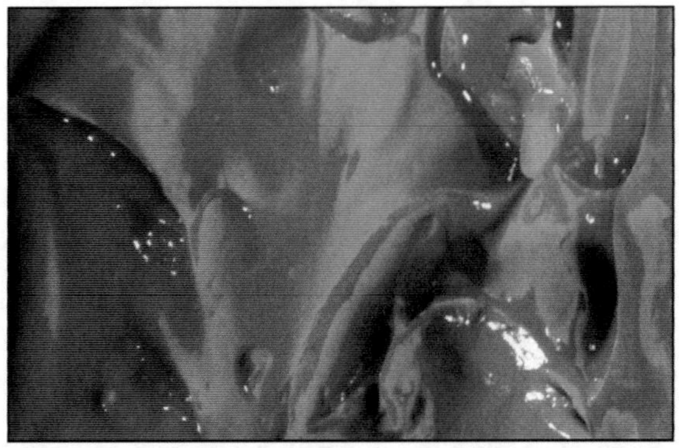

(a)

(b)

(c)

Figure 2. Raw materials; (a) DMS; (b) RPM; (c) POC

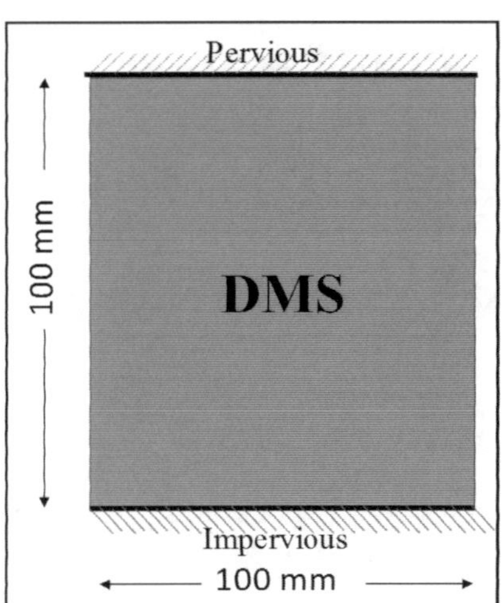

(a)

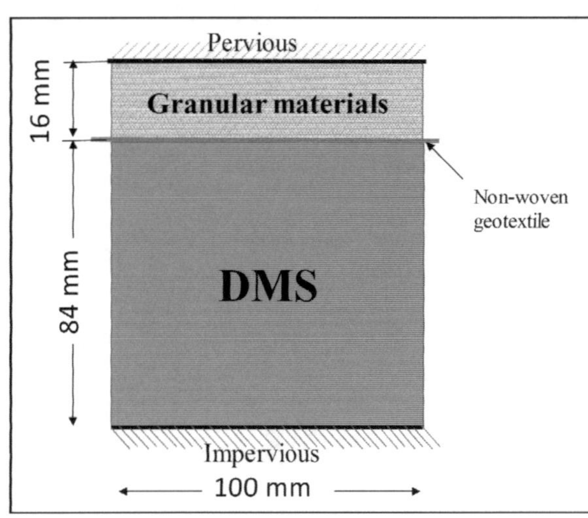

(b)

Figure 3. Schematic cross-section of test configuration on the large oedometer (100 x 100 mm); (a) Control (DMS); (b) Single layer drainage of 50 % granular (RPM or POC)

IOP Conf. Series: Materials Science and Engineering **713** (2020) 012015 doi:10.1088/1757-899X/713/1/012015

3. Results and discussions

3.1 Characterisation of materials

The physical characteristics of DMS are reported in Table 1. The initial water content of DMS is about 218%, which considered as typical for dredged sediments. By using wet sieving and hydrometer method, confirmed that the DMS in Kuala Perlis mainly clay with little bit of silt (silty clay). The proportion of clay about 61%, silt is 38% and fine sand only 1%. The ratio of liquid limit of DMS (wc/LL) is 3.27, higher than previous study at the same dredging site but different time. This indicate that space of time of sedimentation influenced the properties of DMS. The specific gravity, Gs of DMS is 2.68. According to Unified Soil Classification System (USCS), POC and RPM are classified as well-graded gravel and RPM is considered a poorly graded gravel. The Gs of POC are significantly lower than RPM with 2.23 and 2.39 respectively. In this case, it much lower than the typical Gs of virgin aggregates that within in the range of 2.65-2.70. However, the content of bitumen that still coated on the aggregates do influence the density of RPM. Same goes to POC, where the incineration process at the final stage change the physical structure and properties.

As summarised in Table2, it shows the chemical composition of DMS, POC and RPM from the X-ray fluorescence (XRF). The results indicated that all materials contained several kinds of oxides such as Al_2O_3, SiO_2, Na_2O, MgO, CaO and so on. Among all these oxides compound, sodium oxide (Na_2O) occupies the dominant portion of DMS with the percentage of 46.22% and proven to have higher salt content. Followed by silica oxide (SiO_2) and alumina (Al_2O_3) with 30.53% and 13.8% respectively. Chemical compounds in Kuala Perlis reportedly to have similar results where sodium oxide much higher but lower percentage of alumina (8.7%) [15]. On the other hand, Wan Salim [16] reported to have lower in sodium oxide (1.44%). This suggests the activity of dredging change the chemical compound of seabed. POC indeed to have higher silica oxide than RPM with 73% and 68.3% respectively. Figure 4 shows the peaks obtained from analysis of DMS, POC and RPM through the X-ray diffraction (XRD) test. Aforementioned earlier, all 3 materials majorly to have higher quartz components, which a type of silica oxide that found to be prominent.

Table 1. Properties of DMS

Initial water content, w_c	218.07 %	
Specific gravity, G_s	2.68	
Liquid limit, LL	66.5 %	
Plastic limit, PL	55.8 %.	
Plasticity index, PI	10.69	
pH	8.0	
W_C/LL	3.27LL	
Soil classification (USCS)	MH (high plasticity silt)	
	Sand	1%
	Silt	38 %
	Clay	61%

Table 2. Chemical composition of DMS and granular materials

Minerals composition (%)		Materials		
		DMS	RPM	POC
Alumina	Al_2O_3	13.376	15.626	7.895
Calcium oxide	CaO	0.296	1.144	2.907
Potassium oxide	K_2O	0.778	3.529	7.667
Magnesium oxide	MgO	2.199	0.873	3.063
Silicon oxide	SiO_2	30.533	68.274	73.042
Sulphur trioxide	SO_3	0.496	1.914	0.402
Titanium dioxide	TiO_2	0.274	0.222	0.177
Sodium oxide	Na_2O	46.221	5.902	0.436
Iron oxide	Fe_2O_3	2.246	2.157	2.455
Chlorine	Cl	3.411	0.041	0.076
Phosphorus pentoxide	P_2O_5	0.103	0.208	1.794

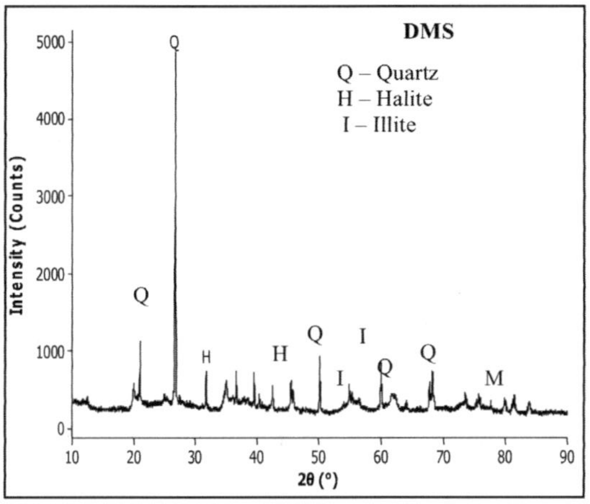

(a)

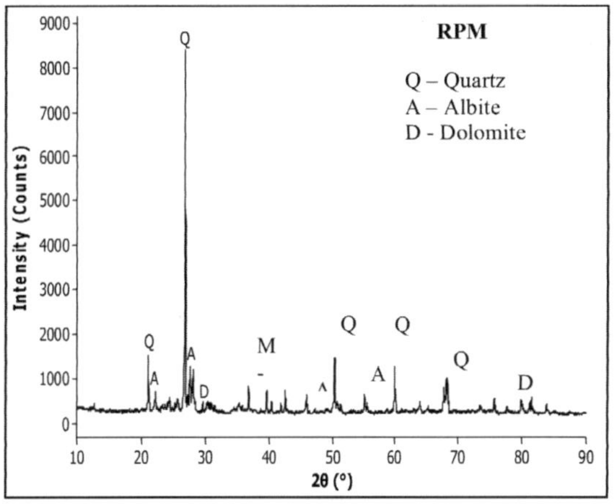

(b)

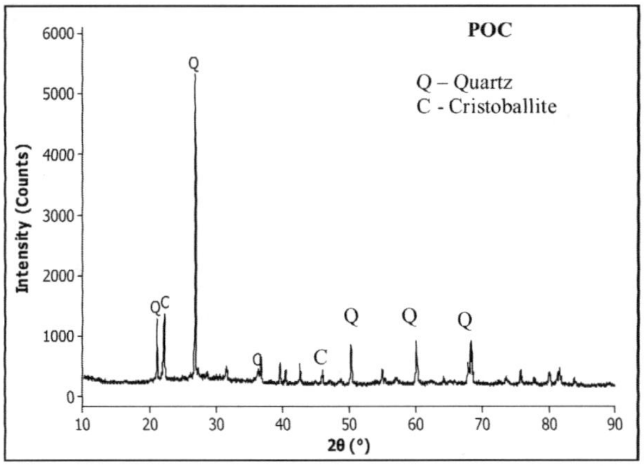

(c)

Figure 4. XRD patterns; (a) DMS; (b) RPM; (c) POC

3.2 Consolidation behaviour

A large-modified oedometer test setup was used to consolidate the DMS samples with 50 % of granular materials (RPM and POC) in single drainage condition. The compression curves, coefficient of consolidation (c_v), volume of compressibility (m_v) and permeability (k) were estimated at the incremental of vertical stresses as the variation of effective vertical stress (σ'_v) in kPa was plotted in figure 5, 6 and 7. The compression curves proven that DMS significantly change with the σ'_v where, the DMS more settle than RPM and POC layering. The water in soil dissipate much faster than DMS only because of the thickness of granular materials influence DMS to consolidate when the increment of σ'_v at each stage. As seen in the figure 6a, m_v of both materials at early stage shows the gap about 5%. However, an overlapping happens at σ'_v = 12.5 kPa to 800 kPa which means that both granular have large voids to let the water in DMS dissipate rapidly. Same goes to figure 7, where the trend in the same condition as m_v. It is noticeable that the k curves of RPM and POC have high k value than the DMS specimens with a k value < 10-9 m/s as impermeable in practice. As a result, the excess pore pressure of dissipation become more slowly because of low permeability of DMS and this will take a long period of time for the DMS to fully consolidate. Overall, the reduction of settlement happens largely with a layer of RPM and POC. This could be stiffening the DMS much faster than DMS alone.

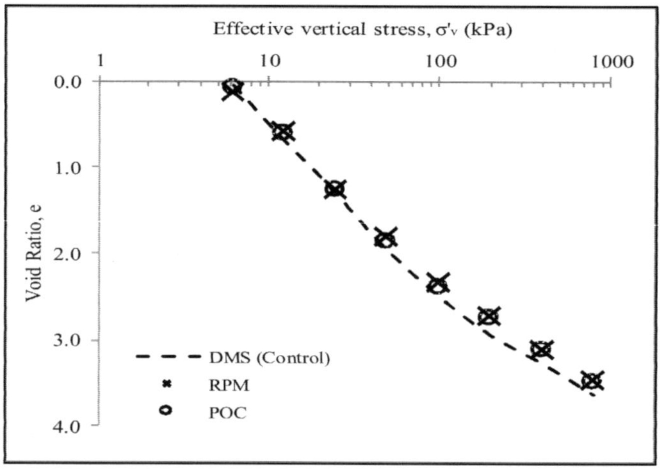

Figure 5. Compression curves

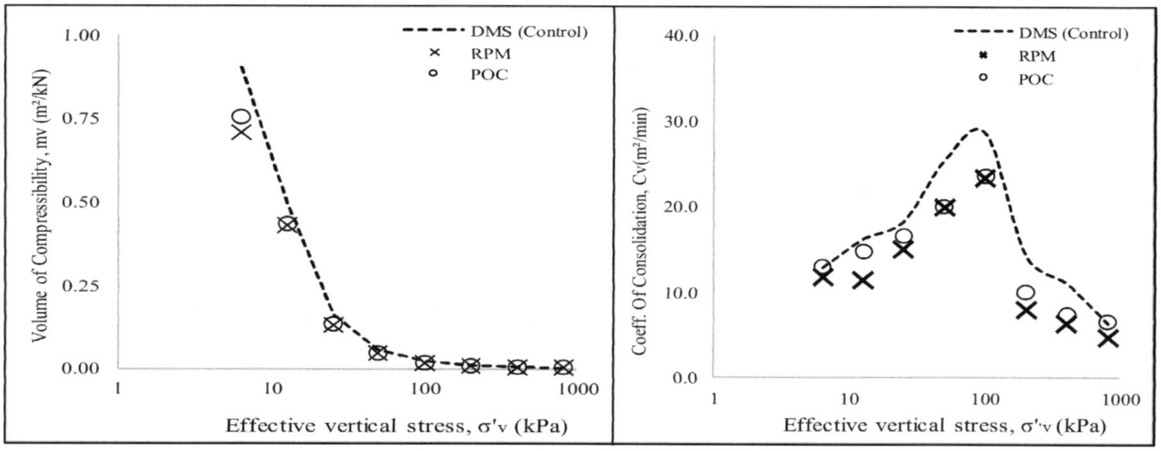

(a) (b)

Figure 6. (a) Volume of compressibility, m_v; (b) Coefficient of consolidation, c_v

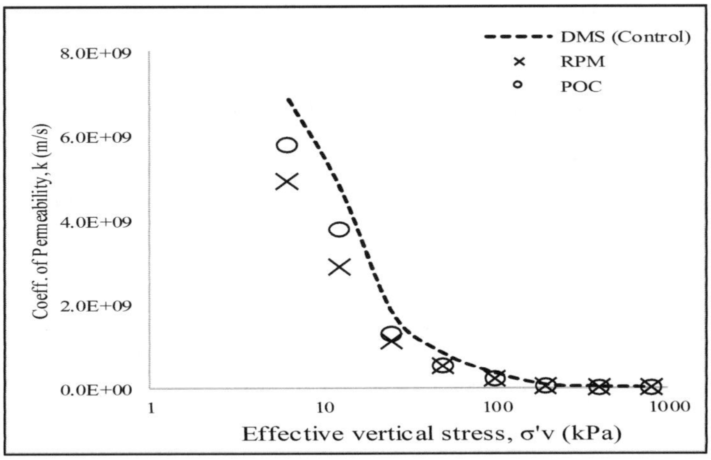

Figure 7. Coefficient of permeability

4. Conclusion

Dredged marine soils (DMS) are generally classified as waste materials in civil engineering applications or known as geo-waste because of poor engineering properties such as low in shear strength, permeability and high compressibility. DMS can be considered as a resource and reusable material for construction purpose unless contamination is found to be excessive. This study investigated the physical and chemical properties of DMS and recycled granular materials (RPM and POC). DMS in Kuala Perlis mainly clay with little bit of silt (silty clay). The proportion of clay about 61%, silt is 38% and fine sand only 1%. The ratio of liquid limit of DMS (wc/LL) is 3.27, higher than previous study at the same dredging site but different time. This indicate that space of time of sedimentation influenced the properties of DMS. all 3 materials majorly to have higher quartz components, which a type of silica oxide that found to be prominent. The water in layer RPM and POC dissipate much faster than DMS only because of the thickness of granular materials influence DMS to consolidate when the increment of σ'_v at each stage. The reduction of settlement happens which can be reuse as backfill materials in civil applications for future works. On other hand, the consumption usage of natural aggregates can be less and it became more sustainable for the environment.

5.0 References

[1] Manap N and Voulvoulis N 2015 Environmental management for dredging sediments- The requirement of developing nations *Journal of Environmental Management* **147** 338-348
[2] Berilgen S A, Kilic, H and Ozaydin, I K 2007 Determination of undrained shear strength for dredged golden horn marine clay with laboratory tests *Proceedings of the Sri Lankan Geotechnical Society's First International Conference on Soil and Rock Engineering* pp. 5-11.
[3] Wang D X, Abriak N E, Zentar R and Xu W 2012 Solidification/stabilization of dredged marine sediments for road construction *Journal of Environmental Technology* **33**(1) 95-101
[4] Bortone G and Palumbo L 2007 *Sustainable Management of Sediment Resources* 1st Ed. Elsevier.
[5] International Association of Dredging Companies (IADC) 2009 *Dredging Materials as A Resource.* Netherlands
[6] Siham K, Bernard F, Abriak NE 2008 Marine dredged sediments as new materials resource for road construction *Journal of Waste Management* **28**(5) 918-928
[7] Shahri Z and Chan C-M 2015 On the Characterization of Dredged Marine Soils from Malaysian Waters: Physical Properties, *Journal of Environment and Pollution,* Canadian Center of Science and Education **4** (3)

[8] Satoh T and Kitazume M 2003 Sea Reclamation with Dredged Soft Soil for Central Japan International Airport *Proceeding of Nakase Memorial Symposium, Yokosuka (Japan)* pp 311-318

[9] Ganesalingam D, Arulrajah A, Ameratunga J, Boyle P and Sivakugan N 2011 Geotechnical Properties of Reconstituted Dredged Mud, *Proceedings from the Pan-AM CGS Geotechnical Conference* pp 1-7

[10] Schlue B F, Kreiter S and Moerz, T 2009 Time-dependent deformation of dredged harbor mud used as backfilling material *Journal of Waterway, Port, Coastal, and Ocean Engineering* **135**(4) 154–163.

[11] Howard I L, Vahedifard F, Williams J M and Timpson C 2017 Geotextile tubes and beneficial reuse of dredged soil: applications near ports and harbours *Ground Improvement, Proceedings of the Institution of Civil Engineers (ICE)* pp. 1-14

[12] Kanadasan J and Razak H A 2014 Mix design for self-compacting palm oil clinker concrete based on particle packing *Materials Design* **56** 9-19.

[13] British Standard Institution (BSI) 1990 BS1377:1990 Methods of test for soils for civil engineering purposes (United Kingdom: British Standard Institution)

[14] Grubb D G, Wazne M, Jagupilla S C and Malasavage N E 2011 Beneficial use of steel slag fines to immobilize arsenite and arsenate: slag characterization and metal thresholding studies *Journal of Hazardous, Toxic and Radioactive Waste* **15** 130-150

[15] Salim W S W, Sadikon, S F, Mohd Salleh S, Mohd Noor NA, Arshad M F and Wahid N 2012 Assessment of physical properties and chemical composition of Kuala Perlis dredged marine sediment as a potential brick material *2012 IEEE Symposium on Business, Engineering and Industrial Applications (Bandung, Indonesia)* pp 188-191

[16] Kaliannan S 2016 *Light solidification of Kuala Perlis dredged marine soil via admixtures of GGBS-cement and sand: 1D compressibility study* Master Thesis (UTHM)

Acknowledgments
The authors would like to thank the relevant laboratories at the University for the technical assistance and support given throughout the project. Special funding by Soil Instruments (M) Sdn. Bhd that enables the paper to be included in GCoMSE2019 is greatly appreciated.

IOP Conf. Series: Materials Science and Engineering **713** (2020) 012016 doi:10.1088/1757-899X/713/1/012016

Study on The Effects of Plastic as Admixture on The Mechanical Properties of Cement-Sand Bricks

Ming Fai Chow[1*] and Muhammad Amin Khalili Rosidan[2]

[1]Institute of Sustainable Energy (ISE), Universiti Tenaga Nasional, Jalan IKRAM-Uniten, 43000 Kajang, Selangor Darul Ehsan, Malaysia
[2]Department of Civil Engineering, College of Engineering, Universiti Tenaga Nasional, Jalan IKRAM-Uniten, 43000 Kajang, Selangor Darul Ehsan, Malaysia
Corresponding author: Chowmf@uniten.edu.my

Abstract. Cement-sand bricks is one of the building materials that being used as alternative for a low-cost house in Malaysia. Nowadays, many researchers have been researching for green and sustainable materials to replace conventional building materials. In this paper, plastic bottle has been studied as a new alternative admixture in cement-sand brick production. The objective of this research is to determine the effect of plastic size on mechanical properties of cement-sand bricks. The lab testing used for this research are compressive strength and water absorption tests. Based on the findings, the bigger size of plastic added into the brick mixture, the higher compressive strength can be achieved. This is due to the plastic can increase the structural integrity of the cement-sand brick. Also, the bigger size of plastic added into the mixture, lower rate of water absorption was obtained. This may be due to plastic help in reducing the volume of void of the cement-sand brick. Further studies are needed to improve the cement-sand brick production with plastic as admixture material.

1. Introduction

Urbanization process associated with population growth has significantly increases the building construction and pollutant build-up at the catchment surface area [1 - 4]. Plastic pollution has been a very worrying issue over the years. For more than 50 years, global consumption and production of plastics have been gradually increasing. In 2013, about 299 million tons of plastics were produced, 4 percent more than 2012, and has shown significant growth over the years. By 2015, plastic production had reached 297.5 million tons in 2015 according to reports done by Global Industry Analysts. As majority of the plastic that been produced and consumed will end up in landfills and ocean every year, steps to minimize the plastic pollution are urgently needed. One of the solutions that can be done is by producing plastic waste-based building materials. Green materials have been highly promoted for reducing the pollution and enhancing the environmental quality at the urban city [5 - 9].

Nowadays, many modifications have been made in the procedure for manufacturing and composing the brick. Many waste materials such as tannery sludge [10], ceramic dust waste [11], waste marble sludge [12], construction and demolition waste [13], plastic bottle [5], agricultural solid waste [14] and degraded municipal solid waste [15] have been used to produce various type of bricks. Mansour et al. [5] investigated the use of waste plastic bottles as an alternative sustainable building material. The block of bottles filled with air showed slightly higher strength than the other types of bottle filling and has proved to have structural stability with a high factor of safety. The block of bottles filled with air can be used for construction as either a bearing wall for a roof slab or a partition (in a multi- story building) or with a factor of safety of 5.8 in one story building. The thermal properties for different type of infill materials have also been examined. The simulation results showed that plastic bottles filled with air construction model showed much better thermal insulation properties than the conventional block construction. Therefore, using plastic bottles as an infill building material, not only reduce their waste disposal problem, but is considered an acceptable thermal insulation material and can rise the environmental awareness.

In today's era, brick is considered one of the building materials sought after for use in the construction of various types of building structures. However, depending on the brick composition and the manufacturing procedure either through molding, combustion, and other processes its properties is been observed especially in its compressive strength. Furthermore, the brick has its own class and type in various international standard codes depending on the severity of environmental condition and the importance of structures. Thus, a comprehensive study of brick composition and properties and various factors related to the production process is indispensable for standardization. To enable a variety of industry, mining solid municipal waste in the brick industry, a better understanding of the various waste materials as brick composition will help in achieving sustainable development goals [8]. Therefore, this research is aimed to determine the compressive strength and water absorption rate of brick samples mixed with various sizes of plastic sheet.

2. Methodology

In this study, the mechanical properties of cement-sand brick with plastic as its admixture is investigated. The experiment or lab testing used for this research are compressive strength test to find out the sample compressive strength and water absorption test to find out the water absorption rate. There are 36 sample have been prepared with different mix ratio and plastic size to compare the findings for all samples. The materials used for this research are cement, aggregates, water and plastic bottle that being cut into different sizes. For aggregates, the maximum size needed is 19 mm and 9.75 mm for minimum size. For sand, the maximum size needed is 2 mm and minimum size of 0.0625mm. All samples have same dimension, but different sizes of plastic used as admixture as shown in Table 1. The plastic bottle is cut into size of 2x2 cm, 3x3 cm and 4x4 cm. There are 4 batch mixes that consist of control mix (no plastic), 2x2 cm mix, 3x3 cm mix and 4x4 cm mix that produce about 9 samples for each mix. Mix ratio that used in this study is 1:6 (cement: sand) for control sample and 1:5.5:0.5 (cement: sand: plastic). The materials that have been mixed are put into brick mold (210mm x 92 mm x 56 mm) to shape it into rectangular brick sample. After done with molding, the brick sample is let it air dry for 24 hours. After being dried for 24 hours, the brick samples then undergo curing process for 7, 14 and 28 days. After that, all samples undergo compressive strength test whereby the maximum amount of compressive load a material can bear before fracturing is measured and water absorption test whereby the weighed of specimens after submerged in water is weighed and compared. These tests were conducted according to the American Society for Testing and Materials (ASTM) Standards. Equation (1) is used to determine the water absorption rate for the brick samples.

Table 1. Mix Design Ratio

Sample/ Plastic Size	Volume	Cement	Sand	Plastic Bottle	Water
	m^3	kg	kg	kg	kg
Control	0.0011	2.55	15.3	-	1.275
2x2 cm	0.0011	2.55	14.0	1.275	1.275
3x3 cm	0.0011	2.55	14.0	1.275	1.275
4x4 cm	0.0011	2.55	14.0	1.275	1.275

$$\textit{Percentage of water adsorption (\%)} = \frac{W_2 - W_1}{W_1} \times 100\% \quad (1)$$

Where: W_1 = *Wet weight of the sample (kg)*
W_2 = *Dry weight of the sample (kg)*

3. Results and discussions

Figures 1(a) – (d) show the results for maximum load of the control cement-sand brick samples and brick samples with different plastic size admixtures. The value of maximum load applied is the highest value of load that sample brick can sustain before failed or cracked. The value of compressive strength is obtained by dividing the maximum applied load with the area of sample brick. The compressive strength for brick samples with different plastic sizes for 28 days curing time is shown in Figure 2. The compressive strength of brick samples is increasing except for brick sample with 3x3 cm plastic. The compressive strength of sample bricks being added with 4x4 cm plastic is higher than the control sample and higher than the standard value of compressive strength of bricks that being used in construction which is 3.5 N/mm^2.

(a)
(b)
(c)

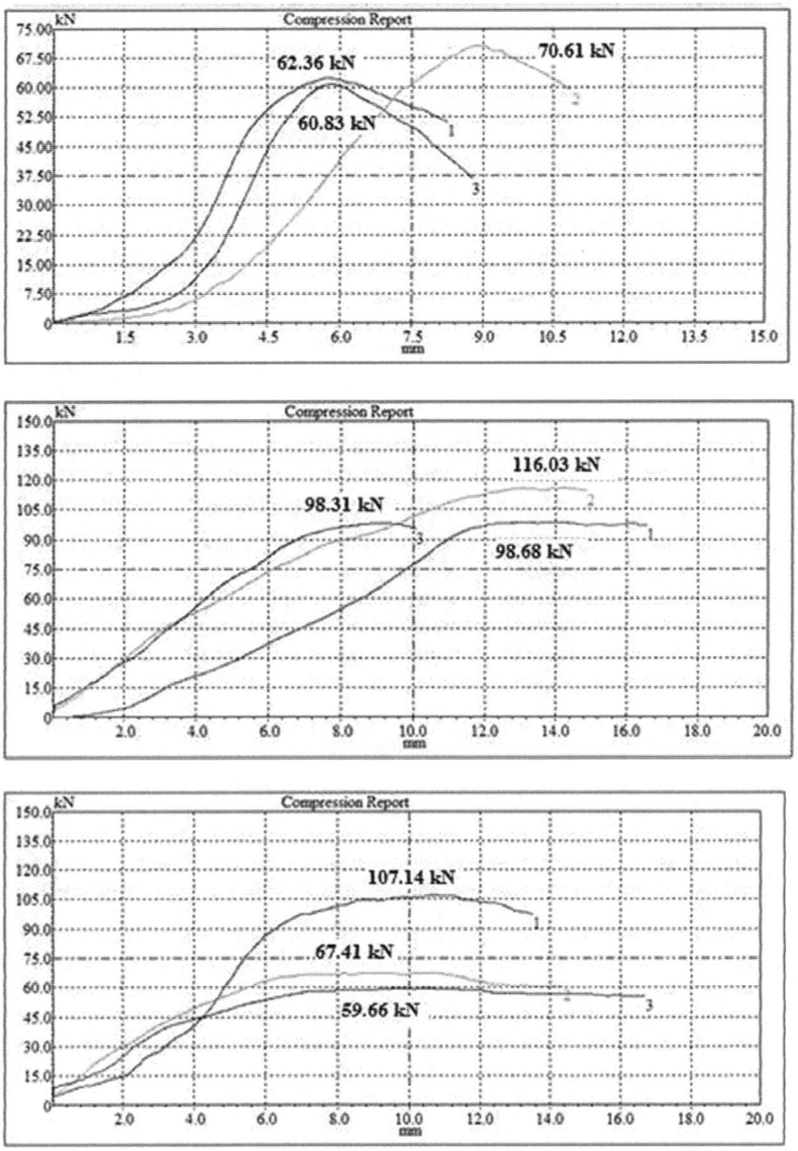

(d)

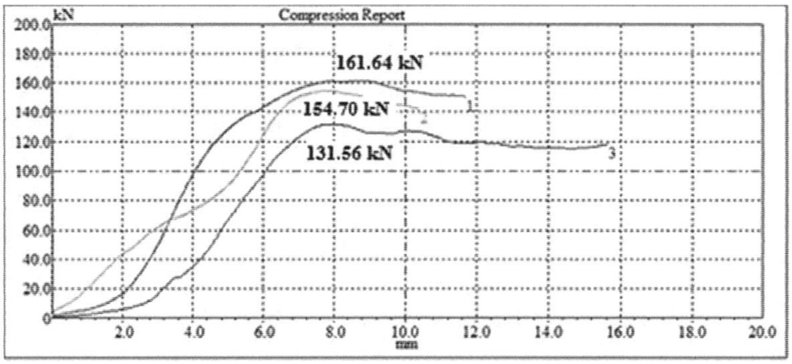

Figure 1. The maximum loads applied for (a) control samples, (b) samples with 2x2 cm plastic sheet, (c) samples with 3x3 cm plastic sheet, and (d) samples with 4x4 cm plastic sheet,

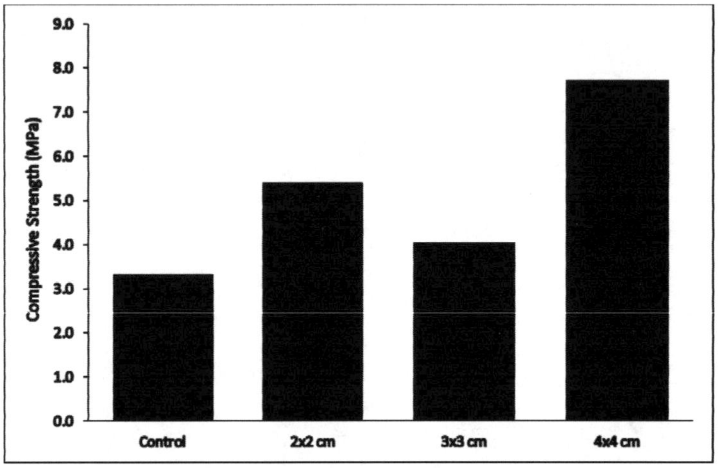

Figure 2. Compressive strength for brick samples with different plastic sizes for 28 days curing time

According to the standard, the percentage of water absorption should not exceed 20% of the original weight. From the results in the Table 2, all sample bricks have percentage of water absorption value less than 20% of its original weight with the highest value of 4.95% for brick sample consists of 3x3 cm of plastic and lowest value of 3.54% for brick sample consists of 2x2 cm of plastic.

Table 2. Water absorption test result

Sample	Weight of specimen during dry as W_1 (g)	Weight of specimen during wet as W_2 (g)	Percentage of water (%)
Control	2046.02	2129.82	4.10
2x2 cm	2231.57	2310.50	3.54
3x3 cm	2016.41	2116.20	4.95
4x4 cm	2098.34	2178.77	3.83

Based on the compressive strength test, we can observe that the size of plastic added into mixture can affect the characteristics of the brick. The bigger size of plastic added into the mixture, the higher compressive strength it can achieve. This is because the plastic added as mixture helps in increasing the structural strength of the bricks. From the water absorption test, it can be seen that the bigger size of plastic added into the mixture, the lower water absorption rate obtained as the plastic reduces the volume of voids in the concrete. It also found that the brick samples become lighter compared to the conventional cement-sand bricks as the plastic reduces cement-sand ingredients quantity.

4. Conclusion

In this study, plastic bottle has been used as a new alternative admixture in cement-sand brick production. The objective of this research is to determine the effect of plastic size on mechanical properties of cement-sand bricks. Based on the findings, the bigger size of plastic added into the brick mixture, the higher compressive strength can be achieved. This is due to the plastic can increase the structural integrity of the cement-sand brick. Also, the bigger size of plastic added into the mixture, lower rate of water absorption was obtained. This may be due to plastic help in reducing the volume of void of the cement-sand brick. Further studies are needed to improve the cement-sand brick production with plastic as admixture material.

5. References

[1] Chow M F and Bakar M F A 2019 Environmental benefits of green roof to the sustainable urban development: A review *Lecture Notes in Civil Engineering* (Switzerland: Springer Nature) **9** pp 1525-1541
[2] Haris H, Chow M F, Usman F, Sidek L M, Roseli Z A and Norlida M D 2016 Urban Stormwater Management Model and Tools for Designing Stormwater Management of Green Infrastructure Practices *IOP Conference Series: Earth and Environmental Science* vol 32(1) (United Kingdom : IOP Publishing) 012022
[3] Chow M F, Yusop Z and Toriman M E 2013 Level and transport pattern of faecal coliform bacteria from tropical urban catchments *Water Science Technology* **67**(8) 1822-1831
[4] Chow M F, Yusop Z and Abustan I 2015 Relationship between sediment build-up characteristics and antecedent dry days on different urban road surfaces in Malaysia. *Urban Water J.* **12**(3) 240-247
[5] Mansour A, Mansour H and Ali S A 2015 Reusing waste plastic bottles as an alternative sustainable building material. *Energy for Sustainable Development* **24** 79–85
[6] Chow M F, Abu Bakar M F, Sidek L M and Basri H 2017 Effects of substrate types on runoff retention performance within the extensive green roofs *J. Eng. and App. Sci.* **12**(21) 5379-5383
[7] Chow M F, Bakar M F A and Wong J K 2018 An overview of plant species and substrate materials or green roof system in tropical climate urban environment. *AIP Conference Proceedings* (USA: AIP Publishing) **2030** 020004
[8] Murmu A L and Patel A 2018 Towards sustainable bricks production: An overview *Constr. Build. Mater* **165** 112–125
[9] Chow M F and Yusop Z 2019 Contributions of dry and wet weather runoffs to annual pollutant loading in tropical urban catchments *Lecture Notes in Civil Engineering* (Switzerland. Springer Nature) **9** pp 1511-1521
[10] Juel M A I, Mizan A and Ahmed T 2017 Sustainable use of tannery sludge in brick manufacturing in Bangladesh *Waste Management* **60** 259–269
[11] Amin S K, El-Sherbiny S A, El-Magd A A M A, Belal A and Abadir M F 2017 Fabrication of geopolymer bricks using ceramic dust waste *Constr. Build. Mater.* **157** 610– 620
[12] Munir M J, Kazmi S M S, Wu Y F, Hanif A and Khan M U A 2018 Thermally efficient fired clay bricks incorporating waste marble sludge: An industrial-scale study *Journal of Cleaner Production* **174** 1122–1135
[13] Seco A, Omer J, Marcelino S, Espuelas S and Prieto E 2018 Sustainable unfired bricks manufacturing from construction and demolition wastes *Constr. Build. Mater.* **167** 154–165
[14] Kizinievi O, Kizinievi V, Pundiene I and Molotokas D 2018 Eco-friendly fired clay brick manufactured with agricultural solid waste *Archives of Civil and Mechanical Engineering* **8** 0–9
[15] Goel G and Kalamdhad A S 2017 Degraded municipal solid waste as partial substitute for manufacturing fired bricks *Constr. Build. Mater.* **155** 259–266

Acknowledgments
The authors would like to thank the Chair of Energy Economics grant (Project No. 2018003KETST) provided by Energy Commission Malaysia and Tenaga Nasional berhad (TNB) and supports provided by Innovation & Research Management Center (iRMC) Universiti Tenaga Nasional.

IOP Conf. Series: Materials Science and Engineering **713** (2020) 012017 doi:10.1088/1757-899X/713/1/012017

Influence of Dolomite and Granite Waste Content on The Properties of Artificial Marble

H Abdul Samad[1] and R Abd Rashid[1]

[1] Mineral Research Centre, Department of Mineral and Geoscience Malaysia, Jalan Sultan Azlan Shah, 31400 Ipoh, Perak, Malaysia
Corresponding author: hamizah@jmg.gov.my

Abstract: The main objective of this work is to characterize and investigate the effect of mineral dolomite and granite waste content on the physical and mechanical properties of artificial marble. Five different samples of artificial marble were prepared with different proportions of fillers. In this study, granite waste is used a filler by substitute the dolomite content from 10 wt% to 40 wt%. Test specimens were evaluated for water absorption analysis, flexural, compression and Barcol hardness properties. It is found that the optimum 10 wt% of granite waste in the composition has increase the flexural and compression strength and reduces the water absorption rate, while the substitution of granite waste does not change significantly the Barcol hardness properties.

1. Introduction

Artificial marble also known as synthetic marble or cultured marble and some of called as marble composite. It is formed from a mixture of polymer resin, mineral fillers, catalyst and pigments. Generally, natural marble is highly porous, permeable and have higher tendency to cracking, chipping and discoloring. Extra process and high maintenance are thus required to prevent such problems. Furthermore, the complexity of the processes such as quarrying, slab cutting and transporting has resulted in increasing the production cost of the natural marble. To overcome these difficulties, artificial marble is produced artificially and the usage is increasing due to depleting supply of natural marble.

Artificial marble is a versatile material created in various natural color, design and patterns which mimic the look of natural marble, granite and quartz. Various techniques like casting, isostatic pressing, and dry casting are used to cast the artificial marble. In artificial marble composition, filler is used as an aggregate while resin is acts like glue hence strengthening the marble. Artificial marble has a wide range of application includes as floor and wall tiles, solid surface kitchen countertops, sinks, bathtubs, bathroom vanities and etc. In most of the artificial marble manufacturing industries, imported alumina tri-hydrate (ATH) filler, limestone or marble dust/powder/particles are commonly used as the main raw material.

In this study, dolomite is introduced as the main filler to substitute the limestone and marble dust/powder/particles. Dolomite is a sedimentary carbonate rock mineral which contains of calcium and magnesium carbonate with chemical composition of $CaMg(CO_3)_2$. In Malaysia, dolomite is the most abundant carbonate mineral after limestone which mainly used as an aggregate for road construction and soil fertilizer. Dolomite has a significant potential to be explored as it shares similar properties with limestone. Therefore, the application of dolomite in many industries should be studied and carried out to increase the diversity.

Besides dolomite, granite aggregates is the main mineral resources available in the country. Granite was quarried as a dimension stone; crushed stone or aggregate for road construction, railroad foundations and for building applications likes paving, monuments, countertops and floor tile. The used of granite waste as filler in the production of artificial marble is being studied in the present work due to the high production of crushed granite that generates a dumping of waste. A million tonnes of waste from quarry industries are being released from crushed activity, cutting, processing and grinding. In dry season, the granite powder or dust dangles in the air, flies and deposits on vegetation and crop. All these conditions are significantly affecting the environment and local ecosystems. By considering the existing available of local mineral resources and quarry industrial waste, this study offers an alternative material in manufacturing of artificial marble by combination of dolomite and granite waste as a filler in the matrix of polyester resin. Hence, the objective of this work is to characterize and investigate an influence of both materials on the properties of artificial marble.

2. Materials and methods

2.1 Materials

The local dolomite was supplied by dolomite quarry from Perlis, while granite waste was collected in Keramat Pulai, Perak. Dolomite and granite waste were grounded into powder form using planetary ball milling with a speed of 250 rpm for 15 minutes for mineral characterization. Both raw materials have been characterized using X-ray diffractometer (XRD) (D8 Advanced, Bruker, German) and X-ray fluorescence (XRF) spectrometer (Shimadzu XRF-1700). In this study, polyester resin is used as the polymer matrix and methyl ethyl ketone peroxide (MEKP) was added to be functioned as hardener with the amount of 1 wt% from resin content. Figure 1 shows the images of dolomite and granite waste used in this study.

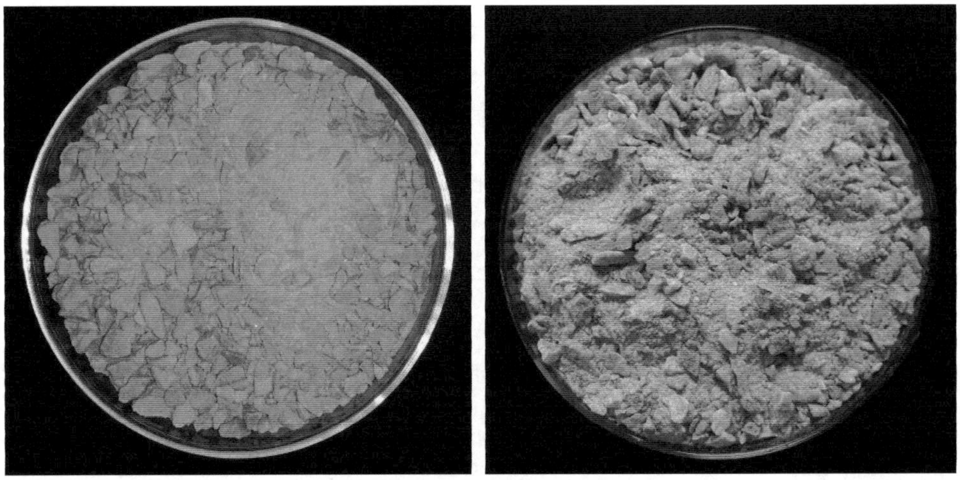

Figure 1. (a) dolomite (b) granite waste.

2.2 Preparation of artificial marble

The origin size of dolomite used in the preparation of artificial marble is 2.00 mm while granite waste contained of powder and dust have a maximum aggregate size of 2.36 mm. The mixture of polyester resin and fillers were well mixed at 200 rpm using a planetary mixer about 10 to 15 minutes and stirred until the fillers is evenly dispersed and wetted by the resin. Then, MEKP is added to initiate the polymerization reaction and the mixture was cast into the mold and left for curing for a few hours or until fully set. A total of 5 samples have been prepared from the base ratio of 70:30 (filler:matrix resin) in weight %. Table 1 shows the mix proportions of artificial marble.

Table 1. Mix proportions of artificial marble.

No	Sample	Resin (% wt)	Fillers	
			Dolomite (% wt)	Granite waste (% wt)
1	D70 (control)	30	70	0
2	D60GW10	30	60	10
3	D50GW20	30	50	20
4	D40GW30	30	40	30
5	D30GW40	30	30	40

** Notes: D=Dolomite; GW=Granite waste

2.3 Characterization of artificial marble

Sample of artificial marble was set for physical and mechanical analysis. The water absorption test was conducted according to ASTM D785. A square test specimen with dimension size of 12.7 mm x 12.7 mm x 12.7 mm was prepared. The weight of the specimen is measured in air (W_1) and then the specimen was immersed in distilled water for 24 hours. On removal from water, the specimen need to wipe off and specimen weighed again (W_2). Water absorption was calculated as below equation:

$$\text{Water absorption (\%)} = \left(\frac{W_2 - W_1}{W_1}\right) \times 100 \quad (1)$$

Flexural test of the cured marble was performed according to ASTM D790. In order to determine the flexural strength, the dimension of 25.4 mm x 12.7 mm x 8 mm rectangular bars specimens was prepared. The test was conducted under 3-point loading at a cross-head speed of 0.2 mm/min and the support span length is 20 mm. The compression test also was performed according to ASTM D695. The dimension size of marble specimen is 12.7 mm x 12.7 mm x 12.7 mm. Both flexural and compression test are performed using the Instron Universal Testing Machine (UTM, Model 3366, Instron).

Barcol hardness test is performed using a portable Barcol Hardness Impressor in order to determine the hardness by indentation of material through the depth of penetration on the indenter. The dimension size of marble specimen is 50 mm x 50 mm x 8 mm as recommended in ASTM D258. Data is expressed as Barcol numbers given by the instrument which the reading on a 0 to 100 scale. A 10 readings of measurement were indicates and the average value was calculated.

3. Results and discussion

3.1 Characterization of raw materials

Figure 2 shows the XRD pattern of dolomite and granite waste. The main peak observed in the dolomite sample corresponds only to mineral dolomite. While, XRD analysis on granite waste reveals the presence of mineral quartz, albite, microcline and calcite as the main mineral. The chemical composition of the dolomite and granite waste is listed in Table 2. The main composition of dolomite is calcium oxide (CaO) and magnesium oxides (MgO) with the contents are 30.28% and 20.08%, respectively. While, granite waste shows the main oxide presented are silicon dioxide (SiO_2) and alumina oxide (Al_2O_3) with the composition are 75.44% and 12.47%, respectively.

The carbonate content of dolomite is fundamental in determining the chemical purity of the resources and is the basis of chemical grade dolomite classification. Table 3 is listed the types of carbonate rock based on calcium carbonate ($CaCO_3$) content. The value of $CaCO_3$ can be determined as the following calculation:

$$\text{\% CaO x Conversion Factor} = \text{\% } CaCO_3\text{; conversion factor is 1.78} \quad (2)$$

IOP Conf. Series: Materials Science and Engineering **713** (2020) 012017 doi:10.1088/1757-899X/713/1/012017

From the calculation, the obtained value of $CaCO_3$ in dolomite sample is 53.89% and classified as pure dolomite rock. This result is in agreement with the XRD analysis which the main phase found only dolomite mineral. From this analysis and based on the calcium carbonate content, dolomite is found to have a potential to replace partially the properties of limestone and marble and granite waste is combined with dolomite to increase the physical and mechanical properties of artificial marble.

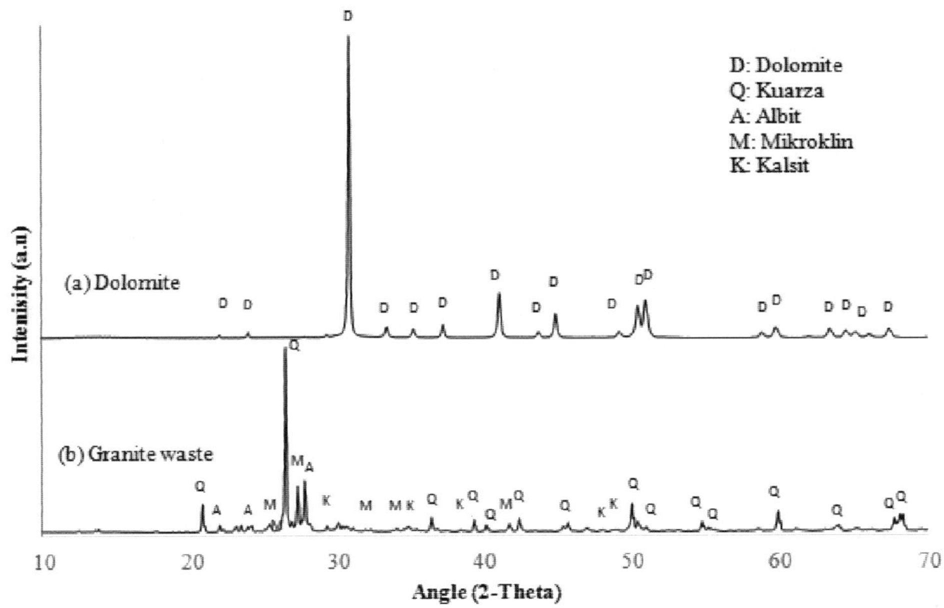

Figure 2. XRD pattern of (a) dolomite (b) granite waste.

Table 2. Chemical composition of dolomite and granite waste.

Sample	Oxides (%)					
	SiO_2	Al_2O_3	CaO	MgO	Fe_2O_3	K_2O
Dolomite	0.63	0.20	30.28	20.08	0.24	0.03
Granite Waste	75.44	12.47	1.21	-	1.27	6.18
Sample	Oxides (%)					LOI (%)
	Na_2O	TiO_2	MnO	Cr_2O_3	ZrO_2	47.36
Dolomite	1.10	0.02	0.02	0.04	-	47.36
Granite Waste	3.19	0.14	-	-	0.10	-

Table 3. Types of carbonate rock based $CaCO_3$ content.

$CaCO_3$ content (%)	**Types of carbonate rock**
< 58.85	Dolomite
58.85-77.14	Calcitic Dolomite
77.14-95.43	Dolomitic Limestone
95.43-97.71	Magnesian Limestone
>97.71	Limestone

IOP Conf. Series: Materials Science and Engineering **713** (2020) 012017 doi:10.1088/1757-899X/713/1/012017

3.2 Characterization of artificial marble

Figure 3 shows the marble specimens in water bath during water absorption test while Figure 4 represents the result of water absorption versus composition of artificial marble with different filler of dolomite and granite waste contents. The water absorption was varies between 0.270% to 0.328% and the value water absorption does not change significantly between control sample and the marble sample content of granite waste at 10 wt% (D60GW10). As can be seen, there is generally a slight increase in water absorption with replacement of granite waste content from 20 wt% to 40 wt%. It might be due to the non-homogeneous particle size of granite waste which formed in the powder, fine dust and large particles. As the granite waste content increased, the powder and fine dust in the mixture could contributed to increase the viscosity of mixture and lead to produce a pores. However, the value found in the present work is acceptable as literature reported the range of water absorption for artificial marble in industries is from 0.09% to 0.40%.

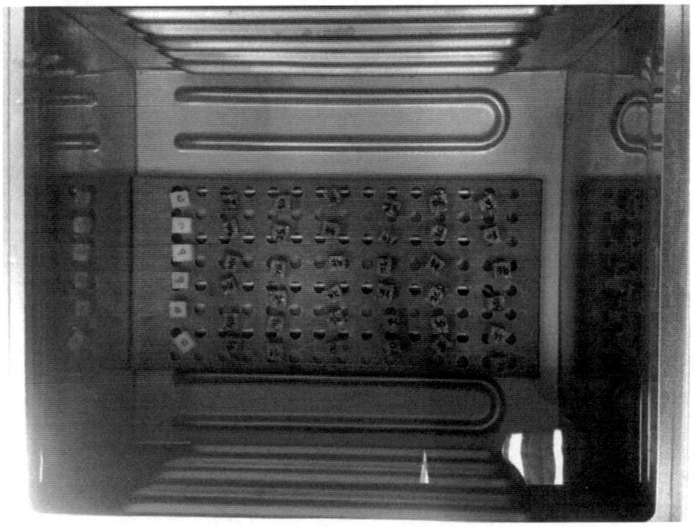

Figure 3. Marble specimens soak in water bath for water absorption test (from top view).

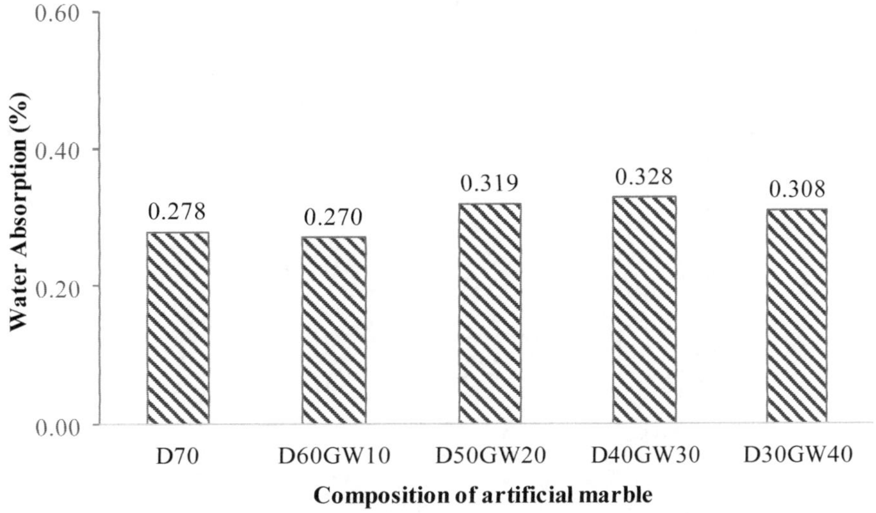

Figure 4. Water absorption of artificial marble.

Figure 5 shows the result for flexural strength versus composition of artificial marble with different filler of dolomite and granite waste contents. It was seen that flexural strength increase with replacement of granite waste by 20 wt% and among all the samples, D60GW10 was found to be the highest strength. This was believed due to high hardness of granite waste that contained of mineral quartz in the composition. Unexpectedly, artificial marble decreased with increasing granite waste content until it reached a maximum value at 40 wt%. The decrease of flexural strength can be explained by powder and fine dust of granite waste particles, which exhibited a higher surface area and subsequently resulting agglomerates. An agglomeration increases the average particles size by the combination of powder particles and acted as stress concentrations or weak point that breaks when stress is applied, hence decreasing the flexural strength.

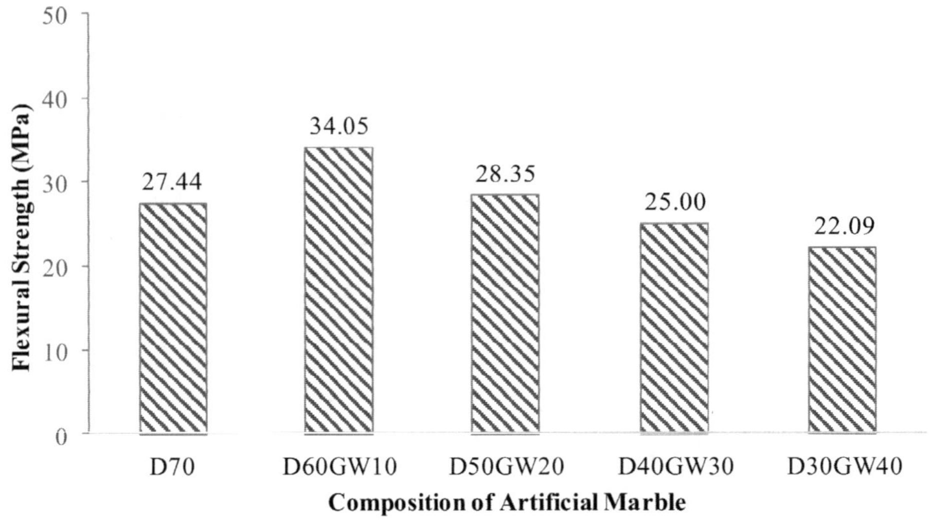

Figure 5. Flexural strength of artificial marble.

Figure 6 shows the compression test images of artificial marble during testing and Figure 7 shows the result for compression strength versus composition of artificial marble with different filler of dolomite and granite waste content. The compression strength value of marble samples containing of granite waste was determined between 83.11 MPa and 89.93 MPa. These values were lower than the value of the control sample, which was 93.65 MPa. The decrease may be attributed to the increasing viscosity of the mixture which caused the increase of pores formation during the casting. This subsequently introduced a weak point in the sample, thereby decreasing the compression strength. This result can be supported by the increase of water absorption rate composed of granite waste from 20 wt% to 40 wt% as observed in Figure 4.

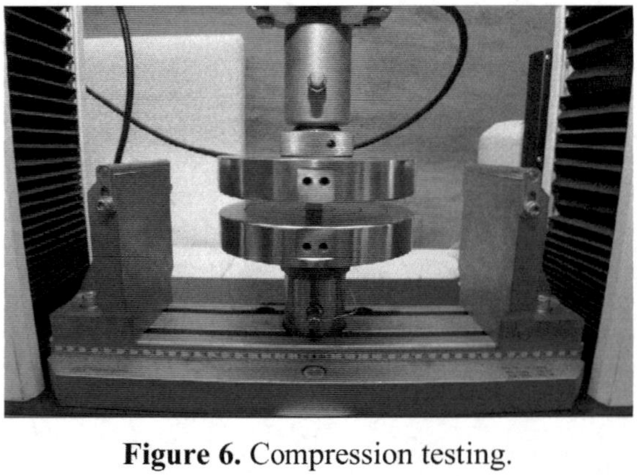

Figure 6. Compression testing.

IOP Conf. Series: Materials Science and Engineering **713** (2020) 012017 doi:10.1088/1757-899X/713/1/012017

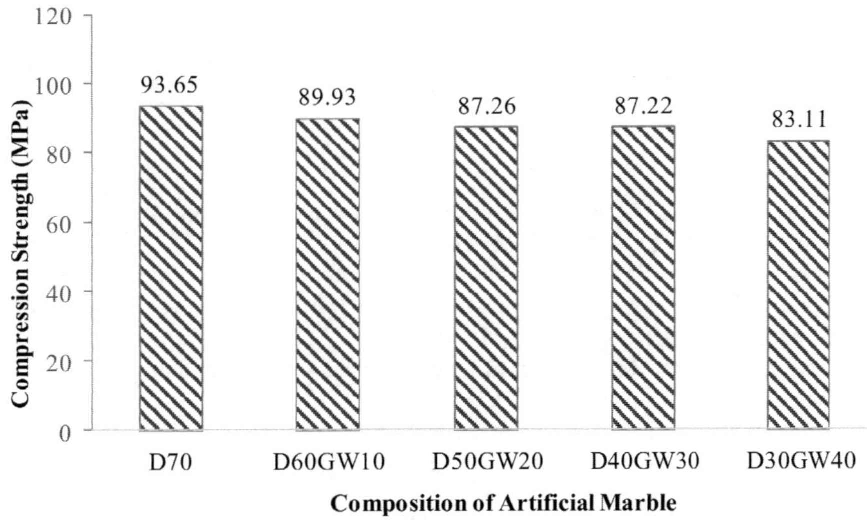

Figure 7. Compression strength of artificial marble.

Figure 8 shows the Barcol hardness result of the artificial marble versus composition of artificial marble with different filler of dolomite and granite waste content. From the figure, it is found that Barcol hardness slightly decreases with the replacement of granite waste until maximum at 40 wt%. Apparently, that replacement of granite waste does not change significantly the Barcol hardness properties. However, the value of Barcol hardness found in this study is in the range of the commercial value measured in industries, which was between 45 HBa to 65 HBa.

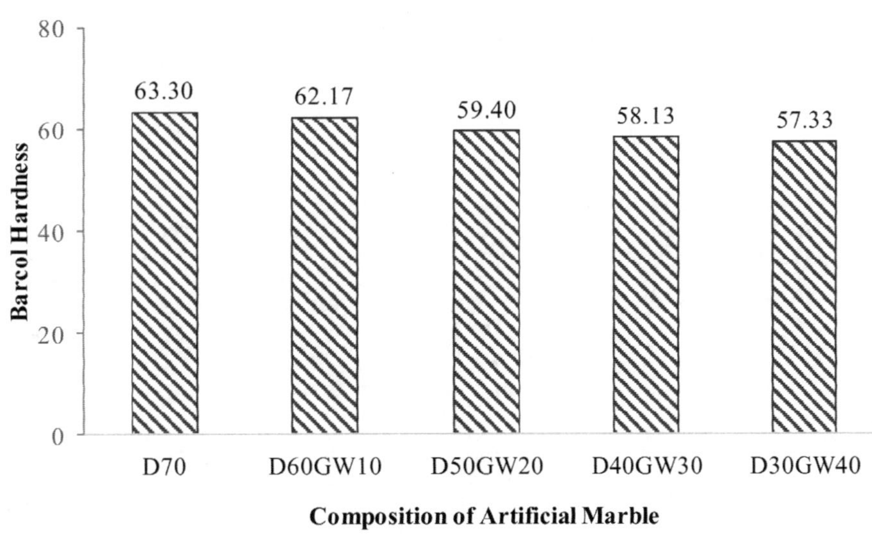

Figure 8. Barcol hardness of artificial marble.

4. Conclusion

The present study focusing on improving the physical and mechanical properties of artificial marble composed with dolomite and granite waste filler. The test result shows that, the water absorption value was increases and the flexural and compression strength slightly decreases as the content of granite waste replacement increases to the maximum at 40 wt%. The decreases of the mechanical properties can be related to the powder and fine particle in granite waste that contribute to the increase of mixture viscosity. However, the optimum 10 wt% of granite waste replacement gives better properties which demonstrated a low water absorption, and higher flexural and compressive strength. Dolomite and granite waste seems have a good potential to utilize as filler in the manufacturing of artificial marble. However, some recommendations for the future studies are suggested such as morphology analysis and method of development by using a vacuum mixer to reduce the existing pores.

5. References

[1] Bera P, Guptha N, Dasan K P and Natarajan R 2012 Recent developments in synthetic marble processing *Rev. Adv. Mater. Sci* **32** 94-105

[2] Bilgin A, Yuceturk G, Arslan S, Bilgin N and Polat S 2013 The effects of mineralogical and petrographical features of The Lake District Rocks (Isparta, Turkey) on the quality of artificial marbles, *Epitoanyag* **65** 21-25

[3] Aruniit A and Tall K 2011 Infuence of filler proportion on mechanical and physical properties of particulate composite *Agron. Res. Biosyst. Eng.* **1** 23-29

[4] Lee M Y, Ko C H, Chang F C, Lo S L, Lin J D, Shan M Y and Lee J C 2008 artificial stone slab production using waste glass, stone fragments and vacuum vibratory compaction *Cem. Concr. Compos.* **30** 583-587

[5] Gokdai D, Borazam A A and Acikbas G 2017 Effect of marble: Pine cone waste ratios on mechanical properties of polyester matrix composites *Waste Bimass Valor* **8** 1855-186

[6] Fernanda S S, Carlos E G R, Ruben J S R 2018 Physical and mechanical characterization of artificial stone with marble calcite waste and epoxy resin *Mat. Res.* **1** 1-5

[7] Hamdan A and Fakharuddin A 2016 *Penilaian Sumber Dolomit Negeri Perlis Indera Kayangan, Jabatan Mineral Dan Geosains Malaysia* 2016 (No. Laporan JMG.KPP (MPI) 01/2016)

[8] Carlos E G R, Ruben J S R, Carlos S C and S N M 2014 Production of synthetic ornamental marble as a marble waste added polyester composite *Materials Science Forum* **775-776** 341-345

[9] Borsellino C, Calabrese L and Bella G D 1921 Effects of powder concentration and type of resin on the performance of marble composite structures *Constr. Build. Mater.* **23** 1915-1921

Acknowledgement

The authors would like to thank dolomite and granite quarry in Perlis and Perak, respectively for providing the raw materials. Authors are also gratefully acknowledged Director of Mineral Research Centre (MRC) and all staffs from Advanced Materials Technology Section for their tireless and valuable assistance to complete this study.

IOP Conf. Series: Materials Science and Engineering **713** (2020) 012018 doi:10.1088/1757-899X/713/1/012018

Comparative Evaluation on the MOE between EN, BS and ASTM of Concrete Containing PET

Muhammad Amzar Othman[1], Shahiron Shahidan[1]*, Shamrul Mar Shamsudin[1], Noorwirdawati Ali[1],Sharifah Salwa Mohd Zuki[1] Faisal Sheikh Khalid[1], Fadzli Mohamed Nazri[2]

[1] Faculty of Civil and Environmental Engineering, Universiti Tun Hussein Onn Malaysia, 86400, Parit Raja, Batu Pahat, Johor, Malaysia
[2] School of Civil Engineering, Engineering Campus, Universiti Sains Malaysia, 14300 Nibong Tebal, Penang, Malaysia
Corresponding author: shahironshahidan@gmail.com; m.amzar10@gmail.com

Abstract. The wide use of plastic bottles has led to a drastic increase in the disposal of plastic bottles. Reusing waste plastic bottles in concrete may be beneficial to the environment and help ensure economic viability at the same time. Previous researchers had investigated the application of waste materials in the construction field by using polyethylene terephthalate as one of the solutions to address this environment threat. In this research, the aim of the study is to investigate the modulus of elasticity of concrete material incorporated with polyethylene terephthalate. To achieve the objective of the study, the modulus of elasticity test was conducted to determine the Modulus of Elasticity (MOE). A comparison of MOE between Eurocodes (EC), British Standard (BS) and American Society for Testing and Materials (ASTM) Standards was conducted. At the end of this study, the value of modulus of elasticity resulting from the plastic bottle concrete material was expected to be classified as the same category as common concrete. Besides, the plastic bottle concrete material was also expected to possess similar properties as other conventional materials. Optimistically, the results of this research resulting from the MOE concrete plastic bottles that are not classified as a similar category as conventional concrete recorded a difference of 4.26% to 59.29%. It can be concluded that shred plastic bottles have a lower feasibility and do not have the potential to be used in concrete mixes.

1. Introduction

In this era of modernization, the advancement of science and technology is increasingly vibrant with the emergence of new research. Polymer-based technology is increasing with recent researches resulting in the production of high quality plastic materials. Although the plastic-based products are increasingly on the market and widely used, the use of plastic materials has led to excessive plastic quantities and it still no effective solution to eliminating this waste material without polluting the environment. Therefore, there are several other alternatives to reuse this wasted material and to avoid further contaminating the environment.

Plastic bottles are used more commonly known as 'Polyethylene terephthalate or PET bottles' are increasingly used in construction work. PET bottles are also suitable for use with other beverages and durable [1]. Various waste materials can be recycled such as tires, plastic, waste residues and so on. Removal of plastic bottles using waste incinerators cause environmental pollution with the release of carbon gases derived from plastic bottle burning. In addition, there is the same process of pyrolysis, in which plastic bottles or plastic materials are processed into re-oil. Both combustion and pyrolysis processes are more expensive than the recycling and uneconomical processes.

This report will focus on the use of plastic bottles to be recycled and used in concrete for the construction work. Reuse of waste plastic bottles in the concrete may have two environmental advantages and at the same time ensure economic viability. In this case, an attempt can be made to control environmental pollution and to save natural resources by using plastic bottles disposed in cement concrete [2]. Further testing and investigations are needed to determine the suitability used for concrete production in the construction industry. Therefore, this study relates to sustainable development due to the use of waste, plastic bottles to produce concrete in construction. Therefore, waste plastic bottles are recycled to replace the fine aggregate into the concrete. With this study, concrete plastic bottles can be compared with that of normal concrete in terms of strength and modulus of elasticity. As a result of the study, it is to decide that concrete plastic bottles are suitable for use in a construction.

2. Material and Procedure

2.1. Material preparation
In this research, the materials used included Ordinary Portland Cement Type 1 (OPC: Type 1) which was based on MS EN 197-1:2014 with a grade of 42.5 N, Fine Aggregates (FA) size of (0.075-5) mm, Coarse Aggregates (CA) size of (5-20) mm, Superplasticizers (SP), water and recycled PET fibers with 50 mm long and 5 mm width.

2.2. PET fiber preparation
Before being cut into fiber form, polyethylene terephthalate (PET) plastic bottles were collected and cleaned. The recycled PET plastic bottles were collected from areas around Universiti Tun Hussein Onn Malaysia and residential colleges located at Taman Universiti. The process of collecting the plastic bottles took around one months. After the bottles were collected, they were cleaned and dried to get rid of any impurities. Finally, the recycled PET pieces were cut into the desired size and shape which was 50 mm in length and 5 mm in width. On the Figure 1 (a) to Figure 1 (d) show the steps of cutting the PET bottles into desired fibers.

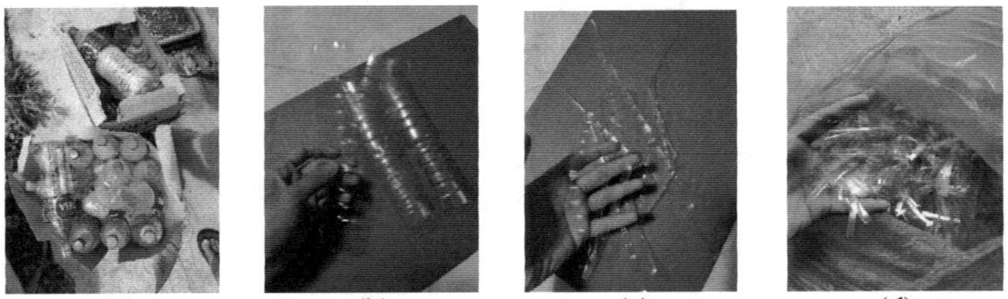

(a) (b) (c) (d)

Figure 1. The preparation of recycled PET fibers: (a) Collecting the recycled PET plastic bottles, (b) Cleaning and drying the recycled PET plastic bottles, (c) Cut recycled PET bottles into smaller pieces, (d) Cut the recycled PET pieces into desired size and shape.

2.3. Mix Design
The process that conducted to selected the most suitable ingredients of concrete and to determine their relative quantities to achieve the desired strength is mix design. In this study, the concrete proportioning was designed using the Concrete Mix Design Form Table 1 and Trial Mix Table 2 shows the proportion of the materials needed in this study based on DOE method.

Table 1. Concrete Mix Design Form

Stage	Item	Value
1	1.1 Characteristic Strength	30 MPa at 28 days
		Proportion defective 5%
	1.2 Standard Deviation	8 MPa
	1.3 Margin (k = 1.64)	1.64 x 8 = 13.12 Mpa
	1.4 Target Mean Strength	30 + 14 = 44 Mpa
	1.5 Cement Type	OPC
	1.6 Aggregate Type: Coarse	Uncrush
	Aggregate Type: Fine	Uncrush
	1.7 Free Water/Cement Ratio	0.48
2	2.1 Slump or Vebe Time	Slump 50 -100 mm
	2.2 Maximum Aggregate Size	20 mm
	2.3 Free Water/Cement Ratio	195 kg/m
3	3.1 Cement Content	195/0.48 = 406.25 kg/m^3
4	4.1 Relative Density of Aggregate (SSD)	-
	4.2 Concrete Density	2400 kg/m^3
	4.3 Total Aggregate Content	2400 - 410 – 195
		= 1795 kg/m^3
5	5.1 Grading of Fine Aggregate	70%
	5.2 Properties of Fine Aggregate	30%
	5.3 Fine Aggregate Content	1795 x 0.30 = 539 kg/m^3
	5.4 Coarse Aggregate Content	1795 x 539 = 1255 kg/m^3

Table 2. Mix design of concrete

Quantities	Cement (kg)	Water (kg)	Fine aggregates (kg)	Coarse aggregates (kg)	PET fibers (kg)
Per m^3	410.00	195.00	540.00	1255.00	420.00
0.014 m^3	5.74	2.73	7.56	17.57	5.88
Normal	5.74	2.73	7.56	17.57	0.00
0.5% PET	5.74	2.73	7.56	17.57	0.21
1.0% PET	5.74	2.73	7.56	17.57	0.43
1.5% PET	5.74	2.73	7.56	17.57	0.64
2.0% PET	5.74	2.73	7.56	17.57	0.85
TOTAL	**28.70**	**13.65**	**37.80**	**87.85**	**2.13**

2.4. Specimens and Test

In this research, the total of 30 specimen were prepared are cylinder mould with size (100 x 200) mm. Five types of mixes were prepared where the control specimens prepared with 0% volume of fibres followed by 0.5%, 1.0%, 1.5% and 2.0% of recycled PET fibres added into the mix. The concrete properties were tested after a curing period of 7 days and 28 days respectively. Compressive test was conducted as specified in the test method BS 1881-116:1983, Part 116: Method for the determination of compressive strength and modulus of elasticity of concrete cylinder [3] and ASTM C469: Standard Test Method for Static Modulus of Elasticity and Poisson's Ratio of Concrete in Compression [4].

3. Result and Discussion

The analysis of the results obtained from the data collected from the laboratory tests. An analysis was done according to the parameters used in controlling the effect of the percentage of recycled PET fibres in concrete under curing conditions of 7 days and 28 days respectively. In this part, it consists of result for slump test, compressive strength test and modulus of elasticity.

3.1 Slump Test

Figure 1 showed a decrease of workability was recorded when adding PET to the mix. Based on the concrete mix design by using the DOE (Department of Environmental) method, the slump range target of this mixture was between 50 mm to 100 mm. The normal concrete in the experiment showed that the slump of concrete was 75 mm which lied between the target range of 50 mm and 100 mm and the slump kept decreasing with the addition of fibre into concrete mixture. The reason behind this decrease is due to present of PET fibres in concrete causes more friction between the particles and this leads to less workability in the mixtures. Besides, the high content and large surface area of the fibres can easily absorb the cement paste thereby increasing the viscosity of the concrete mixture. As the PET content increases, the plasticity and consistency of fresh concrete will decrease. The effects of recycled PET fibres on the slump test between a normal concrete mixture and a mixture with PET fibres are shown in Figure 2.

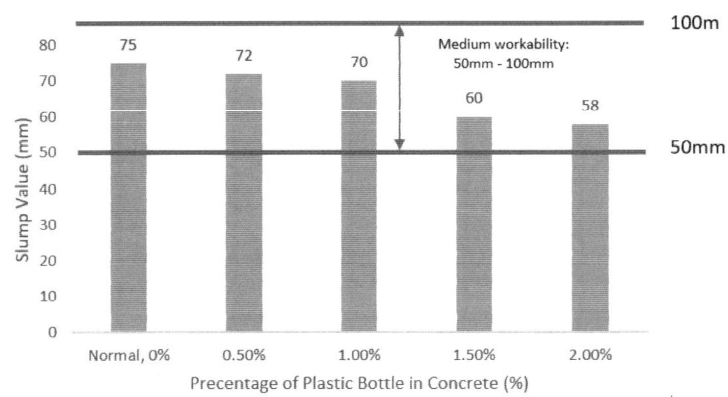

Figure 2. Slump value of shred plastic bottles in concrete

3.2 Compressive Strength

According to the test results from Figure 3 the compressive strength of plastic bottles concrete decreased when there was an increase in the percentage of shredded plastic bottles in the concrete mix. The compressive strength values for normal concrete at 7 days and 28 days were 19.13N/mm² and 20.9N/mm², respectively. The highest compressive strength value of 10N/mm² was obtained by the concrete sample containing 0.5% of shredded plastic bottles at 7 days whereas the concrete sample containing 1.0% of shredded plastic bottles achieved the highest compressive strength value of 15.73N/mm² at 28 days. These findings also concurred with the results reported by previous researchers. It was evident that the long fibres reduced the compressive strength more than the short fibres at a given fraction. Short fibres get mixed thoroughly with the other constituents and therefore a mesh-like structure was not formed as in the case of long fibres [5]. In general, the rate of reduction in strength was found to decrease with the increase in the percentage of recycled PET fibres and also the length of PET fibres.

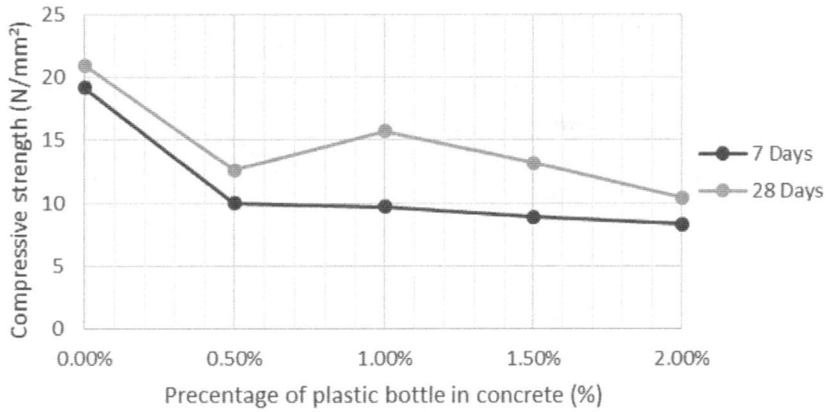

Figure 3. Compressive strength of shred plastic bottles in concrete

3.3 Modulus of Elasticity between EC, BS and ASTM

According to the findings presented on Figure 4, the findings clearly show that the modulus of elasticity of the concrete specimens containing varying percentages of shredded plastic bottles was the highest for all curing ages according to the Eurocode. Overall, at the curing age of 7 days, the MOE value based on the Eurocode was higher than the British Standard and the ASTM standard in the range of 6.73% to 18.21% and 46.02% to 86.3%, respectively. Besides, the MOE value based on the Eurocode was also higher than the British Standard and the ASTM standard in the range of 0.81% to 11.91% and 30.02% to 64.87% respectively at the curing age of 28 days. This finding indicated that during the structural analysis for buildings, the assumption on the modulus of elasticity of concrete is vital. It is known that the standard equations are established based on experimental results obtained for normal strength concrete and the standards generally propose empirical equations for predicting the modulus of elasticity depending on the compressive strength of concrete [6]. The comparison of MOE suggested by the standards indicated that the uses of the existing equations are appropriate for plastic bottle concrete. It also suggests that the MOE value for each standard depends on safety factors, cost effectiveness and sustainability.

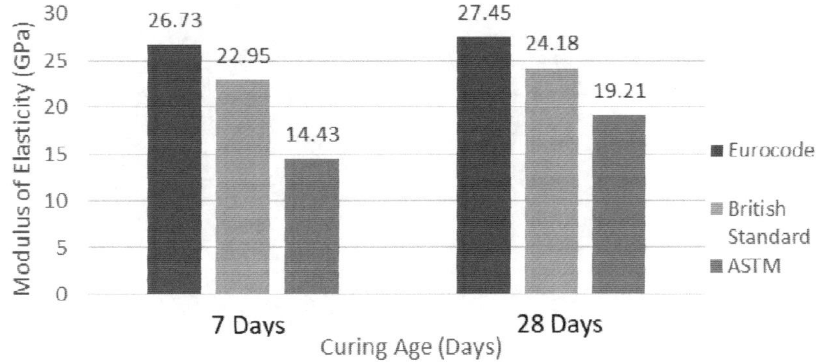

(a)

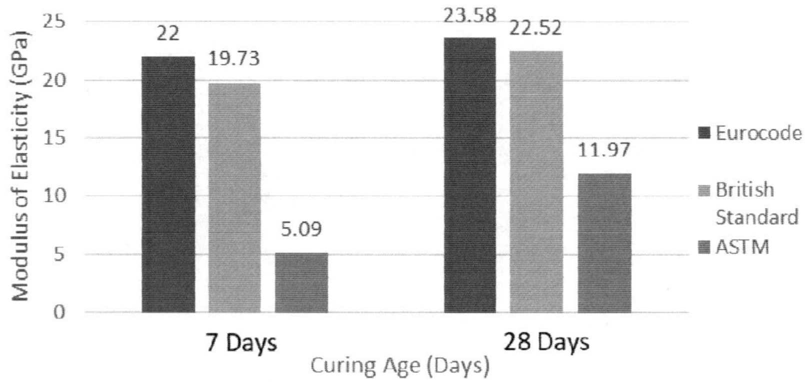
Modulus of Elasticity (GPa)
25
20
15
10
5
0
22
19.73
5.09
23.58
22.52
11.97
7 Days
28 Days
Curing Age (Days)
Eurocode
British Standard
ASTM

(b)

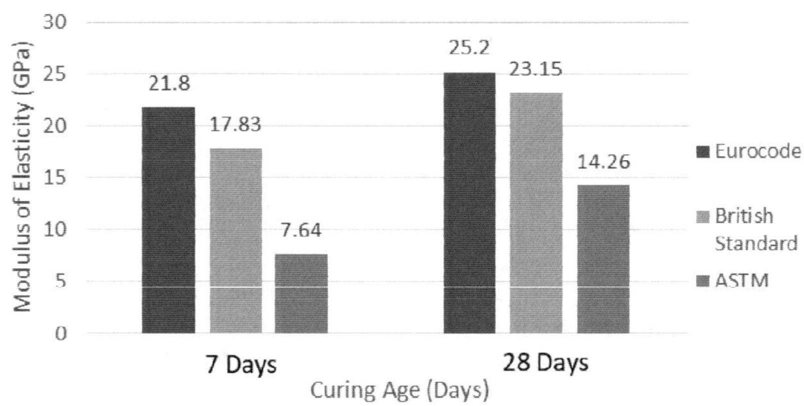
Modulus of Elasticity (GPa)
30
25
20
15
10
5
0
21.8
17.83
7.64
25.2
23.15
14.26
7 Days
28 Days
Curing Age (Days)
Eurocode
British Standard
ASTM

(c)

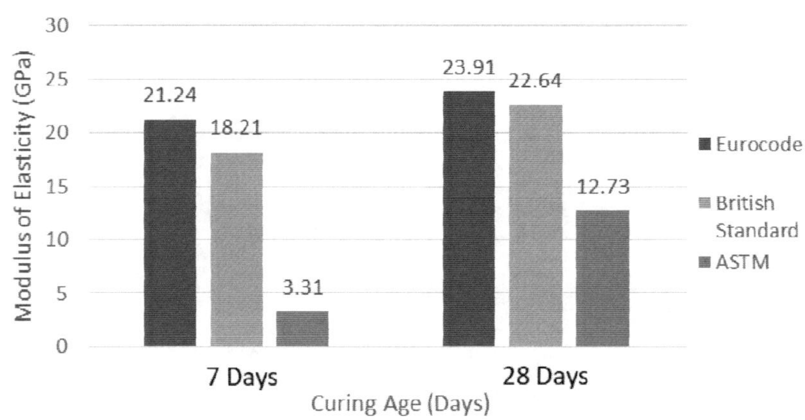
Modulus of Elasticity (GPa)
30
25
20
15
10
5
0
21.24
18.21
3.31
23.91
22.64
12.73
7 Days
28 Days
Curing Age (Days)
Eurocode
British Standard
ASTM

(d)

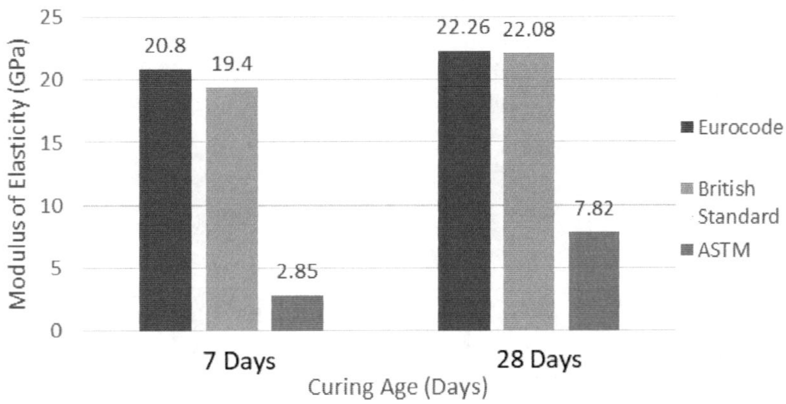

(e)

Figure 4. Comparison of modulus of elasticity of plastic bottles concrete between Eurocode, British Standard and ASTM at (a) 0%; (b) 0.5%; (c) 1.0%; (d) 1.5%; (e) 2.0%

4. Conclusions

In this research, the conclusion of investigate the potential applications of the shred plastic bottles as concrete material in this study were achieved. The following recommendations are on the physical properties and chemical properties of shred plastic bottles need further investigation especially the interconnection between fine and coarse aggregates and cement paste which may affect the concrete's modulus of elasticity.

5. References

[1] Eurocode 2 2014 Design of concrete structures EN1992-1-1 Part 1-1: General Rules and Rule for Building (London: British Standard)

[2] Bui N K, Satomi T and Takahashi H 2018 Recycling woven plastic sack waste and PET bottle waste as fiber in recycled aggregate concrete: An experimental study *Waste Management* **78** 79–93

[3] British Standard BS1881-121: 1983: Testing Concrete. Method for determination of static modulus of elasticity in compression (London:British Standard)

[4] ASTM 2002 Standard Test Method for Static Modulus of Elasticity and Poisson's Ratio of Concrete in Compression (Astm C469) 04 1–5

[5] Shahidan S, Ranle N A, Zuki S S M, Khalid F S, Nazri F M and Ridzuan A R M 2018 Concrete Incorporated with Optimum Percentages of Recycled Polyethylene Terephthalate (PET) Bottle Fiber *International Journal of Integrated Engineering* **10**(1) 1–8

[7] Leman A S, Shahidan S, Senin M S and. Ramzi Hannan N I R A 2016 .Preliminary Study On Chemical And Physical Properties Of Coconut Shell Powder As A Filler In Concrete *IOP Conf. Ser. Mater. Sci. Eng.* vol 160 (United Kingdom: IOP Publishing) pp 12059

[8] Ramzi hannan N I R, Shahidan S, Maarof M Z and Ali N 2016 Physical and Chemical Properties of Coal Bottom Ash (CBA) from Tanjung Bin Power Plant *IOP Conf. Ser. Mater. Sci. Eng.* vol 160 (United Kingdom: IOP Publishing) pp 12056

[9] Senin M S, Shahidan S, Leman A S and Hannan N I R R 2016 Analysis of Physical Properties and Mineralogical of Pyrolysis Tires Rubber Ash Compared Natural Sand in Concrete material *IOP Conf. Ser. Mater. Sci. Eng)* vol. 160 (United Kingdom: IOP Publishing) p. 12053

[10] Senin M S, Shahidan S, Leman A S, and Hannan N I R R 2016 Properties of Cement Mortar Containing Rubber Ash as Sand Replacement *IOP Conf. Ser. Mater. Sci. Eng.* vol 160 (United Kingdom:IOP Publishing) p. 12055

[11] Shahidan S, Leman A S, Senin M S and Hannan N I R R 2017 Suitability of Coconut Shell Concrete for Precast Cool Wall Panel-A Review *MATEC Web Conf.* vol 87 (1005) (Paris: EDP Science) pp 1–6

[12] Leman A S, Shahidan S, Yusuf M Y, Mohd Zuki S S and Misnon N. A 2017 Workability and Compressive Strength for Concrete With Coconut Shell Aggregate Behaviour Of Concrete With Coconut *MATEC Web Conf.* vol 87(1017) (Paris: EDP Science) pp 1–6

Acknowledgments
The author would like to thank Ministry of Education Malaysia (KPM), Universiti Tun Hussein Onn Malaysia, Grant MTUN KPM K122 and Grant Industry M007 for making this important research viable and effective.

IOP Conf. Series: Materials Science and Engineering **713** (2020) 012019 doi:10.1088/1757-899X/713/1/012019

Durability Evaluation of Cement Board produced from Untreated and Pre-treated Empty Fruit Bunch Fibre through Accelerating Ageing

P Peter [1], N M Z Nik Soh[2] Z A Akasah[3], and M A Mannan[4]

[1,2,3] Faculty of Civil and Environmental Engineering, Universiti Tun Hussein Onn Malaysia, 86400 Parit Raja, Batu Pahat, Johor, Malaysia.
[4] Department of Civil Engineering, Faculty of Engineering, Universiti Malaysia Sarawak, 94300 Kota Samarahan, Sarawak, Malaysia.
Corresponding author: nikzaini@uthm.edu.my

Abstract. Durability issue known that natural fibre tends to degrades in cementitious matrix particularly when in hot and humid exposure for a period of time. This experimental approach evaluates the potential of EFB cement board's durability in accelerated ageing condition. Initially, the EFB fibres are pre-treated for 24 hours by being soaked in 1% NaOH solution and oven-dried at 103±2°C. Cement boards of 1300 kg/m^3 density are fabricated in the laboratory, with 2.5:1:2 ratio of ordinary Portland cement, EFB fibre and water with 3% $CaCl_2$. Accelerated ageing assessment of hot water immersion for 10 days, 30 days and 60 days also wet dry cycle for 5 cycle, 15 cycle and 30 cycle at 70°C temperature are then conducted. The tested samples reveal finding ranges in modulus of rupture (MOR) of 3.98 – 9.11 MPa; modulus of elasticity (MOE) of 1056 – 4699 MPa; internal bonding (IB) of 0.28 – 0.53 MPa and thickness swelling (TS) of 1.66 – 9.25%. In addition, the field emission scanning electron microscope (FESEM) showed degradation mechanism of EFB cement board of before and after the accelerated ageing assessment. Conclusively, the findings showed that, accelerated ageing approach for durability assessment of cement boards could be done in quantifying the durability of samples in shorter time.

1. Introduction

It is generally known that most countries cultivated oil palm as a major plantation and in Malaysia, oil palms are the country's largest and most valuable crops thus further developments for a long-term strategy are needed for its potential as a material in structural and non-structural applications [1–5]. The natural fibre composite is known as more cost-effective, material with low energy consumption, environmentally friendly and sustainable [6, 7]. On top of that, it is important to take into account the adaptation of biodegradable composites, not simply for costs and availability but also for the consideration of occupants' health factor in the long-term service of the building [8,9,10]. Substantially, various research and development activities globally as well as in Malaysia, are looking towards discovering an alternative source and an innovatively reliable raw material from natural fibres and the oil palm empty fruit bunch fibre (EFB) is a fascinating discovery in Malaysia to specifically explore its durability potential as a composite material [11]. However when exposed to various ageing condition, the fibre itself experiences severe degradation in the cement matrix thus causing premature deterioration to the cement composite material [12, 13]. Amongst such climate factors as temperature, humidity, solar radiation, precipitation as well as pollutants throughout the year; both high temperature and humidity are the factors that most greatly affects the durability performance of building materials especially for a prolonged outdoor application [14-15].

2. Relevant previous studies

According to several studies on EFB fibres, chemical composition contents of lignin, hemicellulose and cellulose are about 13-43.51%, 17-47.7% and 22.90-57.8% respectively [16–18]. Due to the cell structure, natural fibre has limitation of its utilization as cement-based materials because of weathering condition [7,13]. Therefore, the fibres are necessary to be pre-treated and in alkaline pre-treatment with sodium hydroxide (NaOH) solution, studies reported concentration ranges from 0.2–6% solution has the ability to remove impurities and most residual oil on the EFB fibre surface [2,5,9,17,20–24] regardless of whether it is soaked for a period of 30 minutes or up to 24 hours. Whereas in the thermal treatment of fibre by oven-drying approximately at about 100°C–150°C, the composite results in 60% less thermal conductivity and also yield acceptable properties for a residential building material [22]. It is proved by the study of sisal fibre composite of Wahab [5] and Wei [9] that the thermal treatment improved cellulose crystallinity that provides high initial strength as well as maintained the reinforcement of fibre cement matrix in long period of time.

The causes of fibre degradation resulted into the reduction of fibre strength and its flexibility [25, 26]. As reported by Wei, natural fibre in pore solution of cement matrix experienced a linear decrease when exposed after 10 days in accelerated ageing and deteriorated completely in 2 years. Thus, further investigation on natural fibre in exposure of aggressive environment and various humidity for a long-term stability outdoor application is crucial [8-9]. In studies of durability by natural weathering condition, it usually takes period of approximately 3 months up to 12 months and reported that at 10 months exposure, the mechanical properties of hybrid reinforced composite showed better values compared to conventional concrete [25]. Likewise at 6 months of exposure, findings showed poor durability in toughness due to fibre-matrix adhesion [14]. Other than that, a complete reduction value of MOE and MOR also found [28]. In contrast to accelerated ageing condition in a laboratory, numerous totals of cycles and days at various temperatures are studied. According to Wei [9], 70°C or higher are the most effective temperature of accelerating condition to evaluate durability in hot soak immersion for days and in wet-dry cycle. At similar temperature, David [12] reported that ageing in lime saturated hot water is effective to evaluate the degradation of a specimen and the wet-dry ageing effective in evaluating the stability of fibre in cement-matrix. In order to evaluate the durability of a composite in shorter time, it is reported that the toughness of a composite is reduced with the increasing number of soaking and drying cycles [30] where the average values of MOE, MOR and IB are lower than composite without accelerated ageing condition [31].

Furthermore, most manufacturing of natural fibre composite material used calcium chloride ($CaCl_2$) as chemical accelerator to improve composites' mechanical and physical properties. In study by Onuorah et. al. [18], usage amount of 3% $CaCl_2$ is reported to have the highest obtained mechanical properties compared to 1%–2% $CaCl_2$. In conjunction with Kishar et. al. [19], it is reported that cement accelerator of $CaCl_2$ showed higher compressive strength in a cement paste and has lower porosity compared to samples without addition of $CaCl_2$. Although there are research studies done in regards to fibre pre-treatment and manufacturing improvement methods, rarely reported on the durability aspects particularly of EFB fibre as a composite material. Thus, significant assessment on durability potential of EFB cement board is obviously necessary depending on the type of materials used and methods involved.

3. Experimental materials and method

The EFB fibres were obtained from the Global Seed Sdn. Bhd. factory in Kawasan Perindustrian Simpang Renggam, Johor. It was compacted in blocks of bunches that comprised of long EFB fibres and weighed approximately 2 tonnes. The compacted blocks were loosened to be extruded by the crusher machine for the chipping process at Timber Fabrication Laboratory, Universiti Tun Hussein Onn Malaysia to obtain the short EFB fibres.

The sieving processes was done at Malaysia Palm Oil Berhad, Bangi where it separates the dust from the required fibre sizes for cement board fabrication in which the sieve machine has nominal openings of 4.76 mm (4 mesh), 2.83 mm (7 mesh), 1.41 mm (14 mesh) and 0.177 mm (80 mesh). The lengths distribution of the EFB short fibres are approximately 80 mesh, 14 mesh and 7 mesh of EFB fibres of about 8–10 mm, 10–13 mm and 10–18 mm respectively. As mentioned in previous studies [20,21,24], the most effective amount for EFB fibre is 2% NaOH concentration yet it is not discovered as the amount for the durability of EFB cement board properties specifically. Thus, the fibre is pre-treated by soaking for 24 hours in the concentration of 1% NaOH solution. For heat treatment, the fibres are oven-dried at a temperature of 103±2°C for 24 hours, along with being soaked in water at ambient temperature for 30 minutes. After completion of the pre-treatments period, the EFB fibres are washed in running tap water and then dried under the hot sunlight [5, 17]. Substantially, the size of 14 mesh is used for fabrication in this study where it achieves the provisions of Malaysian Standards 1984 similar to the cement board reported by Onuorah et. al [18]. However, only the properties of cement board were reported. Thus, this research attempts to continue by assessing the durability of the material as subjected to an accelerated ageing condition as recommended by Wei [9].

3.1 EFB cement board fabrication

The 350 mm × 350 mm × 12 mm cement board samples of targeted density 1300 kg/m^3 are prepared for a total of 36 samples. The mixture's material of ordinary Portland Cement Type 1, calcium chloride ($CaCl_2$), water and EFB fibre are used to fabricate the cement board. The mixing ratio of cement:EFB fibre:water is 2.5:1:2, $CaCl_2$ is 3% based on cement weight and is used in the form of powder diluted in water while the water-cement ratio is 40% [18,27]. The mixtures are mixed in the mixer machine for 5 minutes at 60–80 rpm. It is then pre-formed by hand lay-up method unto a wooden mould through a wire mesh. Initially, a polythene sheet place unto the metal plate mould is underneath the plywood. The complete mixture is then pre-pressed with another polythene sheet upon it with the metal plate mould placed on the upper part before being transferred to a 50 ton/m^2 cold compressor machine. For approximately 5-7 minutes, the cement board mixture is compressed at 180 mm/min pressure rate. Upon being removed from the machine, the board are let to cure for 24 hours at laboratory ambient temperature then be de-moulded. The EFB cement board is cured for 28 days in the laboratory before being subjected to an accelerated ageing assessment as referenced by Wei [9].

Figure 1. Fabrication processes of EFB cement board.

3.2 Experimental tests of accelerated ageing

The totals of samples are cut each to be tested for MOE, MOR, IB and TS properties and to be subjected to an ageing process as recommended by previous research [9]. The durability test is conducted at 70°C for accelerated ageing conditions of 10 days, 30 days and 60 days and 5 cycles, 15 cycles and 30 cycles for hot water immersion and wet-dry cycle, respectively.

Table 1. Durability assessment of EFB cement board.

Properties	Sample	Ageing (cycles/days)																				
		Control			5 C*			15 C			30 C			10 D*			30 D			60 D		
		U* T	T* N	T* H	U T	T N	T H	U T	T N	T H	U T	T N	T H	U T	T N	T H	U T	T N	T H	U T	T N	T H
Mechanical	*MOR *MOE	*5	*5	*5	*5	*5	*5	*5	*5	*5	*5	*5	*5	*5	*5	*5	*5	*5	*5	*5	*5	*5
	*IB	*5	*5	*5	*5	*5	*5	*5	*5	*5	*5	*5	*5	*5	*5	*5	*5	*5	*5	*5	*5	*5
Physical	*TS	*5	*5	*5	*5	*5	*5	*5	*5	*5	*5	*5	*5	*5	*5	*5	*5	*5	*5	*5	*5	*5

Notes; C = cycles, D = days, UT = untreated EFB, TN = 1% NaOH pre-treated EFB, TH = heat pre-treated EFB, MOR = modulus of rupture, MOE = modulus of elasticity, IB = internal bonding, TS = thickness swelling, 5= number of samples, Total samples = 105

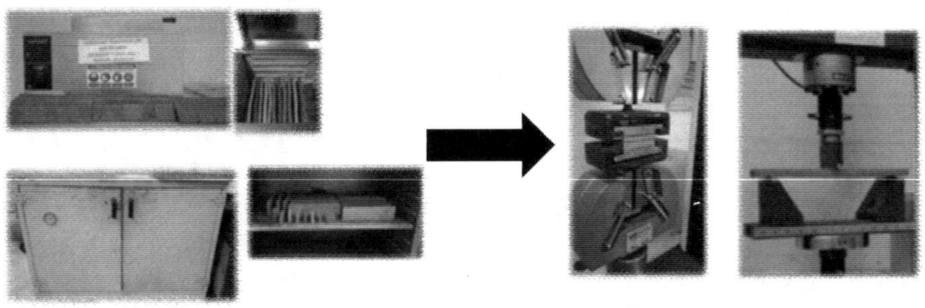

(a) Water bath and ventilated oven (b) UTM INSTRON 3369

Figure 2. Experimental test assessment and set up.

For the accelerated ageing condition of hot water immersion, the samples are immersed completely in a water bath machine. Whereas for the wet-dry cycle, the samples are immersed in water for 15 hours and then oven-dried for 25 hours for each cycle. The Universal Testing Machine (UTM) INSTRON 3369 shown in figure 2 is used for the experimental test of MOE, MOR and IB properties for before and after ageing assessment.

4. Result and analysis

The cement board produced by heat treated EFB fibre showed higher MOE value followed by cement board produced by NaOH treated EFB fibre where in figure 3, showed values range in 2879 – 4699 MPa and 1056 – 3459 MPa respectively. In ageing through warm water immersion, declined value reported approximately 49.58% and 37.15% for NaOH and heat treated EFB fibre as cement board in 10 days to 30 days respectively. Whereas in 5 cycle to 15 cycle of wet-dry cycle, declined value reported about 32.76% and 20.71% for NaOH and heat treated EFB fibre as cement board respectively.

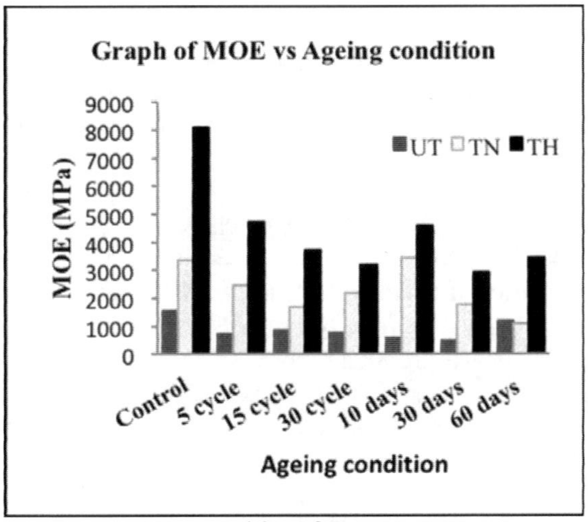

(a) MOE

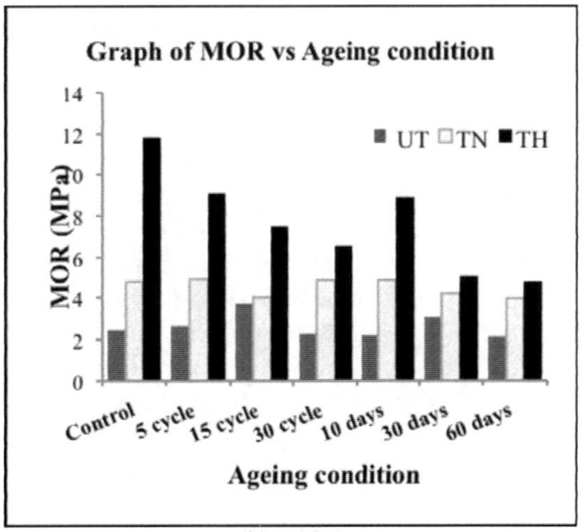

(b) MOR

Figure 3. Graph of MOE and MOR

In figure 3b, the MOR value obtained shows a gradual declined for ageing of wet-dry cycle and warm water immersion. Value ranges are 3.98 – 4.94 MPa and 4.77 – 9.11 MPa for the NaOH and heat treated EFB fibre as cement board respectively. From the obtained result for MOE and MOR for accelerated ageing condition, warm water immersion as well as wet-dry cycles is observed to be the aggressive ageing for the composite as it strength declined, similarly stated by Wei [9].

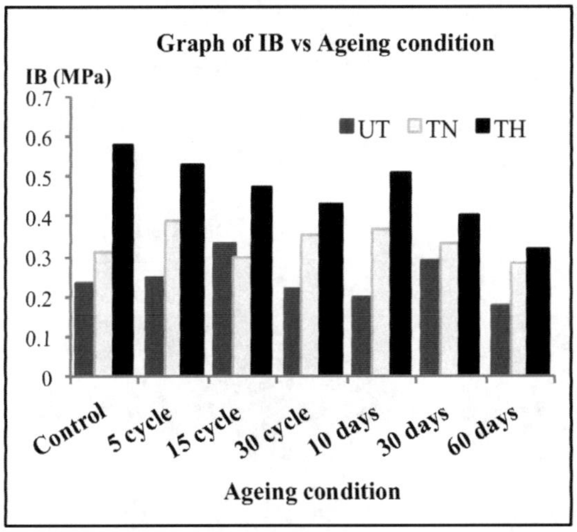

(a) IB

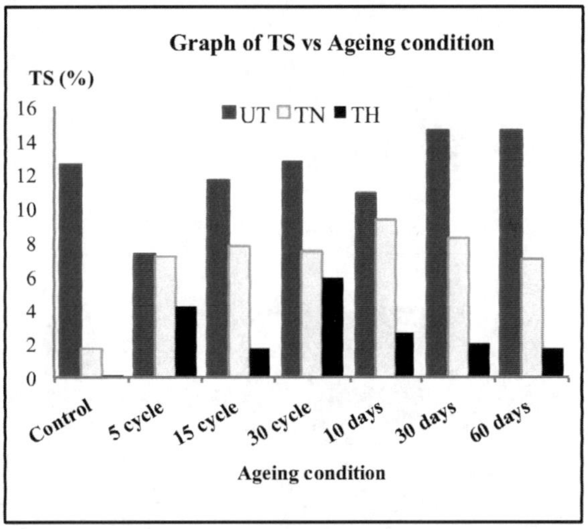

(b) TS

Figure 4. Graph of IB and TS

Similarly observed for the IB in Figure 4, where its value slightly decreased for both accelerated ageing to the ranges of 0.28 – 0.39 MPa and 0.32 – 0.53 MPa for NaoH and heat treated EFB fibre as cement board respectively. In spite of the utmost properties obtained for MOE, MOR and IB, the untreated fibre and followed by NaOH treated EFB fibre as cement board showed slightly higher TS compared to the heat treated EFB cement board in both accelerated ageings where observed values of 7.27 – 14.55%, 6.94 – 9.25 % and 1.66 – 5.81 % respectively, shown in figure 4b.

IOP Conf. Series: Materials Science and Engineering **713** (2020) 012019 doi:10.1088/1757-899X/713/1/012019

4.1. Field Emission Scanning Electron Microscope (FESEM) analysis

The microanalysis of EFB fibre and cement board produced from EFB fibre performed on JEOL JSM-7600F field emission scanning electron microscope. Initially, a fine layers of gold sputter-coated the EFB cement board in order for it to be conductive under the FESEM machine. The microanalysis is crucial to evaluate the effect of accelerated ageing condition on the degradation mechanism of EFB fibre in a cement matrix.

Figure 5a shows the impurities attached on the surface of untreated EFB fibre where it affected the bonding properties between EFB fibre and cement matrix. After the pre-treatment with NaOH in figure figure 5b and heat thermal in figure 5c, the impurities are removed. Compared to NaOH pre-treatment of the fibre, the residual and impurities that attached on the fibre are seen to be mostly removed when pre-treated with heat thermal. The cement board produced from NaOH and heat pre-treated EFB fibre shown in figure 5e and 5f respectively before subjected to the accelerated ageing condition. It is seen that the interfacial bonding between fibre and cement matrix are better compared to untreated EFB fibre as cement board.

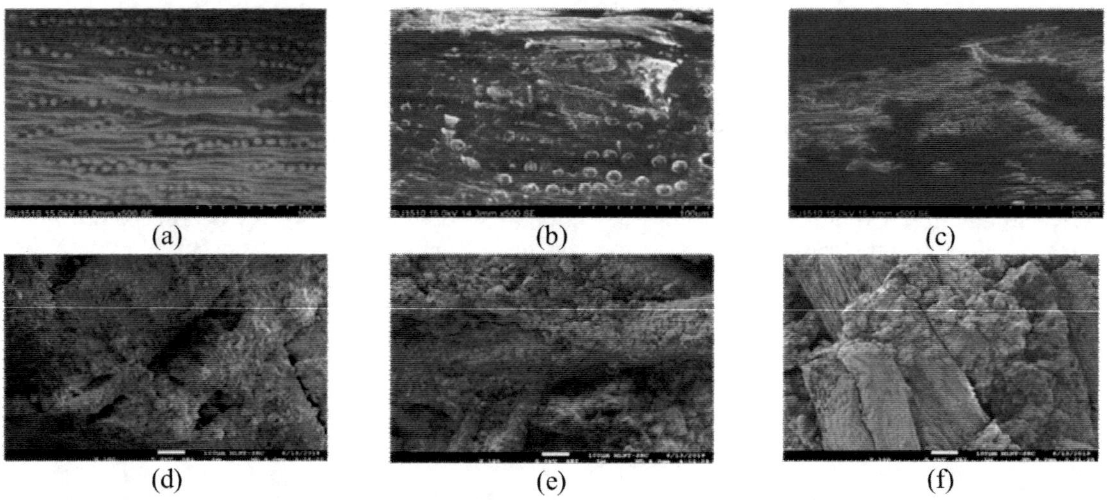

Figure 5. FESEM images of EFB single fibre and EFB cement board without accelerated ageing

Figure 6. FESEM images of 15 wet-dry cycles and 30 days hot water immersion

Whereas in figure 6, shows the FESEM images of the cement board produced by untreated fibre as well as fibre pre-treated with NaOH and heat thermal after the accelerated ageing condition. As can be observed from the figure, interfacial failure between fibre-cement-fibre can be clearly seen after 15 wet-dry cycles and 30 days of hot water immersion. Wei [9] has reported that, the durability of fibre cement based composite is able to evaluate in a reasonably shorter time by the accelerated ageing assessment. Both wet-dry cycles and hot water immersion shows aggressive effect on the reduction strength of the EFB cement board.

5. Conclusion

Based on the experimental results, both hot water and wet-dry ageing greatly affect the degradation mechanism of EFB fibre in cement matrix with increases number of cycles and durations. In terms of MOE, MOR and IB, it was found that significantly value decreases for cement board produced from untreated and pre-treated EFB fibre through both ageing especially wet-dry cycle. Whereas for TS properties, the untreated EFB-cement board has highest percentage of swelling compared to NaOH and Heat treated EFB fibre. Furthermore, the microstructure analysis showed that after 15 cycles and 30 days of ageing, the fibre encounter severe degradation in the cement matrix. This also shows an acceptable approach to determine durability of the EFB cement board within a reasonable short time. In summary, the fibre pre-treatment is the most crucial method in the fabrication of the cement board. Upon durability assessment, the cement board achieves the ultimate properties required for MOE, MOR and IB. However, the TS value is slightly higher amongst the three EFB cement boards due to moisture absorption. Therefore, various ageing condition for durability in accelerated ageing should be conducted on the EFB cement board to explore more of its durability potential as a construction material.

6. References

[1] Dungani R, Karina M, Subyakto, Sulaeman A, Hermawan D and Hadiyane A 2016 Agricultural Waste Fibers Towards Sustainability and Advanced Utilization: A review, *Asian Journal of Plant Sciences* **15** 42–55

[2] Jawaid M, Abdul Khalil H P S and Abu Bakar A 2011 Woven hybrid composites : Tensile and flexural properties of oil palm-woven jute fibres based epoxy composites *Material Science and Engineering A*. **528** 5190–5195

[3] Mat Rasat M S, Wahab R, Sulaiman O, Moktar J, Mohamed A, Tabet T and Khalid I 2011 Properties of Composite Boards From Oil Palm Frond Agricultural Waste *BioResources* **6** 4389–4403

[4] Sreekala M S, Kumaran M G, Joseph S, Jacob M and Thomas S 2000 Oil palm fibre reinforced phenol formaldehyde composites: influence of fibre surface modifications on the mechanical performance *Applied Composite Materials* **7(5–6)** 295–329

[5] Wahab R, Mat Rasat M S, Samsi W, Mustafa M T and Mohd Don S M 2017 Assessing the suitability of Agro-Waste from Oil Palm Empty Fruit Bunches as Quality Eco-Composite Boards *Agricultural Science* **9(8)** 237–247

[6] Hazman A H 2016 *Mechanical and Physical Performance of Empty Fruit Bunch Cement Board (EFBCB) with Different EFB Fibre Treatments* Master Tesis (Universiti Tun Hussein Onn Malaysia)

[7] Onuaguluchi O and Banthia N 2016 Plant-based natural fibre reinforced cement composites : A review *Cement and Concrete Composite* **68** 96–108

[8] Ramesh M, Palanikumar K and Reddy K H 2017 Plant fibre based bio-composites: Sustainable and renewable green materials *Renewable and Sustainable Energy Reviews* **79** 558–584

[9] Wei J 2014 *Durability of Cement Composites Reinforced with Sisal Fibre* PHD Thesis (Graduate School of Arts and Science, Columbia University).

[10] Xie X, Zhou Z, Jiang M, Xu X, Wang Z and Hui D 2015 Cellulosic fibers from rice straw and bamboo used as reinforcement of cement-based composites for remarkably improving mechanical properties. *Composites Part B: Engineering* **78**

[11] Meysam Khoshnava S, Rostami R, Ismail M, Rahmat A R, & Ezekiel Ogunbode B 2017 Woven hybrid Biocomposite: Mechanical properties of woven kenaf bast fibre/oil palm empty fruit bunches hybrid reinforced poly hydroxybutyrate biocomposite as non-structural building materials *Construction and Building Materials* **154** 155–166

[12] David De Klerk M 2015 *The Durability of Natural Sisal Fibre Reinforced Cement-based Composites* Master Thesis (Stellenbosch University)

[13] Wei J, Ma S and Thomas D G 2016 Correlation between hydration of cement and durability of natural fiber-reinforced cement composites *Corrosion Science* **106** 1–15

[14] Ariawan D, Salim M S, Mat Taib R, Ahmad Thirmizir M Z and Mohammad Ishak Z A 2017 Durability of alkali and heat-treated kenaf fiber/unsaturated polyester composite fabricated by resin transfer molding under natural weathering exposure *Adv Polym Technol*. 1–15

[15] Hradil P, Toratti T, Vesikari E, Ferreira M and Häkkinen T 2014 Durability considerations of refurbished external walls *Construction and Building Materials* **53** 162–172

[16] Abdul Khalil H P S, Jawaid M, Hassan A, Paridah M T and Zaidon A 2012 Oil Palm Biomass Fibres and Recent Advancement in Oil Palm Biomass Fibres Based Hybrid Biocomposites *Composites and Their Applications* pp 187–220

[17] Abdullah N and Sulaiman F 2013 The properties of the Washed Empty Fruit Bunches of Oil Palm *Journal of Physical Science* **24(2)** 117–137

[18] Onuorah E O, Okeke C A, Nwabanne J T, Nnabuife E L C and Obiorah S O M 2015 The effects of production parameters on properties of single and 3-layer cement-bonded composites made from oil palm empty fruit bunch and tropical hardwood sawmill residue *World Journal of Engineering* **12(6)** 577–590

[19] Kishar E A, Ahmed D A, Mohammed M R and Noury R 2013 Effect of calcium chloride on the hydration characteristics of ground clay bricks cement pastes *Beni-Suef University Journal of Basic and Applied Sciences* **2** 20–30

[20] Izani M A N, Paridah M T, Anwar U M K, Nor M Y M and H P S 2012 Composites : Part B Effects of fiber treatment on morphology , tensile and thermogravimetric analysis of oil palm empty fruit bunches fibers *Composite* **45(1)** 1251-1257

[21] Ibrahim Z, Abdul Aziz A, Ramli R, Jusoff K, Ahmad M and Jamaludin M A 2015 Effect of Treatment on the Oil Content and Surface Morphology of Oil Palm (Elaeis Guineensis) Empty Fruit Bunches (EFB) Fibres *Wood Research* **60(1)** 157–166

[22] Lertwattanaruk P and Suntijitto A 2015 Properties of natural fiber cement materials containing coconut coir and oil palm fibers for residential building applications *Construction and Building Materials* **94** 664–669

[23] Valášek P, Ruggiero A and Müller M 2017 Experimental description of strength and tribological characteristic of EFB oil palm fibres/epoxy composites with technologically undemanding preparation *Composites Part B* **122** 79–88

[24] Jumhuri N, Hashim R, Sulaiman O, Wan Nadhari W N A, Mohd Salleh K, Khalid I and Razali M Z 2014 Effect of treated particles on the properties of particleboard made from oil palm trunk *Materials and Design* **64** 769–774

[25] Tian H and Zhang Y X 2017 Ageing effect on tensile and shrinkage behaviour of new green hybrid fibre-reinforced cementitious composites *Cement and Concrete Composites* **75** 38–50

[26] Wei J and Meyer C 2015 Degradation mechanisms of natural fiber in the matrix of cement composites *Cement and Concrete Research* **73** 1–16

[27] O Ogunsile B and A Adepegba J 2015 Cement Bonded Particle Board from Musa paradisiaca Stalk *The Pacific Journal of Science and Technology* **16(1)** 12–20

[28] Homkhiew C, Ratanawilai T and Thongruang W 2014 Effects of natural weathering on the properties of recycled polypropylene composites reinforced with rubberwood flour *Industrial Crops and Products* **56** 52–59.
[29] Fidelis M E A, Toledo Filho R D, Silva F A, Mechtcherine V, Butler M and Hempel S 2016 The effect of accelerated aging on the interface of jute textile reinforced concrete *Cement and Concrete Composites* **74** 7–15
[30] Viviane da Costa Correia, Sergio Francisco Santos, Gonzalo Mármol, Antonio Aprigio da Silva and Curvelo H S J 2014 Potential Of Bamboo Organosolv Pulp As A Reinforcing Element In Fiber–Cement Materials *Construction and Building Materials* **72** 65–71
[31] Garzón N, Sartori D, Zuanetti I, Barbirato G, Ramos R, Fiorelli J and Savastano Jr H 2012 Durability Evaluation of Agro-Industrial Waste-Based Particle Boards Using Accelerated Aging Cycling Tests *Key Engineering Materials* **517** 628–634

Acknowledgments
The authors would like to convey their appreciation to laboratories of Faculty of Civil and Environmental Engineering and Microelectronics & Nanotechnology-Shamsuddin Research Centre of Universiti Tun Hussein Onn Malaysia, and to the Contract Research Grant Scheme UTHM (U522) for financing in conducting this research study.

IOP Conf. Series: Materials Science and Engineering **713** (2020) 012020 doi:10.1088/1757-899X/713/1/012020

Influence of varying density on the mechanical and physical properties of Medium Density Fibreboard (MDF) containing kenaf

Noorli Ismail[1,2], Muhammad Azzim Mohmed Jomali[2], Hasniza Abu Bakar[1,2*], Tong Yean Ghing[1,2], Nik Mohd Zaini Nik Soh[1,2] and Raden Maizatul Aimi Mohd Azam[3]

[1]Jamilus Research Center and Sustainable Construction, Faculty of Civil and Environmental Engineering, Universiti Tun Hussein Onn Malaysia (UTHM), 86400, Batu Pahat, MALAYSIA.

[2]Faculty of Civil and Environmental Engineering, Universiti Tun Hussein Onn Malaysia (UTHM), 86400, Batu Pahat, MALAYSIA.

[4]Civil & Structural Engineering Branch, Public Works Department (JKR), 50480 Kuala Lumpur, MALAYSIA

Corresponding author: hasniza@uthm.edu.my

Abstract. Generally, Medium Density Fibreboard (MDF) is made from rubberwood, wood fiber and pine wood with combination of non-woody plant and agricultural residues in producing furniture and laminated floor. The usage of kenaf as main production of MDF had attracted attentions from researchers due to its anatomy of physicals and high content of organic compound namely are cellulose, hemicellulose and lignin. Due to these advantages, kenaf showed the huge potential in producing MDF with the bigger scale of production. In present study, nine (9) panel boards of MDF with sizes of 350 (length) x 350 (width) x 12 (thickness) in millimetres were fabricated based on three (3) different target densities started from 0.7 g/cm^3 to 0.8 g/cm^3. A total of three (3) samples for each target density were prepared for the measurements. The proportion of 65:15 which was described as ratio of kenaf particles to urea formaldehyde (UF) were selected and measured from the total volume of panel board. Testing of Modulus of Rupture (MOR), Modulus of Elasticity (MOE), Internal Bonding (IB) and Thickness Swelling (TS) were conducted in the present study. All measurements were carried out according to the *BS EN 310:1993*, *BS EN 319:1993* and *BS EN 317:1993.* Results showed the increment of MOR, MOE and decrement of IB and TS were associated with increment of target density. Thus, target density of 0.8 g/cm^3 for MDF is the best selection due to high MOR and MOE and low IB and TS corresponding to other two (2) target densities. The decrement of panel thickness also capable to achieve high density associated with high MOR and MOE. Therefore, the significant outcome shows the kenaf plant have the potential as alternative material to produce the Medium Density Fibreboard (MDF).

1. Introduction

Kenaf also known as *hibiscus cannabinus L* which is non-woody fiber plants and can be classified as Mallow family [1-2]. Subsequently, due to the high yield production which was about three to five times greater than other types of fiber plant [3-4] and growth well in tropical climates with high rainfall [2], kenaf then was introduced in Malaysia. Furthermore, collaboration between few research development institutions in Malaysia namely are Malaysian Agriculture Research and Development (MARDI),

Universiti Putra Malaysia (UPM), Malaysia Rubber Board (MRB) and Malaysian Palm Oil Board (MPOB) had been made in order to promote the usage of kenaf in Malaysian production industries. These collaboration among different institutions in Malaysia helps to reduce the usage of timber, thus it can decrease the deforestation rate in Malaysia [6].

Generally, the kenaf crop can be used to produce furniture, fibreboard, pulp, paper products, bio-composites, automotive panel and their components as an alternative materials [5]. Moreover, kenaf morphology also has high cellulose, holocellulose and lignin contents [7-10] that leading to have good bonding between the layers if it was formed as a panel [11]. Due to these characteristics, kenaf shows a good potential as main contents in producing MDF which consists of dry formed product of several panels combined together with a synthetic resin binder. The conventional MDF averagely was able to provide tensile strength between 283 MPa and 800 MPa and elastic modulus (MOE) from 200 MPa to 600 MPa [34]. However, the manufacturer of MDF [12] stated that the mechanical properties of MDF was varies depending on the panel thickness. Table 1 shows the increment of density in MDF's panel capable to achieve high mechanical properties of MOR and MOE. While, the decrement of panel thickness increased the thickness percentages of swelling which was indicated that MDF should be avoided contact with the water sources.

Table 1. Mechanical and physical properties of MDF subjected to different thickness [12]

Mechanical Properties	**Unit (mm)**	**>2.5-4.0**	**>4.0-6.0**	**>6.0-9.0**	**>9.0-12.0**	**>12.0-19.0**	**>19.0-30.0**
Thickness tolerance	mm				± 0.2		
Average of density	kg/m^3	800-900	780-840	760-800	720-760	700-740	650-690
Board moisture	%				5-8		
Modulus of Rupture (MOR)	MPa	0.65	0.65	0.65	0.60	0.55	0.55
Modulus of Elasticity (MOE)	MPa	NA	NA	NA	NA	1050	1000
Internal bonding (IB)	MPa	NA	NA	NA	NA	850	700
Swelling of thickness (TS)	%	35	30	17	15	12	10

2. Literature Review

Previous studies showed that the rubberwood and wood fibers were commonly used to produce MDF and sometimes with combination of others non-woody plant and agricultural residues [13-18, 21-25]. Besides, pine wood was also used as main contents of MDF [26-27]. However, full usage of non-woody plant such as kenaf is encouraged to replace the demand of wood plant due to affordable price and high yield production. Up to the present, MDF made of kenaf showed significant findings [11,19,21-23,28] in enhancing modulus of rupture, modulus of elasticity and internal bonding associated with low thickness swelling which was the main properties in producing MDF. Study of [11] confirmed that the kenaf's MDF recorded the increment of modulus of elasticity and modulus of rupture of 51% and 64%, respectively corresponding to MDF made from rubberwood. While, [21] determined high internal bonding corresponding to rubberwood and other combination materials in MDF panel. Subsequently, [28] stated that the application of kenaf in MDF showed lower thickness swelling than MDF produced from bagasse and wheat straw [29,30]. However, the previous studies of [11,19,21-23,28-30] were conducted with target density of MDF between 0.7 g/cm^3 and 0.73 g/cm^3. None of studies was published with high target density for panel board of MDF containing kenaf. Therefore, three (3) target densities were introduced namely 0.7 g/cm^3, 0.75 g/cm^3 and 0.8 g/cm^3 to investigate the potential of MDF made from kenaf with high density.

3. Materials and Methodology

Kenaf plant and urea formaldehyde (resin binder) were supplied by Timber Fabrication Laboratory, Faculty of Civil and Environmental Engineering, Universiti Tun Hussein Onn Malaysia (UTHM), Malaysia. The making of MDF was started by preparing the kenaf particles followed by board manufacture prior to the testing which was described at following sub-section. Testing performed in the present study were MOR, MOE, IB and TS at Light Structures Laboratory, Faculty of Civil and Environmental Engineering, Universiti Tun Hussein Onn Malaysia (UTHM), Malaysia.

3.1. Preparation of kenaf particles

The raw matured kenaf required to chip and crush using grinding machine to obtain kenaf particles prior to the panel board production. The process of preparing the kenaf particles was shown in Figure 1. This process was made in order to ensure the particles can be easily mixed up with urea formaldehyde.

Figure 1. Process performed to produce kenaf particles by preparing (a) raw kenaf followed by (b) chipped and crushed using Grinding Machine and (c) blended with Urea Formaldehyde (UF) prior to the MDF production.

3.2. Preparation of board manufacture

Nine (9) panel boards of MDF with sizes of 350 (l) x 350 (w) x 12 (t) in millimetres were fabricated based on three (3) different target densities which included 0.7 g/cm^3, 0.75 g/cm^3 and 0.8 g/cm^3. For the each target density, a total of three (3) specimens were prepared for the testing purposes. The kenaf particles of 65% and urea formaldehyde (UF) of 15% were blended using spray gun as shown in Figure 2(a). Table 2 tabulates the target densities of panel board for Specimen 1 was equivalent to total weight of 1029 g when target density of specimen was multiplied with volume of panel board. Thus, the weight of UF required was 237.46g which was measured from ratio of 15/65 multiplied with total weight of board panel, 1029g. While total weight of kenaf particles, 668.85g was required based on 65% proportion multiplied with total weight of board panel. Upon the completion of blending process, the mixture of kenaf particles and UF was manually formed into the wooden mould prior the compaction as shown in Figure 2(b).

The panel board of MDF was pressed using Cold Press Machine at temperature 150°C and pressure of 3 MPa for 6 minutes as displayed in Figure 2(c). Then, Fig. 2(d) shows the MDF were kept at temperature of 25 ±3°C in the oven with 65% of relative humidity for 24 hours followed by curing process for 7 days. The process conducted in preparing the board panel of MDF was referred to Figure 2 for better understanding.

Table 2. Total weight of kenaf particles and urea formaldehyde (UF) based on different target density in making MDF

Specimen No	Target density (ρ) (g/cm^3)	Volume of panel board (V) (cm^3)	Weight of UF (15%) (g)	Weight of kenaf particles (65%) (g)	Weight of panel board (ρ x V) (g)
1	0.70	1470	237.46	668.85	1029
2	0.75	1470	254.42	716.63	1102.50
3	0.80	1470	271.38	764.40	1176

Figure 2. (a) Blending process between kenaf particles and urea formaldehyde (UF) with selected proportion ; (b) Kenaf particles were manually filled in the wooden mould ; (c) Compaction was performed using Cold Press Machine ; (d) MDF was conditioned in the oven for 24 hours with selected temperature and relative humidity.

3.3. Experimental Testings

After the curing process was completed, the main board panels of 350 (l) x 350 (w) x 12 (t) in millimetres were cut into four (4) sizes was used for the testing purposes as shown in Table 3. The size specimen of 290 (width) x 50 (length) x 12 (thickness) was prepared for testing of MOR and MOE as accordance to BS EN 310:1993. While BS EN 319:1993 was used as the guidance for testing of IB and TS on the specimen with sizes of 50 (width) x 50 (length) x 12 (thickness). The board panel was bonded to loading block using epoxy glue as displayed in Figure 3(a) prior the testing. All the testing of MOR and MOE was determined using Universal Testing Machine. Accordance to BS EN 319:1993, determination of swelling in thickness was conducted after immersion in water. Figure 3(b) shows the vernier clipper was used to measure thickness of specimen before and after water immersion with time interval of 24 hours. Then, the different thickness of panel board was divided with thickness before immersion and expressed as percentages.

Table 3. Measurement of size specimen accordance to specification of standard [31-33]

Testing performed	Size specimen (width x length x thickness)	Specification of the standard
Modulus of rupture (MOR) Modulus of elasticity (MOE)	290 x 50 x 12	BS EN 310: 1993
Internal bonding (IB) Thickness swelling (TS)	50 x 50 x 12	BS EN 319: 1993

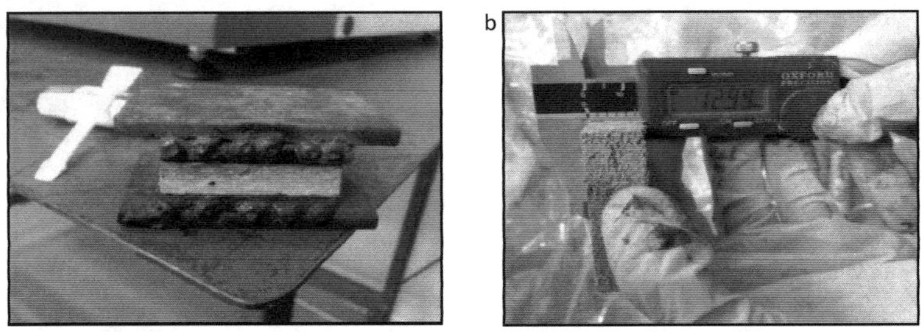

Figure 3. (a) The specimen of MDF was glued and bonded to loading block prior internal bonding test; (b) Vernier clipper was used to measure thickness of swelling on kenaf's MDF

4. Results and discussions

The result of MOR, MOE, IB and TS were plotted by classifying each parameters into different target densities as shown in Figure 4 and Figure 5. Figure 4 records MOR and MOE of panel with density of 0.8 g/cm^3 increased doubled corresponding to panel density of 0.7 g/cm^3 associated with decrement of IB and TS as seen in Figure 5. Furthermore, the average of MOR, MOE and TS recorded for target density of 0.7 g/cm^3 was in similar range as prescribed by [12] with thickness between 12 mm and 19 mm as stated in Table 1. While, the range of MOR, MOE and IB recorded in this study was also similar to MDF's panel which was consists of hundred percent kenaf in study of [21] with target density 0.7 g/cm^3. Similar pattern was also found in study of [28] which was recorded similar TS at about 11.45% with application of kenaf stick in MDF production in target density between 0.72 g/cm^3 and 0.73 g/cm^3. Small percentage of TS showed higher consistency and provides good stability in interconnection between MDF's panel [23]. Overall, the increment of target density of MDF indicated the increment of MOR, MOE and decrement of IB and TS. Thus, target density of 0.8 g/cm^3 for MDF is recommended due to high MOR and MOE and low IB and TS corresponding to other two (2) target density.

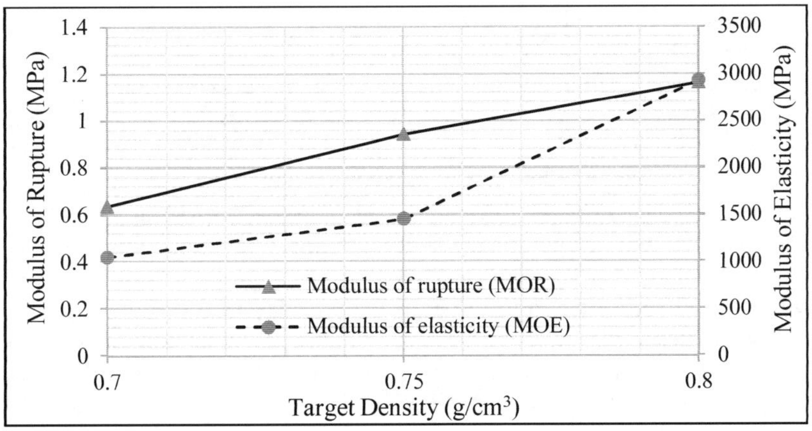

Figure 4. Modulus of Rupture (MOR) and Modulus of Elasticity (MOE) of Medium Density Fibreboard (MDF) containing kenaf at different target density

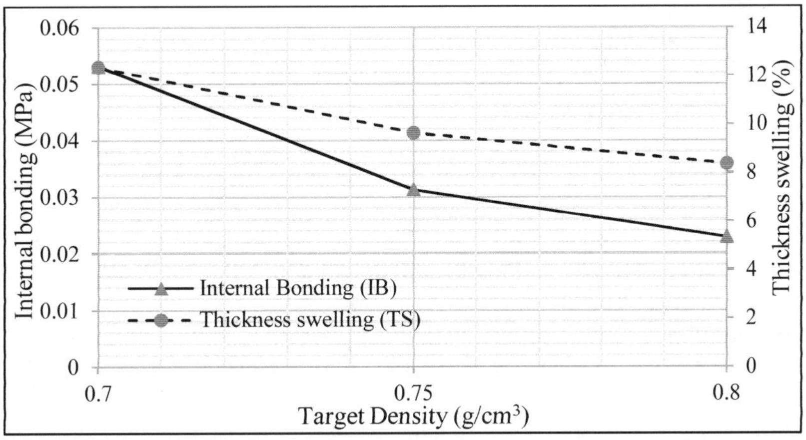

Figure 5. Internal bonding (IB) and thickness swelling (TS) of Medium Density Fibreboard (MDF) containing kenaf at different target density

5. Conclusions

Finding showed that the increment of target density of MDF indicated the increment of MOR, MOE associated with decrement of IB and TS as described in [12]. Therefore, target density of 0.8 g/cm^3 for MDF is recommended due to high MOR and MOE and low IB and TS compared to other two (2) target densities. In conclusion, Kenaf (Hibiscus Cannabinus L) can be pontentially be used as an alternative material in producing Medium Density Fibreboard (MDF) which able to produce high mechanical properties and low physical properties.

6. References

[1] H'ng P S, Khor B N, Tadashi N, Aini A S N and Paridah M T 2009 Anatomical structure and Morphology of new kenaf varieties. *Asian Sci. Res.* **2(3)** 161-166

[2] Faruq G, Alamgir M A, Rahman M M, Motior M R, Zakaria H P, Marchalina B, and Mohamed N A, 2013 Morphological characterization of kenaf (hibiscus cannabinus l.) in Malaysian tropical environment using multivariate analysis *Animal & Plant Sciences* **23(1)** 60-67

[3] Han J S and Rymsza T A 1999 Determining the minimum conditions for soda-anthraquinone pulping of kenaf bast, core, and whole stalk fibers *Second Annual American Kenaf Society Conference* (USA) pp 230-235

[4] Lemahieu, P J , Oplinger E S and Putnam D H 2003 Kenaf. In: Alternative Field crops Manual. retrieved on 2 April 2019 from _http://www.corn.agronomy.wisc.edu/FISC/Alter natives /Kenaf.htm

[5] National Kenaf and Tobacco 2018 (Official Portal:National Kenaf and Tobacco Board) retrieved on 2 October 2018 from http://www.lktn.gov.my/index.php/en/

[6] Harun J, Paridah M T and Khalina A 2009 Kenaf- Its establishment and journey towards energizing the wood-based and biocomposite industry in Malaysia. *In: International Conference on Prospect of Jute and Kenaf as Natural Fibres* (Dhaka-Bangladesh) pp 25-30

[7] Voulgaridis E, Passialis C and Grigoriou A 2000 Anatomical characteristics and properties of kenaf stem (Hibiscus cannabinus) *IAWA* **21(4)** 435–42

[8] Alireza, A, Jalaluddin, H, and Mohd Nor M Y 2003 Pulping and papermaking properties of Malaysian cultivated kenaf (Hibiscus cannabinus) *In: Proceeding of the second technical review meeting on the national project kenaf research project* (Serdang) pp 108–115

[9] Abdul Khalil H P S, Ireana Yusra A F, Bhat A H and Jawaid M 2010 Cell wall ultrastructure, anatomy, lignin distribution, and chemical composition of Malaysian cultivated kenaf fibre *Ind Crops Prod.* **31(1)** pp 113–21

[10] Zaini L H, Jonoobi M, Paridah M T and Karimi S 2013 Isolation and characterization of cellulose Whiskers from Kenaf (Hibiscus cannabinus L.) bast fibers. *Biomater Nanobiotechnol* **4** pp 37–44

[11] Paridah M T, Nor Hafizah A W, Zaidon A, Azmi I, Mohd Nor M Y and Nor Yuziah M Y 2009 Bonding properties and performance of multi-layered kenaf board *Tropical Forest Sci.* **21(2)** 113– 22

[12] Dongwha Malaysia 2018 (Wood Based Panels) retrieved on 29 July 2018 from http://www.dongwha.com/myen/product/product1_01.asp

[13] Grigoriou A, Passialis C, and Voulgaridis E 2000 Kenaf core and bast fibre chips as raw material in production of one-layer experimental particleboard. *Holz Roh-Werkstoff* **58** pp 290–291

[14] Wang D and Sun X S 2002 Low density particleboard from wheat straw and corn pith. *Ind Crops Prod.* **15** pp 43–50

[15] Pandey A and Nema P K 2004 Development of particleboard from soybean husk without resin. IE (I) J – AG 2004 **85** pp 5–9

[16] Kalaycioglu H, and Nemli G 2006 Producing composite particleboard from kenaf (Hibiscus cannabinus L.) stalks *Ind Crops Prod.* **24** pp 177–80

[17] Zheng Y, Pa Z, Zhang R, Jenkins B M and Blunk S 2007 Particleboard quality characteristics of saline jose tall wheatgrass and chemical treatment effect. *Bioresource Technol* **98** pp 1304–1310

[18] Migneault S, Koubaa A, Nadji H, Riedl B, Zhang S Y and Deng J 2010 Medium-density Fiberboard produced using pulp and paper sludge from different pulping processes *Wood and Fiber Science* **42(3)** pp 292-303

[19] Rokiah H, Norhafizah S, Othman S, Sugimoto T, Hiziroglu S, and Sato M 2010 Effect of particle geometry on the properties of binderless particleboard manufactured made from oil palm trunk. *Mater. Des.* **31** pp 4251–4257

[20] Garcia-Ortuno T, Andreu-Rodriguez J, Ferrandez-Garcia M T, Ferrnadez-Villena,M, and Ferrandez- Garcia C E 2011 Evaluation of physical and mechanical properties of particleboard made from gaint reed (Arundo donax L.) *Bio Res* **6(1)** pp 477–86

[21] Juliana A H, Paridah M T, Rahim S, Nor Azowa I, and Anwar U M K 2012 Properties of particleboard made from kenaf (Hibiscus cannabinus L) as function of particle geometry *Materials & Design* **34** pp 406-411

[22] Aisyah H, Paridah M and Sahri M 2012 Influence of thermo mechanical pulping production parameters on properties of medium density fibreboard made from kenaf bast *Applied Sciences* **12(6)** pp 575-580

[23] Norul Izani M A, Paridah M T, Mohd Nor M Y and Anwar U M K 2013 Properties of medium-density fibreboard (MDF) made from treated empty fruit bunch of oil palm *Tropical Forest Science* **25(2)** pp 175–183

[24] Alpar T and Marko G 2017 Energy Grass as Raw Material for MDF Production *Acta Silv. Lign. Hung.* **13(1)** pp 69–79

[25] Gul W, Khan A and Shakoor A 2017 Impact of Hot Pressing Temperature on Medium Density Fiberboard (MDF) Performance, *Advances in Materials Science and Engineering* **2017 (**4056360)

[26] Kartal S, Nami, Green and Frederick 2003 Decay and termite resistance of medium density fiberboard (MDF) made from different wood species. *International biodeterioration & biodegradation* **51** pp 29-35

[27] Akbulut, Turgay, Hiziroglu, Salim, Ayrilmis and Nadir 2014 Surface absorption, surface roughness and formaldehyde emission of Turkish medium density fibreboard *Forest Products* **50(6)** pp 45-48

[28] Das A K, Mondal L R and Shams M I 2015 Physical and mechanical properties of kenaf (Hibiscus cannabinus) MDF *Asian Applied Science* **8(3)** pp 204-209

[29] Markessini E., Roffael E., and Rigal L., (1997), Panels from annual plant fibers bonded with urea-formaldehyde resins, *Proceedings of the 31st particleboard/composite material symposium*, (Washington, USA) pp 147-160

[30] Hosseinabadi Z, Faezipour H M, Latibari J and Enayati A 2008 Properties of medium density fiberboard made from wet and dry stored bagasse *Agriculture Science* **10** 461-470

[31] BS EN 310:1993 1993 Wood-based panels. Determination of modulus of elasticity in bending and of bending strength

[32] BS EN 319:1993 1993 Particleboards and fibreboards. Determination of tensile strength perpendicular to the plane of the board

[33] BS EN 317:1993 1993 Particleboards and fibreboards. Determination of swelling in thickness after immersion in water

[34] Wambua P, Ivens J and Verpoest I 2003 Natural Fibres: Can They Replace Glass in Fibre Reinforced Plastics? *Composites Science and Technology* **63(9)** 1259-1264

Acknowledgments

The author would like to acknowledge Timber Fabrication Laboratory and Light Structures Laboratory, Faculty of Civil Engineering and Environmental, Universiti Tun Hussein Onn Malaysia (UTHM), Malaysia.

IOP Conf. Series: Materials Science and Engineering **713** (2020) 012021 doi:10.1088/1757-899X/713/1/012021

The Influence of Moisture Content, *pH* and Liquid Limit on Electrokinetic-treated Dredged Marine Sediments Stability by using Calcium Chloride as an Enhancement Agent

Nurul Syakeera Nordin[1,2] **and Chee-Ming Chan**[1]

[1] Faculty of Engineering Technology, Universiti Tun Hussein Onn Malaysia, 86400 Parit Raja, Batu Pahat Johor
[2] Research Centre for Soft Soils, Universiti Tun Hussein Onn Malaysia, 86400 Parit Raja, Batu Pahat Johor
Corresponding author: chan@uthm.edu.my

Abstract. Reusing the dredged marine sediment (*DMS*) that was spawned from maintenance dredging work will reduce its disposal effect towards the environment. Furthermore, reusing the *DMS* could optimize the storage capacity of existing confine disposal facilities (*CDF*) and can prolonging its lifespan. Treating the *DMS* with an adequate pre-treatment is the best way to manage this sediment in sustainable solution. The purpose of this research is to study the influence of moisture content (w), *pH* and liquid limit (*LL*) towards the *DMS*'s stability after treated with electrokinetic (*EK*) method. Stainless steel plate was used as the electrode, while distilled water (*DW*) and calcium chloride ($CaCl_2$) were applied as the enhancement agents. The aforementioned enhancement agents are electrically injected into the *DMS* which causing flow of the solutions through the pores in *DMS* under 50 V/m of applied direct current. The *EK*-treated *DMS* shows increment in strength and it was influenced by w, *pH* and *LL*. The dry zone area was created near the anode and wet zone was created near the cathode. The application of $CaCl_2$ create a better result in physical properties of *DMS* compared to pure systems. The *EK*-treated *DMS* hence could be reuse as a geomaterial in construction activities. Largely, *EK* has significantly improve the quality of *DMS* even though the strength increase observed was not homogeneous throughout the specimen.

1. Introduction

Mosavat *et al.* [1] have demonstrated the various effects on the soil due to EK treatment. Figure 1 summarises the chemical changes in the clay with the alteration in the physical and chemical characteristics of the clay soil. In their studies, the authors have investigated the performance of EK towards kaolinite by using stainless steel electrodes. The authors highlight when a direct current (DC) electric potential is applied to soil, it will stimulate the migration of electricity, pore fluid, ions and fine particles. These migrations occurred across the soil towards the oppositely charged electrodes. Henceforth producing a combined effect of a chemical, hydraulic and electrical (*CHE*) gradient [1-2].

The authors have mentioned there are various complex electrochemical effects regarding the application of electric potential. Those effects are the electrolysis reaction, electric potential difference, electric current flow and leading to production of electrophoresis and electromigration. The electrolysis reaction at the anode and cathode are as equation (1) and (2).

Anode:	$2H_2O - 4e^- \rightarrow O_2\uparrow + 4H^-$	$E_0 = -1.229$	(1)
Cathode:	$2H_2O + 2e^- \rightarrow H_2\uparrow + 2OH^-$	$E_0 = -0.828$	(2)

The oxidation process occurred at the anode and it generate an acidic font. While reduction process occurred at the cathode and generating a base font. The standard reduction electrochemical potential (E_0) is the measure of the reactants in its own standard states [1][3]. Electrochemical effect creates an electric potential difference that lead to generation of electroosmosis (EO). EO is a flow of pore fluid across the soil profile. It is known as the primary phenomena in EK that responsible for species transport through the porous media [1][4-6].

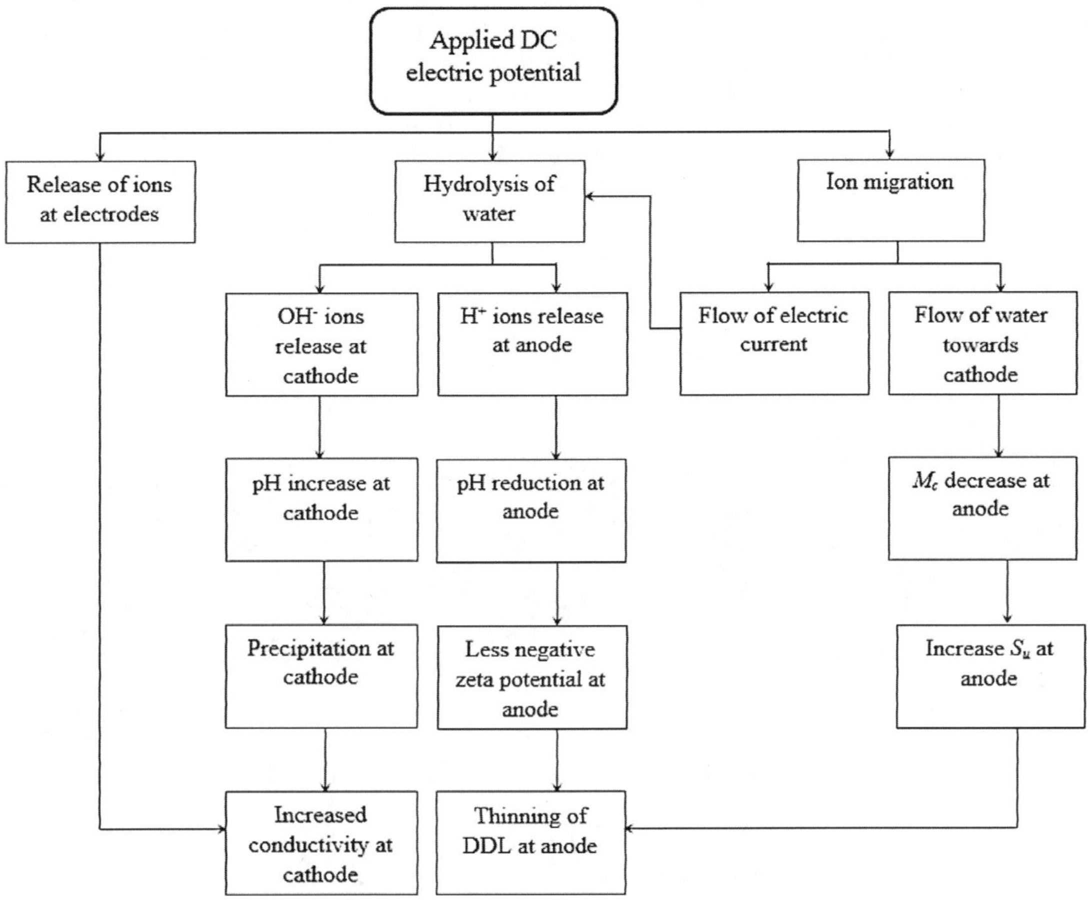

Figure 1. Summary mechanisms of EK treatment [1].

In year 2014, Mosavat *et al.* [1] traces the development of undrained shear strength (S_u) towards *EK* treated soil. In their case study of *EK* treatment towards kaolinite, the authors identify the strength development at five different normalized distances from anode and at five different time intervals. The S_u was measured directly using a hand vane shear apparatus. In their study, two types of enhancement solution were used to explore the *EK* performance. These enhancement solutions were permitted through the soil via anode. Two types of enhancement solution are hydrated lime and saline solution. Hydrated lime was selected due to its extensive use as a soil stabilizer and soil treatment agents. The saline solution was prepared by mixed 0.1 N $CaCl_2$ with 1:100 sodium silicate (Na_2SiO_3) solutions. The combination of $CaCl_2$ and Na_2SiO_3 will result in the formation of hydrated calcium silicate. This mixture is generally used as grouting material. Silicate grouts has demonstrated the advantages in being reliable, safety and ease of use, environmental acceptability and adaptability over a wide range of applications. In their research, the authors claim a change in the electro-osmotic potential occurred across the tank with the development of differential zeta potential in soil under the influence of the electric field. It is also associated with the changes in soil *pH* and the changes in the chemical concentrations of the soil pore solution. Jayasekera and Hall [7] mentions the location of wet and dry regions are reliant on the mineralogical properties of the soil. Its properties include the permeability characteristics, clay type and content, diffuse double layer (*DDL*) characteristics, etc.

2. Experimental work

2.1 Material

The studied material was dredged from Kuala Perlis Jetty, Perlis of Peninsular Malaysia in year 2016. Selection of dredging site was made based on Malaysian Marine Department (Jabatan Laut Malaysia, JLM) dredging schedule. The sediments (figure 2) was dredged at depth 4 to 6 m from sea level using backhoe dredger (figure 3). The properties result of the *DMS* are reported in table 1. The *w* is about 240.74 % (3.36 *LL*), dried in an oven at 100 ºC (BS 1377-2:1990:3.2). The particle density (G_s) value of *DMS* is 2.36 Mg/m^3, measured using a small pycnometer method (BS 1377-2:1990:8.3). The liquid limit (*LL*) value is 71.70 % by using a cone penetrometer method (BS 1377-2:1990:4.3). The plastic limit (*PL*) and plastic index (*PI*) value for *DMS* are 40.06 % and 31.64 %, respectively (BS 1377-2:1990:5). The *DMS* consist of 2.5 % of sand, 80.5 % of silt and 17 % of clay. The *DMS* was classified in high plasticity silt, *MH* for its soil classification (Unified Soil Classification System) [8-9].

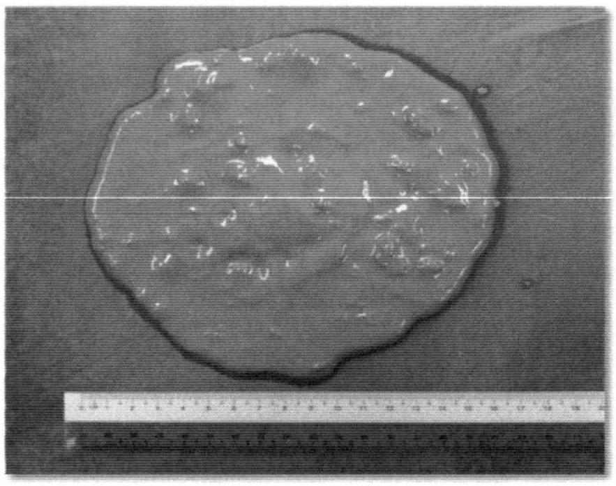

Figure 2. *DMS*

Figure 3. Backhoe dredger.

Table 1. Properties of dredged marine sediments.

Natural water content, *w* (%)	240.74		
LL (%)	71.70		
PL (%)	40.06		
PI (%)	31.64		
w/LL	3.36 *LL*		
G_s (Mg/m^3)	2.36		
Particle size distribution (%)	Sand: 2.5	Silt: 80.5	Clay: 17
Soil classification	High plasticity silt, *MH*		

2.2 EK experimental set-up

The *EK* cell (figure 4) was made from transparent acrylic plate (15 *mm*) and was divided into three compartments. Each compartment was separated with perforated walls. The usage of transparent acrylic plate was to prevent short circuiting and monitoring the *DMS* level during the experiment. Those three compartments are the specimen cell and two electrolyte reservoirs (anode and cathode compartments). The internal dimension for specimen cell and electrolyte reservoirs were 280 × 165 × 413 *mm* and 40 × 165 × 413 *mm*, respectively.

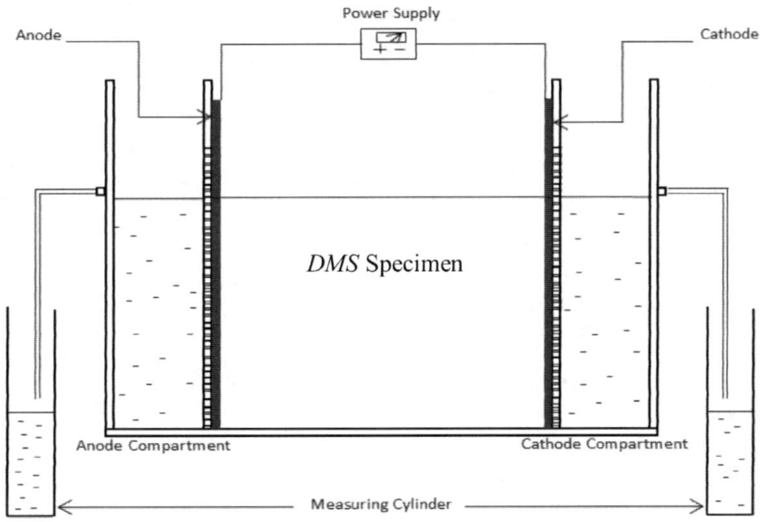

Figure 4. *EK* cell.

Two types of *EK* experimental setups were erected with different enhancement agents in electrolyte compartment. Those setups were namely as EK_0 (*DW - DW, natural system*) and EK_1 ($CaCl_2$ *- DW* system). Each electrolyte reservoir was filled with 1000 *mL* of catholyte/anolyte. A summary of *EK* experimental setups is listed in table 2. The experiment was conducted in 14 days. The period was considered from the previous study by Nordin and Ahmad Tajudin [10-11]. In his analysis, the volume of electrolyte used were monitored every 24 hours and the author identifies that the volume was constant after 224 hours (9.3 days). When *EK* experiment was terminated, the treated specimen was sectioned into six equal parts at each 45 *mm* from the anode. Those treated specimens were investigated through its *w* (BS 1377-2:1990:3.2), S_u (BS 1377-7:1990:3), *LL* (BS 1377-2:1990:4.3) and *pH* (BS 1377-3:1990:9).

Table 2. The summary of EK experiment.

Parameters	EK_0	EK_1
Applied voltage (V/m)	50	
Electrodes	Stainless steel	
Anode compartment electrolyte	*DW*	1.0 *M* $CaCl_2$
Cathode compartment electrolyte	*DW*	
Experimental duration	14 d	

3. Results & Discussion

3.1 EK effect on w and S_u variations

The correlation between *w* and S_u of *EK* treated *DMS* is shown in figure 5. The graph presents the intercorrelations of six measurements of *w* - S_u at different normalized distance from the anode. These data are divided into three zones for analysis. The first zone was 0 – 95 *mm* (near the anode), second zone 95 – 190 *mm* (middle area), and third zone was 190 – 280 *mm* (near the cathode) from the anode. These results indicate decreasing *w* with increasing S_u. It is interesting to note that a wet region was created in the cathode area, while dry region in the anode area. A comparison of these *EK* phases reveals that the strength development is not homogenous but varies accordingly to the distance from the anode. Another striking result to emerge from the data is the high increment of S_u in EK_1. It shows that the application of $CaCl_2$*-DW* in the *EK* system has significantly improved the strength of *DMS* compared to EK_0 (*DW–DW*).

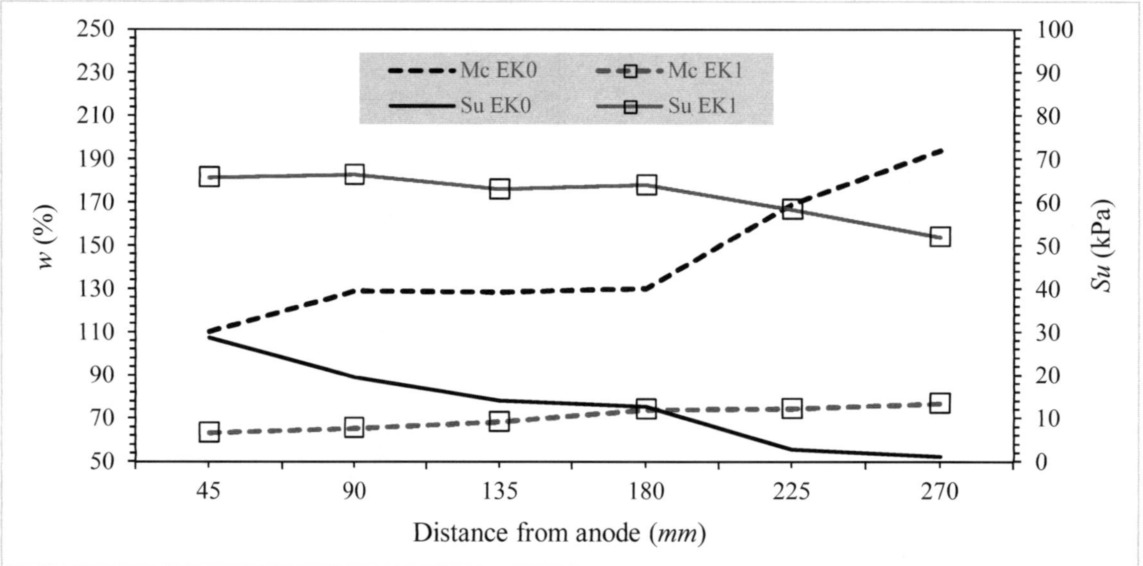

Figure 5. Variations of *w* and S_u for *EK*-treated *DMS*.

Asavadorndeja and Glawe [12] mentioned in their study that the decrease in *w* provides some contribution to the increase in strength. In the third zone, the decrease in S_u and increases in M_c are similar in all *EK* phases. It is encouraging to compare this observation with Liaki *et al.* [13] who found similar trend of M_c - S_u relationship near the cathode area. The authors claimed that the circumstances are in accordance with classical soil mechanics theory. A high increment of S_u of EK_1 is attributed to the introduction of calcium ions into the *DMS*. The injected of calcium ions from the anode were moved into the *DMS* by electroomosis process. The positive performance of calcium ions as the injecting chemical in *EK* has been proven since 1960s by Esrig and Gemeinhardt. They have investigated the performance of calcium ions in *EK* stabilization towards an illitic clay. The results showed an improvement in the strength compared at the same water content and it increased more than one order of magnitude [14].

The results corroborate with the postulations of Mosavat *et al.* [1] who found similar condition in *EK*-treated soil, where the driest region occurred in the first zone and the wettest region occurred in the third zone. It was further claimed that consolidation took place in the anode area, with the reduction in *w* producing *EK* cementation effect. Referring to Mosavat *et al.* [1] and Jayasekera & Hall [7], the location of wet and dry regions is highly dependent on the mineralogical properties of the soil, such as permeability, soil type and diffuse double layer (*DDL*) characteristics.

3.2 Relationship between w and Su

Figure 6 presents the relationship between *w* and S_u of *EK*-treated and untreated *DMS*. There is a significant positive relationship between strength and moisture parameters of untreated *DMS* (R^2 = 0.9892). These relationship of untreated *DMS* were measured by using a maintain load test. The untreated *DMS* was consolidated to a different *w*, henceforth measure its strength by using a laboratory vane shear test. Hence, the exponential line of untreated *DMS* provides an indication of experimental variability in identifying the processes that affected the strength development in treated *DMS* [12][14]. This finding corroborates with the ideas of Liaki *et al.* [13], who proposed that the data that lie above and/or shift to the right of the exponential line denote soil strengthening. Whilst, for data that lie below and/or shift to the left denote soil weakening. The authors declared that the plotted data that lie a small distance away from the exponential line might be attributed to chemical changes arising on the surface of the soil particles (e.g. *DDL*). On the other hand, data lying further away from the exponential line were attributed to chemical reactions occurring in the soil (e.g. compound precipitation).

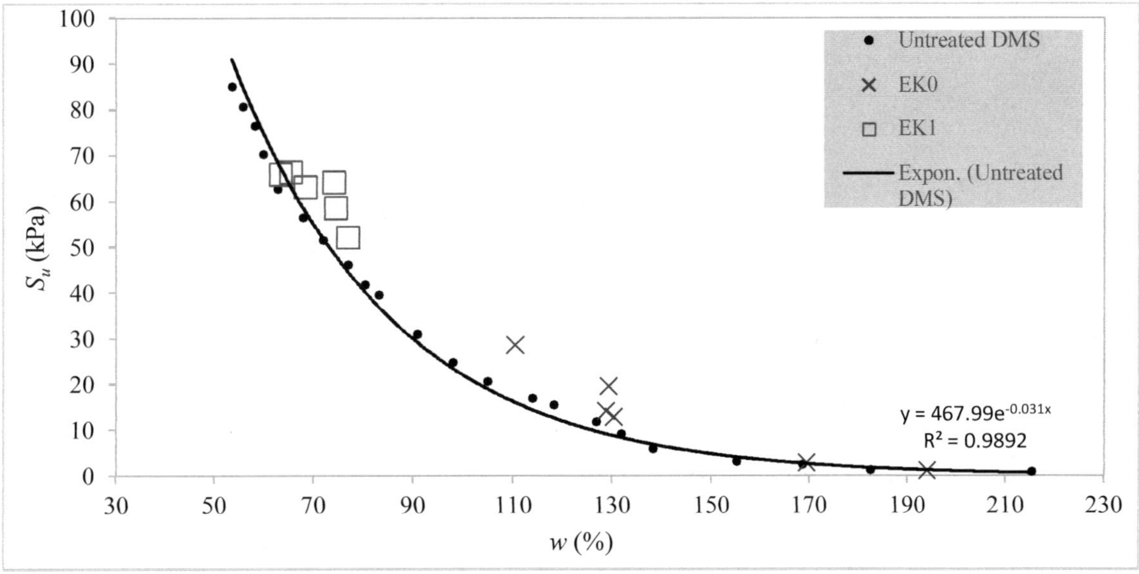

Figure 6. Relationship between w and S_u of untreated and *EK*-treated *DMS*.

Most of the plotted data of *EK*-treated *DMS* shows an improvement and lie above the exponential line. This study has found that generally right after the termination of *EK*, the treated *DMS* near the cathode developed as a wet region compared to other sections. This attributed to the electroomosis process which carried water molecules from the anode towards cathode. The present study confirms previous findings by Liaki *et al.* [13] in the strength development of *EK* treated *DMS*. The strength development in *DMS* is mainly affected by the *DDL*. This is due to the small distance of plotted data from the exponential line and it is also reinforced by the pH value of *EK*-treated *DMS*. Where, only for the further distance away from the exponential line of plotted data are affected by the precipitation process and required a high pH value (alkaline condition). A further study could assess by increasing the molarity or concentration of chemical used in order to investigate the strength development that affected by the compound precipitation process.

3.3 EK effect on pH and S_u variations

The relationship between *pH* and S_u of the *EK*-treated *DMS* is presented in figure 7. As can be seen from the figure, pH at the first zone (near the anode) were reported significantly more acidic in all *EK* phases than the other two zones (middle area & near the cathode). The correlation between *pH* and S_u is suggestive of the strength development being significantly affected by the pH value. The *pH* in all *EK* phases displayed increment towards the cathode and remained in alkaline condition near the cathode area. This trend is similar to previous studies where acidic (pH < 7) and alkaline condition (pH > 7) were observed to develop in the anode and cathode regions of *EK*-treated soil, respectively [12-13].

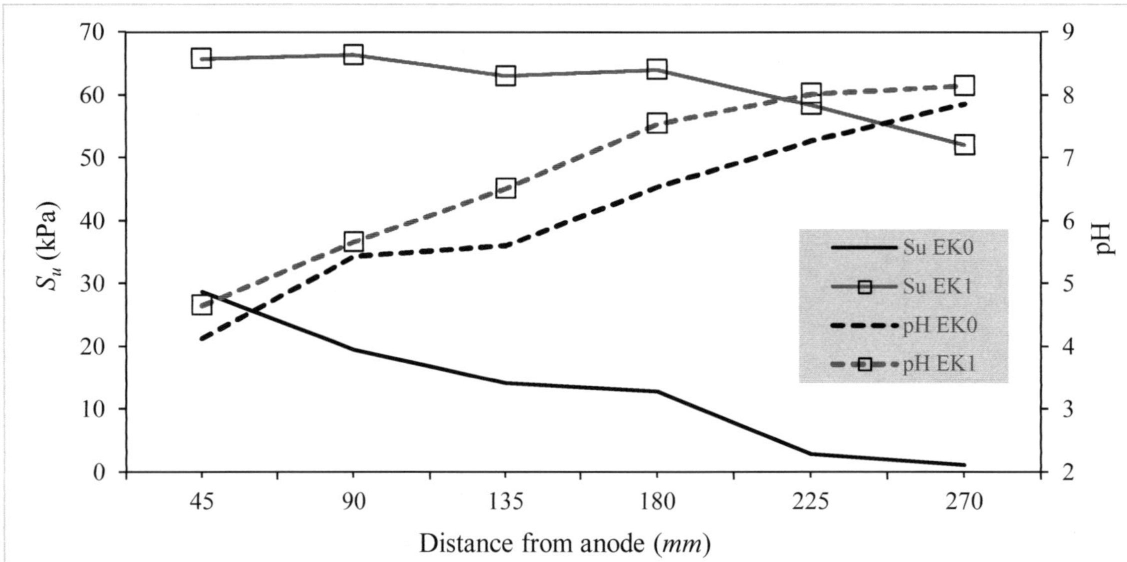

Figure 7. Variations of *pH* and S_u for *EK*-treated *DMS*.

In 1999, Ozkan *et al.* [14] reported that acidic soil conditions can cause ionization of soil, dissolution of basic precipitates, desorption of surface cation species into the pore fluid and leading to a flocculated soil structure. The present results corroborate with the findings by Ozkan *et al.* [14], where the acid front in the anode area recorded the highest strength in the EK-treated *DMS.* Asavadorndeja & Glawe [12] postulated that the formation of cementing agents in *EK* stabilization is mainly dependent upon the soil's *pH* and the availability of free calcium ions. The application of cations as an enhancement agent and the processes involved explains the ions would migrate into the soil through the processes of electromigration and osmotic advection, hence improving the soil's strength. One of the mechanisms that provides the greatest contribution in increasing the soil's strength is precipitation. As mentioned before, *DDL* is the major mechanism that underscores strength development in *DMS* compared to precipitation [13]. There are several possible explanations for this result primarily near the cathode area that merely affected by the precipitation. Some authors have speculated that precipitation reaction normally occurred when pH of soil solution more than seven [12]. Henceforth, it shows that the precipitation reaction is limited to the cathode region.

3.4 EK effect on pH and LL variations

Figure 8 shows the relationship between *LL* and *pH* relative to distance from the anode. It is clearly indicated that the *LL* values for *EK*-treated *DMS* is *pH* dependent. Predominantly, the post-*EK LL* values decreased with increased distance from the anode. At first zone, the *LL* for *EK*-treated *DMS* rose from the untreated *DMS* value (71.70 %). Second zone two acts as a 'transition zone' between the *LL* and *pH*, while third zone illustrates the post-*EK DMS* pH rose from the untreated *DMS* value (6.80).

According to Liaki *et al.* [13], the results of an acidic environment and higher *LL* values near the anode may attributed to flocculation in the *DMS* structure. Besides, the compression of the *DDL* may occur as there are more positive ions (i.e. H^+) existent by implication from the *pH* values. Henceforth, this circumstance supporting the finding in S_u value which clarified the higher strength in anode were reinforced by a flocculated structure in *EK*-treated *DMS.*

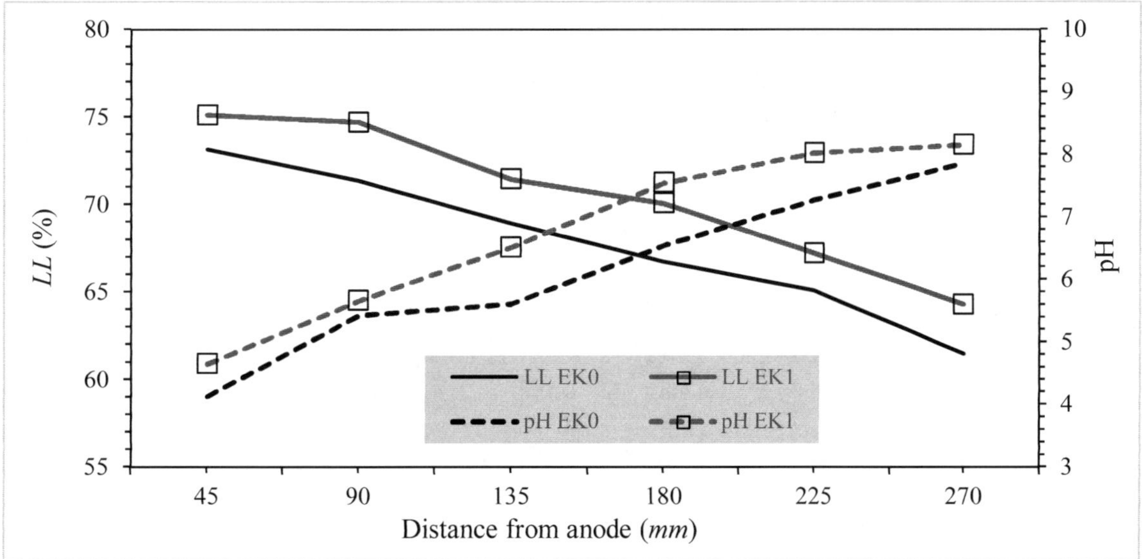

Figure 8. Variations of *pH* and *LL* for *EK*-treated *DMS*.

Additionally, the application of stainless steel electrode may have indirectly raised the *LL* at first zone. This is due to the greater number of higher valency positive ions (i.e. Fe^{+}) emerging into the system near the anode. This could also be affected by the thinning of *DDL* via cation exchange where the higher valency ions replace the lower valency of ones [13]. In year 1983, Casagrande postulated that the increase in *LL* mainly near the anode is affected by the electroosmosis process. On the other hand, Mitchell (1993) observed that the increasing cation valence tends to increase the *LL* of non-expansive minerals. Thus, these findings are corroborates as observed by several past research works that clarified the impact of the type of electrode used. The findings of this result have an important implication to support the physicochemical changes in *EK*-treated *DMS.*

4. Conclusions

An *EK* treatment towards *DMS* shows an improvement in enhancing its S_u. It is also developing the driest and wettest zone in the vicinity of anode and cathode, respectively. The application of $CaCl_2$ as stabilizers create a better result in physical properties of *DMS* compared to other systems. The grey colour of *DMS* turns to reddish brown in the vicinity of anode. This is due to the release and migration of iron oxides that affected from *EK* processes. The iron oxides that adsorbed onto the *DMS* surface contributed to a cementation effect between sediment particles. This cementing compound has improved the strength of the *DMS*. The *w*, *pH* and *LL* values of *EK* treated *DMS* are correspond with the S_u value in all system [15].

5. References

[1] Mosavat N, Oh E and Chai G 2014 Laboratory investigation on physico-chemical changes in kaolinite during electro-kinetic treatment subjected to enhancement solutions, *EJGE* **19** 1215 1233

[2] Jayasekera S 2008 *An investigation into modification of the engineering properties of salt affected soils using electrokinetic* University of Ballarat

[3] Iyer R 2001 Electrokinetic remediation, *Particulate Science and Technology* **19** 219-228

[4] Acar Y B, Gale R J, Alshawabkeh A N, Marks R E, Puppala S, Bricka M and Parker R 1995 Electrokinetic remediation: basics and technology status *Journal of Hazardous Materials* **40** 117 – 137

[5] Casagrande L 1949 Electro-osmosis in soils, *Geotechnique* **1**(3) 159 – 177

[6] Mitchell J K and Soga K 2005 *Fundamentals of Soil Behavior.* 3rd ed. Berkeley: University of California
[7] Jayasekera S and Hall S 2007 Modification of the properties of salt affected soils using electrochemical treatments *Geotech Geol Eng.* **25** 1-10
[8] Nordin N S and Chan C M 2017 Undrained shear strength of low dosage cement-solidified dredged marine soils (DMS) for reclamation works *International Journal of GEOMATE* **13**(35) 180-186
[9] Nordin N S and Chan C M 2018 Evolution of shear strength with varying cement dosages in dredged marine soils *MCRJ Special Issue* **5**(3) 65-72
[10] Nordin N S 2015 *Physicochemical properties of batu pahat clay using electrokinetic stabilisation method* Master Thesis (Universiti Tun Hussein Onn Malaysia)
[11] Ahmad Tajudin S A, Nordin N S, Marto A, Madun A, Zainal Abidin M H, Jefferson I and Mohammad Azmi M A 2016 The monitoring and cementation behavior of electrokinetic stabilisation technique on Batu Pahat marine clay *International Journal of GEOMATE* **11**(26) 2581-2588
[12] Asavadorneja P and Glawe U 2005 Electrokinetic strengthening of soft clay using the anode depolarization method *Bulletin of Engineering Geology and the Environment* **64** 237–245
[13] Liaki C, Rogers C D F and Boardman D I 2010 Physico-chemical effects on clay due to electroigration using stainless steel electrodes *J. Appl. Electrochem.* **40** 1225-1237
[14] Ozkan S, Gale R J and Seals R K 1999 Electrokinetic stabilization of kaolinite bu injection of Al and PO_4^{3-} ions *Ground Improvement* **3** 135-144
[15] Jayasekera S 2015 Electrokinetics to modify strength characteristics of soft clayey soils: a laboratory based investigation *Electrochimica Acta* **181** 39-47

Acknowledgement

Technical support by RECESS (Research Centre for Soft Soils) at the University is duly acknowledged. Special thanks to Soil Instrument (M) Sdn. Bhd. For funding inclusion of the work in GCoMSE2019.

IOP Conf. Series: Materials Science and Engineering **713** (2020) 012022 doi:10.1088/1757-899X/713/1/012022

Behaviour of Post-Tensioned Kempas Timber Beam with Two Tendon Types

W mushina[1], D Y Eng Chuan[2], j mushina[3], N Abd Ghafar[2] and K Boon[2]

[1] Shatra technical institute, Southern technical university
[2] Department of Structural and Material Engineering, Univeristi Tun Hussein Onn Malaysia, Parit Raja
[3]Shatra municipality, Thi-qar municipality, Shatra Iraq
Corresponding author: eng.wesam16@gmail.com, david@uthm.edu.my, norhayati@uthm.edu.my, koh@uthm.edu.my

Abstract. The post-tensioning technology is commonly used for concrete structures especially bridges. This technology is applied for timber elements in many countries effectively. In Malaysia the timber post-tensioning still not known yet. This research aim to propose the post-tensioning technique for timber beams. The timber type is Malaysian Kempas and the device used to apply the post-tensioning force is hydraulic jack. Three types of specimens were tested using the bending test (four-point bending test) for a simple support beam. two type of specimens are post-tensioned timber beams with black threaded rod bar (PTBR) and post-tensioned timber beams with silver threaded rod bar (PTSR). All the specimens' length is 1.2meter and the dimension of timber is 90mm x 40mm. The increment of moment capacity of post-tension timber beam compared to the timber beam the moment resistance increment is 1.42 times. The moment resistance of the post-tensioned timber beam had average increment is 27.4% - 31.8%. The other finding is the failure mode of specimens it is changed from brittle tension failure to ductile failure (compression failure) the cause of this behavior is the role of the post-tensioning rods.

1. Introduction

Pre-tensioning and post-tensioning forces are well known in structural elements. They are widely used in many systems and with different materials. Especially concrete, as they can modify the bending stresses by reducing the tensile stresses. This property of tensioning is also useful in wooden structures. Although wood has a very high tensile strength when it is defect-free. Its tensile strength in real structures is less than its compressive strength due to the presence of defects intrinsic to this material. The effects of applying a pre-stressing force on timber beams has been studied and analysed several times Lantos [1]. Used a mono-wire cable for pre-stressing with the aim to reinforce low-grade wood. The result was fewer variable failure loads with only a modest increase in strength. Peterson [2] attempted to increase strength and stiffness by using a pre-stressed high-strength steel strip bonded to the beam tensile surface, with moderate success. Reinforcing glulam beams with bonded pre-stressed FRP laminates in the tension zone has recently been proposed as a mean of utilizing low-quality timber by improving its structural performance [3].

Stress-laminated deck plates are typically used in bridge decks. A pre-stressing force is applied perpendicular to the grain and increases the shear friction between the individual pieces of timber [4] More recently, a pre-stressed LVL-concrete composite beam was constructed and tested by Deam et al [5]. The pre-stressing tendons had little effect on stiffness or strength, but they did reduce the defection from the permanent load, particularly when the tendons were draped. New Zealand and Australia recently developed frames of unbounded post-tension timber to construct long-span structures in multi-storey building [6]. The laminated veneer lumber (LVL) and wire strands were the main material for that system. This system can resist the load from earthquake by generating moment –resisting frames [7].

Researchers used steel bars at Italy to glue laminated (glulam) timber beam as pre-stressed tendons. This tendon (steel rod) used in pre-stressing timber beams increased the stiffness, ductility and ultimate load of timber compared to normal timber [8]. The steel or carbon FRP plates used by early researcher to apply pre-stressing in timber beam to increase strengthen of timber [9]. These researches of the reinforcement systems didn't success to use as construction market at Italy. The reason of that is the long-term of efficiency and reliability for these systems still do not have enough proven experimental data [8]. McConnell et al. research [10] considers the glulam post-tensioned timber act as reinforced beam with addition eccentric force (post-tensioning force). Figure1 explain the stress diagram that happened in post-tensioned glulam timber, the post-tensioning force could neutralize the tensile stress resulting due the load.

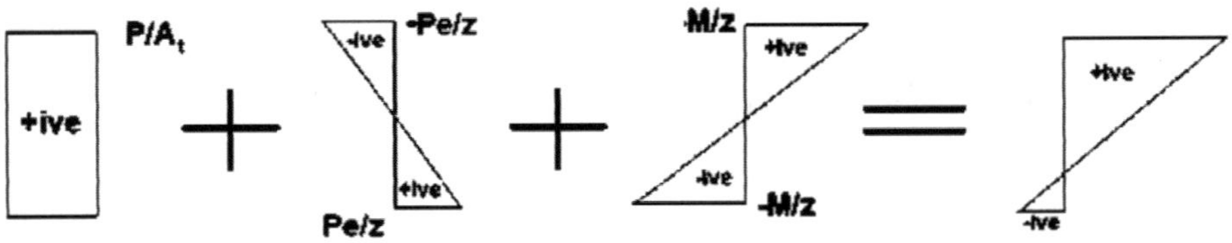

Figure 1: Stress profiles of unreinforced and tension reinforced glulam timber [10]

This research aim to propose the post-tensioning technique for timber beams. The type of timber is Malaysian Kempas and hydraulic jack used to apply the rod bar force that present the post-tensioning force to the timber beam. Two type of threaded rod bar used as tendons silver color and black color.

2. Materials and methods

2.1. The specimens detail

The details of cross section for the three type of specimens is shown in Figure 2. The timber beam is the first specimen. The post-tensioned timber beam is second specimen. The post-tension timber concrete composite beam is the third specimen. The top layer of concrete is laid on the top of post-tension timber beam.

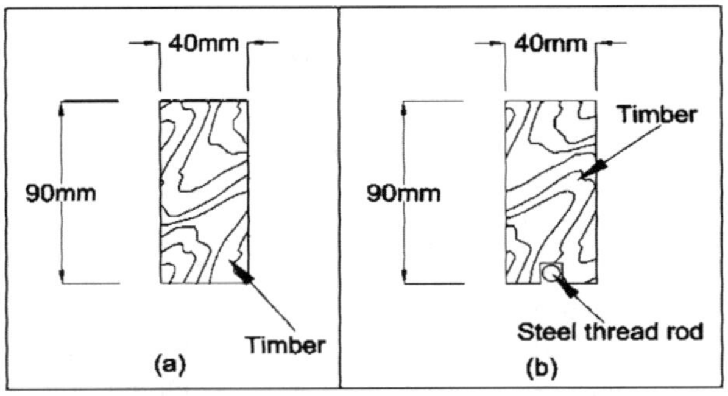

Figure 2. Cross section details: (a) Timber only, (b) Post-tensioned Timber

3 series with 4 specimens for each series are prepared with acronym TB as notation for timber beam. For post-tensioned timber beam with black rod the notation is acronym PTBR and for post-tensioned timber beam with silver rod Acronym PTSR as notation.

2.2. Post-tensioning method and preparation for specimens post-tensioned timber beams.

Post-tensioning by Pre-stressing Jack method is the standard method of post-tensioning in concrete structures. Similar method is used in the case of timber post-tensioning. The main difference is that low post-tensioning force required in case of timber post-tensioning. Which lead to use threaded rod bar as tendons. The components used are the device (pump and jack) of capacity 200 KN. extra fittings are required to fix the timber specimen and transfer the load from device to the specimen. These fitting are pair of C channel covered by plate to fix the specimen during the process, two end plates with size (40x110mm. thickness 10mm) fixed with screw. Two type of rod bars black color (8.85 mm) and silver color (9.4mm) with nuts Figure 3 (b) and (c) [11].

The total components are shown in Figure 3 and 4. Chair steel bracket at the timber beam end to transfer the load from the jack to the beam, Figure 4. Show the position of two dial gauges to measure deflection of timber and displacement of rod bar.

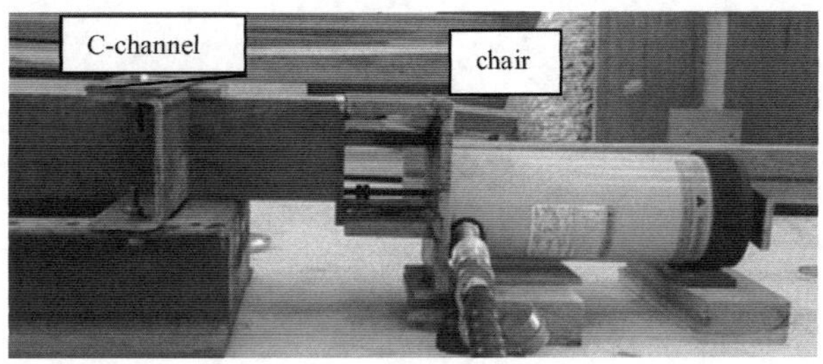

(a)

(b)

(c)

Figure 3. a) the chair position during the post-tensioning process b) black rod bar c) silver rod bar

Figure 4. a) The arrangement of jack and pump during post-tensioning process

Figure 4. b) The dial gauge location

2.3. The bending test procedure
All the specimens had been carried out the four-point bending test to determine the bending strength of the specimen, bending modulus of elasticity, the material flexural stress-deflection response. This test usually carried out using Universal Testing Machine (UTM) according to BS EN 408:2010 code [12]. The arrangement of the bending test is shown in Figure 5. The I-beam used at two points over the span to ensure apply symmetrical load in the test, refer Figure 5. The load rate set to 5.0 mm/min until the specimen failure. The maximum deflection occurs at the middle of span. Fore that the (LVDTs) linear variable differential transformers are located at midpoint of the span to record the specimen deflection.

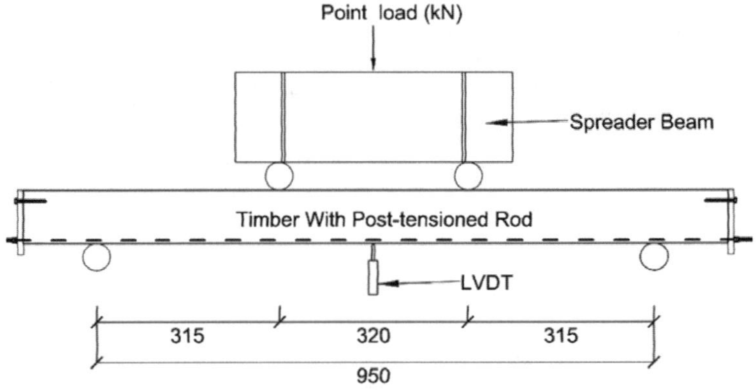

Figure 5. The arrangement of four-point bending test

3. Preliminary tests
The supplementary tests concerned the timber beam are (4 point bending test) and the compression test. The threaded rod bar test is tensile test for the two types black color and silver color. The detail of test specimens and results were presented in Table 1. The modulus of elasticity for black rod bar is 205.769GPa and silver rod bar is 150.133GPa

Table 1: the preliminary tests results in (MPa)

The test type	Number of specimens	Size mm	Results	
			Average value	COV%
4 point bending test	5 timber	1000x40x90 (l x w x h)	86.5	7.6%
Compression test	2 timber	40x90x380 (l x w x h)	65.8	0.1%
Tensile test	2 Rod bar silver	100x9.4 (l x diameter)	Yield stress	Ultimate stress
			298 4%	318.5 1%
Tensile test	2 Rod bar black	100x8.85 (l x diameter)	Yield stress	Ultimate stress
			519 1.5%	538 1.7%
			3.11	2.74

4. Tensioning force estimation

The theoretical post-tensioning force is obtained from the deflection formula used in case of pure bending moment acting on the end beams. This formula is calculated using moment area method and its value $\Delta = \frac{M \times L^2}{8EI}$ [13]. The post-tensioned force (PT) is calculated by two methods theoretically and experimentally. The theoretical method using the deflection equation of pure bending moment at the end $\Delta = \frac{M \times L^2}{8EI}$, where $M = PTxe$ then $\Delta = \frac{PT \times e \times L^2}{8EI}$ and it was related to the PT upward deflection. The experimental method depend on rod bar deformation ($\Delta_{\text{rod bar}}$) and rod bar modulus of elasticity $E_{\text{rod bar}}$, the $PT = E_{rod\ bar} x \epsilon x A$. The results of the two methods are listed in Table 2.

Table 2: Post-tensioning force estimation theoretically and experimentally

Series references	Upward Deflection (mm)	Estimated PT Force (kN) $PT = \frac{\Delta \times 8EI}{eL^2}$	Estimated PT Force (kN) $PT = E_{rod\ bar} x \epsilon x A$
PTBR-1	3.84	22.3	22.2
PTBR-2	2.66	15.4	17.7
PTBR-3	4	23.2	22.8
PTBR-4	3	17.4	19.0
PTSR-1	2.84	16.5	15.0
PTSR-2	2.86	16.6	15.1
PTSR-3	3	17.4	15.5
PTSR-4	3.38	19.6	16.7

5. The Effect of Post-Tensioning Method on Timber Beam Behavior

The behavior of timber beam and post-tensioned beam is presented in Table 4. The F_u is the ultimate load during the timber beam bending test. The M_{exp} is the collapse bending moment value which calculated from the equation $M_{exp} = \frac{F_u \times a}{2}$. The timber beam bending strength fb that calculated by equation $f_b = \frac{3F_u a}{bh^2}$. The equation for modulus elastic, E is $E = \frac{F_u a}{48I\,\delta_{max}}(3L^2 - 4a^2)$ which depend on the specimen length L, experimental collapse bending load F_u and the deflection at mid-span in collapse condition, δ_{max}. The average bending stress of timber beam is 86.5MPa and the average modulus elastic for timber beam is 8961.0Mpa. The mode failure of all the specimens for timber beam is tensile crack started at tension edge. For specimens PTBR, the moment improvement had experimental average 27.4% compared to the timber beam bending strength. Next, for specimens PTSR, the bending strength improvement is 31.7%. The failure type of the post-tensioned beam observed after the bending test show two types. First type is shown in Figure 6(a) display anchorage longitudinal shear failure and the steel end plate is torn out. Second type is shown in Figure 6(b) display the compression with longitudinal shear failure occurs at the mid span.

Table 3. Experimental data results with bending strength increment of timber beam and post-tensioned beam.

Specimen Ref.	F_u (kN)	M_{exp} (kNm)	PT upward def. (mm)	Δ_{max} (mm)	Increment (%)
TB- Average	30.4	4.8		21.2	
PTBR-1	34.2	5.5	2.84	43.7	11.67
PTBR-2	38.3	6.1	2.66	37.8	25.06
PTBR-3	40.1	6.4	4	22	30.94
PTBR-4	43.5	7.0	3	26.2	42.04
Average	39.0	6.2		32.4	27.4
Std. Dev.	3.9	0.6		10.1	
COV %	9.9	9.9		31.0	
PTSR-1	39.2	6.3	2.84	22.3	28.00
PTSR-2	43.3	6.9	2.86	26.7	41.39
PTSR-3	40.0	6.4	3	25.0	30.61
PTSR-4	39.0	6.2	3.38	26.2	27.35
Average	40.4	6.5		25.1	31.8
Std. Dev.	2.0	0.3		2.0	
COV %	4.9	4.9		7.9	

(a)

(b)

Figure 6. Typical failure modes of the PT beams silver rod: (a) at the anchorage, (b) at the mid span

A comparison between timber beam and post-tensioned timber beam bending test results are shown in Figure 7. It is obvious that post-tensioned timber show higher bending capacity and more ductility. The data results of all the tests for the two types of rod bars are shown in Figure 8a) and the close results of tests for both types are shown in Figure 8b). the specimens which is close to average value are shown in Figure 8c) it can see that the silver rod bar give small increment than black rod bar.

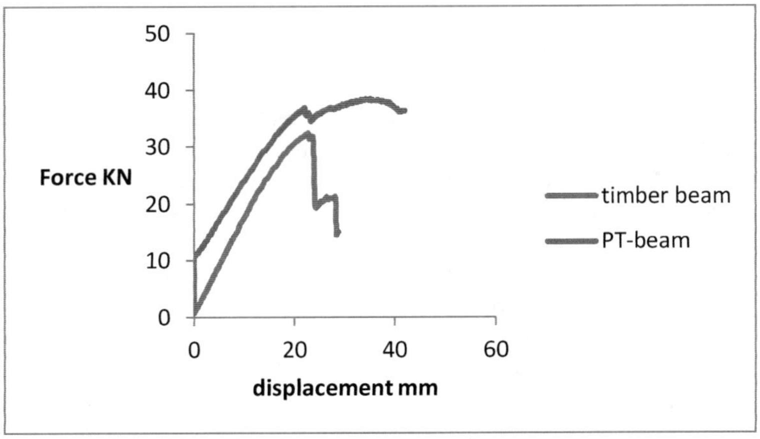

Figure 7. Comparison between bending test for timber beam and PT beam

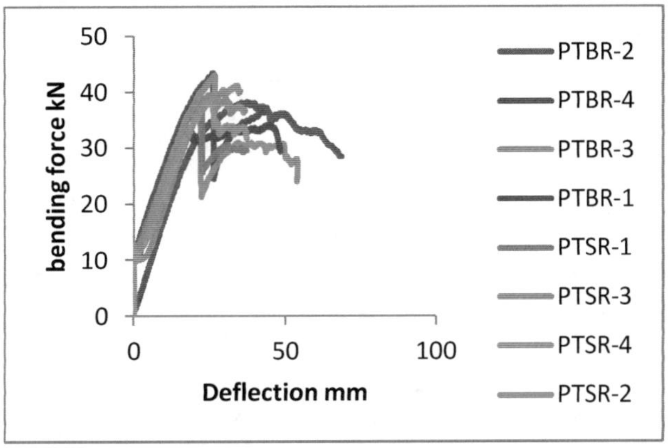

a)

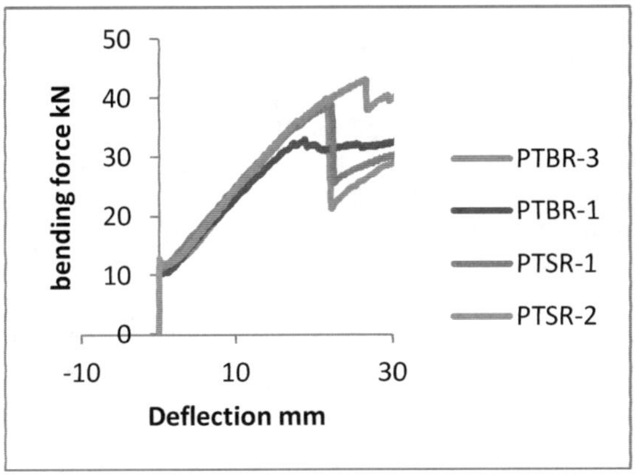

b)

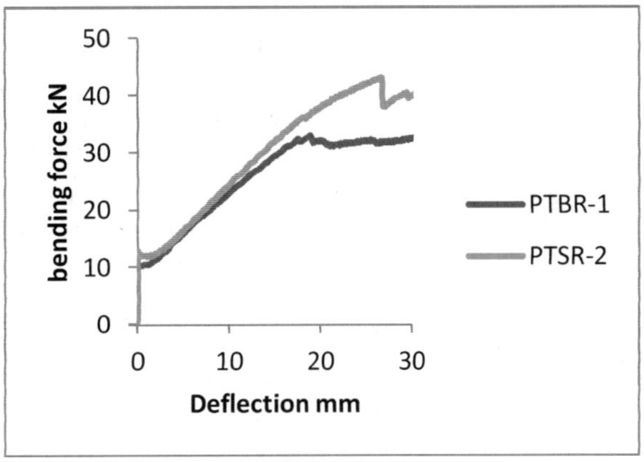

c)

Figure 8. A comparison between a) all data results b) the close results c) the typical results

6. Conclusions

In term of post-tensioning timber the average increment of moment resistance compared with timber beam capacity is 27.4% for black color rod bar and 31.8% for silver rod bar. The common failure mode of beam after bending test it is observed that a compression with longitudinal shear failure occurs at mid span. The other failure type in less degree is anchorage longitudinal shear failure and the steel end plate is torn out. The bending test results for post-tensioning timber beams show the critical specimens beam is PTBR-4 and PTSR-2. Both of the bending strength increment is 42% compared to timber beam bending capacity, the average experimental bending capacity for silver rod bar is 6.5kNm which is higher than black color type 6.2kNm with variation value 6.1-7.0 kNm depending on the timber properties and defects presence in the timber. The analytical of stress distribution, show reduction in the resultant stress value due to the post-tensioning stress that result from post-tensioning force eccentricity. The maximum negative displacement that can be achieved is 4mm. the post-tensioning fitting parts, end anchorage locking nut, end plate and the timber at the end support were efficient during post-tensioning process.

7. References

[1] Lantos G 1964 Reinforced and post-tensioned glue laminated beams under development at TRADA laboratories *Civil Engineering (London)* ***59***(690) 86-87

[2] Peterson J 1965 Wood beams prestressed with bonded tension elements *Journal of the Structural Division* **91**(1) 103-120

[3] Triantafillou T C and Deskovic N 1992 Prestressed FRP sheets as external reinforcement of wood members *Journal of Structural Engineering* **118**(5) 1270-1284

[4] Ritter M A, Geske E A, Mason L, McCutcheon W J, Moody R C and Wacker J 1990 Performance of stress-laminated bridges *Wood Design Focus* **1**(3) 12-16

[5] Deam B L, Fragiacomo M and Gross L S 2008 Experimental behavior of prestressed LVL-concrete composite beams *Journal of Structural Engineering* **134** 801-809

[6] Buchanan A, Palermo A, Carradine D and Pampanin S 2011 Post-tensioned timber frame buildings *Struct. Eng.* **89** 24–30

[7] Palermo A , Pampanin S , Carradine D , Buchanan A H , Dal Lago B , Dibenedetto C, Giorgini S and Ronca P 2010 Enhanced performance of longitudinally post-tensioned long-span LVL beams, *11th World Conf. Timber Eng. 2010, WCTE 2010 (Italy)* vol 1 pp 449–459

[8] De Luca V and Marano C 2012 Prestressed glulam timbers reinforced with steel bars *Construction and Building Materials* **30** 206-217

[9] Kliger R, Al-Emrani M, Johansson M and Crocetti R 2007 Strengthening glulam beams with steel or CFRP plates In *Proc. Asia-Pacific Conf. FRP Struct.(APFIS 2007)* pp 291-296

[10] McConnell E., McPolin D and Taylor S 2014 Post-tensioning of glulam timber with steel tendons *Construction and Building Materials* **73** 426-433

[11] Mushina W, Yeoh D, Mushina J, AbdGhafar N and Boon K 2019. Kempas Timber Un-Bonded Post- Tensioning Solution New Approach *International Journal Of Civil Engineering And Technology (IJCIET)* **10**(03) 634–647

[12] EN, B. (2012). 408: 2010+ A1: 2012. *Timber structures. Structural timber and glued laminated timber. Determination of some physical and mechanical properties* (London: British Standard)

[13] Pathak D 2012 *hand book to SCC junior mechanical recruitment exam guide* (Disha Publication: Agra college) pp A-13

IOP Conf. Series: Materials Science and Engineering **713** (2020) 012023 doi:10.1088/1757-899X/713/1/012023

Vibration Behaviour of Natural Timber and Timber Concrete Composite Deck System

Jalal Mushina [1], **NorHayati Abd Ghafar** [2], **David Yeoh** [2], **Wissam Mushina**[3], **Koh Heng Boon**[2]

[1] Shatra municipality, Thi-qar municipality, Shatra Iraq
[2] Department of civil engineering, UTHM University, Parit Raja Malaysia
[3] Shatra Technical Institute, Southern Technical University, Shatra Iraq
Corresponding author: jalalmushina@yahoo.com, noryati@uthm.edu.my, david@uthm.edu.my, eng.wesam16@gmail.com, koh@uthm.edu.my

Abstract. Timber–Concrete Composite (TCC) structure consists of timber joists or beams effectively interconnected to a concrete slab cast on top of the timber members. This type of structures has applications in multi-storey buildings and short-span bridges particularly in developed countries such as America, Canada, Australia and New Zealand. In Malaysia, these timber concrete composite (TCC) fields are extremely uncommon compared to European country. This paper is investigated the TCC deck system as a new slab structure element that could be used in Malaysia. The use of TCC deck system with Malaysian Kempas timber and normal weight concrete is advantageous in providing a lighter structure, more sustainable and possibly a smaller foundation. Therefore the main objective of this research is to study the vibration behavior of kempas timber (natural timber) in timber deck and TCC deck by experimental modal testing. The vibration data results were conducted by modal experimental test. Then the model is verified by finite element using SAP 2000 software. The connector type that used is screw in TCC deck. The size of each deck is 3.6m x 0.8m, the thickness of concrete is 65mm and the timber beam unit with dimension 3.6 x 0.09x0.04 m. The specimens of timber and TCC deck were tested using an electrodynamic shaker. The data were analyzed using ME Scope software to determine the natural frequency and mode shape of the timber and TCC deck system. The modal experimental results for all tests were more than 8 Hz, it accepted since the vibration limitation for timber floor is 8 Hz. The TCC deck frequency due to shaker excitation is higher than only timber deck which means that the TCC improve vibration resistance property compared with only timber deck vibration resistance.

1. Introduction

The vibration is a movement that occurs either on the ground or in the structure of the building and it affects the structure of the building and can cause discomfort to the user, vibrations should be dealt with because of the extreme vibration will cause damage to the structure of the building. Human rhythmic activities such as walking, running, dancing, jumping and aerobic are among the causes of vibration on the floor surface. Even during this activity, the occurrence of vibrations caused by humans and the dynamic burden produced by humans. The vibration susceptibility problems of using full timber floors can be solved by combining timber with concrete[1].Timber-concrete composite (TCC) construction may be used to improve the performance of structural timber floor systems[2].The vibration caused by the higher harmonics of the activity frequency can also cause discomfort and excessive deflections in the floor panels[3].The natural frequency and damping of LCC beams decreased if the span length was

IOP Conf. Series: Materials Science and Engineering **713** (2020) 012023 doi:10.1088/1757-899X/713/1/012023

increased EC5[4]. For vibration to be acceptable, the natural frequency of the designed flooring systems must be higher than the forcing frequency[5].This study was conducted to show the effect of concrete topping on timber deck (TCC deck) on the vibration resistance property. The electrodynamic shaker is used as excitation source to report the timber and TCC deck vibration data which express the deck behavior. These tests determine the frequency value at specific mode that reflect the behavior of timber and TCC deck. An electrodynamics shaker has been conducted on surface of timber and TCC deck system. The raw data from shaker testing was exported has been made by using ME Scope software. The other objective of SAP 2000 finite element modelling is regard to transformation from the actual deck to a finite element TCC deck, as well as mesh refinement.

2. Material properties

A-Timber: The type of timber used for this project is Kempas (Scientific name Koompassia malaccensis), it was natural timber in Malaysia. The dimensions of this timber beam was of 40 mm (width) x 90 mm (depth) x 3600 mm (length) and the density of timber was 850 kg/m3 and elastic modulus (G) 12.5 GPa.

B-The shear connectors: The connectors used for this study was screws. It has 68.70 mm length, 5.47 mm diameter and the spacing between the screws was 250 mm from design.

C-The concrete: The design of the concrete mix was based on design of experiments British Standards. The target characteristic strength of the concrete for this project is 35 MPa. The w/c ratio is 0.4 and the targeted slump is in the range of 70-120 mm. TCC floor deck is a composite in which a kempas timber was represented as a deck (lamination timber) with 3600 mm length, 800 mm width and 90 mm depth of timber and layer of concrete on top with 65 mm thickness, the two materials are bound together by screw connector.

3. Modal testing

Modal testing is the form of vibration testing of an object whereby the natural frequencies and mode shapes of the timber and TCC under test are determined.

3.1 Grid line for the specimens

The grid lines were drawn along the surface of the specimen for marking point to place the accelerometers, becomes easier for deck of timber and TCC which located at 400mm along the length and 200mm along the width, the grid line was made based on the area of the study specimen.
The accelerometers nodes were located at intersection points of grid lines and the electrodynamics shaker is located on the top of the floor deck to apply the vibration force, each overlapping point needs to be numbered from one to the last point of the grid as shown in Figure 1.

Grid number on the specimen can be started from either direction that is through vertical or horizontal directions as long as it is easy to read and understood during electrodynamics shaker test. Each grid line had been marked with a number to facilitate the process of retrieving data. The accelerometer placed at each numbered grid line starting from the first row. A wax will be used to make sure an accelerometer stick between plate and the deck. Make sure an acceleremoter stick neatly and the connection wires not messy to reduce an error occur during the testing. There are nine rows for vertical direction and 5 rows for horizontal for one deck as shown in Figure 1. Due to limited number of accelerometers, the accelerometers were roving to capture all data from the points on the specimen. Furthermore, one accelerometer was placed as a reference point. The electrodynamic shaker also connected to the data logger. For the shaker, the controller software that been used was spektra software package. Furthermore, the shaker placed at one point and no need to changed places like the accelerometer. The vibration responses were recorded by accelerometers, which are attached along the specimens to measure the vibration response from the specimens and were connected to a data recorder that was connected to the computer.

In this paper study four tests of vibration, the first test is case (A), it means the single timber deck, The second test is case (B), it means the single TCC deck, The third test is case (C), it means the double TCC deck arrange along the width (side to side) and the fourth test is case (D), it means the double TCC deck arrange along the length as shown in Figure 2, 3, 4 and 5 respectively, there were a concrete joint between the two TCC decks in case (C) and case (D) as shown in Figure 4 and 5. The supports were used in these cases, steel I section support.

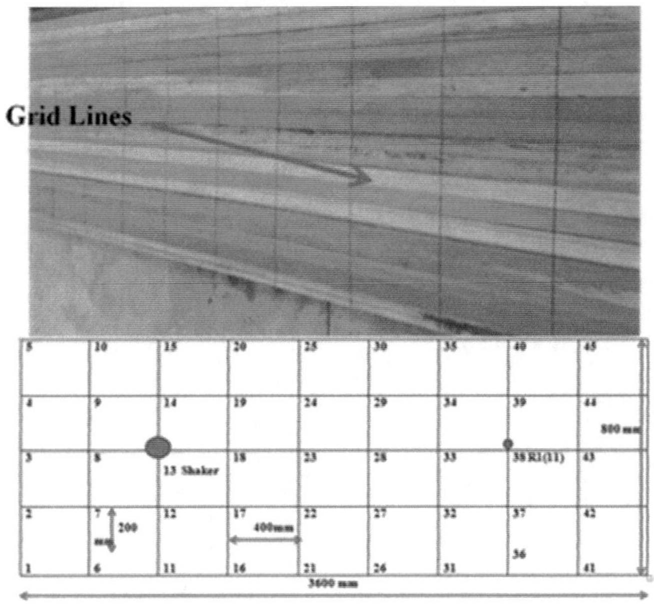

Figure 1. The Grid lines and Grid number on the specimen were drawn along the Deck of timber and TCC

Figure 2. Case (A) single timber deck and TCC

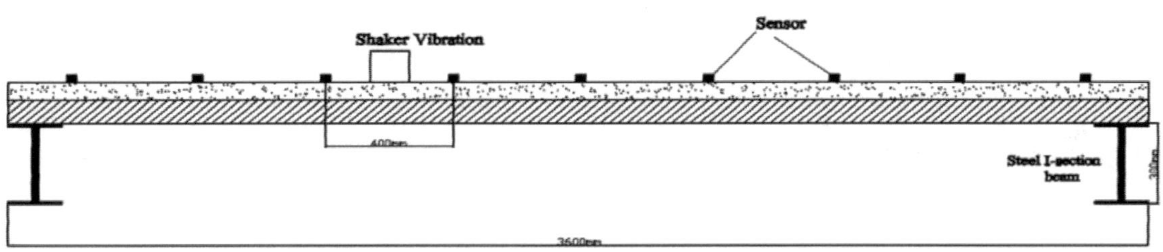

Figure 3. Case (B) single TCC deck

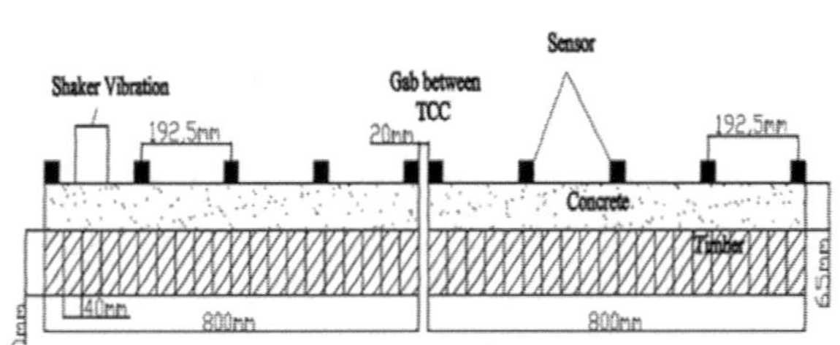

Figure 4. Case (C) double TCC deck arrange along the width (side to side)

Figure 5. Case (D) double TCC deck arrange along the length

3.2. Testing arrangement for Specimens

Twelve accelerometers were mounted at different locations of surface the deck floor to measure their responses. The accelerometer is a device tool that was placed on a specimen to record the response data during the experimental modal testing, using electrodynamics shaker. The accelerometers were connected to data logger to record all the response data as shown in Figure 6 and 7. The computer needs to have the IMC Wave software to generate the data from accelerometer. The equipment that was used during the test has been identified so there was no error during the installation.

Electrodynamic shaker is a tool used to vibrate the specimen. The shaker test was to determine the vibration result at the timber and TCC deck floor specimens. The frequency range for the shaker has been set from 1 Hz to 100 Hz. The time for shaker to running is only for 2 minute 20 seconds. This shaker was connected to the data logger and the data obtained was inserted into the IMC for analysis of the graphic form that occurs when the shaker vibrates as shown in Figure 7, it show the procedure of shaker tool setup.

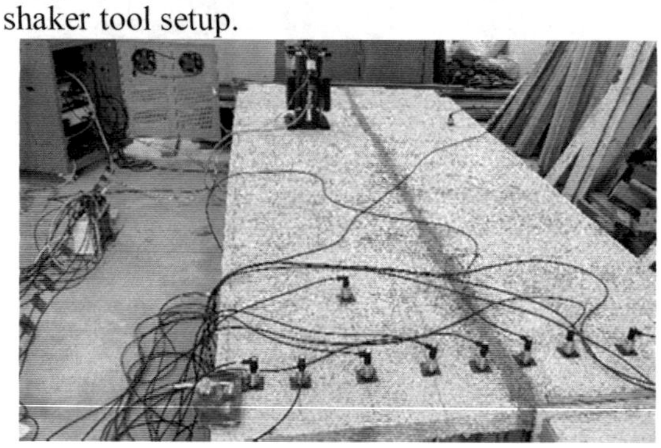

Figure 6. The accelerometers were connected to data logger to record all the response data

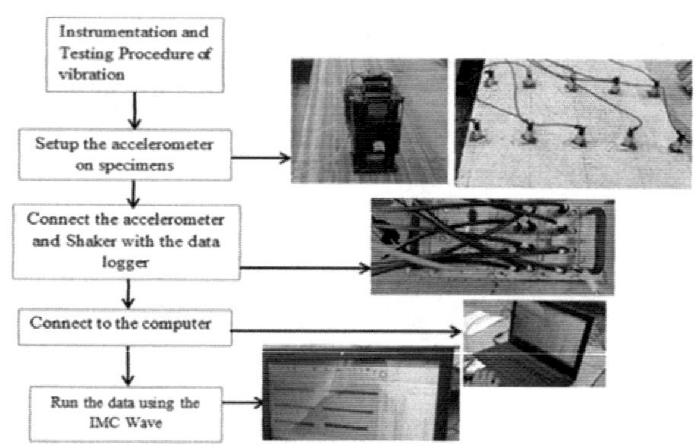

Figure 7. Procedure of shaker tool setup

3.3. Testing of procedure for Specimens

This test was done by using electrodynamic shaker test for the case (A), case (B), case(C) and case (D) in order to determine the natural frequency and mode shape.

The first test is case (A) single timber deck, the vertical force excitation was produced by an electrodynamic shaker, which was placed on top of the specimen at point (13) numbered grid line and placed an accelerometer at one node as the reference point at (38) as shown in Figure 1 and 8. Place an accelerometer at node 1 – 5 follow on the grid. Repeat changing the position of an accelerometer to the node 6 – 10, 11 – 15 until 41-45.

The second test was case (B) single TCC deck, it was same the first test (case A) as shown in Figure 8.The third test was case (C) double TCC deck arrange along the width (side to side), the vertical force excitation was produced by an electrodynamic shaker, which was placed on top of the specimens at point (23) numbered grid line and placed an accelerometer at two nodes as the reference point at (29) and (73) as shown in Figure 9. The fourth test was case (D) double TCC deck arrange along the length , the an electrodynamic shaker was placed on top of the specimens at point (13) numbered grid line and placed an accelerometer at two nodes as the reference point at (38) and (82).The shaker test was to determine the vibration result at the timber deck specimen. The accelerometers were attached at these node points to measure the vibration response from the specimens and were connected to a data recorder (Data logger) that was connected to the computer.

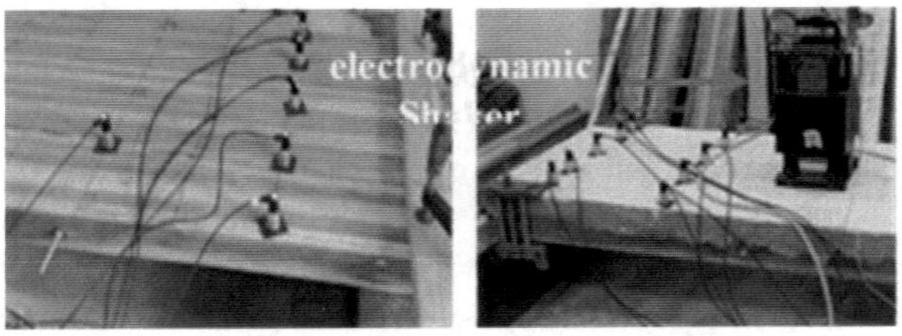

Figure 8. Electrodynamic shaker, which was placed on top of the case (A) and Case (B) specimens at point (13) numbered grid line

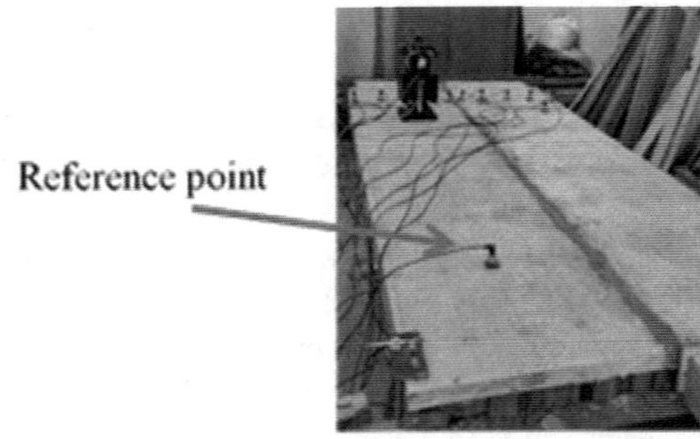

Figure 9. Case (C) double TCC deck arrange along the width, electrodynamics shaker, which was place at point (23) numbered grid line

3.4. Results of test for Specimens

An electrodynamic shaker test has been done, the natural frequency of the first mode shape for case (A), case (B), case(C) and case (D).The raw data of shaker test has been converting in ME Scope to get natural frequency and mode shape of all cases. The results of natural frequency for case (A) was 12.4 Hz, for case (B) was 12.9 Hz as shown in Table 1, while natural frequency of case (C) was 13.7 Hz and 17.5 for case (D) as shown in Table 2 The values of natural frequency were accepted for all cases, since the fundamental natural frequency of residential timber floor must be 8 Hz as a serviceability vibration limitation as states in Eurocode 5.

Table 1. Natural Frequency and mode shape of case (A) single timber and case (B) single TCC deck

Mode	Case (A) Single timber deck	Case (B) Single TCC deck
Mode 1	Frequency =12.4 Hz 	Frequency =12.9 Hz 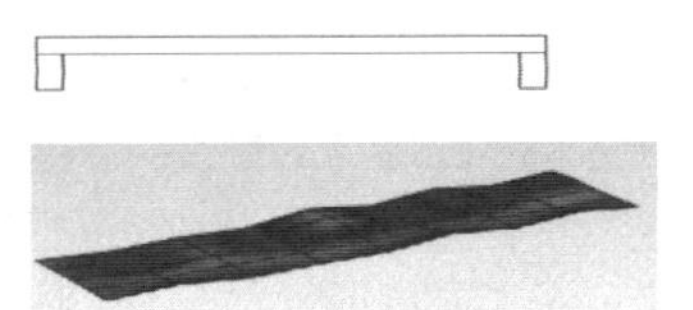

Table 2. Natural Frequency and mode shape of case (C) double TCC deck arrange along the width and case (D) double TCC deck arrange along the length

Mode	Case (C)	Case (D)
Mode 1	Frequency =13.7 Hz 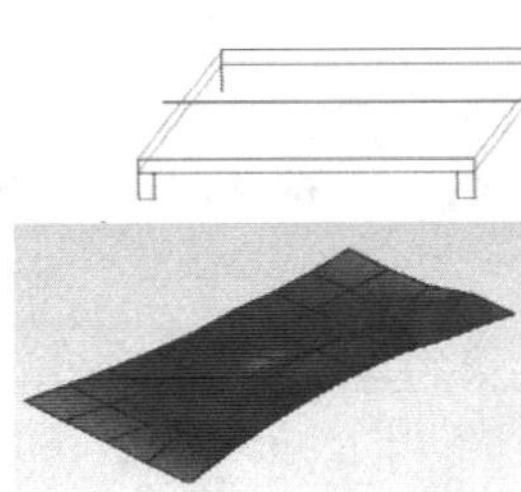	Frequency =17.5 Hz 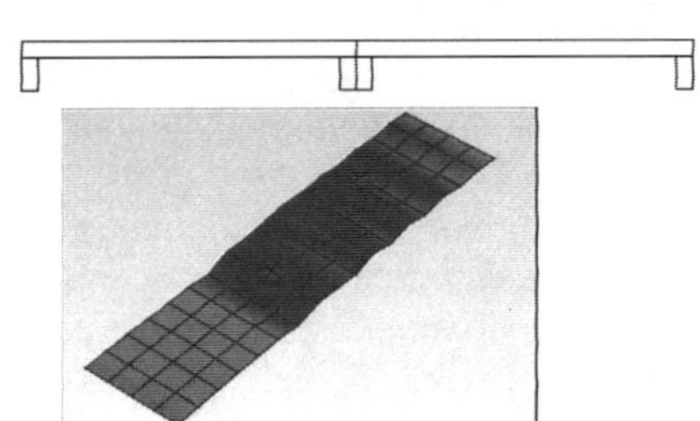

IOP Conf. Series: Materials Science and Engineering **713** (2020) 012023 doi:10.1088/1757-899X/713/1/012023

4. element method (FEM)

4.1. Modelling procedure of TCC

The first step of the modelling was to insert the material properties of concrete and kampas timber into the SAP 2000. For this study, concrete grade 35 was selected for use. However, the Kempas timber properties needed to be manually inserted. The Kempas is an orthotropic material, which means that the modulus of elasticity of kempas timber is 12.5kn/mm^2.In the modelling, the shell element represented the concrete slab, the beam element represented the kempas joists, and the link element represented the screw. For the link element, the shear connector stiffness value was inserted 46.9 kN/mm. The value 46.9 kN/mm was obtained from the push-out test. The rigid link was fixed at all directions. The details that needed to be inserted into the SAP 2000 are listed in Table 3, which covers the shell and beam elements.

Table 3. Details which needed to be inserted into the SAP 2000 software package

Material Properties			
Concrete		Kempas timber	
Mass, m	2400 kg/m^3	Mass, m From test	850 kg/m^3
Modulus of elasticity, E	31 kN/mm^2	Modulus of elasticity, E1 From test	12.5 kN/mm^2
Poisson ratio, U	0.3	Modulus of elasticity, E2 [6]	2.037 kN/mm^2
Shear modulus, G	9.56 kN/mm^2	Modulus of elasticity, E3 [6]	0.9 kN/mm^2
Concrete compressive strength, f'c	35 N/mm^2	Poisson ratio, U [6]	U12 = 0.369, U13 = 0.618, U23 = 0.428
		Shear modulus, G [6]	G12 = 1.075 kN/ mm^2 G13 = 1.0105 kN/mm^2 G23 = 0.1075 kN/mm^2
Shell element			
Thickness	65 mm		
Beam element			
Cross section	40 x 90 mm		
Link element			
Direction, U1	Fixed	Direction, U2 (From test)	46.9 kN/mm

The TCC model was then ready to be analyzed via modal analysis, including for natural frequency and mode shape. Modal Analysis is the study of the natural characteristics of the structures which includes frequency and mode shapes. It is use to design all types of structures including automotive structures, aircraft structures, computers, rackets and etc[7].Modal Analysis simplifies the vibration response of a complex structure into a set of single degree of freedom systems that can easily be understood[8]. It has become a regular in finding modes of vibration of a machine or structure.

4.2. Modelling result of TCC

The results of frequency from the finite element for the case (B) single TCC deck system which the cross section area of each TCC deck is 3.6 x 0.8 m, the thickness of the timber deck is 90 mm and thickness of the concrete is 65 mm, was 12.8 Hz, for case (C) double TCC deck arrange along the width was 13 Hz and case (D) double TCC deck arrange along the length was 14.5 Hz as shown in Table 4, Figure 10 and 11.

Table 4. Result of single and double TCC deck into the SAP 2000 software package

Floor	FEM (Hz)
Case (B) single TCC deck system	12.8
Case (C) double TCC deck arrange along the width	13
Case (D) double TCC deck arrange along the length	14.5

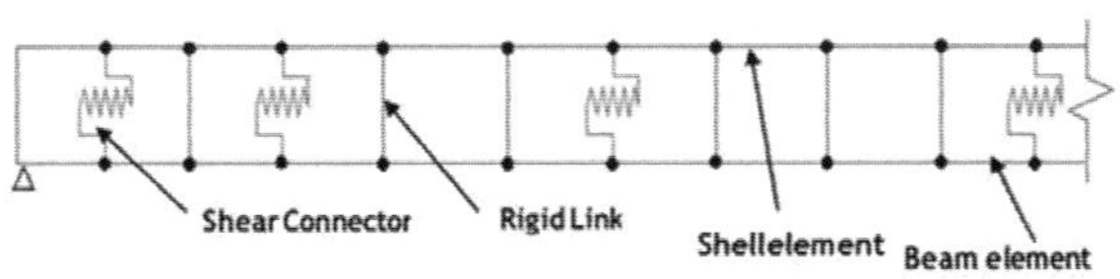

Figure 10. Deck and shell element with shear connector and rigid link

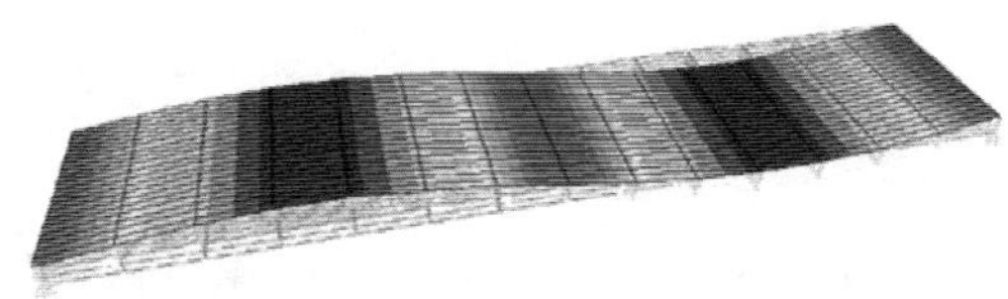

Figure 11. Second mode of A typical 3.6-meter TCC deck-span model from SAP 2000

5. Discusses the results

5.1. Discusses the experimental results for specimens

5.1.1. Effect of timber deck on vibration

The results shown that the natural frequency of the single timber deck (case A) for the first mode shape of timber floor was 12.4 Hz more than 8 Hz limitation of code, this mean there was enhancement when used timber lamination to resistance the vibration force.

These laminations of timber give the deck more stability to resistance the human activates (vibration) because its weight, more sustained and less ductile as the number of timber lamination increased as the beam tends to break at lower displacement.

5.1.2 Effect of topping concrete on TCC deck

The results shown of natural frequency of the case (A) single timber deck for the first mode shape was 12.4 Hz and the natural frequency of case (B) single TCC deck was 12.9 Hz as shown in Table 5.
This mean the experimental result of case (B) was higher than experimental result of case (A) by the differences 0.5 Hz. The different percentage between them is 0.04%. The additional concrete part in TCC deck gives the deck more stability, increased the stiffness of the TCC floor, due to mass increasing. The connector's part in TCC deck is responsible for the degree of composite.

Table 5. Experimental result of natural frequency for single timber and TCC Deck

Floor	Experimental result of case (B) TCC deck (Hz)	Experimental result of case (A) timber deck (Hz)	Different Hz	Percentage of different (%)
Floor deck	12.9	12.4	0.5	0.04

5.1.3. Effect of concrete gap on result of natural frequency for TCC floor deck

Based on the result obtained, the result of case (B) single TCC deck (12.9 Hz) was less than case (C) and case (D), this mean the concrete joint was increased the value of frequency, due to stiffness increasing for case (C) and case (D) as shown in Table 6.

The experimental result for case (D) was 17.5 Hz and it was higher than experimental results for case (C) which was 13.7 Hz by the differences 3.8 Hz. The differences between both ways are more than

27.7%, because the case(D) present real continuous TCC deck due to support existence at middle, while the case(C) present concrete joint connection along the total length of deck. Therefore the natural frequency of case (D) was higher than case(C).

Table 6. Natural Frequency and Mode Shape of case (A) ,(C) double TCC deck arrange along the width and case (D) double TCC deck arrange along the length

Mode	Case (B)	Case (C)	Case (D)
Mode 1	Frequency =12.9 Hz	Frequency =13.7 Hz	Frequency =17.5 Hz

5.2. Discusses the results of modelling
From the result of FEM, we note the frequency of case (B) single TCC deck system was 12.82 Hz. The frequency of case (C) was 13 Hz and case (D) was 14.5 Hz were more than frequency case (B). Also it notes the frequency of case (D) was more than case (C). It can see that the concrete joint in case (C) and case (D) were improved the vibration resistance, through increasing natural frequency from 12.822 Hz to 13 Hz and 14.5, because it make the two panel act as one deck to vibration resistance.

5.3. Comparison between the experimental results and Finite element Modelling.
The fundamental natural frequency is the most important frequency for serviceability in a structural design compared to the higher frequencies.

The comparison between the experimental results and finite element Modelling only focuses on the fundamental frequency, the experimental results for frequency of modal test of case (B) is 12.9 Hz and it is 12.82 Hz for FEM, this mean the experimental result is very little higher than FEM by the differences 0.08 Hz. The different percentage between them is 0.006 %. The small differences between the natural frequencies for experimental results and FEM results for case(A) and case(B) prove that the FEM results from sap 2000 were accepted and the model can be used to extend the results to other cases. For the second case, the experimental results for frequency of modal test of case (C) is 13.7 Hz and is 13 Hz for FEM, also mean the experimental result is little higher than FEM by the differences 0.7 Hz. The different percentage between them is 0.05 %.

For the third case, the experimental results for frequency of modal test of case (D) is 17.5 Hz and is 14.5 Hz for FEM, also mean the experimental result is higher than FEM by the differences 3 Hz. The different percentage between them is 20 % as shown in Table 7. The rigidity of the support systems in the tests probably caused the big differences in case (D). The variable of individual timber beam stiffness of the TCC deck panel may effect the overall vibration behaviour of the specimens. While, the TCC deck systems in FEM modelling have constant stiffness and rigid support compared to real specimen.

Table 7. The differences in natural frequency between vibration test results and analytical modelling

Floor	Experimental (Hz)	FEM (Hz)	Different Hz	Percentage of different (%)
Case (B)	12.9	12.82	0.08	0.006
Case (C)	13.7	13	0.7	0.05
Case (D)	17.5	14.5	3	20

6. Conclusion
The modal experimental results for all tests were more than 8 Hz, it is accepted since the vibration limit for timber floor is 8 Hz according to EC5. The TCC deck frequency due to shaker excitation is higher than only timber deck which mean that the toping concrete in TCC improve vibration resistance property compared with only timber deck vibration resistance, it can concluded that the TCC slab is comfortable solution for vibration due human activities. The laminations timber are given the deck more stability to resist the human activates (vibration) because its weight, more sustained and less ductile as the number

of timber lamination increased as the beam tends to break at lower displacement. The additional concrete part in TCC deck gives the deck more stability, increased the stiffness of the TCC floor, due to mass increasing. The connector's part in TCC deck is responsible for the degree of composite.

The case(D) present real continuous TCC deck due to support existence at middle, while the case(C) present concrete joint connection along the total length of deck. Therefore the natural frequency of case (D) was higher than case(C).From FEM, It can see that the concrete joint in case(C) and case (D) were improved the vibration resistance, through increasing natural frequency from 12.82 Hz single TCC deck (case B) to 13 Hz case(C) and 14.5 case (D), because it make the two panel act as one deck to vibration resistance. The small differences between the natural frequencies for experimental results and FEM results for case (A) and case(B) prove that the FEM results from sap 2000 were accepted and the model can be used to extend the results to other cases.

7. References

[1] Rijal R, Samali B and Crews K 2010 Dynamic performance of timber-concrete composite flooring systems: *Incorporating Sustainable Practice in Mechanics of Structures and Materials* pp 315-319.
[2] Ghafar A 2008 Forced vibration testing on LVL-Concrete composite floor systems: *In 7th fib PhD Symposium in Stuttgart, Germany* pp 1-6
[3] De Silva S S and Thambiratnam D P 2009 Dynamic characteristics of steel–deck composite floors under human-induced loads: *Computers & Structures, 87*(17-18), pp.1067.
[4] Eurocode 5 2004 Design of Timber Structures Part 1-1: *General – Common rules and rules for buildings.* (London: British Standard)
[5] Allen D E and Pernica G 1998 *Control of floor vibration*: Ottawa, ON Canada: Institute for Research in Construction National Research Council of Canada
[6] Amaruddin H I, Hassan R, Amin N M and Malek N J 2014 Finite Element Model of Mortise and Tenon Joint Fastened with Wood Dowel Using Kempas Species: In *InCIEC Springer, Singapore* pp 3-14
[7] Avitabile P 2001 Experimental modal analysis *Sound and vibration* **35**(1) 20-31
[8] Tech Note 2007 Basics of Modal Testing and Analysis: *TN-DSA-003*

IOP Conf. Series: Materials Science and Engineering **713** (2020) 012024 doi:10.1088/1757-899X/713/1/012024

Stiffness and Strength Degradation of Timber Concrete Composite under Fatigue Loading

Kevin J.T Yeo[1] **and David E.C Yeoh**[1]

[1] Jamilus Research Centre, Faculty of Civil and Environmental Engineering, Universiti Tun Hussein Onn Malaysia, 86400 Batu Pahat, Johor, MALAYSIA
Corresponding author: david@uthm.edu.my

Abstract. The design of TCC is often associated with Gamma Method to calculate its effective stiffness as its characteristic strength to predict its SLS and ULS behavior. TCC structures used as bridges are subjected to fatigue loadings during service life. Fatigue loadings causes degradation of strength and softening of TCC structure. Such effect produces damaging consequences on the concrete, timber and especially the shear connectors. The effective stiffness degradation of the TCC structure subjected to fatigue loading within SLS and beyond SLS are studied due to scarcity of past researches on the fatigue behavior of TCC structures. This research was carried out by utilizing Malaysian tropical timber (Koompasiamalaccensis) where its strength properties was tested according to ASTM D198 standard procedures. Not much significant changes of its effective stiffness and connection stiffness is observed when subjected to fatigue loadings within SLS but the degradation process occurs at an instantaneous rate when subjected to fatigue loading beyond SLS. The S-N curve, strain profile and post-mortem analysis generated provides a summarization and indicative prediction for the fatigue life behaviour of the designed TCC.

1. Introduction

TCC structures can be used as an alternative solution to Industrial Building System (IBS) for building floor system and short bridges across monsoon drains for pedestrian and light vehicles instead of steel or reinforced concrete. In general, the concrete layer in TCC structure is designed to resist compression and bending while the timber part resists the tension and bending with the shear force transferred through the connectors [1]. However, the technology of TCC is still new in Malaysia and many researches are required to understand the structural behaviour of TCC structures. The continuous growth of interest in TCC has sparked an interest in its fatigue behaviour. To date, the body of knowledge in the fatigue behaviour of TCC is still not well established due to the scarcity of experimental test data [2].

Fatigue loading is generated by recurring force such as mechanical system, moving traffic or equipment testing on the structure. Hence, fatigue is a state where the material is weakened when fatigue load is applied which leads to microstructural changes. A S-N curve for a particular structural component relates a prescribed number of cycles (N) within a nominal, usually constant magnitude or applied stress range (S) where the component can safely with-stand [3]. The S-N curve has been a fundamental of mate-rail fatigue studies. It not only functions for characterizing fatigue test results but also serve for predicting the fatigue life of the structure or material which involves fatigue damage and stress ratio effect [4].

The fatigue loading imposed on the structure is often complicated in reality. Hence, it is easier to look at constant values while studying the principles of fatigue behaviour as real load conditions are

IOP Conf. Series: Materials Science and Engineering **713** (2020) 012024 doi:10.1088/1757-899X/713/1/012024

difficult to replicate [5]. Depending on the applied stress level, there are many types of fatigue test such as:

- Short-life test: Stress level are designated above SLS of the specimen and some specimens are expected to fail statically.
- Long-life test: Stress level are chosen below SLS of the specimen so that it does not fail after a preassigned number of cycles.

Under some circumstances, complex sequences with varying maximum stress amplitudes are probably required [3].

The present study is aim at conducting a fatigue test by utilizing Malaysian tropical timber- Kempas in the TCC structure to investigate its stiffness and strength degradation when subjected to fatigue loading within SLS (low amplitude) and beyond SLS (high amplitude). The research begins with the preliminary testing of materials such as concrete, shear connectors and the Kempas timber according to ASTM D198 standard procedures so that their strength properties can be applied in the Gamma Method for the design of TCC deck specimen. The S-N Curve derived from experimental results will provide an approximation value on how the stiffness and strength deteriorate under increasing numbers of fatigue loading.

2. Experimental Investigations

Gamma Method is commonly used in the design of TCC structures. In the Gamma method, a reduction factor known as shear coefficients, γ is introduced to measure the degree of interaction from the concrete slab. The value γ ranges from 0 representing no interaction between timber and concrete to 1 signifying full composite action. Preliminary testing for the properties and characteristic strength of timber, concrete and shear connectors are required to be applied in the Gamma Method [6]. The fatigue test of the TCC slab deck will be proceeded once it was designed from the Gamma Method where the effective stiffness and maximum moment resistance was determined.

2.1. Preliminary Testing of Materials

Since Malaysia is known for producing good quality tropical timber, the resources of Malaysia timber should be widely utilized. Kempas (Koompasiamalaccensis) is chosen among a wide variety of tropical timber in Malaysia because of its good characteristic strength and is very suitable to be used as structural timber in construction. The properties testing for Kempas timber such as tensile test, compression test and bending test was done according to the procedures of ASTM D198 where the tensile test specimens have a special tensile grip area and neck-down shape configuration to determine its tensile strength [7].

The stress-strain graph is determined to identify their respective strength properties. Concrete that achieved an average strength of 35MPa was used throughout the experiment. In this experimental research, screws are used as shear connectors as they are simple to install, easily available and high load-bearing capacity. The screws have 70mm long, 7mm in diameter and the spacing between the thread of the screw is 2mm. The push-out test was continued with increasing laminations of timber so that the strength of the shear connectors for the TCC deck can be determined. The results obtained are applied in the Gamma Method for the design of TCC deck.

2.2. Experimental Setup for Fatigue Test

There are 3 TCC specimens constructed for the fatigue test of TCC structure so that an S-N Curve can be obtained. The designed TCC specimens have a span of 3.6m and a total of 10 shear connectors are used for the TCC structure with unequal spacing because the shear force at the edge of the structure is bigger. Figure 1 shows the plan view of the TCC structure while Figure 2 shows the front view of the TCC structure. Figure 3 shows the TCC specimens placed under the actuator for fatigue test.

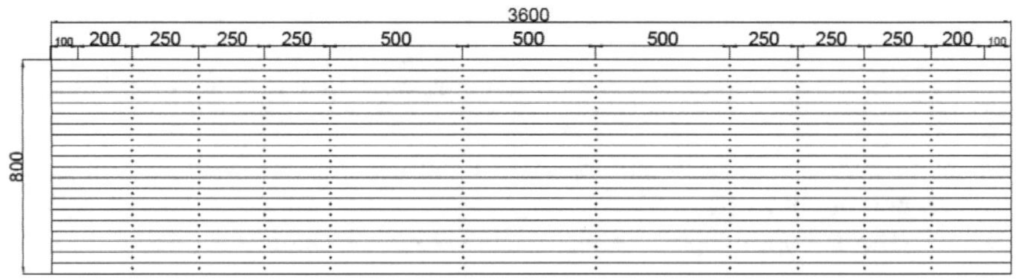

Figure 1: Top view of designed TCC specimen (Shear connectors spacing is centre to centre)

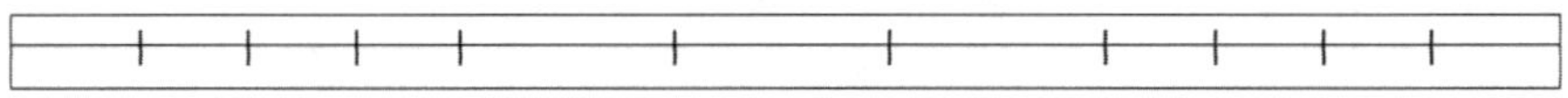

Figure 2: Front view of designed TCC specimen

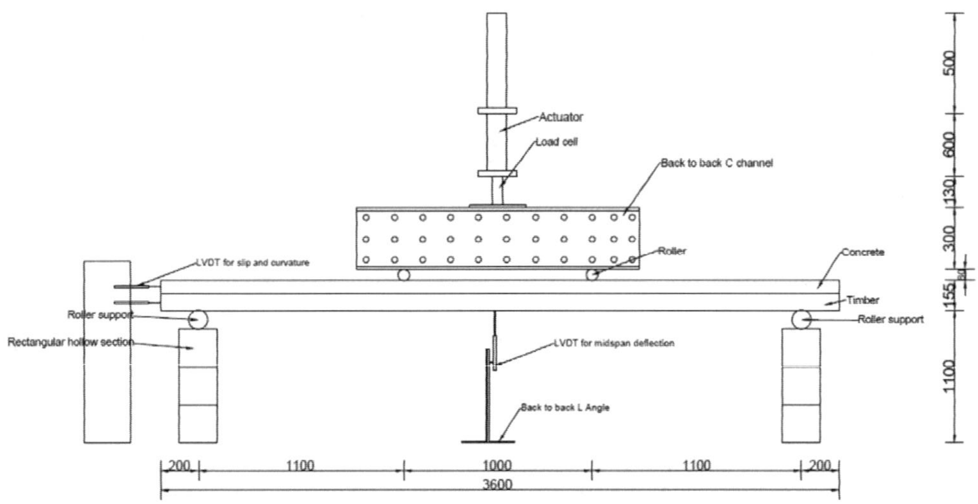

Figure 3: Experimental setup for fatigue test

In Figure 1 and 2, the spacing of shear connectors at the end-span of TCC is smaller because more shear connectors are needed there to provide a better composite action compared to the midspan. In Figure 3, three Linear Variable Differential Transducer (LVDT) was situated at the surrounding of the TCC specimen. One is located at the midspan to detect its midspan deflection and the other two was at the edge of the concrete layer and the timber layer respectively to detect the slip difference of the TCC specimen. The back-to-back C channel act as a spreader beam for the applied loading during the experimental testing. Strain gauges are also attached to the TCC specimens during the fatigue test. Figure 4, 5 and 6 shows the position of strain gauges attached to the TCC specimen to observe the stress distribution that occurs in the TCC.

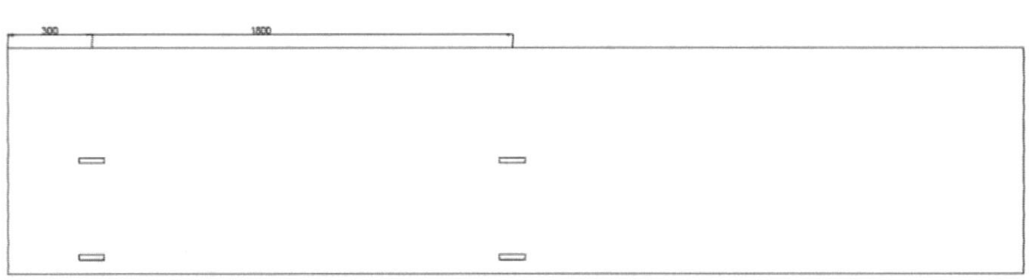

Figure 4: Strain gauge position at top concrete layer

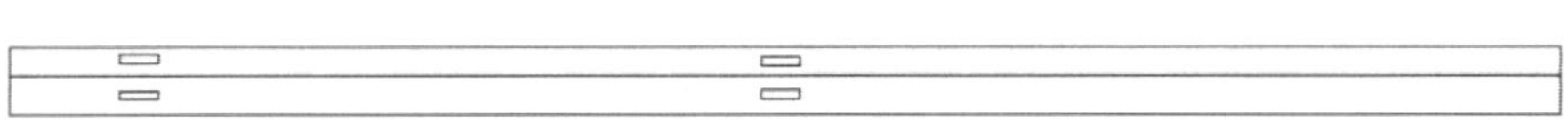

Figure 5: Strain gauge position at front view of TCC

Figure 6: Strain gauge position at bottom timber layer

2.3. Load Protocol for the Fatigue Test

From the TCC specimens constructed, two samples are taken for fatigue loading and one are tested for its actual load bearing capacity. The two samples undergo sinusoidal cycles of fatigue loadings with the frequency of 1 Hz. The amplitude of the fatigue loadings will be different for both specimens with the first one at 10kN minimum and 20kN maximum (within SLS) and the other at 60kN minimum and 120kN maximum (beyond SLS). The low amplitude fatigue loadings are to emulate pedestrian (7km/h) and light vehicle movements (15km/h) while crossing the designed TCC deck.

The load protocol begins with 10 slow cycles with a frequency of 0.1 Hz to obtain a slop in the load-displacement graph where its stiffness after a prescribe number of cycles is obtained from the gradient of the slope. The time required for the TCC specimen to undergo the number of pre-scribe cycles of fatigue loadings can be a very long process. The timing for the prescribed number of cycles required to complete must be estimated and planned out carefully to avoid the actuator from functioning during weekends and public holidays for supervision convenience and preventing over heating of actuator machine. The low amplitude fatigue loading was chosen to be at 10 to 20kN to emulate pedestrian and light vehicles crossing situation It was expected that as the number of fatigue loadings in sinusoidal cycles increases, the degradation of connection stiffness and structural stiffness of TCC will also increase which results in the decrease in load bearing capacity of the TCC deck.

3. Results and Discussions

3.1. Results of Preliminary Testing of Materials

The Kempas timber was tested for its compressive strength, tensile strength and bending strength parallel to grain so that its values can be applied in the Gamma Method. Table 1 below shows the tabulation of results obtained from the preliminary testing of Kempas timber.

Table 1: Summary of mean strength properties in N/mm^2 for Kempas timber (parallel to grain)

Strength Properties (N/mm^2)	Specimen 1	Specimen 2	Specimen 3	Average
Compressive Strength, L=200mm	87.3	87.7	85.8	86.9, (1.1%)
Compressive Strength, L=100mm	91.2	92.4	95.0	93.3, (2.1%)
Tensile Strength	105.3	114.9	104.4	108.7, (5.4%)
Bending Strength	256.4	170.6	189.7	205.6, (21.9%)

Note: Coefficient of Variation (CoV) is given in brackets

From the experimental testing, consistent results are obtained particularly in the compression test and tensile test with a coefficient of variation of 2.1% and 5.4% which means a very low variability in results. For the compression test, it was observed that the specimen with length, L=100mm has an average higher compressive strength than the specimen with length, L=200mm as the latter specimen is slenderer with the same cross section. During the compression test, most of the specimens failed because of the splitting of the timber fibres. Whereas snapping of the timber fibres in the necking part of the specimen occurred in the tensile test. The bending strength of the Kempas timber is very much higher than the tensile and compressive partly because the size of timber specimen (1200mm x 90mm x 40mm) is bigger than the tensile and compressive test specimen with an effective span of 950mm. Hence, there are more timber fibres to support the load. Although the bending strength of the timber is very high, there is still a variance up to 86 N/mm^2 and a 21.9% coefficient of variation due to the fact that timber is a biological material. The tensile test and compression test can achieve more consistent results than the bending due to the smaller size of specimen where the defects and quality of the timber is easier to control. The properties of concrete were tested to ensure it achieve an average strength of 35 MPa in the TCC specimens. Table 2 summarizes the tabulated results for concrete compression test.

Table 2: Compressive strength of concrete after 28 days hydration period

Cube	Compressive Strength (N/mm^2)
1	35.7
2	32.8
3	34.2
Average	34.2

In overall, Table 2 shows consistent results were obtained in the compression test of concrete. On the other hand, the study of the structural performance of shear connectors is classified as another broad range of study in the field of TCC. The strength of the shear connector signifies the maximum load capacity of the specimen while its stiffness is characterized by (Ks) for SLS and (K_u) for ULS which is determined by the secant slip moduli of the maximum load achieved by the specimen in the load-displacement graph at 40% and 60% of its maximum load capacity respectively. Table 3 shows the tabulated results of the push-out test.

Table 3: Values for the properties of shear connectors from experimental push out test

Number of timbers	Connection Stiffness for Serviceability (kN/mm), K_s	Connection Stiffness for Ultimate (kN/mm), K_u	Characteristic Strength of Connection (kN), f_k
1	12.8	11.7	5.3
5	17.4	14.0	12.0
10	24.1	32.1	53.5
20	44.0	45.0	65.0

The properties of shear connectors were tested with increasing laminations of timber until 10 and 20 was derived theoretically as the specimen would be too big to be fitted in the push-out test machine. The results obtained from preliminary testing are applied to the Gamma Method for the design of TCC deck specimens. From the Gamma Method, the designed TCC only has a concrete coefficient,γ_1 of 0.52 which suggest it is not a very good composite action in the concrete layer. The designed TCC has an effective stiffness, EI_{ef} of 5.557 x 1012 N/mm^2 for SLS and 5.601 x 1012 N/mm^2 for ULS. It was predicted that the TCC structure has a maximum moment resistance of 300kNm which is equivalent to a point load of 330kN. However, when the load is increased to a point 183kN, the concrete layer will fail first and eventually the timber part at ultimate load.

3.2. Results of Fatigue Test

The experiment starts with the TCC specimen subjected to low amplitude fatigue loadings of 10 to 20kN until 1 million cycles. The effective stiffness degradation was determined from the gradient of load-displacement graph generated by the 10 slows cycles in the load protocol before the proceeding with the prescribed cycles of fatigue loading. Figure 7 summarizes the effective stiffness degradation for the TCC structure subjected to fatigue loadings within SLS in S-N Graph.

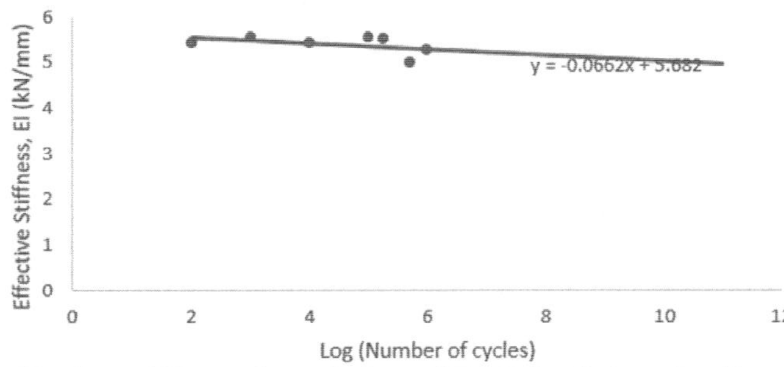

Figure 7: Effective stiffness degradation of TCC for fatigue loadings within SLS

From results of Figure 7, it can be seen that there is a minor amount of stiffness degradation that occurs in the TCC structure when subjected to fatigue loadings within SLS as it completed the total prescribed 1 million cycles of fatigue loading. Slight increase in the effective stiffness was also observed due to the yielding stage of shear connectors during the fatigue loading process. Figure 8 shows the connection stiffness degradation of TCC deck subjected to fatigue loading within SLS in the S-N graph where the connection stiffness is measured by the gradient of the load-slip difference between the timber and concrete tabulated on the load-slip difference graph.

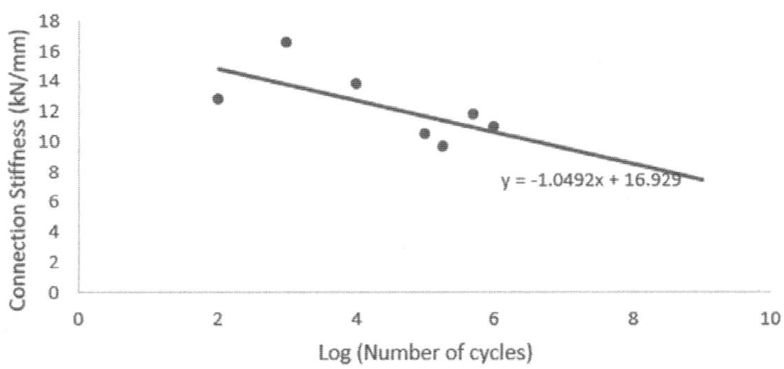

Figure 8: Connection stiffness degradation of TCC for fatigue loadings within SLS

From the results in Figure 8, the connection stiffness degradation for fatigue loadings within SLS is more noticeable compared to the effective stiffness degradation. This phenomenon can be explained with the shear connectors installed in the TCC system has a low yielding strength. From past research on the fatigue behaviour of steel, it was claimed that steel with a higher yield strength has a better performance in fatigue compared to steel with lower yielding strength as their fatigue crack growth rates was compared during the research [8]. As a result, the shear connectors in the TCC deck showed signs of yielding during the early stages of the fatigue loading cycles and its connection stiffness gradually decreases as the fatigue loading in sinusoidal cycles continued. The early yielding and connection stiffness degradation subsequently affects the effective stiffness degradation of the TCC structure but at trivial manner. The overall TCC structure successfully withstand even after subjected to 1 million sinusoidal cycles of fatigue loadings within SLS.

The following TCC specimen was subjected to fatigue loadings beyond SLS where the fatigue loadings were increased to 60kN and 120kN. The fatigue behaviour in terms of stiffness degradation of the TCC when subjected to fatigue loadings beyond SLS occurs at a very instantaneous rate. This situation is observable due to the frequent triggering from the actuator as a sudden and steep increase of midspan deflection was detected while fatigue loadings is applied. Figure 9 shows the effective stiffness of TCC degradation when subjected to fatigue loadings beyond SLS tabulated in the S-N graph and Figure 10 shows the comparison of effective stiffness degradation when subjected to fatigue loadings within SLS and beyond SLS.

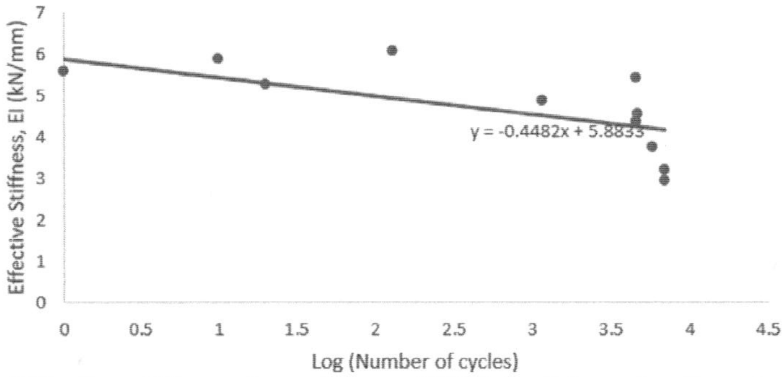

Figure 9: Effective stiffness degradation of TCC for fatigue loadings beyond SLS

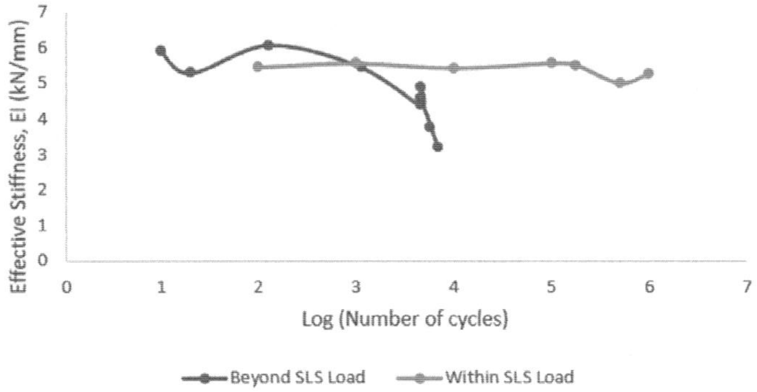

Figure 10: Comparison of effective stiffness degradation of TCC

From the results in Figure 9 and its comparison in Figure 10, it was observed that a steep change in effective stiffness occurs almost immediately when the TCC structure is subjected to fatigue loadings beyond SLS and there is not much of significant changes in stiffness when the TCC structure is subjected to fatigue loadings within SLS even when 1 million sinusoidal cycles of fatigue loadings was completed. This condition can be explained from the perspective of Gamma Method as it can predict the maximum load capacity of the TCC structure. The fatigue loading up to 120kN may be only 40% of the maximum load capacity of the TCC structure but it was almost 70% of the failure load at the concrete layer and also 80% for the shear connectors at the TCC structure. Hence, before the time the TCC reached 6900 cycles of fatigue loading, the concrete layer had already failed and was separated from the timber due to damage of the shear connectors, leaving only the timber to sup-port all the applied loadings and the concrete self-weight. A sharper increase of stiffness degradation was observed at the latter stage of S-N graph when the TCC structure subjected to fatigue loading beyond SLS as it was mainly the stiffness degradation of the Kempas timber. Consequently, there is no need to continue the fatigue loading test as there is already no composite action in the structure. Figure 11 shows the S-N graph for the connection stiffness degradation of TCC deck subjected to fatigue loadings beyond SLS.

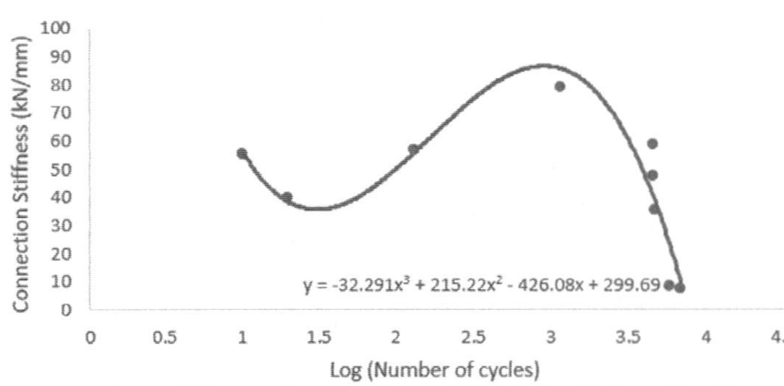

Figure 11: Connection stiffness degradation of TCC for fatigue loadings beyond SLS

From Figure 11, a polynomial curve equation seems to have a better fitting in the results for the connection stiffness degradation of TCC when subjected to fatigue loadings beyond SLS. This is due to the yielding that occurs in the shear connectors. The yielding of the shear connectors would cause the connection stiffness to increase. This leads to an increase in the composite action of the TCC where the timber would displace outwards due to tension and the concrete to displace inwards due to compression when subjected to the loading. The changes of displacement for both components caused an increased gradient in the graph. When the shear connectors break after the yielding, the composite action for the TCC decreases. Hence, both components are displaced outwards due to the tension force which causes the difference of displacement to decrease when the fatigue loading is applied. Resultantly, the gradient of the load-slip difference graph decreases.

3.3. Results of bending test after fatigue loading

Since there is a decrease in the effective stiffness in the TCC structure after subjected to fatigue loading, the maximum load capacity of the TCC structure also decreases. Table 4 shows the comparison of effective stiffness obtained from the Gamma Method calculation with the experimental results. Figure 12 shows the load-displacement graph obtained from the bending test of all the TCC specimens constructed for the fatigue test.

Table 4: Comparison of effective stiffness from Gamma Method with experimental

TCC Specimen	Effective Stiffness (kN/mm)	Accuracy (%)
Gamma Method	5.557	-
Bending (without fatigue)	5.0852	92%, 8% overestimation
Fatigue within SLS (initial)	5.4419	97%, 3% underestimation
Fatigue beyond SLS (initial)	5.8833	94%, 6% underestimation

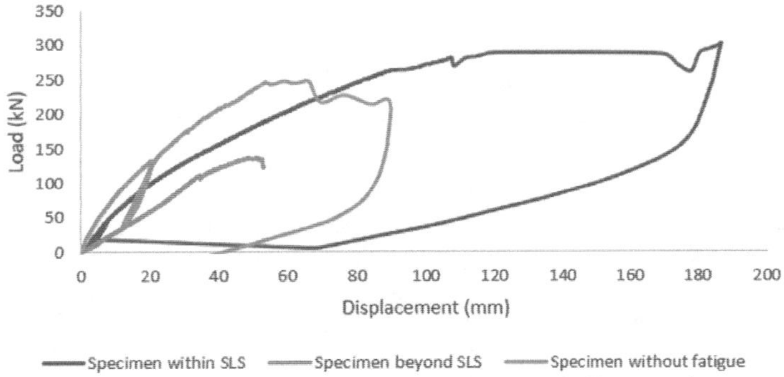

Figure 12: Comparison of load capacity of all 3 specimens

From the results in Table 4 and Figure 12, this research shows good accuracy of the Gamma Method in determining the effective stiffness of the TCC structure up to 90% and also the relationship of the effective stiffness degradation when subjected to fatigue loadings with its load bearing capacity. From the load-displacement graph, the TCC specimen with no fatigue loadings subjected shows higher stiffness while comparing with the other 2 specimens but have a lower maximum load capacity compared to the specimen subjected fatigue loadings within SLS. This reason is due to timber is a biological material and its quality and strength properties is diverse. Similar pattern of results was shown in past researches and also preliminary testing. Therefore, it was not necessary to discuss further on the bending strength of the TCC specimen without fatigue loadings on it as it did not support the hypothesis of this research. Since there is an insignificant amount of effective stiffness degradation that occurs in the TCC specimen subjected to fatigue loadings within SLS, there is not much changes in the load bearing capacity (300kN) when compared with the prediction in the Gamma Method (330kN). For the TCC specimen that was subjected to fatigue loadings beyond SLS, the effective stiffness was greatly reduced due to damage in the concrete and shear connectors, leaving the Kempas timber part to support the fatigue loadings and the concrete self-weight. Resultantly, the TCC specimen subjected to fatigue loading beyond SLS reaches its ultimate limit state (ULS) when loaded up to 130kN. The bending test results also shown the good elastic properties of the Kempas timber in the TCC structure in the load displacement graph in figure 20 as it recovers from the deflection at ULS once the loading was removed. Figure 13 shows the S-N graph generated for comparison of TCC structures load capacity.

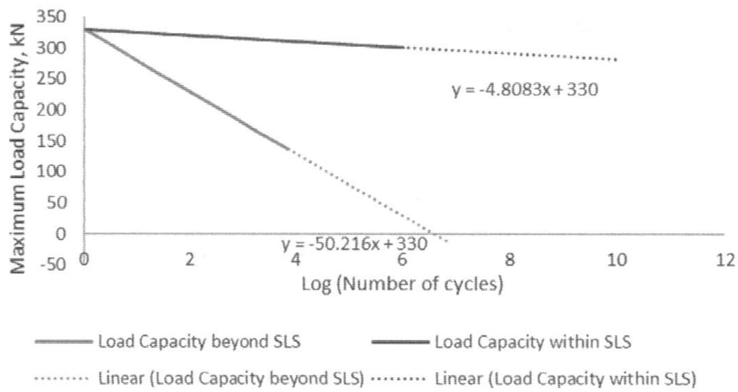

Figure 13: S-N graph for comparison of TCC load capacity

Due to the in-consistency in bending strength of the Kempas timber with a 21.9% coefficient of variation in the preliminary testing, the bending strength of the TCC specimen without fatigue loading did not achieve the desired load capacity as predicted in the Gamma Method. Hence, the value calculated from the Gamma Method was used in the S-N graph shown in Figure 13 as the quality and defects of solid timber (Kempas) is difficult to control when applied in a structural scale specimen. The S-N graph generated provide a stark contrast comparison in the load bearing capacity of TCC structure when subjected to fatigue loading with SLS and beyond SLS.

3.4. Strain Profile Comparison

The strain profile which shows the stress distribution that occurred in the TCC structure was tabulated to support the findings in the fatigue test results and also the hypothesis of the research. The strain profile was emphasized during the bending test of the specimens after subjected to different amplitudes of fatigue loadings. From there, different composite action for each of the TCC specimen was observed. Figure 14 and 15 shows the tabulated strain profile at the midspan of TCC deck for fatigue loadings within SLS and beyond SLS.

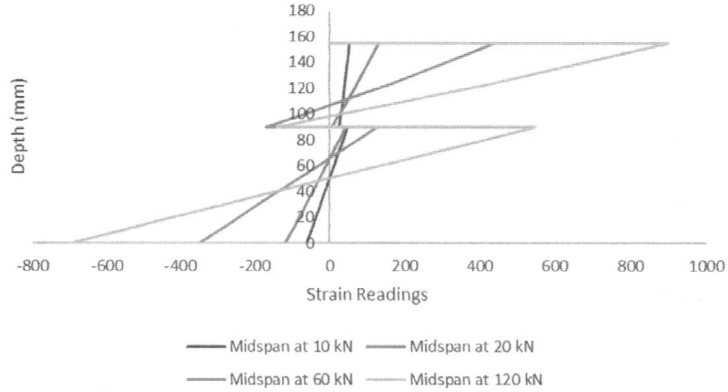

Figure 14: Strain profile at midspan of TCC with fatigue within SLS

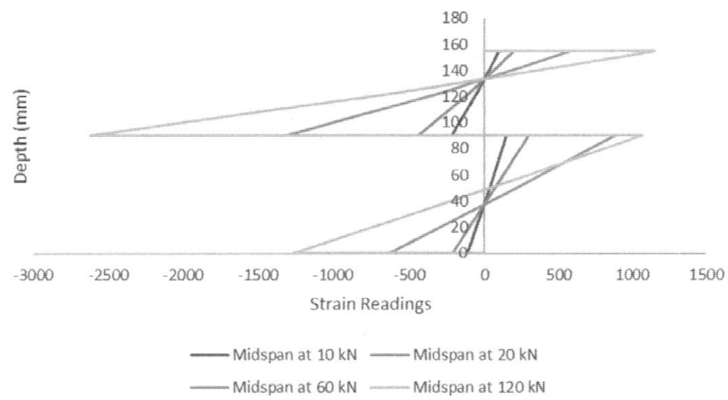

Figure 15: Strain profile at midspan of TCC with fatigue beyond SLS

By comparing Figure 14 and 15, the TCC deck after subjecting to fatigue loading within SLS still has a better composite action during the bending test compared to the TCC deck subjected to fatigue loading beyond SLS. This is due to the shear connectors is still intact in the TCC deck thus still capable of transferring the shear force to the timber part making the concrete part more exposed to compression zone while the timber is more exposed to the tension zone. Contrarily, most of the shear connectors in the TCC deck subjected to fatigue loading beyond SLS was already broken and unable to transfer the shear force from the concrete component to the Kempas timber. Hence, no composite action was observed in its strain profile in Figure 15 as both of the concrete and timber component was subjected to compression and tension zone equally. For both of the strain profile tabulated in Figure 14 and 15, both of the specimens were compared at the load level of 10kN, 20kN, 60kN and 120kN without its load level at ULS because the strain value at ULS load level is too large and would make the earlier mentioned load levels looked insignificant and difficult to compare when tabulated together.

3.5. Results of Post-Mortem Analysis

During post-mortem analysis, the concrete layer of the TCC is hacked away to observe the conditions of the shear connectors after fatigue loadings and bending test. Any deformations or breaking of shear connectors in the TCC specimens were recorded. Table 5 and table 6 shows the tabulated condition of shear connectors where broken shear connectors are marked with an 'X' while '90' suggests that the shear connectors are still in good condition without any deformations. Any value below 90 suggest the degree of bending for the shear connectors.

From the observations in Table 5, not much of damage of shear connectors is observed for TCC specimen subjected to fatigue loadings within SLS with most of the shear connectors still being in good condition. In Table 5, some slight deformations only occurred at the endspan of the TCC structure as the shear force in a uniformly distributed loading is usually larger. Since the fatigue loading within SLS does minimal damage to the shear connectors, there is still composite action in the TCC structure as shown in the strain profile which enables it to have a higher effective stiffness and load-bearing capacity compared to the TCC specimen subjected to fatigue loadings beyond SLS in Table 6. The damage of shear connectors in the TCC specimen subjected to fatigue loadings beyond SLS are more severe. All the shear connectors at the endspan were already broken or deformed, leaving only the 6th row of shear connectors which is situated at the midspan of TCC in good condition. This is mainly because the shear force at the midspan is usually minimal for a uniformly distributed load. The findings in the post-mortem analysis showed good agreement with the fatigue test results and the strain profile tabulated because most of the shear connectors subjected to fatigue loading beyond SLS were destroyed which resulted in a lower composite action, resulting a lower effective stiffness and load-bearing capacity.

IOP Conf. Series: Materials Science and Engineering **713** (2020) 012024 doi:10.1088/1757-899X/713/1/012024

Table 5: Condition of shear connectors for fatigue loading within SLS

Number of Rows	1st row of connectors	2nd row of connectors	3rd row of connectors	4th row of connectors	5th row of connectors	6th row of connectors	7th row of connectors	8th row of connectors	9th row of connectors	10th row of connectors
1	80	90	90	80	90	90	X	90	90	90
2	80	90	90	80	90	90	80	90	90	90
3	X	90	90	X	90	90	80	90	90	X
4	80	90	90	90	90	90	80	90	90	90
5	80	X	90	90	90	90	80	90	90	90
6	80	90	90	X	90	90	90	90	90	90
7	X	90	90	90	90	90	90	90	90	X
8	X	X	90	90	90	90	90	90	90	90
9	80	90	90	90	90	90	90	X	90	90
10	X	90	90	90	90	90	90	90	90	90
11	X	90	90	90	90	90	90	90	90	90
12	X	90	90	X	90	90	90	90	90	X
13	X	90	90	90	90	90	90	90	90	90
14	80	90	90	90	90	90	90	90	90	90
15	80	90	70	90	90	90	90	90	90	90
16	80	90	90	90	90	90	90	90	90	90
17	80	90	90	90	90	90	90	90	90	90
18	80	70	90	90	90	90	90	90	90	90
19	80	90	90	90	90	90	90	90	90	90
20	80	90	90	90	90	90	90	90	90	90

Table 6: Condition of shear connectors for fatigue loading beyond SLS

Number of Rows	1st row of connectors	2nd row of connectors	3rd row of connectors	4th row of connectors	5th row of connectors	6th row of connectors	7th row of connectors	8th row of connectors	9th row of connectors	10th row of connectors
1	X	X	X	X	90	90	90	90	X	X
2	40	X	X	X	90	90	90	90	80	X
3	X	X	X	X	X	90	90	90	80	X
4	40	X	X	X	90	90	90	90	X	X
5	X	X	X	X	90	90	90	90	X	X
6	50	X	X	X	90	90	90	90	X	X
7	50	X	X	X	90	90	90	90	X	X
8	X	X	X	X	90	90	X	X	80	X
9	X	X	X	X	90	90	X	X	X	X
10	X	X	X	X	90	90	X	X	X	X
11	X	X	X	80	90	90	X	X	X	30
12	X	X	80	80	90	90	X	X	X	30
13	X	X	X	80	90	90	X	X	X	30
14	X	X	70	80	90	90	X	X	X	X
15	X	X	X	X	90	90	X	X	X	30
16	X	X	X	X	90	90	90	X	X	X
17	X	X	X	X	90	90	90	90	80	X
18	X	X	X	X	X	90	90	90	80	X
19	X	X	X	X	90	90	90	90	80	X
20	40	50	X	X	X	90	90	90	80	30

4. Conclusion

Based on the experimental results presented on this paper, the following conclusion are deduced:

- The Gamma Method is capable of determining the effective stiffness of the designed TCC structure up to 90% of accuracy.
- The results from the preliminary testing of timber especially in the bending test showed the bending strength of timber has high coefficient of variation because the defects and quality of the solid timber in a structural scale is more difficult to control. Hence, the material properties are based on average results from the specimens tested.
- In overall, the research showed good accuracy in showing the stiffness and strength degradation of TCC structures when subjected to fatigue loadings. The effective stiffness of TCC and its shear connectors does not change significantly when subjected to fatigue loadings within SLS. The effective stiffness of TCC structure and its shear connectors will decrease rapidly when subjected to fatigue loadings beyond SLS because the loadings could possibly be near to the failure load of the concrete layer and shear connectors.
- The TCC technology combines the timber and concrete elements using shear connectors to achieve a better stiffness and higher load bearing capacity but it is still not a homogenous material where the timber, concrete and shear connectors still have their own failure load.
- Aside from fulfilling the objectives of the research which focus on the fatigue behaviour of TCC structure, this paper also provides an overview on what can be improved in future researches such as overcoming the inconsistency of Malaysian tropical timber and a more effective utilization by post-processing it to engineered timber before being used as structural materials.

5. References

[1] Yeoh, D, Fragiacomo M and Carradine D 2013 Fatigue behaviour of timber–concrete composite connections and floor beams *Engineering Structures* **56** 2240-2248

[2] Balogh J and Atadero R 2013 *Fatigue testing of wood-concrete composite beams* Technical Report (United States: Mountain Plains Consortium) pp 1-24

[3] Zhu H 2014 *Experimental Investigations of Residual and Fatigue Capacities of Timber Connections with Glued-In FRP Rods* PhD Thesis (Vancouver: University of British Columbia)

[4] Burhan I and Kim H 2018 SN curve models for composite materials characterisation: An evaluative review *Journal of Composites Science* **2**(38) 1-29

[5] Lokken, N. (2016). *Fatigue of threaded rods subjected axial load* Master Thesis (Oslo: Norwegian University of Science and Technology)

[6] Zhang C 2013 *Analysis of the Timber-Concrete Composite Systems with Ductile Connection* PhD Thesis (Toronto: University of Toronto)

[7] Ahmad Z, Bon Y C and Abd Wahab E S 2010 Tensile strength properties of tropical hardwoods in structural size testing. *International Journal of Basic and Applied Sciences* **10**(03) 1-6

[8] Carvalho D, Silva A L L, Jesus A M P and Fernandes A A 2015 Fatigue behaviour of structural steels comparison of strain life and fatigue crack propagation data *Mecanica Experimental* **25** 67-78

Dynamic Responses of Sandwich Cemboard-Foamed Concrete Panels Subjected to Low-Velocity Impact

Z M Jaini[1], S N Mokhatar[1], J Hadipramana[2] and M Abu Seman[3]
[1] Jamilus Research Center, Universiti Tun Hussein Onn Malaysia, 86400 Parit Raja, Johor, Malaysia
[2] Faculty of Engineering, Universitas Muhammadiyah Sumatera Utara, Medan 20238, Indonesia
[3] Faculty of Civil Engineering and Earth Resources, Universiti Malaysia Pahang, 26600 Pekan, Pahang, Malaysia
Corresponding author: rizuan@uthm.edu.my

Abstract. Sandwich cemboard-foamed concrete panel (SCFCP) has sparked the attention due to green and sustainability requirements in construction of building. SCFCP is essentially made of fiber cement board as core component and foamed concrete as outer layers. Shear connectors act as joint mechanism and create compositeness behaviour. However, investigations on SCFCP remains passive, creating a number of uncertainties particularly in dynamic responses. Therefore, this study intends to investigate the dynamic responses of SCFCP under low-velocity impact. Nine cube specimens were prepared using size of 100 mm x 100 mm x 100 mm, while nine panel specimens were casting using size of 1220 mm x 300 mm x 76 mm. The compression and impact tests were conducted on cube and panel specimens. The compression test revealed that foamed concrete has compressive strength approximately 18 MPa with density of 1600 kg/m^3. Under low-impact of 1.39 m/s, it is noticeable that the impact force decreases from 8.41 kN to 6.73 kN in correspond to the spacing of shear connector. Contradictory, the displacement shows sprang off pattern with value between 87.90 mm to 96.50 mm.

1. Introduction

Either in residential or high-rise buildings, concrete panel is widely used for the entire construction. Concrete panel is one of component in precast concrete which has been extensively utilised as industrialised building system and received good merit from local councils. It can be used for interior element, landscaping, soundproofing and some can be bearing capacity wall with different size, function and design. In term of advantageous, concrete panel offers excellent durability and flexibility solutions. In addition, it has high potential for salvage and relocation due to high initial embodied energy which can be offset by its extended life cycle up to 100 years [1]. The design of concrete panel has been improved to suit with the latest demands. However, engineering specification still an issue that should be considered in the construction, irrespective of whether the concrete panel is in-situ or precast. In Malaysia, concrete panel has been promoted since the last decade to enhance the importance of precast. Unfortunately, the adoption of concrete panel is still low in current state. The lack of standardization and regulation were identified as major factors [2].

Basically, concrete panel can be classified as homogeneous or sandwich system. In recent years, sandwich panel has received most attention as it can be used for various applications and offer good strength. Sandwich panel is one of combining materials possibly to form composite element that enable optimum design to be composed for the particular application [3]. Basically, sandwich panel can be applied as vertical or horizontal structural component. However, Lakshmikandhan el al. [4] defined the sandwich panel as such system that is more suitable for wall. Because of composite element, it has various advantageous such high strength, better crack resistance and ductility as well as good energy absorption [5, 6]. According to Shutt [7], there are three types of sandwich panel depending on the degree of composite action; non-composite, fully composite and partial composite. Basically, the degree of composite action depends on the bond (naturally or using shear connectors) and rate of deformation between core and outer layers. Therefore, the selection of materials to be used in sandwich panel is paramount to ensure it has fully function as desired.

Lightweight materials are basically demanded for the fabrication and installation of sandwich panel as it not only easy to be combined but able to counter the weight penalty. Among the innovation, sandwich cemboard-foamed concrete panel (SCFCP) has sparked the attention due to green and sustainability requirements in construction. SCFCP is basically consists of fiber cement board as core layer and foamed concrete as outer layers, where shear connectors act as joint mechanism. Since SCFCP is categorized as lightweight structure and still a new product, more attention should be given to promote this type of sandwich panel. However, the investigations on SCFCP remain passive and hence create various uncertainties especially that related to dynamic responses. Under high loading such as seismic, impact, explosion and accidental, the structural behaviour, failure mode and energy absorbing are important parameters that must be fully acknowledged [8]. So that, sandwich panel able to maintain its serviceability at any conditions. Moreover, governing factors on the performance of sandwich panel must also be considered in the design criteria.

Therefore, this study intends to investigate the dynamic responses of SCFCP under low-velocity impact. Low-velocity impact is not a new matter as various works have previously been done on concrete panel [9, 10] and sandwich panel [11, 12]. However, there is still no established information on SCFCP. The effects of shear connector, in term of spacing, on the force-time history and displacement time-history as well as the failure mode are of interest. This study contributes to the in-depth understanding about the performance of SCFCP. In the design prospective, the vital intention is to establish the structure that able to withstand the entire feasible loads for imposed period of time. Therefore, it is necessary to investigate the strength, stability and rigidity of the structure to resist force, whether in static or dynamic. In addition, this study able to solve various issues related to the serviceability of SCFCP. Hence, SCFCP can be used confidently in the construction of building. Whereby, SCFCP could be the figurative structural component that can contribute to an extraordinary performance.

2. Materials and specimens

Foamed concrete with density 1600 kg/m^3 was produced using ordinary Portland cement, sand, rice husk ash, foaming agent, superplasticizer and water. Rice husk ash was used as partially sand replacement at 40%. The mix design is based on the prescribed volume as suggested by Abd Rahman et al. [13] and Jaini et al. [14], where the ratios 0.50 for cement-sand, 0.55 for water-cement, 0.07 for foaming agent-cement and 0.05 for water-foaming agent. The targeted compressive strength of foamed concrete is approximately 18 MPa that deemed suitable for structural component. Foamed concrete was used as outer layers, while fiber cement board based on Primaflex as core layer. There are nine cube specimens of foamed concrete ware prepared using mould of 100 mm width, 100 mm length and 100 mm depth. After the demoulded process, cube specimens were placed at the ambient condition for air curing along 7, 14 and 28 days. Meanwhile, nine panel specimens were prepared using size of 300 mm width, 1220 mm length and 76 mm thickness. Three different spacings of shear connectors were used as detailed in Table 1, while the schematic design of SCFCP can be referred in Figure 1.

Table 1. Details of panel specimens.

Spacing of Shear Connector	Label of Panel Specimen	Quantity
100 mm	SCFCP-100	3
150 mm	SCFCP-150	3
200 mm	SCFCP-200	3

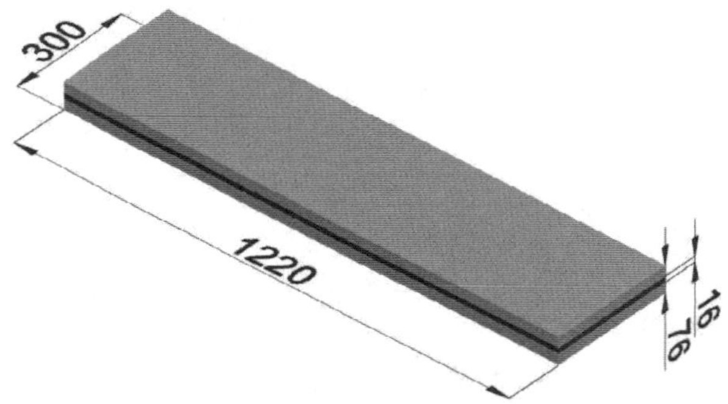

Figure 1. Schematic design of panel specimen SCFCP.

3. Experimental test

The compression test was performance to investigate the compressive strength of foamed concrete. The cube specimens were tested using Ele-Compact Machine 150. The compression test was conducted in accordance to BS EN 12390-2 [15]. After undergone the air curing at 7, 14 and 28 days, cube specimens were cleaned, measured and weighted. A compressive axial load with a specific load rate was applied to cube specimens up to failure, where the load reduces after reaching the maximum value. The maximum load carried by cube specimens was recorded. On the other hand, the setup of impact test requires rolled supports, frame and impactor as can be seen in Figure 2. The effective span of support to support was fixed at 1000 mm. The impact test was performed as accordance to ASTM D7126-15 [16]. The impactor with mass of 94.55 kg and located at the height of 1000 mm was used to initiate the free-fall with constant low-velocity of 1.39 m/s. When the impactor hits the panel specimen, the load cell was used to record the load-time history. On the other hand, the dynamic LVDT was used to capture the displacement-time history. These data are safely stored on data logger.

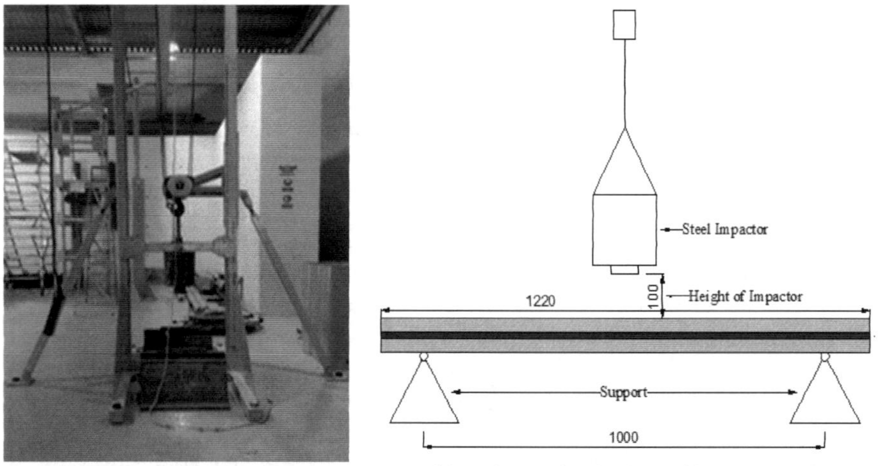

a) Frame of impact test b) Schematic design of impact test

Figure 2. Setup of impact test.

4. Results and discussion

4.1. Compressive strength

Compression test on cube specimens provided the results of compressive strength as can be seen in Figure 3. In general, the compressive strength is significantly increased in corresponds to the curing age. This is the typical trend that can be observed in foamed concrete and normal concrete. It was identified that the compressive strength achieved the targeted characteristic with value of 18.33 MPa, which is slightly over than 18 MPa as required by BS EN 1992-1-1 [17]. Within 7 days, the compressive strength increases around 47.40% followed by 38.90% and 12.71% at the sequent period of 14 and 28 days. The percentage of increment shows better performance than normal concrete. According to Aldrige [18] and British Cement Association [19], the compressive strength of foamed concrete should be in the range of 7.50 MPa to 10.50 MPa. This is particularly for foamed concrete with density of 1600 kg/m^3. However, the compressive strength of this study is slightly lower than obtained by Jaini et al. [20] that also employed rice husk ash as sand replacement in foamed concrete. The type of foaming agent and dilution process during the production of foamed concrete were identified as variance factors.

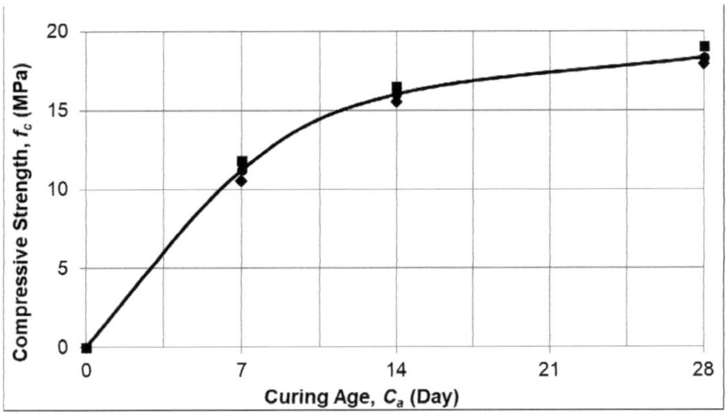

Figure 3. Compressive strength of foamed concrete.

4.2. Impact force and displacement

The results of impact force-time history and displacement-time history are presented in Figure 4 to Figure 6. Basically, these results are combined to produce the impact force-displacement curves. It should be emphases here that impact force-time history and displacement-time history are based on average value. Under the low-velocity of 1.39 m/s and height approximately 1000 mm, the time taken for the impactor to hit the surface of panel specimens was relatively 1.70 to 2.15 split second. It can be observed that the force raised abruptly during the impact stage. This cause the panel specimens experienced excitation state as the force keep oscillate until it behaves static. However, the force remains at the certain value due to the reflected stresses induced by impactor. High energy is transferred from impactor to the panel specimens and create incident stresses that propagate along the thickness and span of SCFCP. Similarly, the displacement also experiences fluctuate condition after imposed by impactor. During which the displacement remains constant at the particular value, the panel specimens were observed break up into two pieces.

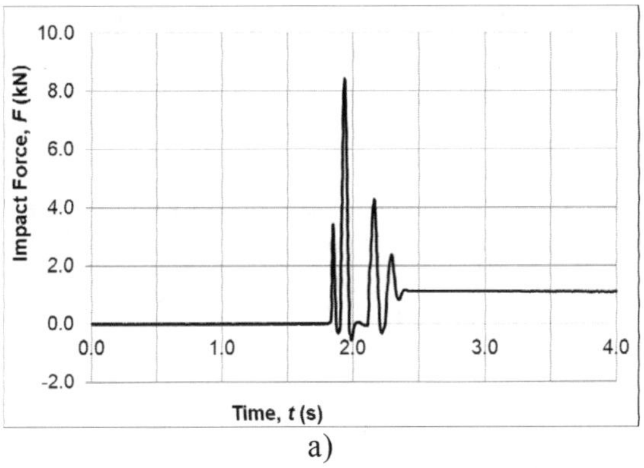

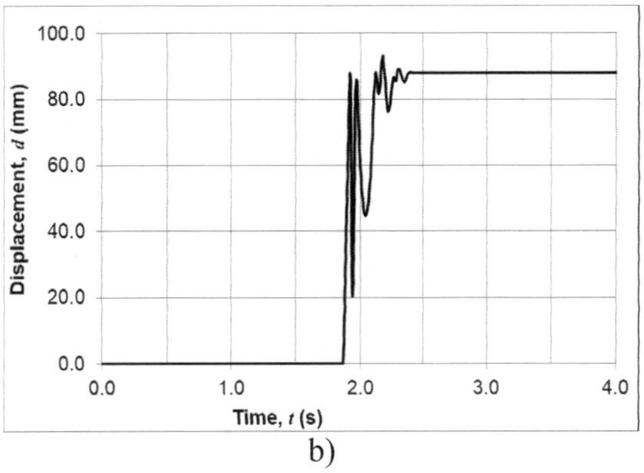

Figure 4. Results for SCFCP-100: a) Impact force-time history and b) displacement-time history.

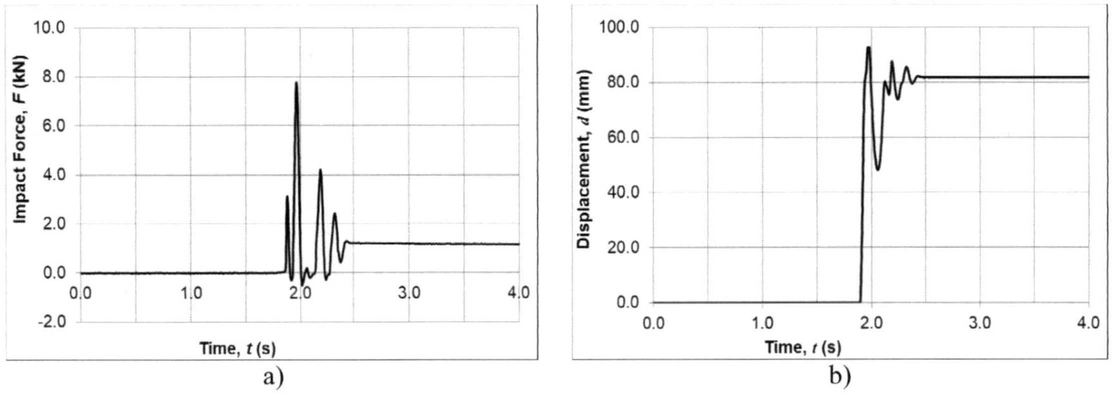

Figure 5. Results for SCFCP-150: a) Impact force-time history and b) displacement-time history.

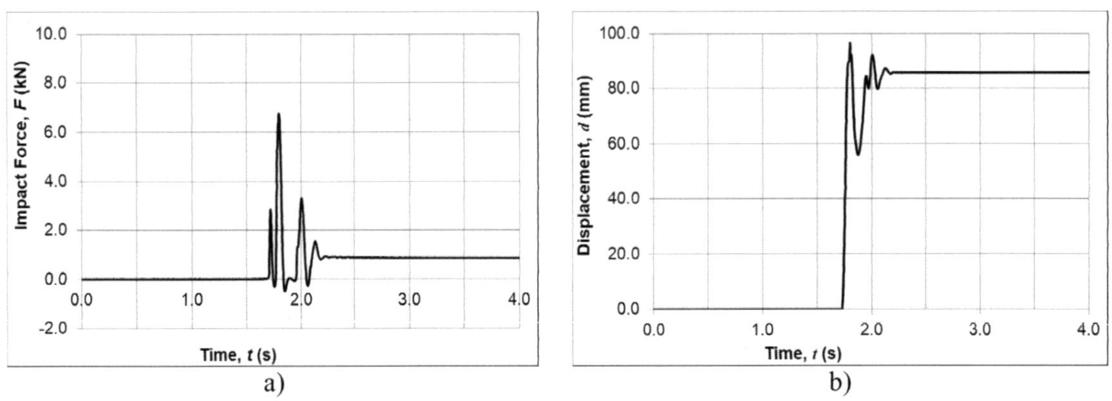

Figure 6. Results for SCFCP-200, a) Impact force-time history and b) displacement-time history.

Figure 7 shows the impact force of SCFCP in corresponds to the spacing of shear connector. The impact force decreases when the spacing of shear connector become bigger from 100 mm to 200 mm. It can be clearly observed that the closest spacing produce high impact force. This happens because of shear connectors contribute to the good stiffness mechanism and subsequently improve the bearing capacity of panel specimens to withstand the loading. As spacing of shear connector is small, hence more bolts can occupy the panel specimens and create almost a perfect bond in between core and outer layers. The relation between impact force and spacing of shear connector can be presented by the best fitted line of inverse linear function, in which the impact force of SCFCP can be predicted by a simple formulation of $F = -0.00168S + 10.155$. Here, F and S are the impact force and spacing of shear connector respectively. The forward and backward forecasts revealed that the impact force for SCFCP with spacing of shear connector at 50 mm, 250 mm and 300 mm are 9.38 kN, 6.26 kN and 5.48 kN respectively.

The displacement of SCFCP against the spacing of shear connector is plotted as shown in Figure 8. The displacement was found become higher as the spacing of shear connector changes from 100 mm to 200 mm. The displacement was recorded at 87.90 mm, 92.60 mm and 96.50 mm for panel specimens SCFCP-100, SCFCP-150 and SCFCP-200 respectively. At the constant thickness and dimension of SCFCP, the displacement is significantly diverse due to the spacing of shear connector. The consumption of input energy by reducing the spacing of shear connector indicates the drop off in displacement. This specifies that the spacing of shear connector at 100 mm able to enhance the strength and geometry property of SCFCP. The presence of bolts expands the rigidity that prevent the bending to be formed excessively although under high impact force. The positive correlation was established by the displacement of SCFCP and the spacing of shear connector. By depending on the spacing of shear connector at 100 mm, 150 mm and 200 mm, the displacement of SCFCP can be depicted as $d = 0.086S + 79.433$ where d is the displacement whilst S is the spacing of shear connector.

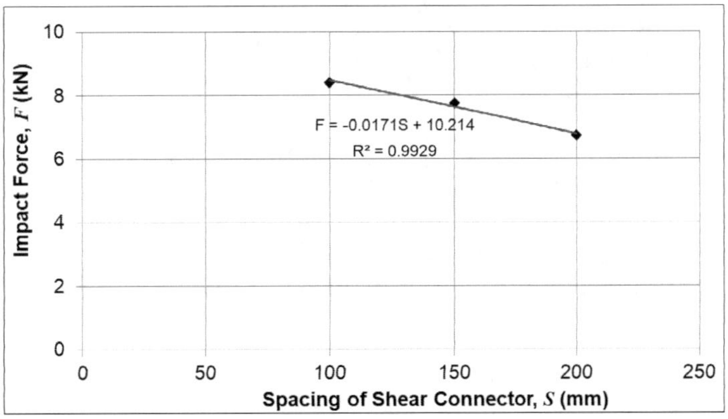

Figure 7. Impact force in corresponds to the spacing of shear connector.

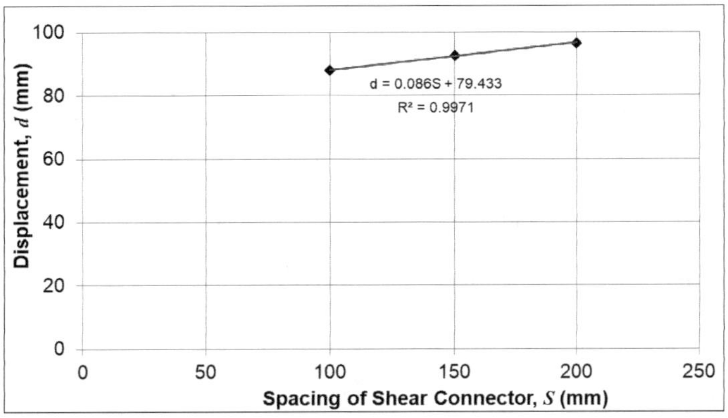

Figure 8. Displacement in corresponds to the spacing of shear connector.

4.3. Failure mode

Under low-velocity impact, the panel specimens were dominated by flexural and horizontal cracks as can be seen in Figure 9. These cracks can be observed occur at the mid-span, exactly under the location of impact. Unlike Yao et al. [21] that discovered the failure mode in compression zone, there is no crater happens at the top surface of panel specimens. The delamination failure in between the core element and outer layers is invisible. However, fracture by spalling appears at the end of impact which is happen in panel specimens SCPCP-150 and SCFCP-200. The spalling led to a phenomenon as similar as shear plug but propagate from beneath to top surface (around 70 degrees) and edge to edge of panel specimens. After the impact, the panel specimens are totally break into two pieces. The separation occurs mainly due to joint failure that indicate the high stresses were transferred from shear connectors to the surrounding area. Noroozi et al. [22] stated that size of bolt and distance arrangement are important factors in observing the failure mode. By understanding the link between the load distribution and configuration of bolt, failure mode can become part of criteria in design of sandwich panel.

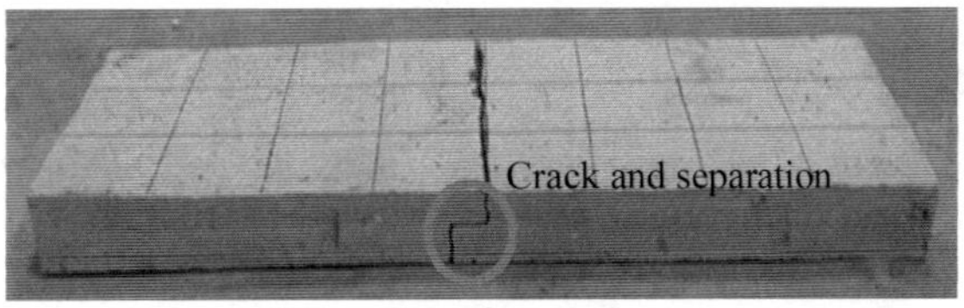

a)

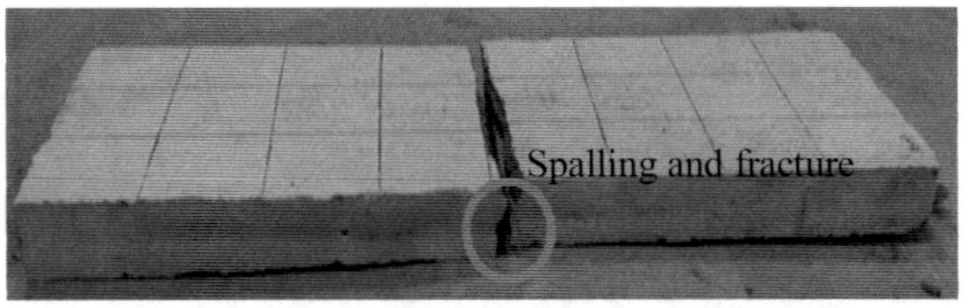

b)

c)

Figure 9. Failure mode of panel specimens, a) SCFCP-100, b) SCFCP-150 and c) SCFCP-200.

5. Conclusion

This study offers a comprehensive investigation on dynamic responses of sandwich cemboard-foamed concrete panel (SCFCP) under low-velocity impact. During the impact test, low-velocity impact was conducted using velocity 1.39m/s that induced from free-fall of impactor at the height of 1000mm. This study was concentrated on the effect of shear connector particularly the spacing at 100mm, 150mm and 200mm. The compression and impact tests pleasant the tremendous results of compressive strength, impact force-time history, displacement-time history and failure mode. In term of impact force and displacement, it was affirmed that SCFCP with the closest spacing of shear connector offers a great strength and serviceability. Meanwhile, failure mode is dominated by the flexural and horizontal cracks. However, this study has deficiency to capture the exact event of failure mode especially that related to the crack propagation and progressive of spalling. It is well understood that the spacing of shear connector plays an important role in governing the dynamic responses of SCFCP as it creates the degree of composite action in which affecting the stiffness mechanism and geometry property.

6. References

[1] Chris R and Downton P 2013 *Your Home - Australia's Guide to Environmentally Sustainable Homes* (Canberra: Department of Climate Change and Energy Efficiency)

[2] Polat G 2008 Factors affecting the use of precast concrete systems in the United States *J. of Construction Engineering and Management* **134** 169-178

[3] Davies J M 1997 Sandwich panels *Thin Walled Structures* **16** 179-198

[4] Lakshmikandhan K N, Harshavardhan B S, Prabakar J and Saibabu S 2017 Investigation on wall panel sandwiched with lightweight concrete *Int. J. of Materials Science and Engineering* **225** 2-16

[5] Insub C, Junhee K and Ho R K 2015 Composite behaviour of insulated concrete sandwich wall panels subjected to wind pressure and suction *Materials* **8** 1264-1282

[6] Soheil S, Amir A and Farhang F 2016 The effect of concrete panel thickness upon composite steel plate shear walls *J. of Constructional Steel Research* **117** 81-90

[7] Shutt C A 1997 *Report Codifies, Details Sandwich Wall Panels* (Chicago: Precast/Prestressed Concrete Institute)

[8] Jimin H and Zhi F F 2001 *Modal Analysis* (Oxford: Butterworth-Heinemann)

[9] Mohamed A Kader and Ahmed F 2019 Improving the resistance of concrete panels to hard projectile impact *Int. J. of Protective Structures* **26** 1-10

[10] Sangkyu L, Gyuyong K, Hongseop K, Minjae S, Gyeongcheol C and Jeongsoo N 2018 Strain behavior of concrete panels subjected to different nose shapes of projectile impact *Materials* **11** 2-15

[11] Meo M, Vignjevica R and Marengo G 2004 The response of honeycomb sandwich panels under low-velocity impact loading *Int. J. of Mechanical Sciences* **47** 1301-1325

[12] Ranjith K P, Kiran K N, Nagasrisaihari S, Chennakesava R B 2018 Damage characterization of sandwich composites subjected to impact loading *J. of Sandwich Structures and Materials* **16** 10-25

[13] Abd Rahman N, Jaini Z M and Mohd Zahir N N 2015 Fracture energy of foamed concrete by means of the three-point bending test on notched beam specimen *ARPN J. Eng. Appl. Sci.* **10** 6562-6570

[14] Jaini Z M, Mokhatar S N, Mohd Yusof A S, Zulkiply S and Abd Rahman M H 2016 Effect of pelletized coconut fibre on the compressive strength of foamed concrete *MATEC Web of Conf.* **47** 01013

[15] BS EN 12390-2 2009 *Testing Hardened Concrete. Making and Curing Specimens for Strength Tests* (London: British Standards Institution)

[16] ASTM D7136-15 2015 *Standard Test Method for Measuring the Damage Resistance of a Fiber-Reinforced Polymer Matrix Composite to a Drop-Weight Impact Event* (West Conshohocken: ASTM International)

[17] EN 1992-1-1 2004 *Design of Concrete Structure Part 1-1: General Rules and Rules for Buildings* (London: British Standards Institution)

[18] Aldridge D 2005 *Introduction to Foamed Concrete, Use of Foamed Concrete In Construction*, (Dundee, United Kingdom)

[19] British Cement Association 1994 *Foamed Concrete: Composition and Properties* (Slough, United Kingdom)

[20] Jaini Z M, Rum R H and Boon K H 2017 Strength and fracture energy of foamed concrete incorporating rice husk ash and polypropylene mega-mesh 55 *IOP Conf. Series: Materials Science and Engineering* **248** 012005

[21] Yao X, Bing L and Kazunori F 2017 Behaviour of reinforced concrete slabs under low velocity impact *ACI Structural J.* **114** 643-658

[22] Noroozi S, Sewell P and Vinney J 2014 Investigation of bolt group configurations on load distribution and joint failure *Mechanics of Advanced Materials and Structures* **21** 422-432

Acknowledgments
Authors would like to thank Universiti Tun Hussein Onn Malaysia and Ministry of Education Malaysia for the continuous supports in term of facilities and financial. This study was conducted under GPPS research grant, Code U696.

IOP Conf. Series: Materials Science and Engineering **713** (2020) 012026 doi:10.1088/1757-899X/713/1/012026

Dynamic Serviceability of Lightweight Composite Deck as Floor System Under Human Excitation

A F Norhalim[1], Z M Jaini[2], N Abd Ghafar[2], M A Majid[2] and A Ahwang[3]

[1] Faculty of Civil and Environmental Engineering, Universiti Tun Hussein Onn Malaysia, 86400 Parit Raja, Johor, Malaysia
[2] Jamilus Research Center, Universiti Tun Hussein Onn Malaysia, 86400 Parit Raja, Johor, Malaysia
[3] Civil and Structural Engineering Branch, Public Works Department, 50582 Kuala Lumpur, Malaysia
Corresponding author: rizuan@uthm.edu.my

Abstract. Lightweight composite deck is widely used as floor system due to the advantage of minimum weight as well as to speedy the construction process. There are numerous investigations on the structural behaviour of lightweight composite deck, but its reliability and understanding on dynamic serviceability remain ambiguous. Therefore, this study intents to numerically investigate the dynamic serviceability of lightweight composite deck as floor system. The lightweight composite deck with dimension of 2900 mm span and 900 mm was modelled in STAAD.Pro. The thickness of lightweight composite deck varied from 100 mm to 225 mm. The isotropic equivalent plate model and effective homogenous criterion were adopted for the geometry and material properties. Dynamic serviceability in term of natural frequency and displacement were analysed in corresponds to the thickness of lightweight composite deck. It is noticeable that by increasing the thickness led to the improvement in the natural frequency. In contrast, the displacement become smaller as the thickness become larger. It was identified that only the lightweight composite deck with thickness 200 mm and 225 mm comply the allowable limit.

1. Introduction

Composite construction has been popular over the last twenty years and has largely accounted for the dominance of steel frames in multi-storey building. This type of construction is structurally efficient because it exploits the tensile resistance of the steel and the compressive resistance of concrete. Applications include commercial, industrial and residential building. As the evolution of construction industry goes by, the current interest on new design of human floor has been discovered. Which is using the lightweight composite deck. This renowned design consists of steel structure and bondeck slab made of lightweight concrete with profiled steel sheet. By using these materials, the construction of floor system evolves significant exploitation since concrete is good in compression while steel shows better performance in tension. It became a new solution in reducing the selfweight of structure and the pedestrian bridge can be constructed in a short time. Besides that, lightweight composite deck is versatile since it can be adapted in many cases, located in between any buildings or elevated from busy road. The good resistance and architectural apparel are also another added value.

Lasheen et al. [1] and Yardim et al. [2] emphasized that lightweight concrete has total weight reduction in term of dead load around 20% to 35%. Even up to 50% as in the latest report by Waldmann et al. [3]. The thickness is normally in the range 100 mm to 250 mm for shallow decking, and in the range 250 mm to 300 mm for deep decking [4]. In the design consideration, the serviceability of composite slab in term of vibration and deflection must be checked to provide safety and to prevent outright structural failure. Penza [5] reported that composite slab with shorter span will give a good performance in vertical shear capacity. There are numerous experimental tests that have been conducted to study the behaviour of composite slab and the comparison with reinforced concrete slab such as by Zhou et al. [6], Holomek and Bajer [7], Mohamed and Sidky [8] and Cifuentes and Medina [9]. It was measured that the load carrying capacity and stiffness for the composite slab and reinforced concrete slab are almost insignificant different. Meanwhile, the dynamic behaviour of composite slab was investigated based on two variables; the thickness and grade of concrete. The findings revealed that by increasing the thickness of composite slab, the natural frequency increases up to 25% [10].

Unfortunately, lightweight composite deck has less stiffness than normal concrete slab. Concurrently, the reduction on selfweight may cause the transient state being dominant, subsequently lead to the high discomfort level. Although, Yardim et al. [2] and Jaini et al. [11] developed lightweight composite deck that achieve acceptable limit of vibration higher than 10Hz, the understanding on real situation for pedestrian bridge is still need to be developed. Lack of prove on its reliability must be tackled to further enhance the acceptance of lightweight composite deck in construction industry, especially for the pedestrian bridge. Since there are demand on the application of lightweight composite deck, but lack of knowledge and established procedures, the numerical modelling using STAAD.Pro will definitely contribute to a rational and robust prediction of vibration responses, including the structural behaviour under ambient and transient states. Moreover, this study will help in the improvement of construction process by taking optimum consideration of thickness, span and others specification of lightweight composite deck.

2. Numerical modelling

2.1. Specification

Figure 1 shows the schematic design of lightweight composite deck that has been implemented from the deck design of floor system in Parlimen Malaysia [12]. The size is 2900 mm span and 900 mm width with various thicknesses from 100 mm to 225 mm. The lightweight composite deck is supported by universal steel beams at both ends with the size of UB 553 x 312 x 150, where the perfect bond was assumed as presence of shear studs. Fixed supports were assigned at each of end-corner of universal steel beams to create the fully constrains at Fx, Fy, Fz, Mx, My and Mz. So that, the movement in lightweight composite deck will significantly affecting the supports and behave globally.

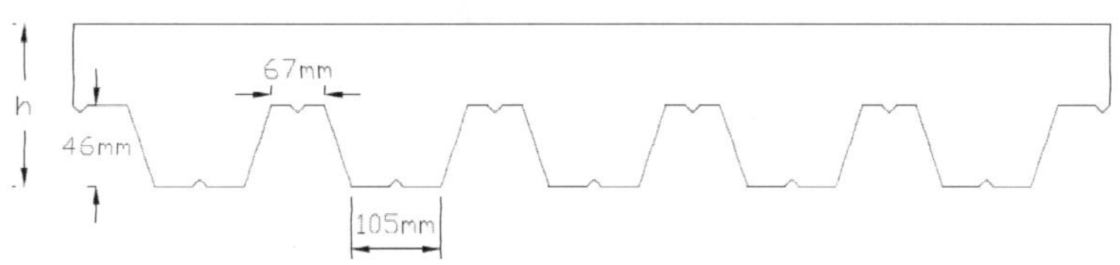

Figure 1. Cross-section of lightweight composite deck

The loads that imposed on the lightweight composite deck are based on the static action and human walking activity. The static action includes selfweight, finishes and services of 1.5 kN/m^2 and variable action for the two-way walkway space. The value of variable action is 3 kN/m^2 as accordance to Class 37 structure category on BS EN 1992-1-1 [13]. On the other hand, human walking activity acts as dynamic excitation in the form of force-time history. Equation (1) as suggested by Gandomkar et al. [14] was used to calculate the load that induced by the human walking activity.

$$F(t) = P_{ai}\cos(2\pi i f_s t) \quad (1)$$

where, P is individual's weight taken as 800 N, a_i is dynamic coefficient for the ith harmonic force component, i is harmonic multiple of the step frequency, f_s is step frequency and t is time in seconds.

2.2. Geometric properties

Numerical modelling in STAAD.Pro was conducted through linear-elastic analysis. Since the lightweight composite deck has the corrugated shape, a special technique so-called isotropic equivalent plate model as established by El-Dardiry and Ji [15] was employed to model the lightweight composite deck. This is due to fact that STAAD.Pro not permitting the corrugated shape to be modelled with the condition of two separated components. In the isotropic equivalent plate model, the corrugated shape was transformed to the flat layer. Therefore, the orthotropic manner of lightweight concrete and profiled sheet was merged to form an isotropic system. Although, the position of neutral axis and second moment of area are changed, the geometric stiffness and mass of lightweight composite deck can be upheld as original. Figure 2 illustrates the transformation of lightweight composite deck from corrugated shape to the isotropic equivalent plate model.

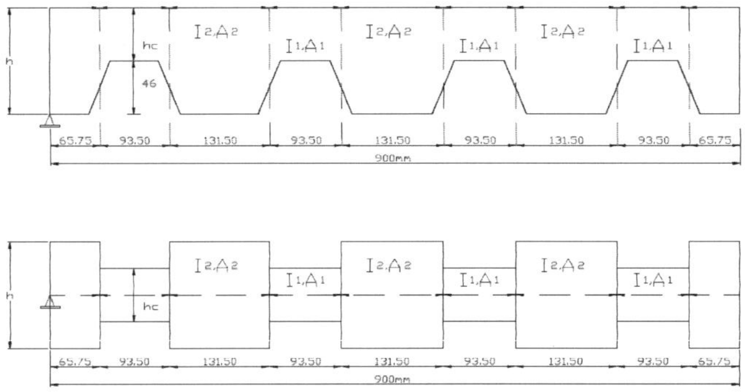

Figure 2. Isotropic equivalent plate model in numerical modelling.

2.3. Discretization

The lightweight composite deck was modelled as plate elements while the universal steel beams were created as beam elements. In term of meshing, STAAD.Pro provides two different selection which are three-noded polygonal element and four-noded quadrilateral element. In this study, the four-noded quadrilateral element was employed to discretize the lightweight composite deck as this type of meshing provide accurate deformation since it can be interpolated to a higher degree. In addition, the four-noded quadrilateral element is more suitable to form uniform meshing. The size of mesh is 20 mm x 20 mm based on convergency obtained from the mesh sensitivity test. Figure 3 shows the discretization of lightweight composite deck in the interface and rendering views.

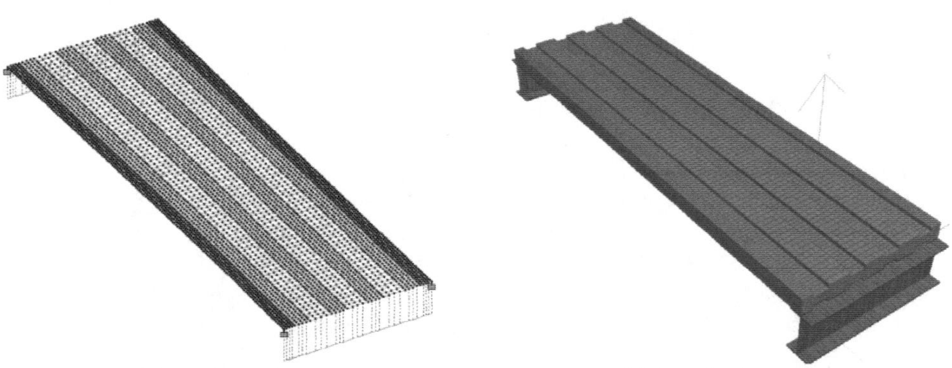

a) Interface view b) Rendering view

Figure 3. Meshing of lightweight composite deck in the interface and rendering views.

2.4. *Material properties*

The lightweight composite deck consists of foamed concrete and profiled steel sheet of ComFlor46. The density of foamed concrete was controlled at 1600 kg/m^3 which found suitable to be used as structural component. Meanwhile, ComFlor46 has 0.9 mm thickness. The material properties of Young's modulus, Poisson's ratio, density, thermal coefficient, critical damping, shear modulus, yield stress, tensile strength, yield strength ratio, tensile strength ratio and compressive strength can be referred in Table 1. The material properties of foamed concrete and profiled steel sheet are based on BS EN 1992-1-1 [13], and the standard manufacturer respectively. When the isotropic equivalent plate model is adopted, the material properties should be changed to the homogenous effective criterion. Therefore, a general rule of mixtures was used to predict the material properties. Therefore, when the thickness changes and hence the material properties alternated as can be referred in Table 2. It should be noted here that the material properties adopted in this study is predominantly by the brittle material model.

Table 1. Material properties of foamed concrete and profiled steel sheet.

Properties	Foamed Concrete	ComFlor46
Young's modulus, E (kN/m^2)	19.2×10^6	210×10^6
Poisson's ratio, ν	0.17	0.3
Density, ρ (kN/m^3)	15.69	76.82
Thermal coefficient, α (/°C)	8×10^{-6}	10×10^{-6}
Critical damping, ζ	0.05	0.03
Shear modulus, G (kN/m^2)	8×10^6	81×10^6
Yield stress, F_y (kN/m^2)	-	280×10^3
Tensile strength, F_t (kN/m^2)	1.84×10^6	0.41×10^6
Yield strength ratio, R_y	-	1.5
Tensile strength ratio, R_t	-	1.2
Compressive strength, F_c (kN/m^2)	20×10^3	-

Table 2. Material properties based on homogenous effective criterion.

Properties	Thickness of lightweight composite deck					
	100 mm	125 mm	150 mm	175 mm	200 mm	225 mm
E (kN/m^2)	21.67×10^6	21.11×10^6	20.80×10^6	20.56×10^6	20.39×10^6	20.25×10^6
ν	0.1717	0.1713	0.1711	0.1709	0.1708	0.1707
ρ (kN/m^3)	16.49	16.31	16.20	16.13	16.07	16.03
α (/°C)	8.03×10^{-6}	8.02×10^{-6}	8.02×10^{-6}	8.01×10^{-6}	8.01×10^{-6}	8.01×10^{-6}
ζ	0.0497	0.0498	0.0498	0.0499	0.0499	0.0499
G (kN/m^2)	8.94×10^6	8.74×10^6	8.61×10^6	8.52×10^6	8.45×10^6	8.40×10^6
F_y (kN/m^2)	3.62×10^3	2.85×10^3	2.35×10^3	2.00×10^3	1.74×10^3	1.54×10^3
F_t (kN/m^2)	1.82×10^6	1.83×10^6	1.83×10^6	1.83×10^6	1.83×10^6	1.83×10^6
R_y	0.0194	0.0153	0.0126	0.0107	0.0093	0.0083
R_t	0.0150	0.0120	0.0101	0.0086	0.0075	0.0066
F_c (kN/m^2)	19.74×10^3	19.8×10^3	19.83×10^3	19.86×10^3	19.88×10^3	19.89×10^3

3. Result and discussions

3.1. *Acceleration-time history*

The overview of plotted acceleration-time history for lightweight composite deck is shown in Figure 4. This acceleration-time history can be converted to the vibration spectrum domain using fast Fourier transform. Therefore, the natural frequency that cause the lightweight composite deck to experience resonant can be easily identified. Figure 5 shows the vibration spectrum domain where the amplitude of vibration is plotted against the natural frequency. The oscillation gives the range of maximum and minimum amplitudes, where it will reduce to the static state. The highest acceleration-time history can be observed at the point of imposed load. For the lightweight composite deck, it was observed that the maximum acceleration is around 1.42 m/s^2 to 4.31 m/s^2 for thickness 100 mm to 225 mm. According to Saidi et. al [16], walking pedestrians can induce considerable vertical and horizontal rhythmic impulsive dynamic loads that are dominated by the pacing rate and by having a higher pacing rate, the vibration responses including frequency, amplitude and duration will be increased.

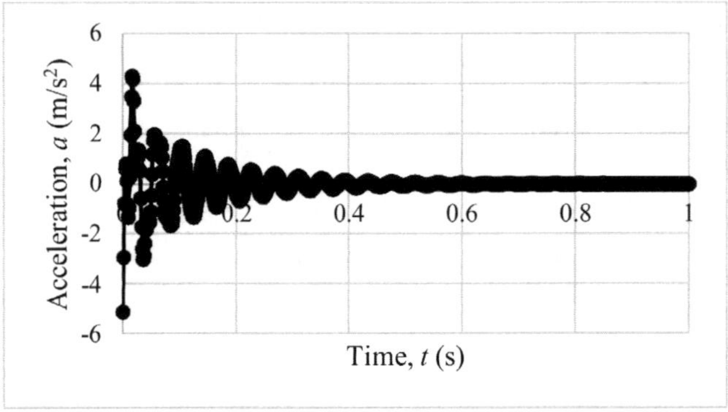

Figure 4. Acceleration-time history of lightweight composite deck.

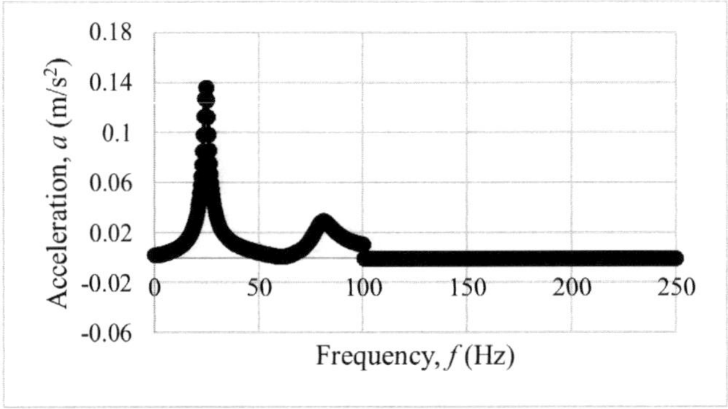

Figure 5. Vibration spectrum domain of lightweight composite deck.

3.2. Natural frequency

It should be emphasized that the natural frequency presented here is based on the first mode shape. In this study, the natural frequency was captured under ambient and transient states. The ambient state is fully depending on the selfweight and service load of lightweight composite deck itself. Meanwhile, the transient state is based on the impose load of human walking activity in the form of harmonic-sinusoidal function. Figure 6 shows the natural frequency in corresponds to the thickness of lightweight composite deck. The natural frequency under the ambient state is plotted together with the natural frequency obtained from analytical approach. It can be observed that the natural frequency of ambient state is slightly close to the analytical approach. The lightweight composite deck with thickness 100 mm has the natural frequency of 23.99Hz and 24.30Hz for the ambient state and analytical approach, respectively. This indicate relatively small difference around 1.28%. Similarly, lightweight composite deck with thickness 125 mm has 1.29% error. Meanwhile, for thickness 125 mm to 225 mm, the natural frequency has the gap less than 7.5%.

It can be observed that by increasing the thickness of lightweight composite deck will eventually lead to the higher value natural frequency. Thickness has direct relationship with the selfweight where it will be increased if the thickness enlarged. Hence, it will increase the moment of inertia and consequently increase the natural frequency. Howard and Hansen [17] agreed that as the equivalent thickness of the floor increased, which corresponds to a greater cross-sectional moment of inertia and cause the first resonance frequency increased. On the other hand, Jaini et al. [11] reported that the natural frequency of foamed concrete composite slab decreases due to the thickness ranging from 100 mm to 150 mm but gradually increased as the thickness reached 175 mm. This can be explained as the presence of the shear bond has been ignored in the modelling process. The lightweight composite deck was modelled as an isotropic equivalent plate model that merged the orthotropic manner of lightweight concrete and profiled sheet to form an isotropic system.

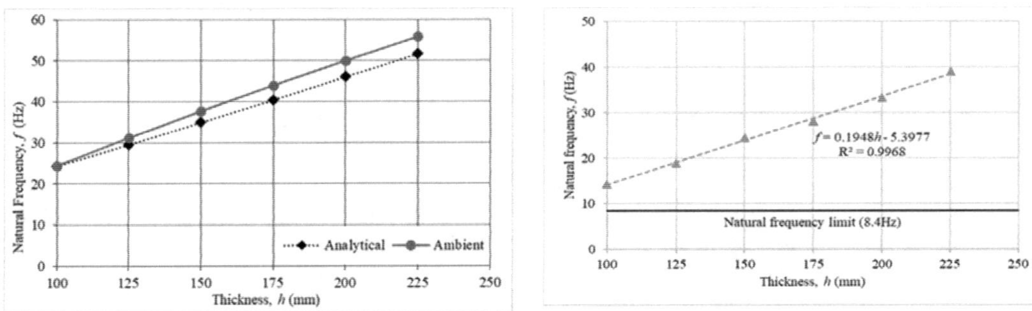

(a) Ambient state (b) Transient state

Figure 6. Natural frequency in corresponds to the thickness of lightweight composite deck.

3.3. Displacement

In the real scenario of design, serviceability aspect become predominant factor that need to be checked carefully. Under dynamic loading, especially vibration, serviceability must be referred to both natural frequency and displacement. Figure 7 shows the displacement in corresponds to the thickness of lightweight composite deck under ambient and transient states. A comparison with analytical approach indicates insignificant difference. The maximum error between ambient state and analytical approach is 37.04%, which happen on the lightweight composite deck with thickness 100 mm. The displacement for lightweight composite deck with thickness 100 mm to 150 mm under ambient state is upper bound of analytical approach. However, the ambient state become lower bound when the thickness increases from 175 mm to 225 mm. It can be observed that the displacement is decreased when the thickness become larger. The deformation of structure and supports begins almost simultaneously. As a result, the energy of external load which is absorbed by yielding supports is fully redistributed to the slab.

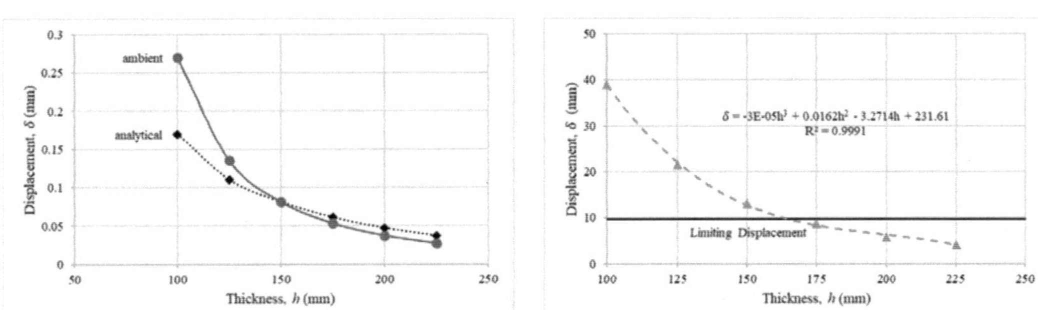

(a) Ambient state (b) Transient state

Figure 7. Displacement in corresponds the thickness of lightweight composite deck.

For a lightweight composite deck that has effective span of 2900 mm, it was computed that the displacement limit is 8.25 mm [13]. Basically, the trend of displacement of lightweight composite deck is almost similar. It can be clearly observed that by increasing the thickness of lightweight composite deck, the displacement is gradually decreased. It is affirmed that the stiffness of lightweight composite deck governs its structural behaviour. By comparing with displacement limit, only lightweight composite deck with thickness 175 mm, 220 mm and 225 mm accomplish the serviceability requirement. The displacement of lightweight composite deck under transient state is 39.01 mm, 21.54 mm, 13.06 mm, 8.62 mm, 5.77 mm and 4.23 mm for thickness 100 mm, 125 mm, 150 mm, 175 mm, 200 mm and 225 mm, respectively. Under human walking activity, the displacement of lightweight composite deck become obviously larger compared to which under the selfweight and service load. The corresponding reaction as the human walking activity generates higher dynamic forces than the quasi-permanent load [13]. The high dynamic forces together with the quasi-permanent load contribute to a higher magnitude of load, thus the reaction forces and vibration responses under human walking activity is higher than under the selfweight and service load.

3.4. *Effective thickness*

In order to determine the effective thickness of lightweight composite deck for the pedestrian bridge, the natural frequency is plotted against the displacement. Figure 8 shows the correlation between the natural frequency and displacement under transient state. In respect to Murray et al. [18] and BS EN 1992-1-1 [13], the lightweight composite deck must fulfil the natural frequency and displacement limits. Principally, the natural frequency must bigger than 10Hz while the displacement must lower than 8.25 mm. Although all thicknesses in lightweight composite deck satisfied the vibration limit, only lightweight composite deck with thickness 200 mm and 225 mm pass the displacement control.

In this study, the displacement results are more dominant than the natural frequency when the lightweight composite deck is subjected under human excitation. All the thicknesses passed the natural frequency limit but not the displacement limit. These corresponding results were influenced by the increasing stiffness as the thickness increased. However, different situation might happen if the span of the lightweight composite deck is longer. The natural frequency and displacement will have an equal role in terms of serviceability stage.

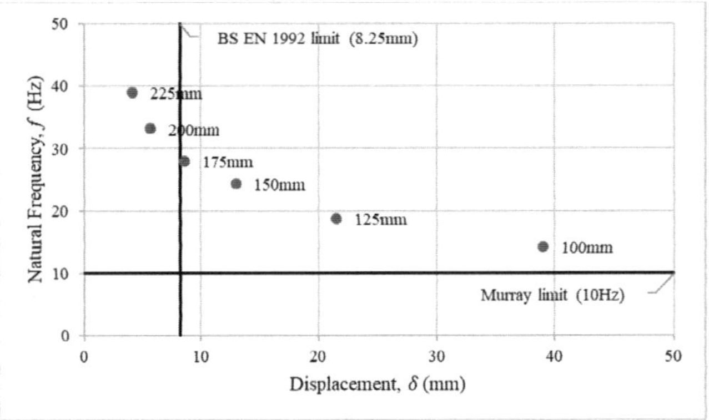

Figure 8. Correlation between natural frequency and displacement.

4. Conclusion

The lightweight composite deck was numerically modelled in STAAD.Pro with various thickness from 100 mm to 225 mm. The isotropic equivalent plate model was employed to construct the lightweight composite deck that originally in the form of corrugated shape. Meanwhile, the material properties of lightweight composite deck were converted from anisotropic to homogenous effective criterion using a general rule of mixtures. The loads that imposed on the lightweight composite deck are based on the static action and human walking activity (in the form of harmonic-sinusoidal function as force-time history). It was discovered that STAAD.Pro able to produce the acceleration-time history and vibration spectrum domain. A comparison of natural frequency and displacement between ambient state and analytical approach provide convincing agreement. It is noticeable that the thickness of lightweight composite deck governs the natural frequency and displacement. When the thickness of lightweight composite deck is increased, the natural frequency is increased as well. On the other hand, the displacement reduces as the thickness become larger from 100 mm to 225 mm.

5. References

[1] Lasheen M, Shaat A and Khalil A 2016 Efficiency of using lightweight concrete slabs in steel-concrete composite beams *XVI Int. Sci. Conf. VSU`2016* Bulgaria pp 181–186

[2] Yardim Y, Waleed A M T, Jaafar M S and Laseima S 2013 AAC-concrete light weight precast composite floor slab *Constr. Build. Mater.* **40** 405–410

[3] Waldmann D, May A and Thapa V B 2017 Influence of the deck profile design on the composite action of slabs made of lightweight woodchip concrete *Constr. Build. Mater.* **148** 887–899

[4] Feldmann M, Heinemeyer C and Volling B 2008 *Design Guide for Floor Vibrations* (Luxembourg: ArcelorMittal) pp 3–5

[5] Penza A 2010 *Composite Slabs with Lightweight Concrete* Master Thesis (Milano: Politecnico Di Milano) pp 81-85

[6] Zhou X, He Y, Jia Z and Nie S 2011 Experimental study on vibration behavior of cold-form steel concrete composite floor *Adv. Steel Constr.* **7(3)** 302–312

[7] Holomek J and Bajer M 2012 Experimental and numerical investigation of composite action of steel concrete slab *Procedia Eng.* **40** 143–147

[8] Sidky A N M, Yasser G and Fahmy M 2012 An experimental investigation of composite floor vibration due to human activities. A case study *HBRC Journal* **8** 228–238

[9] Cifuentes H and Medina F 2013 Experimental study on shear bond behavior of composite slabs according to Eurocode 4 *J. Constr. Steel Res.* **82** 99–110

[10] Venghiac V 2010 The influence of the thickness of the slab and concrete grade on composite floors *Bul. Inst. Polit. Iasi.* **1** 85-92

[11] Jaini Z M, Rum R H M, Ghaffar N H A, Misbah M S A, Desa M S M 2018 Experimental study on vibration responses and energy dissipation of foamed concrete composite slabs *Malaysian Constr. Res. J.* **4(2)** 193-207

[12] Jabatan Kerja Raya 2018 *Constrution Drawing: Proposed Project of Parlimen Building Phase 3B* (Kuala Lumpur: Kuala Lumpur) pp 85-153

[13] Eurocode 2 2014 *EN 1992-1-1 Design Of Concrete Structures - Part 1-1: General Ruels and Rules for Buildings* (London: British Standard Institution)

[14] Gandomkar F A, Badaruzzaman W H W and Osman S A 2011 The natural frequencies of composite profiled steel deck dry board with concrete infill (PSSDBC) system *Latin American J. Solids Struct.* **8(3)** 351–372

[15] El-Dardiry E and Ji T 2006 Modelling of the dynamic behaviour of profiled composite floors *Eng. Struct.* **28** 567–579

[16] Saidi I, Haritos N, Gad E F and Wilson J L 2006 Floor vibrations due to human excitation - damping perspective *Earthquake Eng. Australia* **1** 257–264

[17] Howard C Q and Hansen C H 2002 Vibration analysis of waffle floors *Comput. Struct.* **81** 15-26

[18] Murray T M, Allen D E and Ungar E E 1997 *Steel Design Guide Series 11 Floor Vibrations Due to Human Activity* (America: American Institute of Steel Construction) pp 7-10

Acknowledgments
Authors would like to thank Universiti Tun Hussein Onn Malaysia, Public Works Department (Civil and Structure Engineering Branch) and Ministry of Education Malaysia for the continuous supports in term of facilities and financial. This study was conducted under GPPS research grant, Code U696.

IOP Conf. Series: Materials Science and Engineering **713** (2020) 012027 doi:10.1088/1757-899X/713/1/012027

Shear Performance of Pretensioned Prestressed Concrete Beam with Steel Fibre

Ruqayyah Ismail[1], Lyn Dee Goh[1], Nik Nurul Nadia Nik Hasan[1], Azerai Ali Rahman[2], Fariz Aswan Ahmad Zakwan[1]

[1]Faculty of Civil Engineering, Universiti Teknologi MARA, Cawangan Pulau Pinang, 13500 Permatang Pauh, Pulau Pinang, MALAYSIA.
[2]Faculty of Civil Engineering, Universiti Teknologi MARA, 40450 Shah Alam, Selangor, MALAYSIA.
Corresponding author: gohlyndee147@uitm.edu.my

Abstract. The use of steel fibre has proven to be effective in enhancing the performance of concrete structure. However, its application in the prestressed concrete is not fully understood. Therefore, this study presents the application of steel fibre in the prestressed concrete beam. An investigation on the crack pattern and load-deflection relationship of prestressed concrete beam filled with steel fibre was carried out and a comparison is made with the prestressed concrete beam without the steel fibre. A total of three configurations of beams with size of 150 × 200 × 1200 mm were employed. Steel fibres were added with two different volume fractions of 3% and 5%. Experimental results showed that the control beam experienced shear crack pattern, while the other two beams experienced the flexural-shear crack. In comparison between two different percentages of steel fibre in the beam configurations, more cracks were observed in the prestressed beam that filled with 3% steel fibre compared to the prestressed beam filled with 5% of steel fibre.

1. Introduction

Prestressed concrete members have been used in building structures and infrastructure facilities since the 1960s due to their advantages such as high quality, and durable aesthetic [1] Prestressed concrete is known to have better performance compared to normal reinforced concrete. The application of prestressing concrete benefit the construction as it able to provide more strength for longer span. Reinforced concrete is known with its ability to utilise full-strength capacity in resisting the load directly while prestressing concrete is capable in resisting shear stress due to its pre-compression behavior.

For reinforced concrete beam, shear failure would occur when the length of the beam is three times of the effective depth. Failure mode of shorter spans is due to combination of three elements; shear, crushing, and splitting. For longer spans (plain reinforced concrete beam), the crack occurred easily due to flexural tensile stresses[2]. By prestressing the concrete, the pre-compression are introduced in the structural elements which is subjected to tensile stresses and can be kept under compression during its service time [3]. The external load will generate tensile stresses which is nullified by the pre-compression and could also be a residual compression. Hence, the cracking problem in structure can be avoided and the cross-section can be fully utilised in transferring the load. The used of prestressed concrete could avoid plain concrete's shear failure which commonly happened with sudden collapse and failed without earlier signage [4].

A pull-out behaviour of prestressing strands in reinforced concrete incorporated with steel fibre showed that the addition of the steel fibre into concrete was capable to modify shear performance of the concrete [5]. Yoon and Nishiyama [6] have conducted a test investigating shear failure of deep fibre reinforced concrete beams and found that the shear crack was reduced when utilising a significant content of steel fibre in the element. Furthermore, the additional of steel fibre in concrete was capable to eliminate the occurrence of the concrete spalling as the steel fibre could hold the concrete during the cracking stages [7]. In addition, post cracking response of concrete structural was reported better than conventional element [8]. The claim is strongly supported by Liu et al. [9] with experimental works conducted and demonstrated that the volume of the steel fibre utilised has contributed to the increment of the beam shear strength. As for the use of steel fibre in prestressed concrete beam, with 0.3 and 0.5 percent of additional steel fibre, Yoon and Nishiyama [6] have concluded that the increment amount of steel fibre has led to more crack's appearance. This finding however contradicts the effect of steel fibre in ordinary reinforced concrete beam as reported by previous researchers.

Uncertainty finding of prestressed beam performance with steel fibre has led this study to be initiated. This study is investigating the shear performance of prestressed concrete with additional of steel fibre in the prestressed concrete beam. Therefore, three beams configurations were tested to investigate the shear behaviour and the effect of additional steel fibre to the prestressed concrete beam performance against shear failure. Additional volume fracture of 3% and 5% steel fibres were used in this study and the shear failure was observed and compared with the reinforced concrete beam specimen without steel fibre (control specimen).

2. Methodology

2.1. Specimen Materials

Two main materials used in this study were concrete and steel fibre. The concrete compressive strength was calculated based on the standard BS EN 12390 [10]. Concrete grade 40 MPa was used for all specimens in this study. 0.6% of plasticiser from the weight of cement was added into the concrete mixture for the workability of the mixture. In avoiding concrete segregation, the usage of plasticiser should not be added with maximum amount [11]. While for steel fibre, hooked-end type was used as depicted in Figure 1. The length of the steel fibre is 60 mm and the diameter is in the range of 0.5 mm – 1.0 mm. The proportions of steel fibre used were 3% and 5% of its volume.

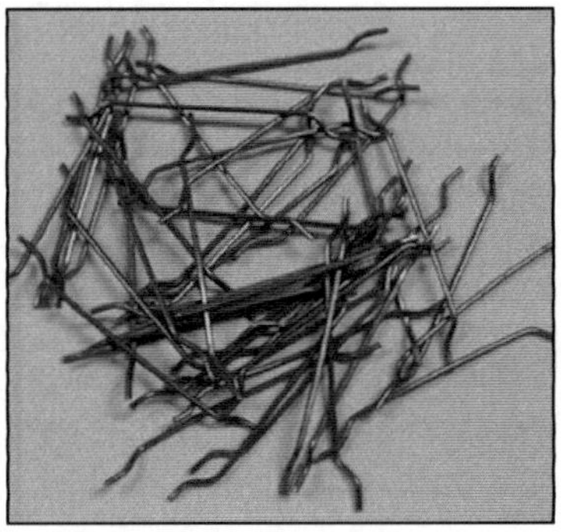

Figure 1. Steel Fibre

2.2. Casting of Beams

All three beam specimens were casted in a formwork of 2250 mm length prestressing bed (Refer to Figure 2). One of the specimens was reinforced concrete beam with four numbers of 12 mm reinforcement bar diameter. The ultimate tensile load for 12.7 mm tendon is 186 kN. For pre-tensioned prestressed concrete, the tendon is tensioned up to 50% from its tendon strength that is approximately 93 kN (2900 Psi). The tendon tensioning was conducted using manual hydraulic jack with 70 mm elongation. The fresh concrete was then placed into the formwork in the prestressing bed and cube moulds. Compacted process was done by hand-rammed and vibrating table. The prestressed concrete specimens were then left for seven days before the formwork was removed. The stressed in the presetressing wires were released from the anchorage ends before the specimens removed from the formwork and they were kept in the laboratory under controlled room temperature. All beams and cubes specimens were tested after a curing period of 28 days. Two-point load testing was carried out for beams and compressive strength test for the cubes.

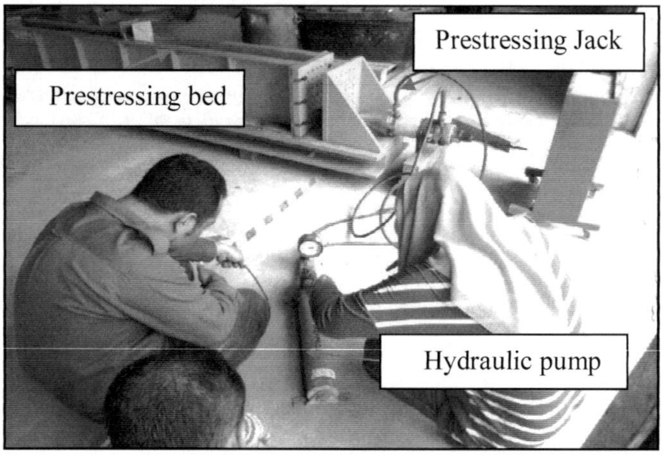

Figure 2. Pre-tensioning Process

2.3. Testing Set-Up

The shear test was conducted using two-point load test method [12]. The load-deflection result of the testing was recorded to differentiate the stiffness of every specimen. The specimen was tested until it reached the ultimate failure. From this study, three linear variable differential transducers (LVDT) were used; left (LVDT 1), middle (LVDT 2) and right (LVDT 3) side of the beam. A 20 mm LVDT was employed. The Figure 3 shows the setting up of two-point load shear testing and the placement of the LVDTs.

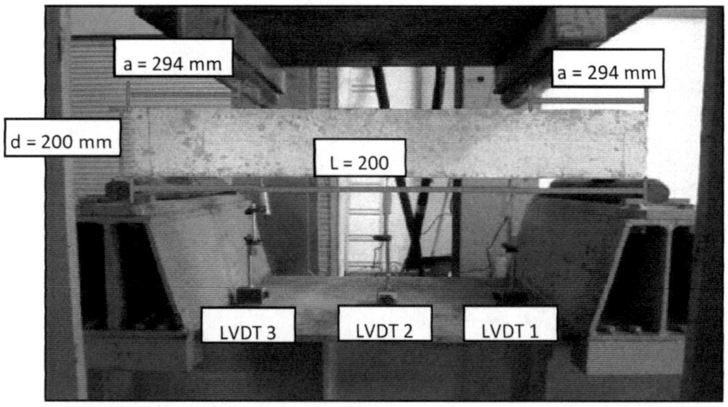

Figure 3. Testing Set-Up of Two Point Bending

3. Results and Discussion

3.1. Concrete Cube Compression Test
Concrete cube compressive strength of 7, 14 and 28 days were recorded for all beam specimens; reinforced concrete beam, prestressed beam with 3% of steel fibre and 5% of steel fibre (refer to Table 1).

Table 1. Cube Compressive Strength

Days	Concrete Cube Compressive Strength (N/mm^2)		
	Control Concrete	3% Steel Fibre	5% Steel Fibre
7	28.0	28.9	29.5
14	33.1	35.4	37.0
28	41.0	43.7	45.7

3.2. Load-Deflection Behaviour
Load-deflection was measured until the beam was failed. The typical load-deflection curves for reinforced concrete beam, prestressed beam with 3% steel and prestressed beam with 5% steel fibre were presented in Figure 4, Figure 5(a) and Figure 5(b), respectively. From the Figure 4, it shows that when load was increased, the deflection is also increased. The highest displacement experienced was recorded by LVDT 3 which was placed at the right support of the specimen where the first crack was observed.

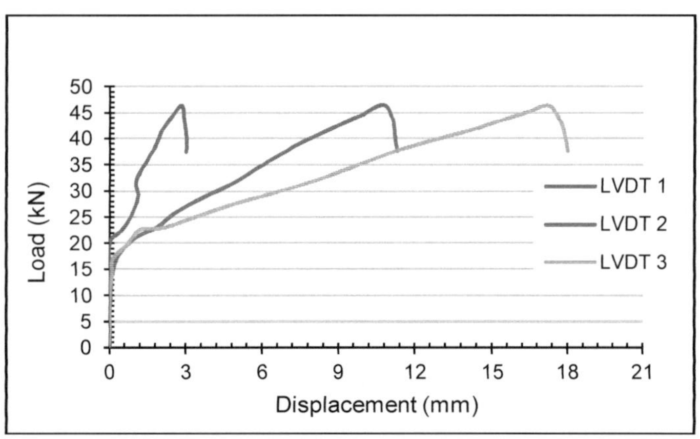

Figure 4. Load-Deflection for Reinforced Concrete Beam

The highest value of the deflection was recorded by LVDT 2 which it was placed at the middle of the beam for beam specimen with 3% of steel fibre (Figure 5(a)). This maximum deflection recorded was 14.02 mm when a 34.32 kN load was applied. Maximum load recorded in the study was 37.58 kN, also by LVDT 2. From Figure 5(b), it shows that, LVDT 2 gave the highest value of deflection compared to the LVDT 1 and LVDT 3. The deflection was higher at the middle position of the beam specimen. The ultimate load experienced by the beam specimen with 5% steel fibre was 44.60 kN.

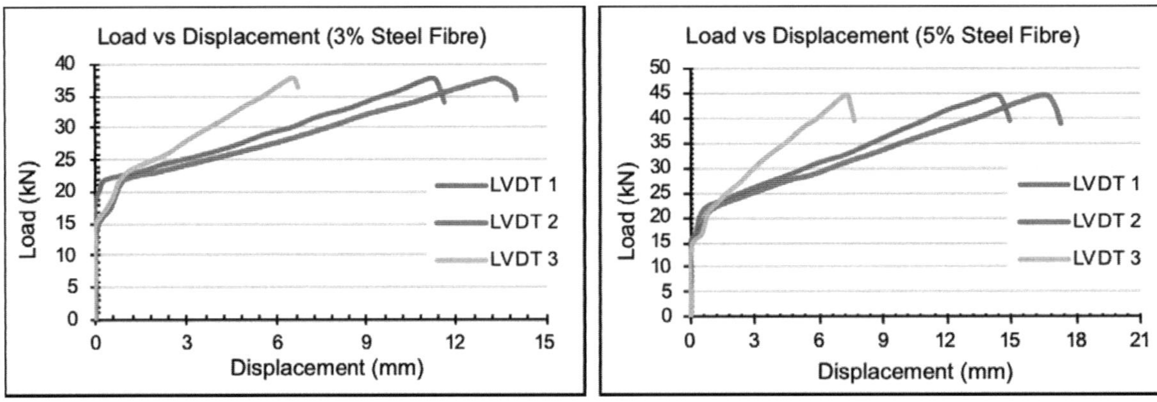

(a) (b)

Figure 5. Load-Displacement for Prestressed Beam with Steel Fibre; a) 3% Steel Fibre b) 5% Steel Fibre

Table 2 shows the results of ultimate load and displacement with percentages difference comparison of each specimen group for the ultimate displacement. The ultimate load for reinforced concrete beam is slightly higher (0.4%) compared to prestressed beam with 5% steel fibre while prestressed beam with 3% steel fibre is recorded lower (18.53%) than reinforced concrete beam specimen. This has showed additional of steel fibre to prestressed beam did not contribute to the strength enhancement of the beam. The ultimate displacement comparison of each LVDTs showed LVDT 1 (left side) experienced the highest increment for prestressed beam with 5% steel fibre (14.24 mm) compared to reinforced concrete beam (2.84 mm). This can be observed from the crack failure happen for prestressed beam with 3% and 5% were more to the left side of the beam. The result contradict with ultimate displacement experienced by LVDT 3 (right side) which reflecting reinforced concrete beam crack failure can be seen at the right side of the specimen.

Table 2. Summary of Results

LVDTs Position	Ultimate Loads (kN)			Ultimate Displacement (mm)			Ultimate Displacement (Different-%)		
	Reinforced Concrete Beam (RCB)	Prestressed Beam with 3% Steel Fibre, (PBSF3)	Prestressed Beam with 5% Steel Fibre (PBSF5)	Reinforced Concrete Beam (RCB)	Prestressed Beam with 3% Steel Fibre, (PBSF3)	Prestressed Beam with 5% Steel Fibre (PBSF5)	RCB - PBSF3	RCB - PBSF5	PBSF3 - PBSF5
LVDT 1	46.4	37.8	44.4	2.84	11.18	14.24	293.0	401.41	27.37
LVDT 2				10.74	13.37	16.58	24.49	54.38	24.00
LVDT 3				17.24	6.52	7.43	62.18	56.90	13.96

3.3. Cracking Pattern

The first crack was observed to occur at the support of the beam. It started from the beam soffit and propagated to the top of the beam. The crack started at 60 mm from the end of the support and continued until 294 mm distance from the support. Figure 6 (a) shows the crack pattern of the control beam which appeared at the right side of the beam.

Crack pattern for specimen with tendon and 3% of steel fibre can be seen in Figure 6 (b). The crack started to form at 390 mm from the right end of the support. From the observation, the crack was categorised as flexural-shear crack as the crack was large and sudden. The crack pattern formed from this specimen is as what has been recorded by Narayanan and Darwish [13]. The study conducted by

Narayanan and Darwish [13] was using un-tensioned reinforcement in the prestressed beam with the 0.75 fraction of steel fibre.

Figure 6 (c) is showing crack pattern formed for prestressed concrete beam with 5% steel fibre. This crack pattern was at 420 mm from the nearest support at left side of the beam and it occurred at the section between the flexural and shear zones. The crack failure mode observed in this beam was categorised as the flexural-shear failure. This beam specimen experienced a sudden crack with wider crack opening. This beam demonstrated more cracks as compared to the prestressed beam filled with 3% of steel fibre. This is due to shear crack increased significantly as the fibre content increased [7]. Increment amount of steel fibre will increase the strength of the beam as the steel fibre will act as bonding material in the concrete. This has been mentioned as well by Yusof et al. [14] whereby when the steel fibre concrete beam was loaded, the steel fibre will perform a bridging effect to the cracks.

(a) (b) (c)

Figure 6. Crack Pattern of Prestressed Beam; a) Reinforced Concrete Beam b) 3% Steel Fibre c) 5% Steel Fibre

4. Conclusion

From this study there are several findings that can be concluded:

i. Reinforced concrete beam specimen in this study has failed in shear crack while for prestressed beams with 3% and 5% steel fibre, the beams failed in flexural-shear crack manner. The reinforced concrete beam showed shear crack failure because it consists of reinforcement bar which resisted the flexural load applied. The prestressed beams with steel fibre showed flexural-shear crack behaviour with large crack occurred and the failure occurred at a very sudden rate compared to the reinforced concrete beam.

ii. More cracks were observed in the prestressed beam with 3% steel fibre compared to the prestressed beam with 5% of steel fibre. This due to 5% steel fibre prestressed beam have higher strength compared to the prestressed beam with 3% of steel fibre.

iii. The ultimate load for reinforced concrete beam is 46.38 kN while for prestressed concrete beam with 3% and 5% steel fibres, the ultimate load recorded were 37.22 kN and 44.63 kN, respectively.

iv. Prestressed concrete beam with 5% steel beam has achieved almost the similar compressive strength as the reinforced concrete beam with 4% differences which shows the existence of 5% steel fibre is sufficient to replace reinforcement in beam and can result in lowering the possibility of shear failure.

5.0 References

[1] Yao D, Jia J, Wu F and Yu F 2014 Shear performance of prestressed ultra high strength concrete encased steel beams *Constr Build Mater.* **52** 194–201

[2] Abi Aghayere 2017 *Reinforced Concrete Design Ninth Edit* (Upper Saddle River New Jersey: Pearson Education Inc) ppp 419

[3] Rajagopalan N 2005 Prestressed Concrete Second Edi. (Alpha Science International, Ltd) pp 508

[4] Cho J-S, Lundy J and Chao S-H 2009 Shear Strength of Steel Fiber Reinforced Prestressed Concrete Beams *Structure Congress 2009 (Austin, Texas)* pp 1058–66

[5] Baran E, Akis T and Yesilmen S 2012 Pull-out behavior of prestressing strands in steel fiber reinforced concrete. *Const. Build Mater.* **28**(1) 362–71

[6] Yoon HJ and Nishiyama M 2012 Behavior of Pretensioned Concrete Beams Using Steel-Fiber Reinforced Concrete *15th World Conf Earthq Eng. 2012 (Lisbon, Portugal)* 1–9

[7] Lim D H and Oh B H 1999 Experimental and theoretical investigation on the shear of steel fibre reinforced concrete beams *Engineering Structure* **21**(10) 937–44

[8] Madan S K, Rajesh Kumar G and Singh S P 2007 Steel Fibers as Replacement of Web Reinforcement for RCC Deep Beams in Shear *Asian Journal Civil Eng.* **8**(5) 479–89

[9] Liu H, Xiang T and Zhao R 2009 Research on non-linear structural behaviors of prestressed concrete beams made of high strength and steel fiber reinforced concretes *Construction Building Material* **23**(1) 85–95

[10] BSEN 12390-3 2009 Testing hardened concrete (London: British Standard Institution BSI) 2009

[11] Mello E, Ribellato C and Mohamedelhassan E 2014 Improving Concrete Properties with Fibers Addition *Int. Sch. Sci. Res. Innov.* **8**(3) 245–50

[12] Omar W and Mohamed R N 2002 The performance of pretensioned prestressed concrete beams made with lightweight concrete *Journal Kejuruteran Awam* **14**(1) 60–70

[13] Narayanan R and Darwish I Y S 1987 Shear in prestressed concrete beams containing steel fibres *Int. J. Cem. Compos. Light Concr.* **9**(2) 81–90

[14] Yusof M and Norazman N 2011 Normal strength steel fiber reinforced concrete subjected to explosive loading *Int. J. Sustain. Constr. Eng. Technol.* **1**(2) 127–36

Acknowledgement

The authors would like to express their deepest appreciation to Universiti Teknologi MARA (UiTM), Cawangan Pulau Pinang for the execution of this research work.

IOP Conf. Series: Materials Science and Engineering **713** (2020) 012028 doi:10.1088/1757-899X/713/1/012028

Strength and Structural Performance of Reinforced Concrete Beam with Artificial Plastic Fibre (APF)

S N Mokhatar[1], N A N A Mutalib[2], A M A Budiea[1], A F Kamarudin[1] and M S Md Noh[1]

[1] Jamilus Research Centre, Faculty of Civil & Environmental Engineering, Universiti Tun Hussein Onn Malaysia, 86400, Parit Raja, Batu Pahat, Johor
[2] Graduate Centre, Universiti Tun Hussein Onn Malaysia, 86400, Parit Raja, Batu Pahat, Johor
Corresponding author: shahruln@uthm.edu.my

Abstract. Behaviour of Reinforced Concrete (RC) in beam structure has captured attention of many researchers. Many past researchers investigated the issue of concrete strength in tensile zone. The present research investigates the effect of waste nylon cable ties fibre that act as an Artificial Plastic Fibres (APF) in normal concrete mix and RC beam element. The concrete strength of 30 MPa have been utilized in this research. The percentage of APF that has been mixed with the concrete paste are 1% and 3%. For strength performance, nine (9) cube specimens and nine (9) cylinder specimens were tested under compressive and splitting tensile test, respectively. In addition, six (6) RC beam specimens were prepared and tested under three points bending test in order to investigate its structural performance. Based on the results, 0%, 1% and 3% of APF in concrete mix produces tensile strength of 1.89 MPa, 3.37 MPa and 2.66 MPa respectively. While, from the compression tests, 50.9 MPa, 47.2 MPa and 24.5 MPa were generated from 0%, 1% and 3% of APF, respectively. From this observations, only 1% of APF was employed in concrete mix to produce RC beam for structural performance investigations. Based on three point bending test, RC beam that consisting 1% of APF produced maximum load about 35.98 kN with 1.8 mm deflection compared with control specimen that can only endured until 27.93 kN of maximum load with 2.37 mm deflection. Further observation reveals that cracking of RC beam consisting 1% of APF reduce and delay the crack formation. Overall, utilization of 1% APF in RC beam successfully improves its strength and structural performance.

1. Introduction

Construction is one of the most dynamic industrial sectors around the world. The construction industry usually uses concrete as the major building materials [1-3]. Despite that, concrete still shows its weaknesses in term of cracking [4] at tensile zone. One of the inventions that can be done to increase the strength and structural performance of concrete beam is by using fibre. Many researchers such as [5-6] and [7] has been conducted study on the performance of concrete beam consisting fibre materials. However, study on efficiency of waste nylon cable ties as plastic fibre in concrete mix is still as yet fully explored. On the other hand, the production of waste in Malaysia is in the average of 30 000 tons with 5 percent were recycled [8]. This waste are also included for non-degradable waste [8]. An existent of this non-degradable waste in large quantities may somehow give bad impact towards environment [9]. Generally, nylon cable ties is one of the wastes that are non-degradable and take the high cost to recycle it back [10]. This material has been produced approximately 4 million tonnes with 13% is the production of chemical fibre production around the world in the year 2000 [10].

In concrete production, combination of steel with concrete helps in increasing the tensile strength capacity of the concrete. Therefore, the function of concrete could contribute for compression region, meanwhile, steel bar resolve problem in tension region [11-14]. Although RC can help in reducing the problem in tensile zone of concrete element, it is still undeniably that steel reinforcement cannot perform effectively in concrete. Steel reinforcement bar will only reduce the crack but not eliminate it [15]. Control or restrain the formation of cracking is significant in order to increase the lifespan of reinforced concrete in term of the durability of concrete structure [16]. In the previous research, many additional ingredient of concrete mix has been investigated such as from local waste product, waste organic materials, waste non-degradable products, etc. There is one literature conducted research on strength properties of nylon fibre and polypropylene fibre in reinforced concrete [17]. This research determines the best type of fibre that can be used to stop/reduce the development of micro-crack before it becomes macro-crack [17]. Based on study by Song et al [17], they found that the nylon fibre has the good strength properties compared to polypropylene fibres in improving the behaviour of concrete. Further, from the study by Nokman and Aziz [18], they applied heavy duty nylon cable ties fibre to solve the problems that related to concrete. They also found that the compressive and tensile strength of concrete can be increased by adding 1% of heavy duty nylon cable ties [18].

Hence, in this study, waste normal nylon cable tie fibre known as Artificial Plastic Fibre (APF) has been proposed as the solution to the above problem by adding this artificial fibre in concrete mix to reduce the tendency of cracking in concrete. Based on the previous research by [18], an additional of this material in concrete can improve the mechanical properties of concrete mix. Basically, this paper highlights the investigation of mechanical properties for concrete consisting of APF. Next, the optimum percentage of APF were added in concrete mix in order to explore the structural behaviour of reinforced concrete (RC) beam consisting this artificial fibre under three points bending test.

2. Samples Preparation and Testing Procedure

Generally, there are three major stages were implemented which includes: a) the preparation of nylon cable ties fibres as artificial plastic fibre known as APF, b) the preparation of concrete mix consisting 1% and 3% of APF, c) the production of RC beam specimens consisting only 1% APF as the optimum percentage of fibre. Compressive strength test were conducted and accordance to BS 1881 Part 116 (1983). All cubes were tested after 28 days of curing. Splitting tensile test is indirect method to obtain the tensile strength of concrete, which the specimen placed in a longitudinal axis and perpendicular to the load. This test conducted after curing at 28 days. The test are accordance to IS: 5816-1970. In this study, the innovation of APF preparation involved process of cutting, cleaning and measure the weight the APF. The length of APF are in the ranges of 30 mm to 50 mm which comply with the normal dimension of existing fibre. To clean up the APF from dust, these artificial fibres are soaked in the water to prevent impurities which may disturbed the bonding interaction with concrete paste. Then, APF are weighed to 1% and 3% as an additional materials in the concrete mix.

2.1. Concrete Mix Design and Specimen Size

Basically, APF acted as additional ingredients in concrete mix which do not affect the normal proportion of main ingredients in concrete mix design. In this research, 1% and 3% of APF were used as additives fibre to concrete mix. The quantities of APF used in this study are tabulated as in Table 1.

Design of normal concrete mix are based on DOE method. The target of concrete strength after 28 days of curing is 30 N/mm^2. The cube size is 0.1 m × 0.1 m × 0.1 m, while, the cylinder height and diameter are 0.3 m and 0.15 m, respectively. For beam specimen, the size is 0.8 m × 0.12 m × 0.15 m.

Table 1. Quantities of APF in kg

	Weight of APF (kg)	
Percentage of APF Sample	1%	3%
Cube	0.10	0.22
Cylinder	0.43	1.30
Beam	1.13	-

2.2. Beam Design

The related specimen details for RC beam are described as in Figure 1. Three points bending test is a testing that measure and monitor the behaviour of specimens after crack initiation and during its propagation. This test revealed the peak load at first crack, ultimate load and maximum deflections of RC beam.

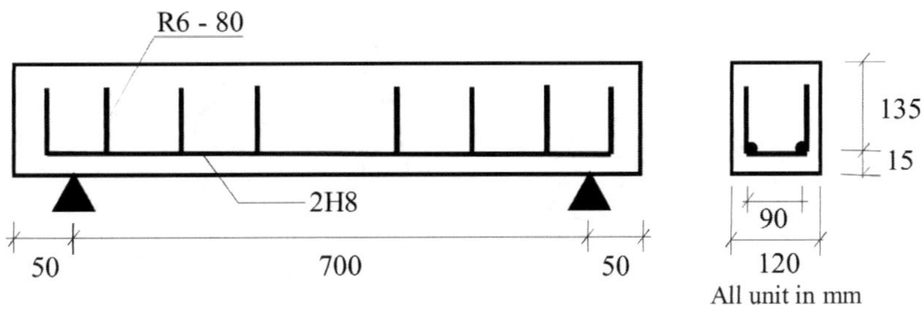

Figure 1. Size and dimensions of steel reinforcement.

3. Results and Discussions

3.1. Compressive Strength Test

Results for the compressive test of the cubes after 28 days are shown in Fig. 2. It can be observed that the usage of APF in concrete has influences the compressive strength. From this results, it can be seen that the presence of APF slightly reduces the concrete strength. However, the percentage difference for 1% APF in concrete mixes as compared with control specimen approximately about 7.3% which are still not gives the major different. However, 3% addition of APF decreases significantly the concrete strength about 51% in comparison to the reference specimen. More amount of APF in concrete are not efficient and suitable under compression state which reflect to the function of fibre can delay the crack formation during tensile.

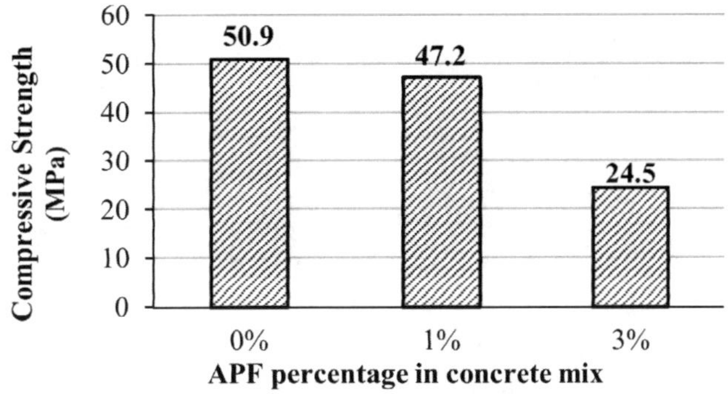

Figure 2. Effect of APF for compressive strength

IOP Conf. Series: Materials Science and Engineering **713** (2020) 012028 doi:10.1088/1757-899X/713/1/012028

3.2. Splitting Tensile Strength Test

The tensile test results for cylinder specimens after 28 days are shown in Figure 3. Generally, the use of APF increases the tensile strength of the concrete owing to the bridging effect of the fibre. It can be seen that concrete containing 1% APF produces highest yet optimum tensile strength about 78% from normal concrete. While, the strength of concrete specimen with 3% APF is increased positively, however for only 41% as compared to control specimen (0% APF). This phenomenon can be reflected to the study by Nokman and Aziz [18] that concluded, the increasing percentage of heavy duty nylon fibre in concrete absorbs more water from concrete mix. Since APF is a hydrophilic material, it easily to absorb water and contributed to the lower water-cement ratio. As a result, APF can be limited until 1% in providing good workability of concrete which decreased result for 3% APF was due to the poor bonding interaction between APF, cement and aggregates that created honeycomb surface. Overall, it is confirmed that only 1% APF are suitable to be added in concrete mix.

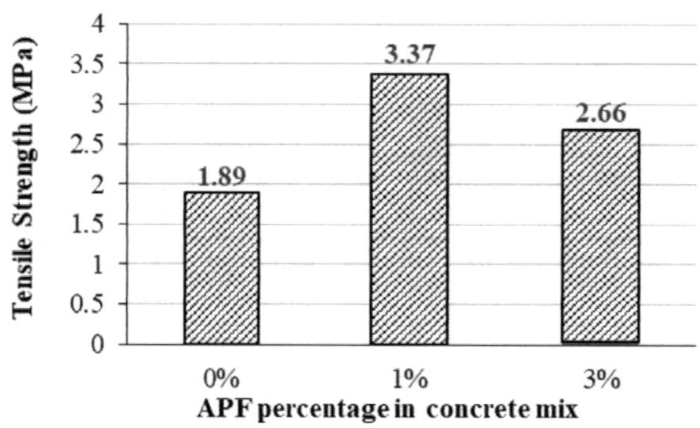

Figure 3. Effect of APF for tensile strength of concrete

3.3. Three Points Bending Test

Concrete mix containing 1% APF were chosen as optimum percentage. For further investigations, normal RC beam that denoted as NRC and RC beam specimens containing only 1% APF were tested under three points bending test. During this test, first crack load, final crack pattern, ultimate load as well as maximum deflections of both type of RC beam were monitored and recorded accordingly.

3.3.1 First Crack Load

To summarize the results, Table 2 shows the average of loading that produced from first crack formation for three specimens of each type of RC beam. Based on graph in Figure 4, it can be found that the additional of 1% of APF in normal concrete beam has been delayed the formation of crack during certain load being applied. It can be also clearly seen that the APF in concrete reduce the crack growth in tensile region. From the graph, the presence of 1% APF in concrete beam specimen produced 18.2 kN loading value as compared to NRC that generated only 16.7 kN. It can be also concluded that the addition of 1% APF in NRC exhibits the higher loading for first crack by reducing the formation of crack that may harm the concrete structure and reinforcement bar inside the beam element.

Table 2. First crack load of beam specimens

Beam	No. Beam	First crack load (kN)
NRC (control)	1	13.08
	2	20.38
	3	16.60
	Average	16.70
RC with 1% APF	1	17.86
	2	17.11
	3	19.62
	Average	18.20

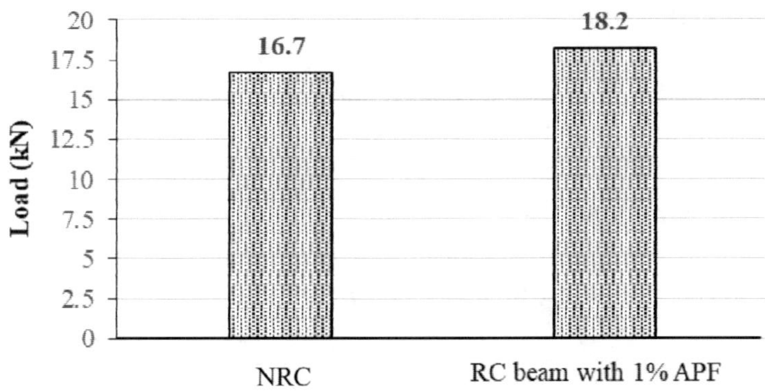

Figure 4. Average of first crack and its loading value for beam specimens.

3.3.2 RC Beam Behaviour (mode of failure)

The crack pattern and failure modes formed after imposed the ultimate load is shown in Figure 5 (a) and 5 (b). It can be seen that many diagonal and flexural cracks are developed for normal reinforced concrete beam that denoted as NRC (refer Figure 5(a)). While, for RC beam consisting 1% of APF generated lesser crack pattern and no severe flexural cracks are observed (refer Figure 5(b)). It is interesting to notice that the presence of 1% APF in concrete can reduce the formation of crack growth. However, this phenomenon can be further verified in the following sub-section for load-deflection graph.

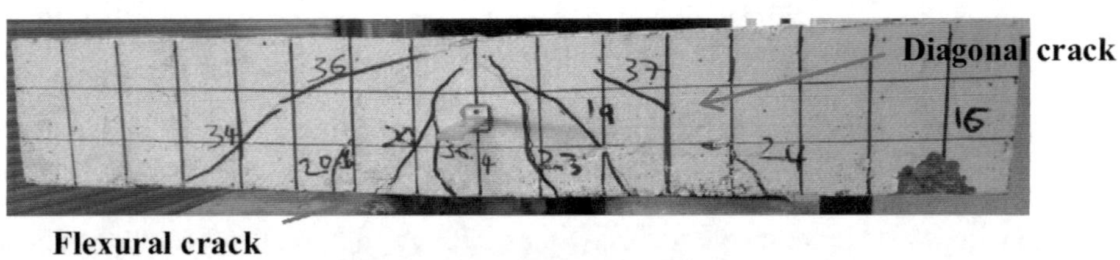

(a) Crack pattern for NRC

(b) Crack pattern for RC beam consisting 1% APF

Figure 5. Mode of failure for RC beam after

3.3.3 RC Beam Behaviour (Load vs. Deflection)

For further observations, the overall load versus deflection curve for both beam specimens has been plotted clearly as shown in Figure 6. From the curve, NRC with 1% APF reaching highest maximum load with low deflection compared with control beam. 1% APF in NRC has increased the maximum load until 35.98 kN with deflection about 1.8 mm. While, for control specimen (NRC), the curve displays the maximum load for about 27.93 kN with 2.37 mm deflection. This curve can be verified the results from sub-section 3.3.1, where the load value trend is comparatively the same. The addition of 1% APF in concrete caused increasing of maximum load but delayed the crack formation. It can be proved from the deflection value for both beam type. Less deflection were recorded for the RC beam consisting 1% of APF.

This result reveals that the characteristic of APF as a fibre on concrete provided positive results towards structural performance of beam element. It is also due to the APF physical characteristic.

Basically, APF has a jagged (rough) physical surface and high toughness produce good bonding interaction with cement paste.

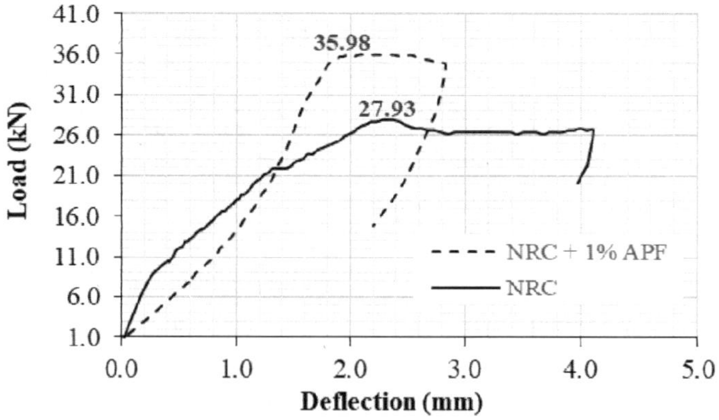

Figure 6. Load verses deflection graph for beam specimens.

4. Conclusions

Based on the analysis and discussion, several conclusions can be made:

i. Different tensile strength of APF does not give different effect towards producing a good performance of concrete.
ii. The innovation of artificial fibre from waste normal nylon cable tie has the capability to replicate the existing steel fibre in industry.
iii. Compressive strength of concrete are not affected significantly by adding APF in concrete mix.
iv. Tensile strength of concrete increased up to 78% by adding 1% of APF in concrete mix.
v. For structural element, 1% of APF in RC beam reduces the problem of severe cracking and crack formation especially at tensile region.
vi. By reusing waste nylon cable ties known as APF as additional substance in concrete mix, the quantity of this waste material disposed at site can be reduced. This would also reduce landfill consumption and promotes cleaner environment for future generation.

5. References

[1] Gartner E M and Macphee D E 2011 A physico-chemical basis for novel cementitious binders *Cement and Concrete Research* **41** (7) 736–749

[2] Mokhatar S N, Sonoda Y, Kamarudin A F, Md Noh M S and Tokumaru S 2018 The effects of pressure dependent constitutive model to simulate concrete structures failure under impact loads *J. of Phys: Conf. Ser* Vol 995 (1) (United Kingdom: IOP Publication) pp. 1 – 11

[3] Mokhatar S N, Zuki S S M, Tan H J, Sonoda Y, Inoue K, Kamarudin A F and Md Noh M S 2018 A fundamental hammering sound test to asees the degree of deterioration in reinforced concrete structure *IOP Conf. Ser: Materials Sc. & Eng.* Vol 431 (12) (United Kingdom: IOP Publication) pp. 1 - 6

[4] Jaini Z M, Boon K H, Mokhatar S N, Mat I, Hazmi H and Hashim N H 2016 Structural behavior of short-span reinforced concrete beams with foamed concrete infill *ARPN J. of Eng. & App. Sc.* **11** (16) 9820 – 9825

[5] Khan M I, Abbas Y M, Fares G 2017 Review of high and ultrahigh performance cementitious composites incorporating various combinations of fibres and ultrafines *J. King Saud Univ. Eng. Sci.* **29** 339–347

[6] Showkat M and Hemant S 2016 Effect of PET fibers on the mechanical properties of concrete. *SSRG Int. J. of Civil Eng.* **3** (12) 1 – 6

[7] Oujiang W, Abdul H Z, Baik S C, Scott D E 1994. Properties of fibre reinforced concrete using recycled fibres from carpet industrial waste. *J of Materials Sc.* **29** (16) 4191–4199

[8] Yahaya N 2007 Solid waste management in Malaysia: policy review, issues and strategies *Asia Solid Waste Manag. Cycle Conf.* 23-28
[9] Alok B, Divyanshu Y, Shreyshi V 2015 Non-biodegradable waste: Its impact & safe disposal. *Int. J. of Adv. Tech. in Eng. & Sc.* **3** (1) 184 – 191
[10] Richards AF 2005 *Nylon fibres, Synthetic Fibres: Nylon, Polyester, Acrylic, Polyolefin* (Cambridge: Woodhead)
[11] Warner R F (Robert Falcon) 1998 *Concrete structures* (Longman, South Melbourne)
[12] Van T K, De B N, De M W, Verstraete W 2010 Use of bacteria to repair cracks in concrete. *Cem. Concr. Res.* **40** 157–166
[13] Sellevold E J and Bjontegaard O 2006 Coefficient of thermal expansion of cement paste and concrete: Mechanisms of moisture interaction. *Mater. Struct.* **39** 809–815.
[14] Marta S 2019 The analysis of failure in concrete and reinforced concrete beams with different reinforcement ratio. *App. Mech.* **89** (5) 885–89
[15] Allen D C and Roper H 1991 Concrete Structures: Materials, Maintenance and Repair (LongmanGroup)
[16] Khan M S 2006 Control of cracking in concrete state of the art transportation research board basic research and emerging technologies related to concrete committee: *Transportation Research Board 500 Fifth Street*.
[17] Song P S, Hwang S and Sheu B C 2005 Strength properties of nylon- and polypropylene-fibre-reinforced concrete *Cement and Concrete Res.* **35** 1546 – 1550
[18] Nokman I A and Aziz Saim A 2016 The compressive and tensile strength of nylon cable tie fibre-reinforced concrete *Struc. & Materials* **1** 328 – 336

Acknowledgment

The authors would like to thank University Tun Hussein Onn Malaysia for the financial support to publish this paper through TIER 1 Code H270.

IOP Conf. Series: Materials Science and Engineering **713** (2020) 012029 doi:10.1088/1757-899X/713/1/012029

Shear Capacity of Reinforced Concrete Beams with Square Cross Section Subjected to Biaxial Bending

R Thamrin[1], **S Haris**[1], **E Dedi**[1] **and E Dalmantias**[1]

[1] Civil Engineering Department, Engineering Faculty, Universitas Andalas, Padang, Indonesia
Corresponding author: rendythamrin@eng.unand.ac.id

Abstract. This paper presents the test result of an experimental study on the shear strength of reinforced concrete beams subjected to biaxial loads. Six simply supported reinforced concrete beams with square cross section were used. Three beams rotated at 22.5 degrees and three beams rotated at 45 degrees to the applied load were tested until failure. All beams were subjected to four-point bending test. The test variable in this study was ratio of longitudinal reinforcement. The values of flexural crack load, diagonal crack load and the ultimate load were observed during the test as well as the midpoint deflection of the beams. Existing empirical equation for shear strength of concrete presented in design codes was used and then compared to the values obtained from the test. Comparison between test results and theoretical shear capacity show that all of equations conservatively estimated the occurrence of shear failure with the values of the test results 25 to 55% higher than the theoretical values depending on the ratio of longitudinal reinforcement. It was also observed that ratio of longitudinal reinforcement influences the angle of diagonal shear crack.

1. Introduction

Shear failure needs to be avoided in the design of reinforced concrete structures to prevent sudden collapse without warning. To avoid collapse due to shear forces, the structure must be carefully designed based on the load acting on it. The direction of the actual load acting on a building structure during an earthquake or similar natural event will be unknown. The direction of the load may be perpendicular to lateral or vertical axes of the cross section but could also be at a certain angle to lateral and vertical axes as illustrated in figure 1(b).

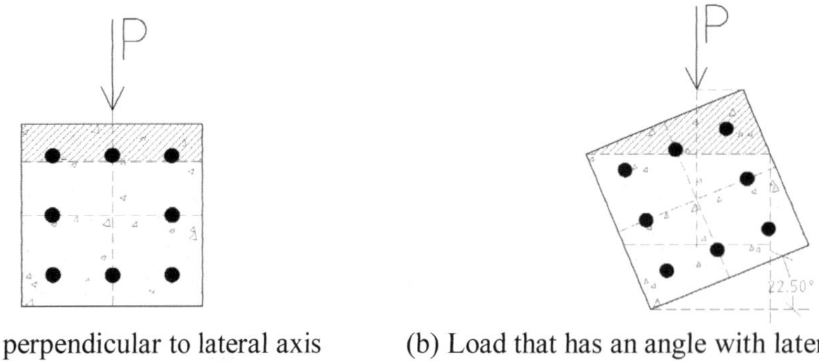

(a) Load perpendicular to lateral axis

(b) Load that has an angle with lateral and vertical axes

Figure 1. The possible direction of load acting on a reinforced concrete member.

Up until now, research carried out on the shear strength of reinforced concrete beams has focused on uniaxial bending, the load only acting in one direction of the lateral axis of the cross-section of the structural elements as shown in figure 1(a). Information on the effect of biaxial shear forces acting with a certain angle to lateral and vertical axes of cross-section in structural elements is limited even though biaxial bending forces in building structures caused by various combinations of loading direction are common [1,2]. Most reinforced concrete codes for building structures do not yet provide design guidelines and formulas for calculating shear force capacity for this biaxial load condition. The behavior of biaxial shear forces acting on structural elements needs to be investigated to determine the effect on the shear strength of reinforced concrete structural elements covering columns and beams under the load of biaxial shear force.

Based on this fact, this study focuses on shear strength provided by concrete of reinforced concrete beams without stirrups loaded with biaxial loads. A numerical analysis was also conducted to obtain the full response of flexural behavior of the rotated cross section.

2. Experimental study

Tests were carried out on six reinforced concrete beams with square cross sections. The simply supported beam was used and loaded monotonically with four point bending. In order to produce a biaxial shear load, lateral axis of cross section was rotated at 22.5 and 45 degrees as shown in figure 2. Figure 3(a) shows the detail of test setup and equipments used while figure 3(b) shows the beam rotated at 45 degree after testing.

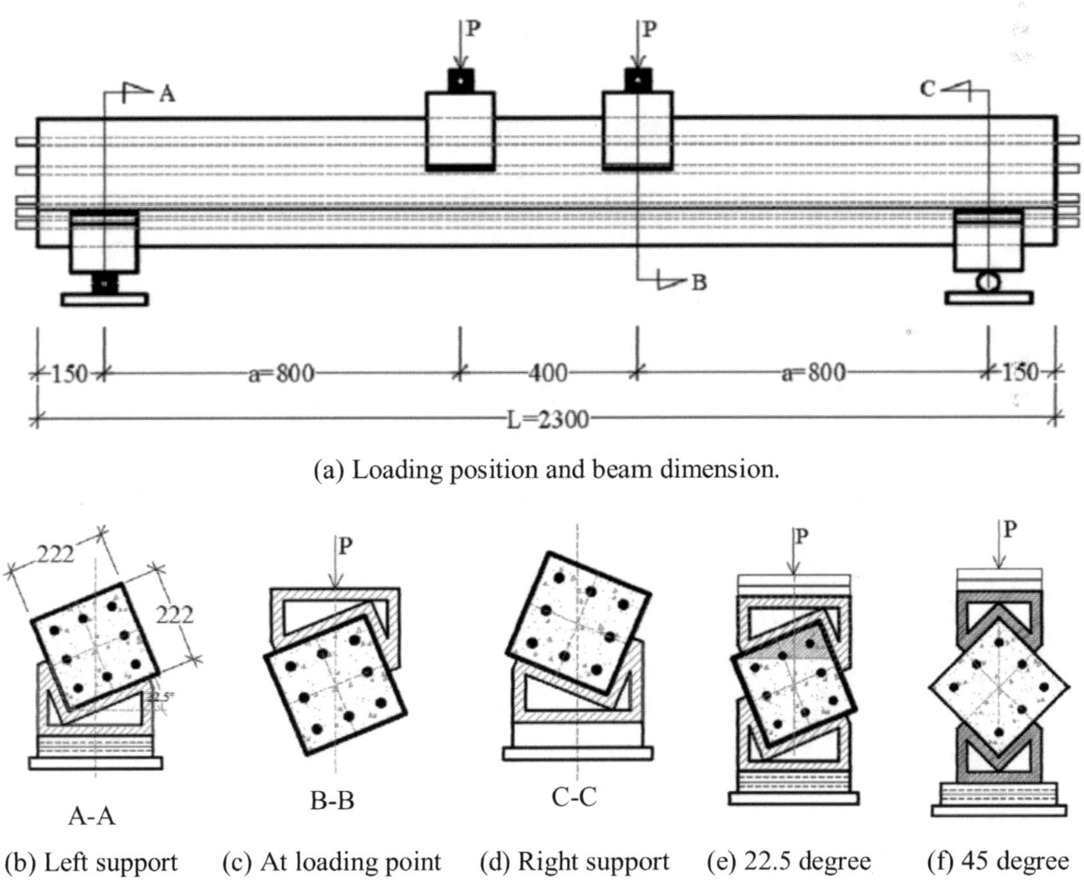

Figure 2. Loading on the tested beam showing rotated cross sections at various selected positions.

As shown in figure 2(a), the distance between supports was 2000 mm, the shear span length 800 mm, the distance between two point loads 400 mm, and the beam end length beyond the support 150 mm. The beam had a square cross section with dimensions of 222 mm x 222 mm as illustrated in figure 2(c). The concrete used had characteristic compressive strength of 25.1 MPa. The test variable in this study was the ratio of longitudinal reinforcement. The longitudinal reinforcement had diameters of 13 mm, 16 mm, or 19 mm with yield strength of 410 MPa, 390 MPa, and 370 MPa, respectively. Table 1 shows the beam data and details of the reinforcement. All beams presented in this report were designed without stirrups. The value of load was measured using a load cell placed on the top of the spreader beam as shown in figure 3(a). In order to obtain deflections, three displacement transducers were positioned at loading points and at midspan of the beam as shown in figure 3(a). The load and displacement that occurred during the test were recorded continuously every 5 seconds with the help of data acquisition system.

(a) Rotation angle 22.5 degree.

(b) Rotation angle 45 degree.

Figure 3. The beams (a) before and (b) after test.

Table 1. Beams data.

Beams Data							Longitudinal Reinforcement	
Beams Identification	fc' (MPa)	b (mm)	H (mm)	d (mm)	a (mm)	α	ϕ_b (mm)	N
G2-01-22.5				185		22.5	13	8
G2-02-22.5				184		22.5	16	8
G2-03-22.5	25.1	222	222	183	800	22.5	19	8
G2-01-45				185		45.0	13	8
G2-02-45				184		45.0	16	8
G2-03-45				183		45.0	19	8

Note: b, d, and H are the dimension before the cross section is rotated.

3. Numerical analysis

In this study, theoretical concrete shear strength for each of the beams was calculated using the empirical equation suggested by ACI 318M-14 [3]:

$$V_c = 0.172\lambda\sqrt{f_c'}bd \quad (1)$$

where fc' is the compressive strength of concrete, b is the width of cross section and d is the effective depth of cross section.

The numerical flexural capacity of the beam was obtained using moment-curvature analysis of reinforced concrete cross section [4,5,6,7]. In this method, the beam cross-section is divided into a number of reinforcement and concrete layers. Each layer is associated with the corresponding stress-strain law. The first step in this method is to obtain strains for each layer with the assumed curvature

IOP Conf. Series: Materials Science and Engineering **713** (2020) 012029 doi:10.1088/1757-899X/713/1/012029

value. The stress values for each layer are obtained by using the strains values obtained from the previous step. The compressive stress-strain law of concrete is obtained from another research report [8] and the tensile stress-strain law of concrete is linear up to the maximum value. A bilinear stress-strain law was applied for the steel reinforcement. Each of the load steps had to satisfy the equilibrium condition of the internal forces, F_i, as expressed in Equation 2. This was solved using an iterative procedure. After this step, the moment-curvature relationship could then be obtained. The moment-curvature analysis of beam cross section is conducted with the help of a computer program.

$$\sum F_{si} + \sum F_{ci} \approx 0 \tag{2}$$

4. Test results and discussion

The results obtained from the test and the numerical predictions for all beams are summarized in Table 2. In this study, due to the absent of stirrups, all beams failed in shear as expected. Typical crack patterns of the tested beams at failure are shown in figure 4 and figure 5. It was observed from the test that the first crack occurred in the zone of constant moment when the applied load reached an average of 8.3. After that, as the load increased, the flexural crack extend to the shear span region which then developed to become diagonal shear cracks at an average load of 41.5 kN. It is revealed from figure 4 and figure 5 that diagonal shear cracks occurred in the shear span region for all the beams tested in this study.

G2-01-22.5

G2-02-22.5

G2-03-22.5

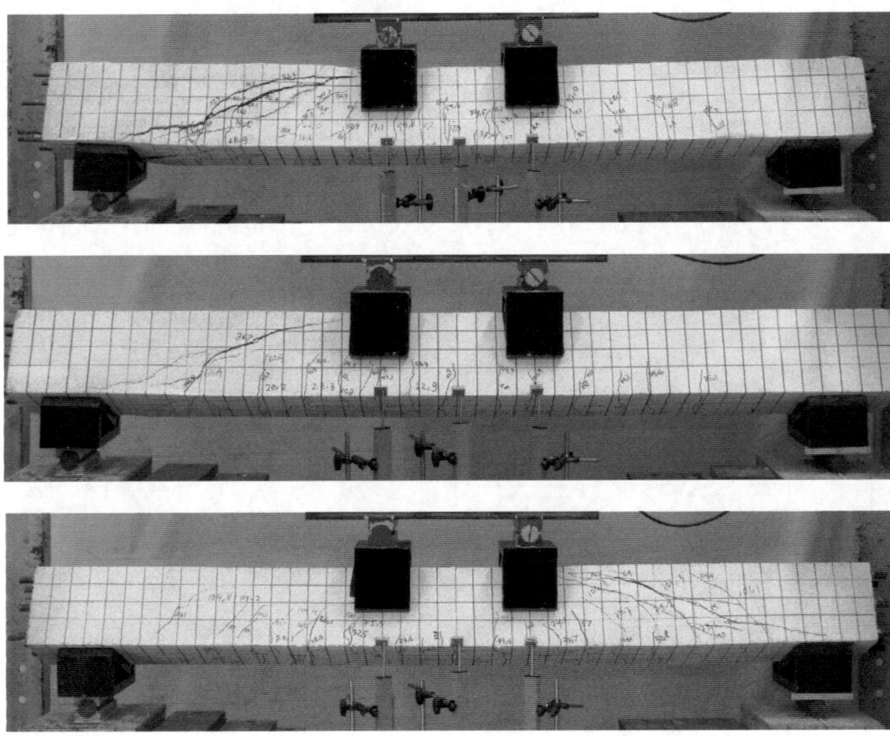

Figure 4. Crack patterns of the beams with rotation angle 22.5 degree.

IOP Conf. Series: Materials Science and Engineering **713** (2020) 012029 doi:10.1088/1757-899X/713/1/012029

G2-01-45

G2-02-45

G2-03-45

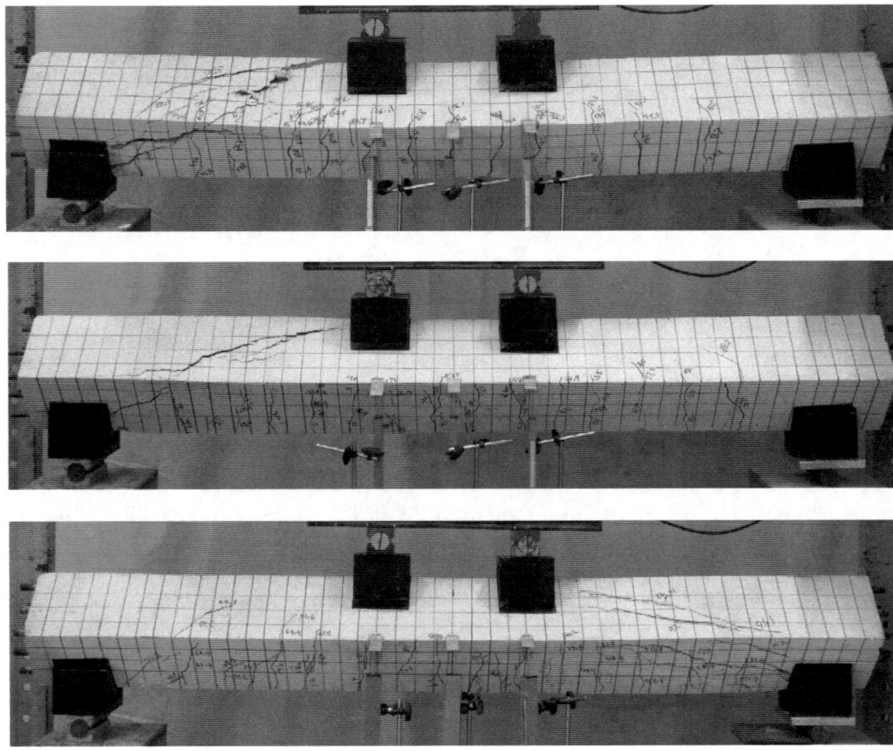

Figure 5. Crack patterns of the beams with rotation angle 45 degree.

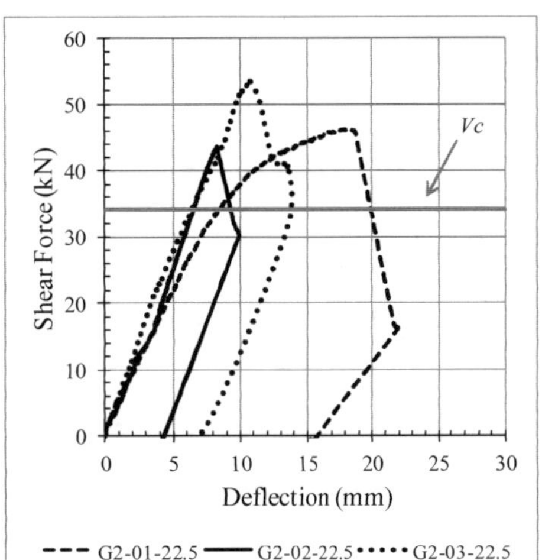

(a) Rotation angle 22.5 degree.

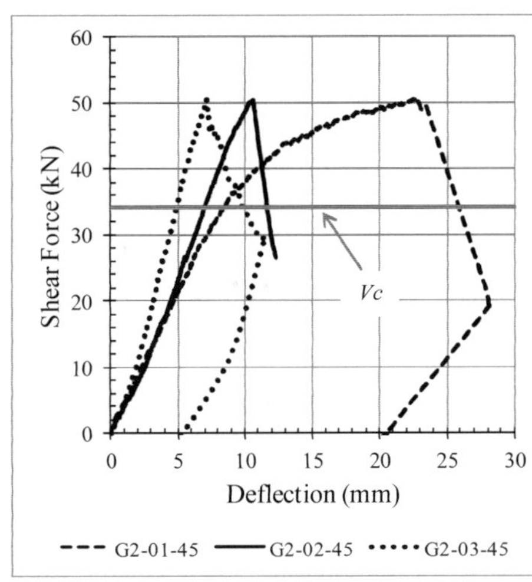

(b) Rotation angle 45 degree.

Figure 6. Shear force vs. deflection of the tested beams.

Table 2. Test results and the numerical predictions.

	Calculated Shear Capacity ACI 318M-14	Analytical Flexural Capacity	Experimental		
			First Crack	Diagonal Crack	Max. Load
Beams Identification	V_c (kN)	V_f (kN)	(kN)	V_d (kN)	V_m (kN)
G2-01-22.5	34.3	42.4	8.0	43.5	46.3
G2-02-22.5	34.1	56.6	6.0	35.0	43.4
G2-03-22.5	33.9	72.0	14.5	37.5	53.4
G2-01-45	34.3	46.3	6.3	42.6	50.7
G2-02-45	34.1	63.0	6.0	44.1	50.4
G2-03-45	33.9	79.8	9.3	46.0	50.4

Immediately after diagonal cracking, all the tested beams failed in shear. As listed in table 1 and shown in figure 6, the measured maximum loads were considerably higher (by 25 to 55%) than the shear strength provided by concrete, V_c, computed using the ACI 318M-14 code. This means that the shear strength design formula provided by ACI 318M-14 is conservative and safe to use for designing structural members that may be subjected to biaxial shear force.

Figure 4 and figure 5 show that the angle of diagonal crack tended to be quite different for each beam. It can be seen that the slope of diagonal crack decreased as the ratio of longitudinal reinforcement increases. This indicates that the ratio of tensile longitudinal reinforcement affected the slope of diagonal crack. This phenomenon has been also reported in previous studies [5].

As can be seen from figure 6, the shapes of the load deflection response of all beams demonstrate suddenly drop after the maximum load. However, the slopes of load deflection curves before the maximum load were significantly different for different beams. The difference in these slopes can be attributed to the difference in the ratio of longitudinal reinforcement.

Figure 7 compares numerical analysis and test results. The purpose of this comparison is to show more clearly that no beams reached flexural capacity because they failed due to shear before reaching bending capacity. Even though figure 7(a) and figure 7(d) show that the tensile reinforcement yielded before shear failure, however none of the tested beams reached the ultimate flexural strength because none of the beams experienced concrete crushing in the compression zone. It was also obtain from the test that beams with 45 degree angle show slightly higher shear strength compare to that beams with 22.5 degree angle.

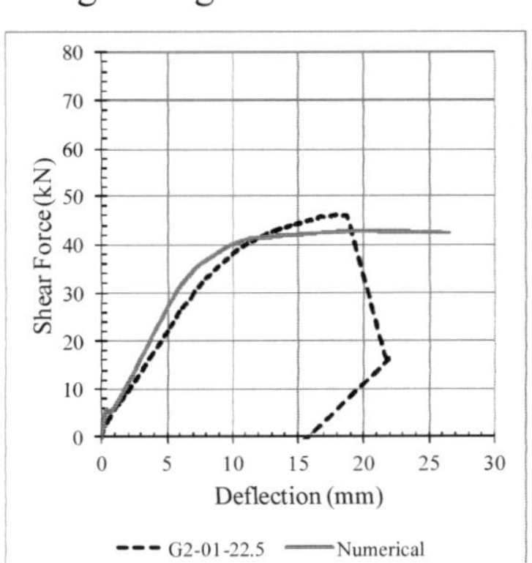

(a) G2-01-22.5

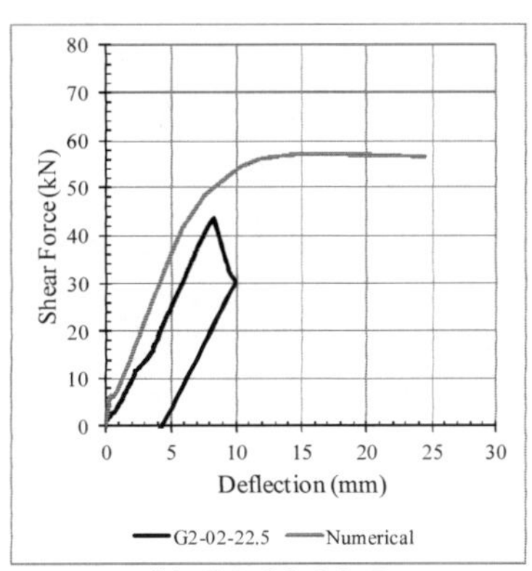

(b) G2-02-22.5

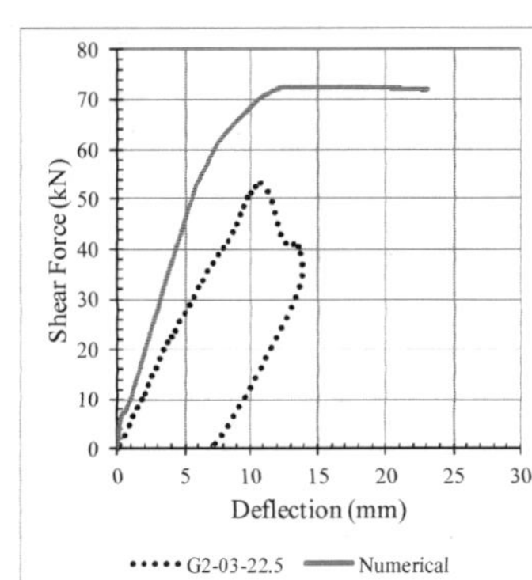

(c) G2-03-22.5

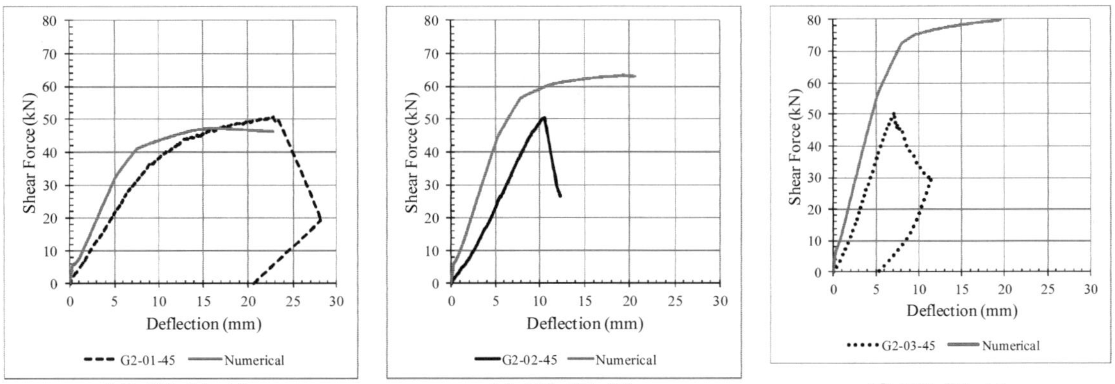

(d) G2-01-45 (e) G2-02-45 (f) G2-03-45

Figure 7. Comparison between numerical analysis and test results.

5. Conclusions

Six reinforced concrete beams with square cross sections designed without stirrups were tested until failure to obtain shear capacity on being subjected to a biaxial shear load. The results indicate that the following conclusions can be written:

- The maximum loads obtained from the test were noticeably higher by 25 to 55% than the theoretical shear strength provided by the concrete, V_c, computed using the ACI 318M-14 equation. This indicates that structural members subjected to biaxial shear force still can be safely designed using the shear strength design formula provided by ACI 318M-14.
- The longitudinal reinforcement ratio significantly affects the diagonal crack angle.
- Numerical flexural strength obtained using layer element method confirms that none of the beams reached flexural capacity because they failed in shear due to lack of shear strength provided in the shear span zone.

6. References

[1] Mark P 2007 No AccessShear-resistant design of biaxially loaded RC beams *Magazine of Concrete Research* **59**(1) 21-28

[2] Tinini A, Minelli F, Belletti B and Scolari M 2016 Biaxial shear in RC square beams: Experimental, numerical and analytical program *Engineering Structures* **126** 469–480

[3] American Concrete Institute 2014 *Building Code Requirements for Structural Concrete and Commentary* (United States: Farmington Hills Michigan)

[4] R Park, and T Paulay 1975 *Reinforced Concrete Structures* (New York: John Wiley)

[5] Thamrin R, Tanjung J, Aryanti R, Nur O F and Devinus A 2016 *Journal of Engineering Science and Technology* **11**(4) 548–62

[6] Thamrin R, Kurniawan R, and Melinda A P 2017 Shear and flexural capacity of reinforced concrete members with circular cross section *Procedia Engineering* **171** 957-964

[7] Thamrin R 2017 Analytical Prediction on Flexural Response of RC Beams Strengthened with Steel Plates *Int. Symposium on Civil and Environmental Eng. 2016 ISCEE 2016 (Melaka)* vol 103 (Paris: MATEC Web of Conference) 02012

[8] Mander J B, Priestley M J N and Park R 1988 Theoretical Stress-Strain Model for Confined Concrete *Journal of Structural Engineering* **114**(8) 1804-1826

Acknowledgements

The authors would like to acknowledge their thanks to the Engineering Faculty, Andalas University for the financial support through Hibah Publikasi in financial year 2019.

IOP Conf. Series: Materials Science and Engineering **713** (2020) 012031 doi:10.1088/1757-899X/713/1/012031

Mechanical Properties of Single Shear Plane of Bolted Steel Connection

A F Kamarudin[1], M K Musa[2], S N Mokhatar[1], T N Tuan Chik[1], S S Mohd Zuki[1], A Abu Bakar[1], J Hadipramana[3] and H H Ahmad Johari[1]

[1] Jamilus Research Centre, Faculty of Civil and Environmental Engineering, Universiti Tun Hussein Onn Malaysia, 86400 Pt. Raja, Bt. Pahat, Johor, MALAYSIA
[2] Faculty of Technology Engineering, Universiti Tun Hussein Onn Malaysia, 86400 Pt. Raja, Bt. Pahat, Johor, MALAYSIA
[3] Program Studi Teknik Sipil, Universitas Muhammadiyah Sumatera Utara Jl. Muctar Basri, Medan 20238, INDONESIA
Corresponding author: fahmy@uthm.edu.my and hanisjoe.hj@gmail.com

Abstract. A 5-storey portable steel tower based on prefabricated frame concept has been built in Jamilus Research Centre, UTHM. It is now undergoing laboratory testing for static and dynamic performance. In this study, the mechanical properties of the connections were investigated via their connection condition between elements (beam and column) and connectors. The frame elements were designed based on S275 steel grade of 77 mm x 77 mm hollow section with 3 mm thickness, and 8 mm thickness of connectors' plate. The connection was subjected to single shear plane action which fastened by bolts. Mild steel (grade 4.6) and high tension (grade 8.8) bolts with 10 mm diameter size were used. Shear, bearing and tensile strengths of the bolts and steel plates were determined via tension testing using 50 tonnes of Universal Testing Machine (UTM). From the results, all bolts and steel plates have satisfied the design strengths requirement of 640 N/mm^2 (high tension bolt), 240 N/mm^2 (mild steel bolt) and 275 N/mm^2 as required by EN 1993-1-8. Meanwhile the ultimate shear force of bolts was found within ranges of 25% (high tension) and 12.5% (mild steel) from their respective ultimate tensile force. Bearing failure was occurred at bolt hole of 3 mm thickness of steel plate but only when it was fastened to high tension bolt. No bearing failure occurred when mild steel bolt was used. It has proven that all bolt spacing and edges distances were sufficient, without tearing failure observed in all testing. Finally, the permissible bearing resistance were checked based on the ratio of experiment against the EN 1993-1-8 relationships, when it has shown to satisfaction if the value is less than 1 as recommended by previous researcher.

1. Introduction

Connections between beams and columns or column to foundation are fragile segments that need to be emphasized in steel frame constructions. The functions of connections are to distribute axial force and moments to columns. Bolted connections are the most preferable in steel constructions because it able to reduce the construction cost since less experience and skilled workers are needed during handling of construction activities.

Geometrical factors of the connections components such as bolt sizes, angle of sections, edge and spacing have given a big impact during preference of materials. Bearing occurs when it has a contact surface area between the shanks of the bolt and the bolt hole. Meanwhile, a bolt is also subjected to a shear and tension resistances that need to be given a great attention in steel constructions. According to

Prinz et al. [1], a bolted beam to column connection is often expected to be fully rigid or fully pinned. However, to be truth, the rigidity of connections is clarified as semi-rigid connection [1]. Besides, the deliberations made for connections rigidity can be used to determine the limits of strength forces for beams and columns in a building system [2].

Bolt is mainly used in connection to resist shear on the surface of plates, and tensile force along the axis of bolts. In a bolt connection, the forces can be transferred into bearing, shear and tension conditions. John and Stephen [3] stated, the force that comes from vertical components, V, act to break off the bolts out of the column, while horizontal component in the case of moment loading, M, acts as to fracture the bolts in tension. There are several types of malfunction in bolt connection which are, plate tearing failure, bearing failure on the plate material, tension failure on the surface area, bolt shearing and any combinations malfunction [3].

Mehran et al. [4] describes most failures in connections are due to plate tearing (refer to figure 1). It occurs when the bolt is at a critical distance or space alongside bolts align with pressure is low [4]. The failure such as plate shear/tear-out and tension failure on the net section mostly fail on plate, and this can be described as catastrophic failure which is sudden failure.

Shearing failure of a bolt shear can be occurred when modulus material of bolt is much lower than the plate which also depending on grade of a bolt. The grade of a bolt is determined when it has greater strength to resist the shearing force more than the capacity of plate load, and its positioning of connection as a whole. In spite of that, there are some factors that influence the shear behaviour. To name a few, sheet thickness, grades of bolts, diameter of bolt, mechanical properties of steel, resistance coefficient and torque applied on bolt [5].

In contrast to bolt bearing, when the failure can be occurred if the modulus material of bolt is much higher than the material of plate. Consequently, the bolt will tilt due to lowest strength of its grade. Meanwhile, the tension failure on the net section can occurs due to the existence of holes that will increase the concentrated of stresses areas on plates [6]. This kind of failure commonly occurs at the joints due to lower net section area of the sheet and thus it can contribute as the fragile section of the plate. These failures have been controlled by the number of bolts and spacing arrangement. Thus, to prevent this failure a proper design with minimum satisfied edge distance and bolt spacing need to be considered [7].

According to American Iron and Steel Institute (AISI) [9], shear and tension testing are used to determine the hardiness and rigidity of the specimens. Standard shear tests can be conducted when two specimens is jointed and placed in tensile testing machine [8]. This tension testing will be loaded till failure. Table 1 shows a guideline to a standard shear test dimension of a specimen with specific width, edge distance, spacing and length. The length of extensometer gauge length, L_g must be short enough to reduce the stretch impact in specimens on the extensometer readings.

This paper is a part of static testing report which involved in portable steel tower investigation built in Jamilus Research Centre, UTHM. In this study, a single shear plane jointed plates (lapping) samples with 3 mm and 8 mm of S275 steel thicknesses is fastened by 10 mm of bolt (high tension and mild steel) with specific edge distance and bolt spacing. The samples were tested under tension testing by using 50 tonnes of UTM machine in order to determine the mechanical properties of the materials and the connections (bolt and connecting plates). Some other discussion was also made on the end distance. Tensile strength, shear strength, bearing resistance and failure modes of the material as well as the connection were reported as the major part in mechanical properties investigations in this study. Allowable strength and acceptable design considerations as provided by previous code of practices and researches in [9][11-13] were also studied and discussed.

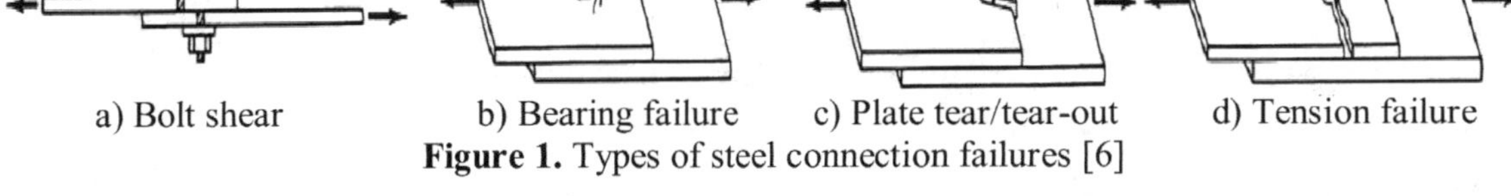

a) Bolt shear b) Bearing failure c) Plate tear/tear-out d) Tension failure

Figure 1. Types of steel connection failures [6]

IOP Conf. Series: Materials Science and Engineering **713** (2020) 012031 doi:10.1088/1757-899X/713/1/012031

Table 1. Recommend dimensions for standard shear test [9]

Fastener diameter, *d* (mm)	**Dimensions of specimen (mm)**				
	Width, w	**Strap length, L_s**	**End distance, e_1**	**Spacing, *p***	**Extensometer gage length, L_g**
≤ 6.5	60	260	30	60	150
> 6.5	10d	200+10d	5d	10d	30.20d
Tolerance	±2	±5	±1	±1	±5

2. Materials and Methodology

All element and connector plates were designed and manufactured from S275 steel grade. 77 mm × 77 mm cross section of square hollow elements section with 3 mm thickness has been cut out into coupon shape (refer to figure 2) for tension testing. 5 mm/min speed was used in tension testing. Similar procedure was carried out on the plate's connector of 8 mm plate thickness for tensile strength. The preparation and testing procedure of tensile specimens was done in accordance to the code of practice of BS EN ISO 6892-1:2016 [11] for tensile strength determination (see figure 3). The specimen preparation of shear and bearing strength on plates and bolts was prepared as given in figure 4. Figure 5 shows the mild steel bolts (Grade 4.6) and high tension bolts (Grade 8.8) with diameter 10 mm which also involved in the tension testing.

The design requirements of the bolts have been referred to MS 146-2006 [12]. The nominal yield strength of the high tension bolt, mild steel bolts and plate must be satisfied to the minimum design requirement of 640 N/mm^2, 240 N/mm^2, and 275 N/mm^2 as required by EN 1993-1-8 [10]. The commentary of the shear test can be referred to American Iron and Steel Institute (AISI) guidelines [9]. A total of eight specimens in single shear plane connection were prepared. The variation of specimen parameters can be differentiated against the bolt types, plat thickness, number of bolts, edge distance and spacing. All steel plates were provided by the strain gauges at the edge distance region, in order to investigate tensile or compression action against the elongation or contractions behaviour. Observation on specific type of failures (refer to figure 1) was also carried out in this study. The recorded yield strength and ultimate strength of bolt and plate from tension testing were tabulated in table 2, whereas figure 6 and 7 show the bolts and coupon failures which taken after testing. Details specification of the each specimen is provided in the Appendix

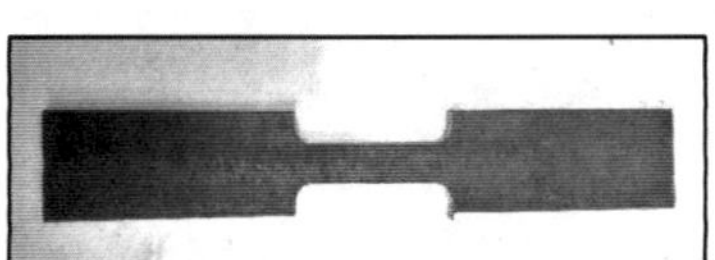

Figure 2. Plate coupon

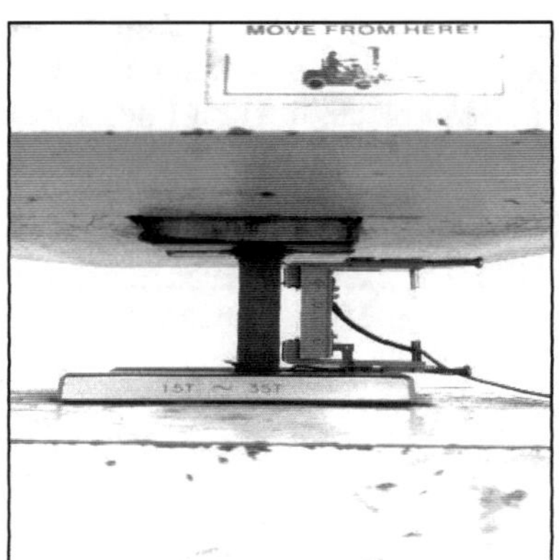

Figure 3. Experimental setup for tensile test on coupun

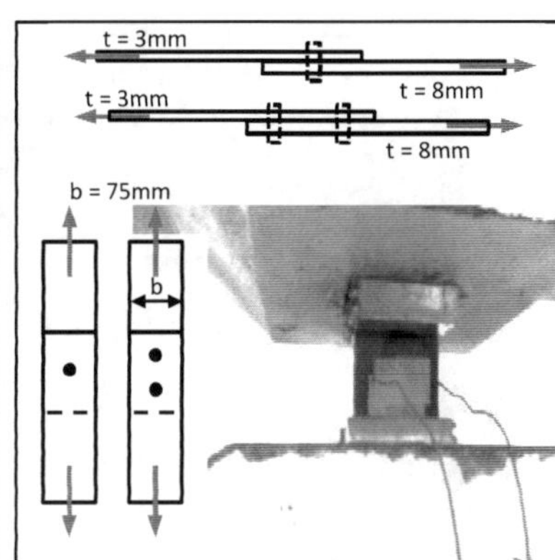

Figure 4. Experimental setup for shear/bearing strengths

a) High tension bold (grade 8.8) b) Mild steel bolt (grade 4.6)

Figure 5. Bolts used in the connections

Table 2. Tensile strengths result of bolts and plates

Materials	Yield strength, N/mm^2	Ultimate strength, N/mm^2
Mild steel bolts	361	395
High tension bolts	642	789
Plates (thickness 3 mm)	333	470
Plate (thickness 8 mm)	311	450

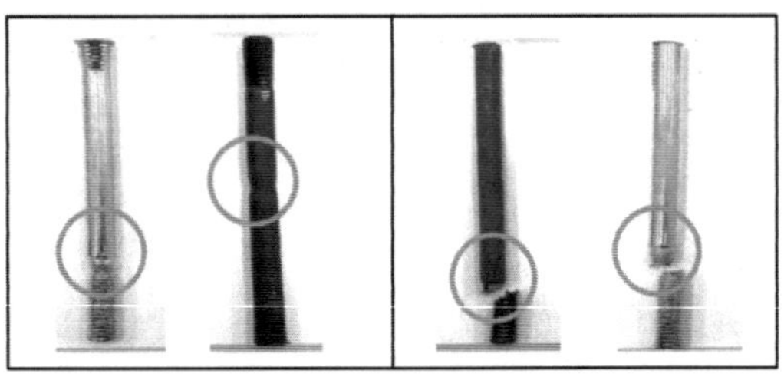

Figure 6. Bolt failures on tensile test

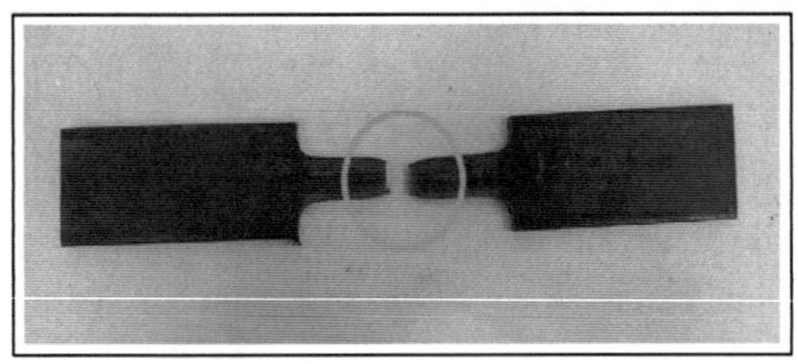

Figure 7. Plate failure of coupon

3. Results and Discussions

3.1 Ultimate shearing force and failure modes of specimens

The ultimate shearing force with respective observed failures has been summarized in table 3. From a sample of testing result on load versus stroke curve of specimen A (see figure 8), the curve has been shown in linear appearance of elastic behaviour which acted by the shearing and bearing resistance of the bolt and plates. According to Mehran et al. [4] taken from an instance of truss connection, the load-stroke curves will implicate the connections of members and bolts as a whole. However, when the bolt is started to tear and unsecure tightly fixed in place, the non-linearity curve is developed then ended by failures.

Table 3. Ultimate shearing force on bolt and type of failures

Specimen	Thickness		Ultimate shearing force (kN)		Failure description
	TP	BP	Mild steel bolt	High tension bolt	
A (two bolts)	3 mm	8 mm	20.02	65.90	• Bolt shearing failure • Plate bearing failure only occurred at 3 mm plate thickness which fastened by high tension bolts • Free damage at 8 mm thickness of steel plate
B (one bolt)			14.00	30.04	
C (one bolt)			14.01	34.02	
D (one bolt)			16.00	31.03	

TP: Top plate, / BP: Bottom plate

IOP Conf. Series: Materials Science and Engineering **713** (2020) 012031 doi:10.1088/1757-899X/713/1/012031

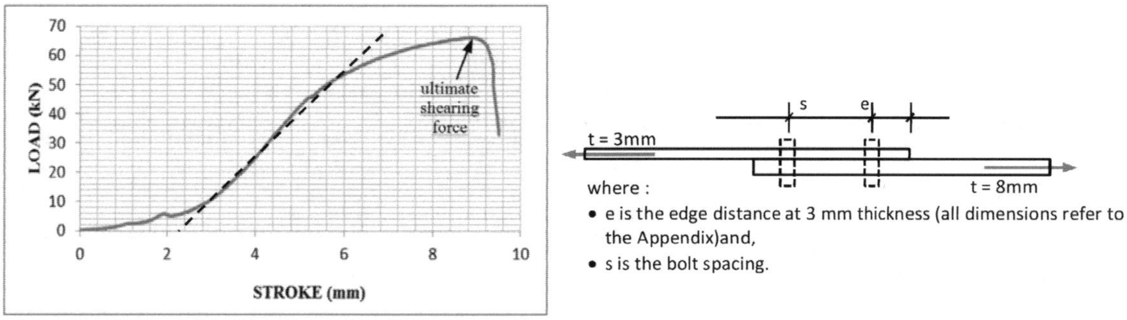

Figure 8. *Specimen A:* Load-stroke curve and double bolted (high tension bolt) diagram

3.2 Strain gauge

Figure 9 shows the strain curves from one of samples marked as specimen C. The strain gauges were attached at the edge distance region of both plates which closer to the bolt hole. The elongation of the plate is occurred along the longitudinal direction of the plates, which inline to the direction of tensile forces. At the first 60 seconds, both steel plates behave in elastic behaviour based on the linear strain curves pattern obtained. At this stage, the plates were able to withstand the external axial load applied which subjected to compression behaviour. After 60 seconds, 3 mm thickness plate is started to deform which then ended by bolt shearing failure after 162 secs. For this specimen, the maximum shear force has archived up to 14.01 kN (refer to mild steel bolt case in table 3). All bolts were failed in shearing. However, in case of high tension bolt, the 3 mm thickness of steel plate was failed in bearing and none of failure observed for 8 mm thickness of steel plate. The strain curve pattern of the rest specimens are attached in the Appendix.

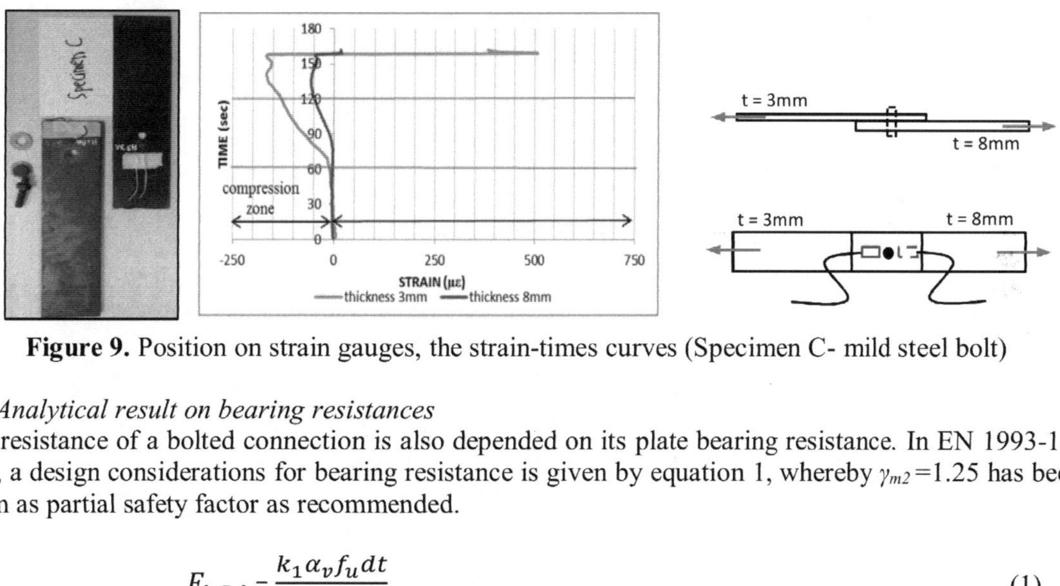

Figure 9. Position on strain gauges, the strain-times curves (Specimen C- mild steel bolt)

3.3 Analytical result on bearing resistances

The resistance of a bolted connection is also depended on its plate bearing resistance. In EN 1993-1-8 [10], a design considerations for bearing resistance is given by equation 1, whereby γ_{m2}=1.25 has been taken as partial safety factor as recommended.

$$F_{b,Rd} = \frac{k_1 \alpha_v f_u d t}{\gamma_{m2}} \tag{1}$$

where:

$F_{b,Rd}$	: bearing resistance,
d	: bolt diameter with d_{hole} = d+1mm
t	: thickness of plate
f_u	: ultimate tensile force = 410 N/mm^2 for S275 t $\geq$ 3mm
γ_{m2}	: safety factor = 1.25
α_v	: $\alpha_b = \frac{e}{3_{dhole}}$ or $\alpha_d = \frac{f_{ub}=800N/mm^2}{f_u}$ (the lesser)

The limitations between design resistances of bolts can be calculated according to the ratio of bearing resistance of specimens obtained from laboratory outcomes ($F_{b,Ed}$) to the bearing resistance relationship as expressed in equation (1). The computed results from this ratio have been summarized in table 4 and 5. For permissible criteria, $F_{b,Ed} \leq F_{b,Rd}$ it must be less than 1 [13], and all tested samples in this study have fulfilled to the acceptable limit.

Table 4. Acceptable limit to bearing resistance of high tension bolt connection

Types of bolt	Load due to bearing, $F_{b,Ed}$ (kN) (Experimental results)	Maximum bearing resistance, $F_{b,Rd}$ (kN) (Manual calculation)	Description ($F_{b.Ed}/F_{b,Rd}$) [13]
High tension bolt	65.90	95.94 (two bolts)	$(\frac{65.90}{95.94}) = 0.69 \leq 1$, **Acceptable**
	30.04	47.97 (one bolt)	$(\frac{30.04}{47.97}) = 0.63 \leq 1$, **Acceptable**
	34.02		$(\frac{34.02}{47.97}) = 0.71 \leq 1$, **Acceptable**
	31.03		$(\frac{31.03}{47.97}) = 0.65 \leq 1$, **Acceptable**

Table 5. Acceptable limit to bearing resistance of mild steel bolt connection

Types of bolt	Load due to bearing, $F_{b,Ed}$ (kN) (Experimental results)	Maximum bearing resistance, $F_{b,Rd}$ (kN) (Manual calculation)	Description ($F_{b.Ed}/F_{b,Rd}$) [13]
Mild steel bolt	20.02	47.97 (two bolts)	$(\frac{20.02}{47.97}) = 0.42 \leq 1$, **Acceptable**
	14.00	24.00 (one bolt)	$(\frac{14.00}{24.00}) = 0.58 \leq 1$, **Acceptable**
	14.01		$(\frac{14.01}{24.00}) = 0.58 \leq 1$, **Acceptable**
	16.00		$(\frac{16.00}{24.00}) = 0.67 \leq 1$, **Acceptable**

3.4 Failure modes of steel plate

According to Zadanfarrokh et al. [8], a ratio between the edge/end distance and bolt diameter can be used to predict the failures types *(e/d)* of the plate which expressed in equation 2 and 3. But, the influence of bearing failure on a plate is mostly occurred to the thin plate compared to the thick plate [14].

Sheet tearing : for $e/d \leq 1.5$ (2)
Sheet bearing : for $e/d \geq 2.25$ (3)

where:

e : edge/end distance
d : diameter of bolt.

Based on the calculated *e/d* ratio of specimens A, B, C and D (refer to table 6), the sheet bearing failure has shown to a great potential when it has exceeded to 2.25. The sheet bearing failure significantly occurred on 3 mm thickness of steel plate when it fastened to the high tension bolt. However, none of a plate failure was observed on specimens with mild steel bolts connection, due to lower resistance of shearing force which makes the bolt to break first before the bearing failure of the steel plate.

Table 6. e/d ratio of specimens A, B, C and D for mild steel and high tension bolts connection

Specimen (3 mm plate thickness)	Minimum end distance and e/d ratio			
	Single bolt connection	e/d	Double bolts connection	e/d
A			40/10	4.0
B	80/10	8.0		
C	116/10	11.6		
D	79/10	7.9		

4. Conclusions

Following are some conclusions made based on the outcomes from laboratory testing and analytical analysis carried out in this study:

i. The tensile strengths of the mild steel bolt, high tension bolt and steel plate have satisfied to the minimum yield strength requirement of 240 N/mm^2, 640 N/mm^2 and 275 N/mm^2 as required by BS EN 1993-1-8. Besides, it has found that the tensile strength of high tension bolt (grade 8.8) has indicated two times higher compared to the mild steel bolt (grade 4.6).

ii. The specimens had experienced compression behaviour at the edge/end distance of the steel plate from the result of strain gauge shown in this region.

iii. The shearing failure of 10 mm diameter of mild steel bolt in single shear plane connection testing is higher if compared to potential of bearing failure of 3 mm thickness of steel plate.

iv. The design consideration on bearing resistance based on the calculations made against the ratio from experimental results to the bearing resistance equation of EN 1993-1-8 have shown to acceptable limit when the values are less than 1.

v. The ultimate shearing forces of both bolt types are found to archive 12.5% (mild steel bolt) and 25% (high tension bolt) from respective ultimate tensile force.

vi. Bearing failure has occurred on the 3 mm thickness of steel plate when 10 mm of high tension bolt was used. This failure has also agreed to *e/d* ratio as recommended by Zadanfarrokh et al. [8]. However, lower shear resistance of mild steel bolt has caused to shearing failure to be occurred first, which slightly different against high tension bolt scenario.

5.0 References

[1] Prinz G S, Nussbaumer A, Borges L and Khadka S 2014 Experimental testing and simulation of bolted bam-column connections having thick extended endplates and multiple bolts per row *Engineering Structures* **59** 434 - 447

[2] Jones S W, Kirby P A and Nethercort D A 1983 The analysis of frames with semi rigid connections- a state-of-the-art report. *Constr Steel Res* **3** p 581-595.

[3] C M John and F F Stephen 2011 Structural Steel Design Fifth Edition (New Jersey: Pearson-Prentice Hall) pp 1-724

[4] Mehran Z, Shelley A and Ronagh H R 2016 An experimental study into the capacity of cold-formed steel truss connections *Journal of Construction Steel Research* **127** 176-186

[5] Yongjiu S, Wang M and Wang Y 2011 Analysis on Shear Behavior of High-Strength Bolts Connection *International Journal of Steel Structures* **11** 203-213

[6] Zadanfarrokh F *1991 Analysis and design of bolted connections in cold formed steel members* PhD Thesis (United Kingdom: University of Salford) 1-394

[7] Hancock G J, Murray T and Ellifrit D S 2001 Cold-formed Steel Structures to the AISI Specification (Boca Raton: CRC Press) p 1-416

[8] Zadanfarrokh F and Bryan E R 1992 Testing and Design of Bolted Connections in Cold Formed Steel Sections. Eleventh International Specialty Conference on Cold-Formed Steel Structures *11th International Specialty Conference on Clod-Formed Steel Structures* (USA) Missouri University of Science and Technology pp 625-662

[9] AISI S100-07 2010 *North American Specification for the Design of Cold Formed Steel Structure Structural Members* (Washington DC: American Iron and Steel Institutes)

[10] Eurocode 3 EN 1993-1-8 2005 *Design of steel structures: Part 1-8: design of joints* (London UK: British Standard Institution)

[11] ISO 6892-1 2016 *Metallic materials – tensile testing. Part 1: Method of test at room temperature* (Geneva: BSI Standards Publication)

[12] MS 146:2006 2016 Hot rolled steel bars for the reinforcement of concrete – specification (Malaysia: Malaysian Standard)

[13] Abidelah A, Bouchair A and Kerdal D E 2012 Experimental and analytical behavior of bolted end-plate connections with or without stiffeners *Journal of Construction Steel Research* **76** 13-27

[14] Ravindra B K and Vishal M V 2014 Experimental Study of Bolted Connections Using Light Gauge Channel Sections and Packing Plates at the Joints *International Journal Advanced Structure Engineering* **6** 105-119

Acknowledgments
The authors would like to acknowledge and expressed gratitude to the financial support by RMC, UTHM under TIER 1 Vot. U837 grant, Faculty of Civil and Environmental Engineering UTHM for the instrumentations, and JRC research members for their great cooperation to the accomplishment of this research.

Appendix: Dimension of specimen A, B, C & D, and strain gauge curves (all dimension in mm).

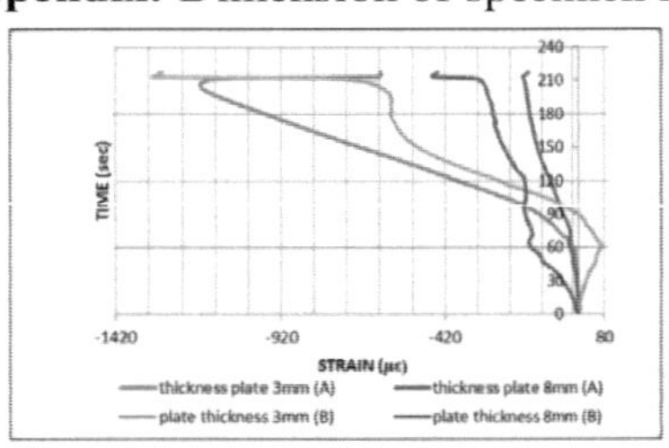

Specimen A – high tension bolt

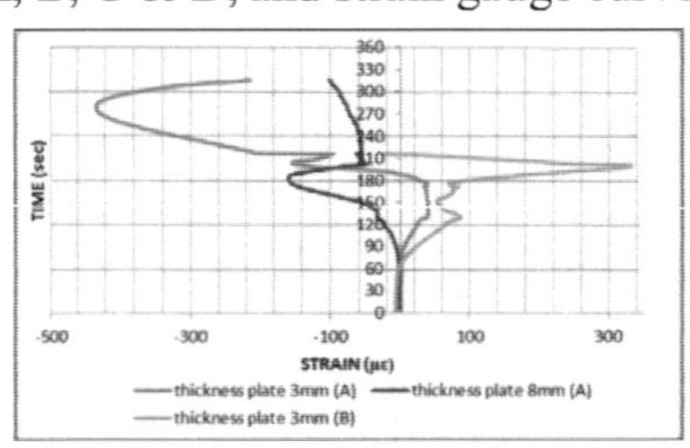

Specimen A – mild steel bolt

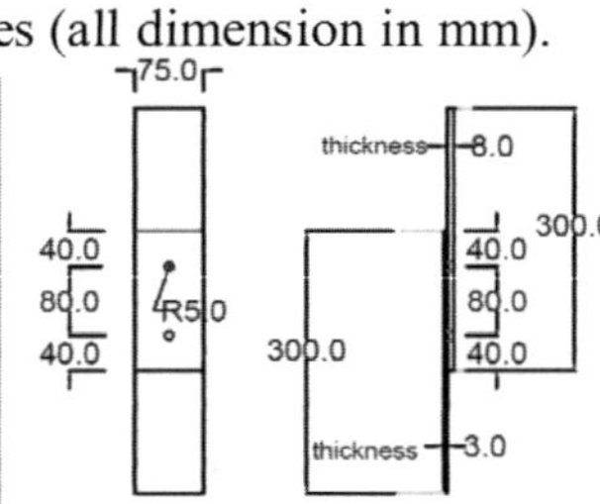

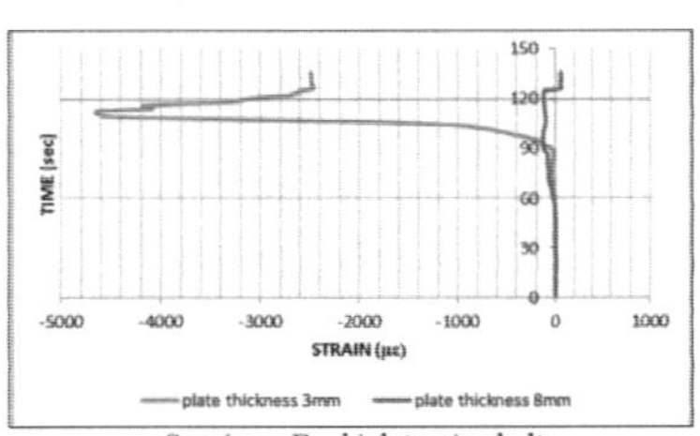

Specimen B – high tension bolt

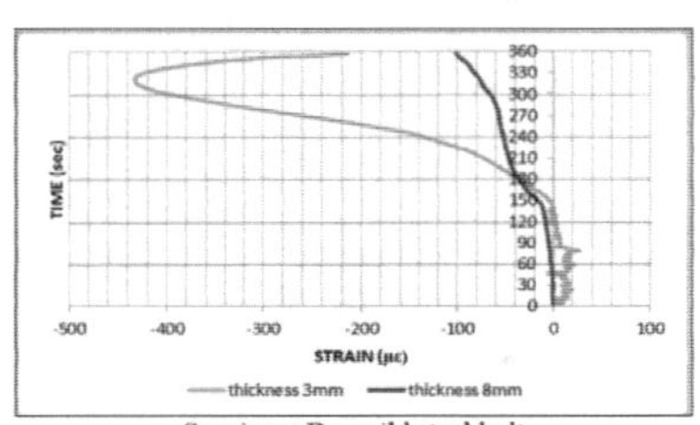

Specimen B – mild steel bolt

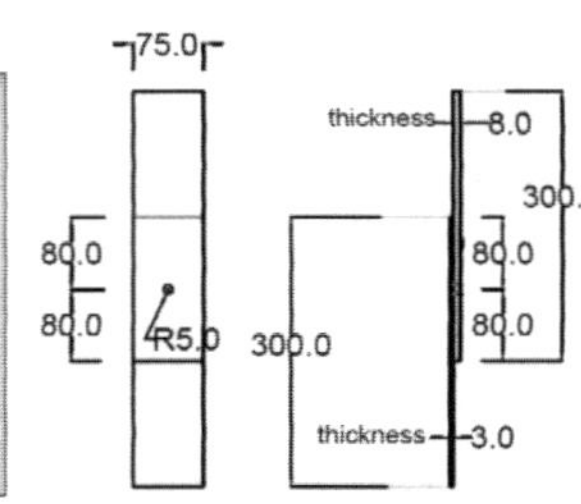

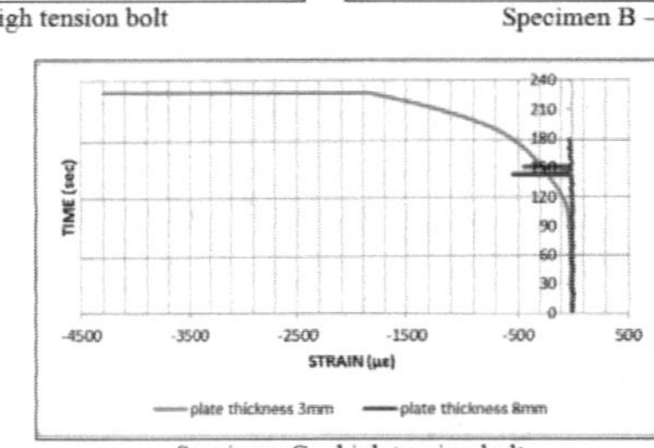

Specimen C – high tension bolt

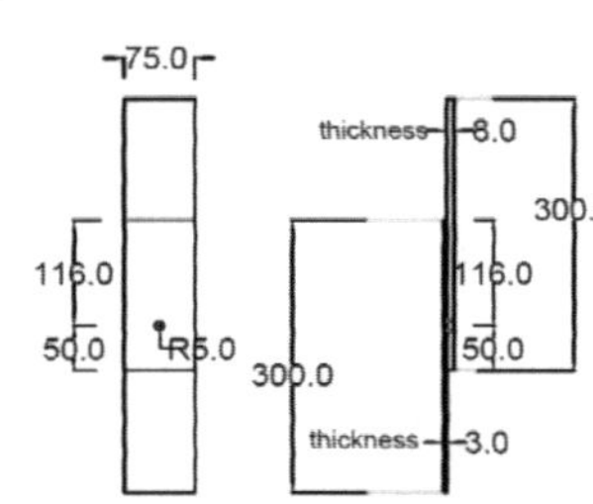

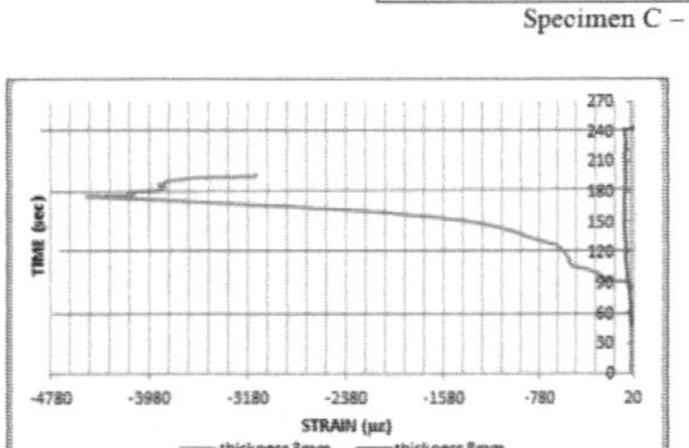

Specimen D – high tension bolt

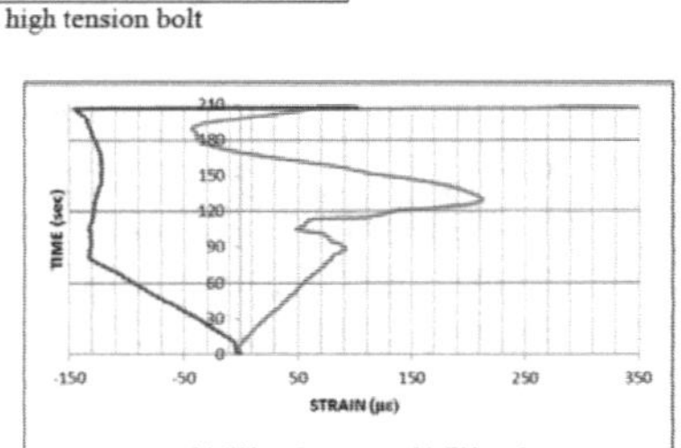

Specimen D – mild steel bolt

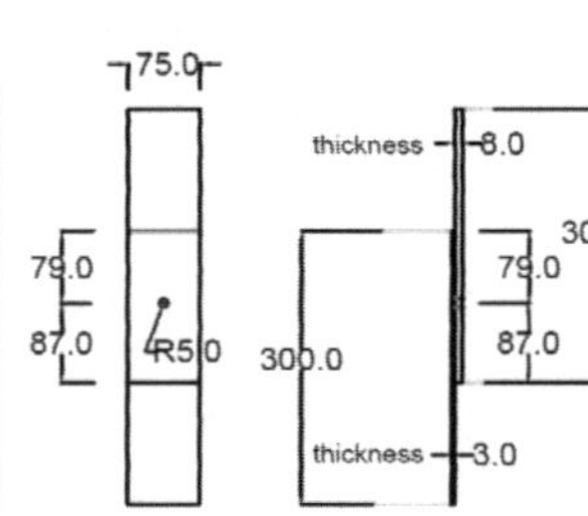

IOP Conf. Series: Materials Science and Engineering **713** (2020) 012032 doi:10.1088/1757-899X/713/1/012032

Finite Element Analysis of Coated (Intumescent Coating Protection) Cellular Steel Beam (CSB) Expose to Fire

F A A Zakwan [1], R R Krishnamoorthy [2], A Ibrahim [2] and R Ismail [1]

[1] Faculty of Civil Engineering, Universiti Teknologi MARA, Cawangan Pulau Pinang, 13500 Permatang Pauh, Pulau Pinang, MALAYSIA
[2] Faculty of Civil Engineering, Universiti Teknologi MARA, 40450 Shah Alam, Selangor, Malaysia
Corresponding author: fariz838@uitm.edu.my

Abstract. Traditionally, standard composite steel beam section acts as a load bearing structural element to sustain an external load. However, in the event of fire, an additional fire load acts on the composite steel beam section. The combined action of the former and latter would accelerate the vertical deformation of the beam. In the case of cellular steel beam (CSB), the vertical deformation predicted to be higher. Due to these circumstances, the structural behaviour of the composite CSB were compromised leading to the critical failure mode of web-post buckling and Vierendeel bending failure. Therefore, it is crucial to evaluate the behaviour of the composite CSB at elevated temperature under both loading action. In this research work, validation process was initiated between the numerical simulation analysis of CSB exposed to the fire with the readily available experimental data work. The validated model was then used to simulate the composite CSB with newly added fire protection material of intumescent coating. From the finite element simulation, the predicted vertical deformation slightly decreased for thicker intumescent coating application onto the beam surface. In conclusion, by applying the intumescent coating, further improvements were predicted for the vertical deformation and subsequently maintains the strength of the composite CSB at elevated temperature.

1. Introduction

Composite steel beam is widely used as the main load bearing structural element for building structures., namely flooring system and roof system. Interaction between concrete slab connected with steel beam creates an increased composite steel beam stiffness against external load at elevated temperature. Fire exposure threat affecting the performance of composite steel beam under fire load. In the last few years, a new type of steel beam was used to accommodate new trends and current needs involving construction building industry. This type of steel beam allows pipelines and conduits to move straight through the web steel beam section. Web opening section were introduced through the web steel beam section of I or H shape. There are several shapes of the web opening available in the market, namely circular, hexagonal, octagonal and sinusoidal shape as illustrated in Figure 1. Commonly, circular web opening shape was the preferred choice due to its shape that can fits circular pipelines and conduits straight the web opening section. This type of steel beam is commonly referred as cellular steel beam (CSB). With the full interaction in the middle of concrete slab and composite CSB, the composite beam increase the bending moment resistance between 50% to 100% and also increase the flexural resistance in comparison with standard steel beam [1, 2]. However, the stiffness of the CSB were slightly reduces due to some parts of the web section were omitted and removed. This scenario leads to several failure

IOP Conf. Series: Materials Science and Engineering **713** (2020) 012032 doi:10.1088/1757-899X/713/1/012032

conditions occurred along the web opening section. Even though the stiffness of the CSB were reduced, the strength of the beam was not compromise.

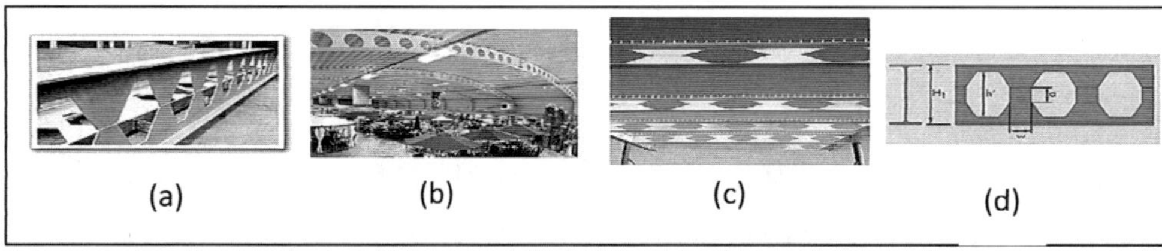

Figure 1. Various shapes of web opening section of steel beam; (a) hexagonal web opening, (b) circular web opening, (c) sinusoidal web opening and (d) octagonal web opening [3–6].

In normal practice for building structures, both passive and active fire protection method were installed and utilized to protect the structural element against unexpected fire hazard. These types of fire protection system provided adequate fire safety during fire risk exposure. Passive fire protection material is classified into a few types of protection, namely board systems, concrete encasement, intumescent coating and spray-on systems [7]. Intumescent coating is widely used for passive fire protection system due it's easier to install the intumescent panting surrounding the structural element as opposed to other types of protection material. For this reason, intumescent coating was selected for this current research. For coated composite CSB, the steel beam stiffness and strength will reduce significantly at elevated temperature. It is believed that the strength of steel beam deteriorated and weaker beyond 550°C of fire exposure and hence critical failure mode occurred before collapse. Vierendeel bending failure mechanism, web-post buckling and vertical deflection are some of the critical failure conditions. Consequently, adopting fire safety material onto the beam could reduce these drawbacks and hence improved the beam stiffness and strength. Thus, a reliable performance-based approach needs to be executed to explore fire resistance behaviour of coated composite CSB when bare to fire. This approach considers important consideration that contribute to fire protection behaviour of the beam, namely fire density, fire distribution, boundary condition and connection between other steel member which were not captured when selecting prescriptive-based approach. The latter design approach can be retrieved from available fire design regulations of BS5950: Part 8, BS EN 1991-1-2, BS EN 1993-1-2 and BS EN 1994-1-2, which were adopted until today. This design approach calculate the fire protection material thicknesses based on fire temperature limit which is up to 550°C when bare to standard fire curve [8]. From then onwards, the depth of the fire safety material was determined straightforward without having to consider important factors as mentioned in performance-based approach previously.

Therefore, a general purpose ABAQUS finite element program were used in present research to replicate and predict the structural behaviour of the coated composite CSB when expose to fire. Numerical simulation model of composite CSB when expose to fire were validated against the available experimental results. Temperature distribution and vertical deformation of composite CSB at elevated temperature were modeled and verified with the previous work conducted by [9–12]. From then onwards, parametric investigations were initiated by applying various intumescent coating thickness to explore the behaviour of the composite CSB at elevated temperature.

2. Cellular steel beam (CSB) and its behaviour

2.1. Cellular steel beam (CSB)

The usage of CSB as the main structural steel member is gaining momentum due several advantages as mention in Chapter 1. In term of self-weight, CSB is much lighter than standard steel beam which reducing the usage of steel material itself. Even though some portion through the web steel beam section of CSB were omitted, the structural strength and integrity is maintained as per standard steel beam. However, several drawbacks need to consider when analyzing the CSB at elevated temperature. Vierendeel bending failure mechanism, web post buckling and large vertical deflection are critically failure mode that has been reported in the literature [2][10][13–15]. Only vertical deformation results were reported in this research while other drawback will be discussed in the next research paper. The predicted maximum temperature and maximum deflection together with the composite CSB member when expose to fire can be located and further discussed. With the supplementary fire safety material of intumescent coating, which is covering the composite CSB, the level of fire safety can be improved. Therefore, different depth of intumescent coating layer was studied in this research. Hence, the nature of coated composite CSB at elevated temperature can improved significantly. In this research, coating thickness of 0.1, 0.3, 0.5, 1.0, 1.5, 2.0 and 5.0 mm were chosen to explore the structural behaviour of the CSB when bare to fire and applied load.

2.2. Prescriptive-based approach

The Prescriptive-based approach is the easiest and safest approach to assess the behaviour of an particular structural component at elevated temperature. This approach is purely referred to standard fire resistance test according to the guidelines stipulated in standard fire codes of BS EN 1991-1-2, BS EN 1993-1-2, BS EN 1994-1-2, BS EN 1363-1, BS 476-20 and ISO 834. The appropriate depth of the fire safety material can be obtained directly form the design regulations. However, there are some factor that needs to be consider when adopting this approach. In this approach, it does not deal with the current nature of fire that differ than standard fire exposure which considering cooling stage. Another reasonable fire nature also needs to be considered in the fire design regulations due to its nature regarding fumes discharge, fire distribution, fire density and fire intensity that affecting structural building when expose to fire [16-17]. In addition, it also does not consider the temperature distribution along the structural element in the course of fire exposure. Consequently, it is exceptional to embrace reliable performance-based technique that able modelled and simulate the performance behaviour of structural member at elevated temperature by considering all crucial factors.

2.3. Performance-based approach

In this more advanced approach, it is quite complicated to analyze the structural behaviour of the structural steel elements. However, by performing performance-based approach by means of numerical simulation, it able to model and simulate complicated structural steel element at elevated temperature. The mechanical properties obtained from the experimental test were used and to be include in the numerical simulation analysis. Two different stages can be implemented in this approach, namely heat transfer analysis (temperature behaviour of structural member) and static analysis (mechanical behaviour of structural member) [18]. Commonly, passive fire safety material was used to insulate the steel beam member by reducing the temperature distribution along the steel member when expose to fire. Hence, intumescent coating material were used in this research to provide full insulation of the composite CSB when expose to fire by reducing the CSB temperature. Due to this scenario, it is expected that the vertical deformation will subsequently reduce due to temperature drop in steel beam member. Therefore, it can predict the most suitable thickness of intumescent layer to be insulated onto the composite CSB. It is crucial and important to adopt a dependable performance based concept, that needs to be acknowledge in the present steel construction [7].

3. Specification of Cellular Steel Beam (CSB)

One sample of bare composite CSB when expose to fire were chosen from the experimental results conducted by Nadjai [9] at University of Ulster, Northern Ireland. The symmetrical unprotected composite CSB A2 (SUC-CSB-A2) was selected and verify against numerical analysis performed in this current study. Two parameters were used for validation purposes, namely maximum temperature distribution and mid-span vertical deflection. The SUC-CSB-A2 sample were bare to a slow heating fire curve as stated in fire design regulations of ISO 834, BS EN 1991-1-2 and PD 7974-1. After validation process has completed, parametric investigations were initiated to figure out the fire protection behaviour of protected SUC-CSB-A2 which is insulated with intumescent coating material. Table 1 presented detailed geometrical properties for composite SUC-CSB-A2 with additional concrete slab on the steel beam surface.

The geometry detailed of the composite SUC-CSB-A2 is depicted in Figure 2. For the first parameter, only one cross-section of fire thermocouples location selected along the beam segment as presented in

IOP Conf. Series: Materials Science and Engineering **713** (2020) 012032 doi:10.1088/1757-899X/713/1/012032

Figure 2(c). Therein are six fire thermocouples attached to the beam section, namely TA1, TA2, TA3, TA4, TA5 and TA6. For the second parameter, the maximum vertical deformation location (DA1, DA2, DA3, DA4 and DA5) were located as shown in Figure 2(d). However, point DA3 were presented with regards to mid-span vertical deformation.

Table 1. Geometrical properties of composite SUC-CSB-A2 and concrete slab.

Steel beam section size	The finished depth of bottom tee steel beam section	Width of upper flange steel beam section	Width of bottom flange steel beam section	Thickness of upper flange steel beam section	Thickness of bottom flange steel beam section	Thickness of web steel beam section
UB 406 x 140 x 39	575 mm	141.8 mm	141.8 mm	8.6 mm	8.6 mm	6.4 mm
Steel grade	**Diameter of circular web opening**	**Thickness of concrete slab**	**Width of concrete slab**	**Normal concrete grade**	**Yield strength steel reinforcement**	
S355	75 mm at 500 mm center to center	150 mm	1200 mm	35 N/mm^2	460 N/mm^2	

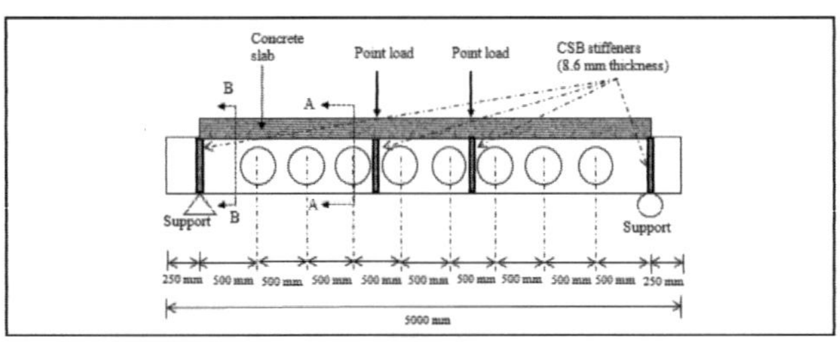

(a)

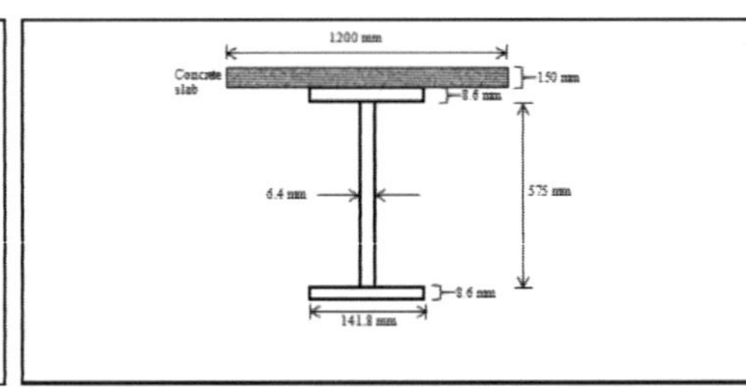

(b)

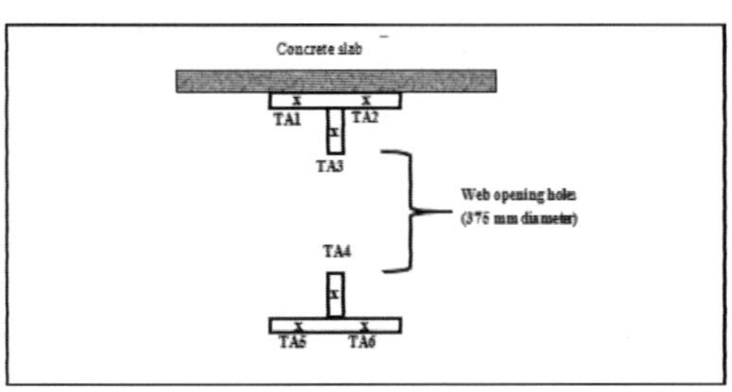

(c)

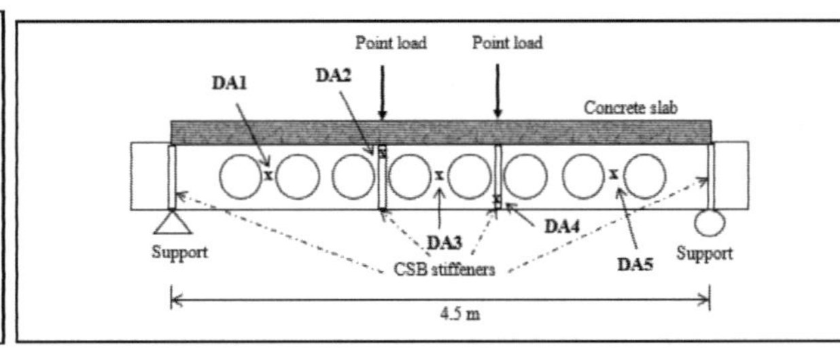

(d)

Figure 2. Composite SUC-CSB-A2 model: (a) front view; (b) cross section B-B view; (c) fire thermocouples position at cross-section A-A; (d) front view location vertical deflection [9][11-12].

4. Numerical modelling

In this non-linear investigation, the stress-strain behaviour for the steel beam and concrete slab were obtained from design codes BS EN 1991-1-1, BS EN 1991-1-2, BS EN 1993-1-1, BS EN 1993-1-2, BS EN 1994-1-11 and BS EN 1994-1-2. For both ends of the coated steel beam, boundary conditions were applied as similar with experimental set up. For experimental program conducted by Nadjai [9], the interaction in the midst of concrete slab and the composite CSB was connected by using shear stud. However, in this current research, shear stud was not considered in the numerical analysis. But instead full interaction constraint was adopted in the middle of the underlying surface of concrete slab and topmost surface of the composite CSB by means of tie constraint to replicate the shear stud function.

Both heat transfer and static analysis were used during the computational finite element analysis program.

During heat transfer analysis stage, a four-node linear heat transfer quadrilateral shell element DS4 model were chosen for bare composite SUC-CSB-A2, concrete slab and intumescent coating. For heat transfer step, the heat from fire exposure were transmitted to the steel beam surface by the process of convection and radiation. The fire exposure started from the bottom fire furnace and equally expose to all surface of the CSB and intumescent coating, which is underneath the concrete slab. The value of the convection used were 25 W/m^2K and 9 W/m^2K for bare and protected steel beam surface as stated in fire design regulations. The radiation value on steel surface was taken as 0.8 as taken from fire design regulation. In addition, the radiation value on intumescent coating surface was taken as 0.825. In terms of steel beam and concrete properties at elevated temperature, the thermal properties and conductivity were taken out the fire design regulations. The specific heat and thermal conductivity for the steel beam and concrete slab used in the numerical analysis were illustrates as in Figure 3. The intumescent coating properties were retrieved from previous research conducted by Krishnamoorthy [19-20].

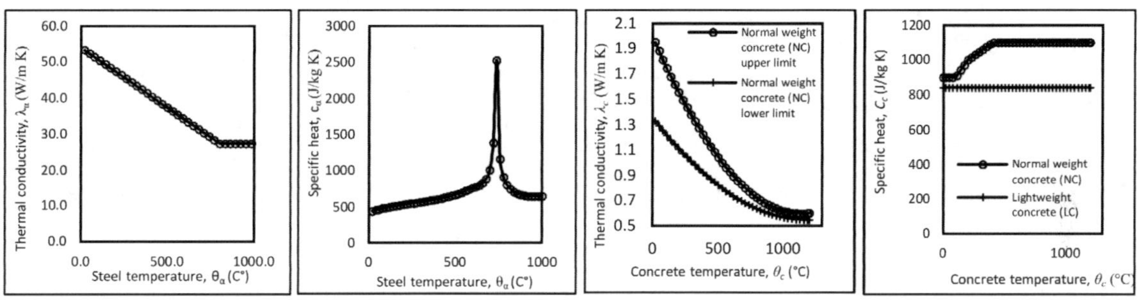

(a) (b) (c) (d)

Figure 3. Thermal properties for ; (a) steel , (b) steel, (c) concrete and (d) concrete [21–23]

For static analysis step, the Newton-Raphson method were employed numerically to execute the material non-linearity equations and geometrically by consolidating the iterative and incremental procedures. An applied load of 90 kN were imposed at two location on the top of concrete slab as illustrates in Figure 2(a) and Figure 2(d). General static analysis was selected to solve the nonlinear behaviour of the composite SUC-CSB-A2 model. A non-linear analysis of finite membrane-strain, fully integrated linear shell element S4 model were chosen for both bare composite SUC-CSB-A2 model and concrete slab. The material nature for steel beam and concrete slab were taken from previous research work [12][20]. In addition, concrete damage plasticity material behaviour was chosen to model the nonlinear plasticity concrete slab behaviour at elevated temperature. The concrete slab and steel beam density are taken as 2400 and 7850 kg/m^3 respectively.

5. Results and discussions

Figure 4 and Figure 5 and shows the predicted and measured maximum temperature distribution at point TA1, TA2, TA3, TA4, TA5 and TA6 of composite SUC-CSB-A2 model. The fire exposure duration for all points is 4700 seconds (78.3 minutes). The overall forecast maximum temperature agrees well with measured maximum temperature at 4800 seconds fire exposure time. From the figures, the forecast maximum temperature is slightly less than the measured maximum temperature for point TA1 and TA2. This behaviour is likely due to point TA1 and TA2 placed in the intermediate section of the top flange beam section which is fully connected to concrete slab. Parts of heat in this section were dissipated through to concrete slab which affecting the temperature distribution through the top flange beam section. Due to this scenario, the outcome from experimental temperature output is different than forecast temperature output. Experimental outcome does not consider the interaction between both steel section and concrete slab, while numerical analysis does capture the effect of the full interaction during simulation analysis procedure. However, TA3, TA4, TA5 and TA6 predicted slightly overestimate value of maximum temperature as compared to experimental data. The behaviour for these four locations were believed caused by the interface reaction when fire exposure hit on the steel beam surface. Convection, radiation and material properties values does have a compelling significance onto the fire safety behaviour of composite CSB at elevated temperature. Meanwhile, Figure 6(a) shows the forecast and measured critical vertical deformation located in the middle of the web section and centre of the composite SUC-CSB-A2 model. From the figures, the forecast critical vertical deflection agrees well with the measured maximum vertical deflection at the end of the 4800fire exposure duration. Point DA3 forecast the critical vertical deflection of 0.21 m as compared to measured vertical deformation of 0.2 m as shown in Figure 6(a). Figure 6(b) shows the forecast critical vertical deflection at point DA3 of composite SUC-CSB-A2 model which is coated with different depth of intumescent layer. The vertical deflection is slightly decreasing when adopting deeper intumescent layer.

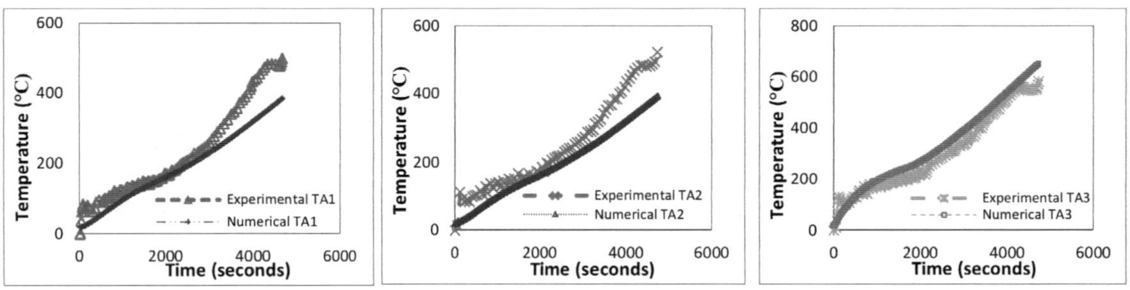

Figure 4. Forecast and measured maximum temperature for point TA1, TA2 and TA3 for composite SUC-CSB-A2 model at the end of 4800 seconds fire exposure.

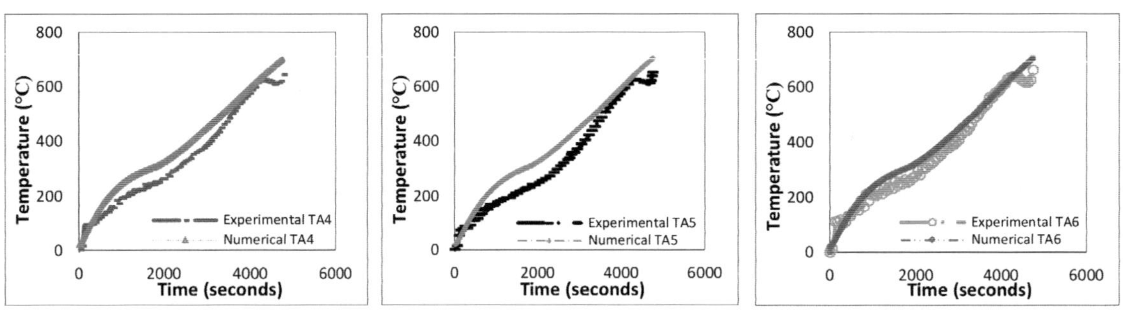

Figure 5. Forecast and measured maximum temperature for point TA4, TA5 and TA6 for composite SUC-CSB-A2 model at the end of 4800 seconds fire exposure.

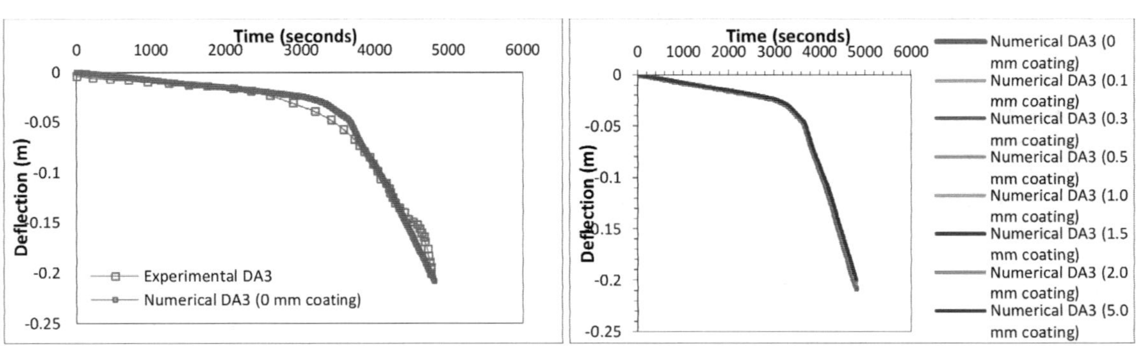

(a) (b)

Figure 6. Predicted and measured maximum vertical deformation for point DA3 for composite SUC-CSB-A2 model (fire exposure 4800 seconds); (a) validation process, (b) applying different depth of intumescent coating layer.

Figure 7(a) illustrates the forecast critical temperature distribution of composite SUC-CSB-A2 model which were obtained through ABAQUS software. Figure 7(b) illustrates the forecast critical vertical deflection of composite SUC-CSB-A2 model when bare to 4700 seconds fire exposure. The beam model was fully covered with 5.0 mm thickness of intumescent layer. Detailed summary of the maximum vertical deflection for all intumescent coating thickness is summarizes in Table 2 as referred to Figure 6 (b). However, there are no significant effect on adopting various intumescent layer onto the composite CSB even though the vertical deflection were slightly drop as presented in Table 2. Adopting deeper intumescent layer onto the protected composite CSB model does slightly contribute in predicting critical vertical deflection drop as compared to bare composite CSB model.

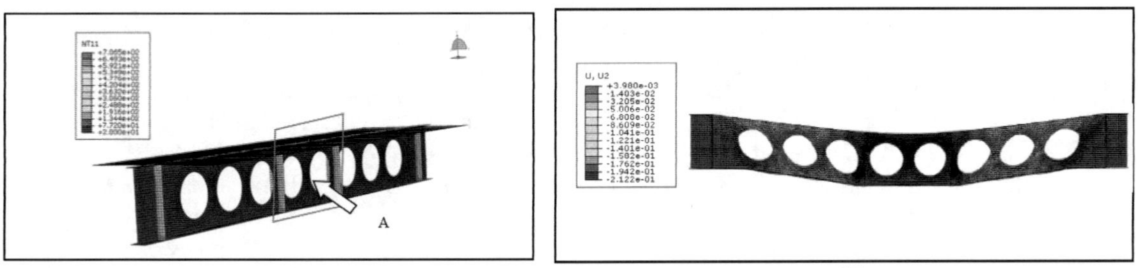

(a) (b)

Figure 7. Composite SUC-CSB-A2 model at elevated temperature (4800 seconds); (a) forecast critical temperature distribution, (b) forecast predicted vertical deflection (with 5.0 mm intumescent coating thickness).

Table 2. Summary of predicted maximum vertical deflection at point DA3 of composite SUC-CSB-A2 model expose to fire exposure of 4800 seconds.

Intumescent coating thickness (mm)	Maximum vertical deformation (mm)	Deflection decrease (mm)	Percentage deflection decrease (%)
No coating	208.599	0	0
0.1	202.453	6.146	2.946
0.3	200.985	7.614	3.65
0.5	200.169	8.43	4.041
1	200.082	8.517	4.083
1.5	199.233	9.366	4.49
2	199.098	9.501	4.555
5	198.859	9.74	4.67

6. Conclusion

The nonlinear structural behaviour of composite SUC-CSB-A2 model under combined fire load and applied load has been investigated and discussed in this paper. A nonlinear finite element model has been created based on composite SUC-CSB-A2 model and verified with available experiment results. The predicted maximum temperature distribution and mid-span vertical deflection of composite SUC-CSB-A2 model has been well justified with experimental results. From the parametric investigation, employing deeper intumescent layer onto the beam model has significantly decrease the mid-span vertical deflection. It shows that adopting thicker intumescent layer does not make a significant difference in terms of mid-span vertical deflection. Therefore, the techniques of applying intumescent layer protection onto the beam model needs to be change to other techniques. Future research related to partial fire protection techniques needs to be investigated in order to obtain its behaviour when expose to fire and under applied load.

7. References

[1] Lawson R M and Hicks S J 2011 *Design of beams with large web openings (SCI P355)* (United Kingdom: SCI)

[2] Bihina G, Zhao B and Bouchaïr 2013A Behaviour of composite steel – concrete cellular beams in fire *Eng. Struct.* **56** 2217–2228

[3] ArcelorMittal 2014 *Angelina TM: Castellated steel beam with sinusoidal openings* Retrived on 5 May, 2019 from http://news.archiexpo.com/press/arcelormittal/angelina-tm-castellated-steel-beam-sinusoidal-openings-55693-188613.html

[4] Gunung Steel Group 2015 *Honey Comb castellated beams* Retrived on 5 May, 2019 from http://www.gunungsteel.com/index.php?option=com_content&view=article&id=76:honey-comb&catid=39&Itemid=171.

[5] SteelConstruction. info 2015 *Thin Film Intumescent Coating* Retrived on 5 May, 2019 from http://www.steelconstruction.info/images/8/80/Intumescent_coating_example_3.png.

[6] Constructalia 2015 *Octogonal castellated beams* Retrived on 5 May, 2019 from http://www.constructalia.com/english/products/structures/steel_sections_and_merchant_bars/cellular_and_castellated_sections/octogonal_castellated_beams_with_octogonal_openings_.

[7] Buchanan A B 2001 *Structural Design for Fire Safety* (New Jersey: John Wiley & Sons Ltd)

[8] Y. C. Wang 2002 *Steel and Composite Structure: Behaviour and Design for Fire Safety* (United Kingdom: Spon Press)

[9] Nadjai A, Vassart O, Ali F, Talamona D, Allam A and Hawes M 2007 Performance of Cellular Composite Floor Beams at Elevated Temperatures *Fire Saf. J.* **42** 489–497

[10] Nadjai A, Petrou K, Han S and Ali F 2016 Performance of Unprotected and Protected Cellular Beams in Fire Conditions *Constr. Build. Mater.* **105** 579–588

[11] Nadjai A 2007 *Behavior of Composite Floor Cellular Steel Beams at Elevated Temperatures* (United Kingdom: SCI)

[12] Bake S 2010 *Cellular Beams at Ambient and Elevated Temperatures* PhD thesis (Manchester: The University of Manchester)

[13] Bake S 2011 Design of beams with large web openings (United Kingdom: SCI) pp. 30–31

[14] Chung K F, Liu C T T H and Ko A C C H 2001 Investigation on Vierendeel Mechanism in Steel Beams with Circular Web Openings *J. Constr. Steel Res.* **57** 467–490

[15] Ellobody E and Young B 2015 Nonlinear Analysis of Composite Castellated Beams with Profiled Steel Sheeting Exposed to Different Fire Conditions *J. Constr. Steel Res.* **113** 247–260

[16] Dwaikat M M S and Kodur V K R 2011 A Performance Based Methodology for Fire Design of Restrained Steel Beams *J. Constr. Steel Res.* **67**(3) 510–524

[17] Parkinson D L and Kodur V K R 2007 Performance-Based Design of Structural Steel for Fire Conditions - A Calculation Methodology *Steel Struct.* **7** 219–226

[18] Bailey C 2015 *One Stop Shop in Structural Fire Engineering* Retrived on 5 May, 2019 from http://www.mace.manchester.ac.uk/project/research/structures/strucfire/Structural FireEngineering/default.htm.

[19] Krishnamoorthy R R and Bailey C G 2009 Temperature distribution of intumescent coated steel framed connection at elevated temperature *Proc. Nordic Steel Construction Conference (Malmo)* vol I (Sweden: Swedish Institute of Steel Construction) pp 572-579

[20] Krishnamoorthy R R 2011 *The Analysis Of Partial And Damaged Fire Protection On Structural Steel At Elevated Temperature* (Manchester: The University of Manchester)

[21] British Standards Institution 2002 *BS EN 1991-1-2:2002 Actions on Structures — Part 1-2: General actions — Actions on Structures Exposed to Fire* (United Kingdom: British Standards Institution)

[22] British Standards Institution 2005 *BS EN 1993-1-2:2005 Design of Steel Structures — Part 1-2: General rules — Structural Fire Design* (United Kingdom: British Standards Institution)

[23] British Standards Institution 2005 *BS EN 1994-1-2:2005 Design of Composite Steel and Concrete Structures —Part 1-2: General Rules — Structural Fire Design* (United Kingdom: British Standards Institution)

Acknowledgments
The authors would like to thank all the staff and assistant engineer in the Computer Laboratory, Faculty of Civil Engineering, University Teknologi MARA, Cawangan Pulau Pinang, Permatang Pauh, Pulau Pinang, Malaysia and Faculty of Civil Engineering, Universiti Teknologi MARA, Shah Alam, Selangor, Malaysia for providing general purpose of ABAQUS finite element program for numerical simulation purposes.

IOP Conf. Series: Materials Science and Engineering **713** (2020) 012033 doi:10.1088/1757-899X/713/1/012033

Vibration Test on Existing Steel Sheet Pile

Sakhiah Abdul Kudus[1], Kunitomo Sugiura[2], Yasuo Suzuki[3]

[1] Faculty of Civil Engineering, Universiti Teknologi MARA, 40450 Shah Alam, Selangor, Malaysia.
[2] Department of Civil and Earth Resources Engineering, Graduate School of Engineering, Kyoto University, Kyoto, 615-8540, Japan.
[3] Department of Civil Engineering, Faculty of Sustainable Design, University of Toyama, Toyama, 930-8555, Japan
Corresponding author: sakhiah@uitm.edu.my

Abstract. A thorough understanding about the vibration of a structure is essential to resolve any vibration problems. In this study, the vibration measurement was conducted on a real plate structure, which is steel sheet pile. The monitored structures are located in Rokkenya-gawa river and Aji-gawa river in Osaka City, Japan. The purpose of this investigation is to obtain information about the overview conditions of this marine structure. A simple vibration measurement as well as thickness measurement have been conducted. There is no serious damage was observed on the monitored structure.

1. Introduction

The vibration characteristics needs to be understood first before any meaningful use in the field of structural health monitoring. The applications of modal testing is a form of experimental evolution conducted to determine the dynamic properties of a structure. In recent years, efforts have been made to combine the modal testing with the analytical methods. Modal testing deals with a real structure directly, and utilizes the analytical method to verify both methods. There are several methods applicable for excitation of the structure and normally are pure random, burst random, period swept sine and hammer impact. The use of hammer impact has a poor signal to noise ratio [1]. The dynamic characteristics of the shaker may be combined with the test structure when the shaker is used to conduct modal testing. Use of the dynamic properties of structure such as natural frequency and mode shape to evaluate the damage has extensively been studied [2-5].

Vandepitte and Olbrechts [6] studied the dynamic behaviour of the test structure when the structure was excited by different excitation methods [6]. Three methods of excitation were considered: hammer excitation; shaker with stringer excitation; and internal shaker excitation. The results obtained show that the loading methods do influence the results obtained. Thus, the loading effect is one of the important parameter to be considered. The study also concludes that the inertial shaker does not only add mass to the system, but may also increase the stiffness of the complete system.

In vibration analysis, free vibration is the natural response of a structure due to impact or displacement. The responses obtain are based on properties of the structure and the vibration data acquired from examination of the mechanical properties of the monitored structure. Structural vibration is generally measured by an electronic sensor which converts the vibration motion into the electrical signal. By analysing the signal, the nature of the vibration can be obtained. The dynamic measurements

IOP Conf. Series: Materials Science and Engineering **713** (2020) 012033 doi:10.1088/1757-899X/713/1/012033

are normally conducted using an accelerometer, where the measurement of vibration response is conducted by mounting the accelerometer to the surface of the structure.

The damage assessment based on vibration technique was selected mainly due to merit of this technique which have the ability to inspect large region in short time, low implementation cost, relatively sensitive towards the local changes in the structure, low energy consumption and repeatability. This study aimed to investigate the vibration characteristic of existing sheet pile exposed to marine environment using vibration test.

2. Methodology

The use of corrugated plate for retaining walls and embankments has received worldwide acceptance due to high strength, lower cost and ease of installation. Figure 1 shows the schematic diagram of the steel sheet pile embedded in soil backfill. Normally, a steel tie rod is used to tie back the sheet pile. The tie rod system is categorised as the most economical sheet piling tie back. The front side of steel sheet pile was facing to the water river side while another side of steel sheet pile is facing to the backfield, which indicates that the backside of this structure is inaccessible for any maintenance or inspection work. The assessment of this type of structure is challenging because of limited access to the structure. The steel sheet pile is only accessible on the river side, and the other side of sheet pile was embedded in the backfill.

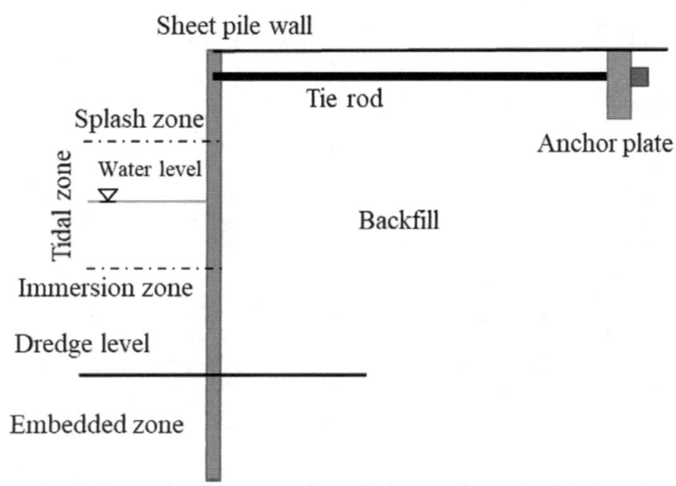

Figure 1. Schematic diagram of steel sheet pile embedded on backfill.

Figure 2 summarizes the testing procedure for modal testing. Normally, the data from modal testing is further analysed for estimation of frequency response function (FRF) to obtain other modal parameters like mode shape and modal damping. A physical structure has an infinite number of mode shape together with resonance frequencies. However, in this work, mode shape and modal damping were not possible to be obtained from the vibration test conducted on steel sheet pile because of limited vibration measurement. In order to obtain the mode shape and modal parameters of the monitored structure, large number of discrete measurement response points need to be recorded on the surface of the structure. In addition, improper selection of a discrete point where the measurement is conducted on nodal points leads to inaccurate characterization of mode shape. The FRF is describe as the relationship between two points in the input-output as a function of frequency.

Figure 2. Basic vibration measurement for modal testing.

IOP Conf. Series: Materials Science and Engineering **713** (2020) 012033 doi:10.1088/1757-899X/713/1/012033

3. Results and discussion

The vibration measurement was conducted on a real plate structure, which is steel sheet pile. The purpose of this investigation is to obtain information about the overview conditions of this marine structure. A simple vibration measurement as well as thickness measurement has been conducted. Figure 3 depicts the site on the Aji-gawa river. The site visit was done on February 2016. The steel sheet pile was used as permanent support. The thickness measurement was conducted on several points and the losses of thickness is about 0.4 mm. The intact thickness of this wall is 10 mm. Thus, it can be concluded that no serious corrosion damage happened here from the thickness measurement data table 1 and table 2 show details of the structure age and maintenance schedule on steel sheet pile located at Aji-gawa river and Rokkenya-gawa river respectively. The structure is almost 50 years old but no serious damage was observed. This was related to the properly scheduled maintenance provided by the authorized authority.

Figure 3. The preliminary investigation of health condition of steel sheet pile located at Aji-gawa river, Osaka.

Table 1. Construction and maintenance details of Aji-gawa river sluice water.

Year of construction	1970 (March)
Age of structure (during the inspection)	46 years
Type of plated structure	Steel sheet pile
History of coating and/or painting	1998 and 2013
Galvanic protection	2018 (plan)
Others	-

Table 2. Construction and maintenance details of Rokkenya-gawa river flood gate J1.

Year of construction	1969 (March)
Age of structure (during the inspection)	47 years
Type of plated structure	Steel sheet pile
History of coating and/or painting	2013
Galvanic protection	2013
Others	-

The vibration test was conducted using an impact hammer, vibration shaker and accelerometer. An accelerometer is a simple mechanical device used in the engineering field to measure the acceleration force. The force in this work is induced via vibration shaker and impact hammer. Figure 4 details the location of excitation and the position of the accelerometer. Two accelerometers were placed in a fixed position 1 m apart and excitation was made in the middle between the both of accelerometer. The frequency of excitation varies from 50 Hz up to 400 Hz using the vibration shaker. Signal analysis can be divided into time and frequency domain, and each domain provides a different view and insight into the vibration.

Fast Fourier transform (FFT) is used to convert an original signal normally in the time domain to representation in the frequency domain. Figure 5, Figure 6, and Figure 7 represent the frequency domain data of the sheet pile when excited by the vibration shaker with frequencies of excitation of 50 Hz, 100 Hz and 150 Hz respectively. The analysis involves imposing excitation on the plate structure and determining the frequency at which the structure resonates when the excitation and the vibration response match. The data was captured by the accelerometer from Channel 1 located at the bottom part of the sheet pile. From the results obtained, the peak of 66.47 Hz might represent one of the resonance frequencies of this structure. In addition, the modal testing result for pile excited by the impact hammer was presented in Figure 8. The pile was excited by several impacts. This section summarize how the impact wave was generated and the vibration response obtained. In order to understand the vibration of the structure, the steady vibration is necessary as compare to only one or small number of impacts. The steady vibration can be induced from the actuator and vibrator.

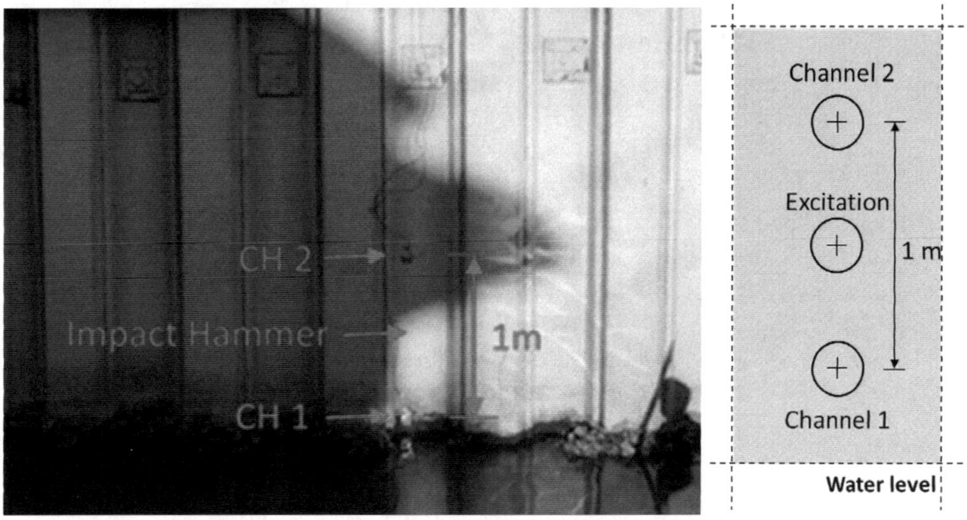

Figure 4. The details of excitation and measurement point on steel sheet pile structure located at Rokkenya-gawa.

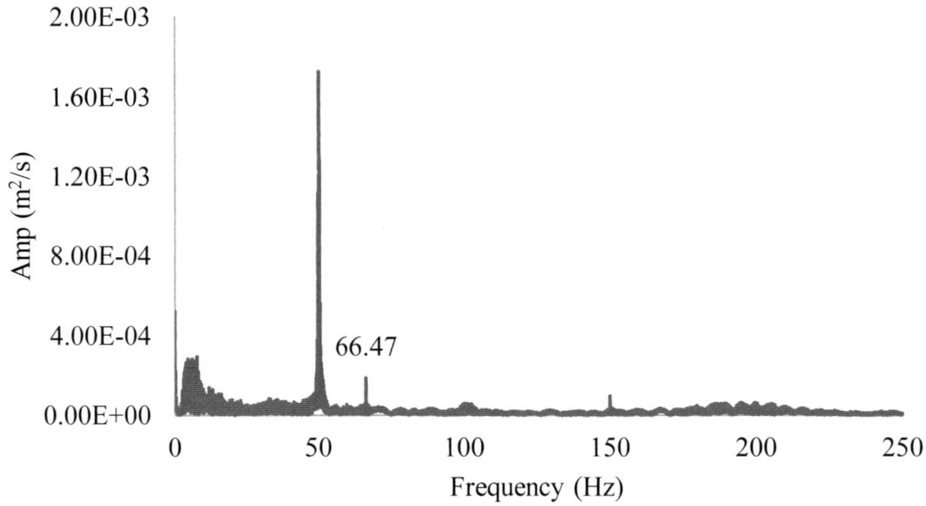

Figure 5. The FFT result from accelerometer Channel 1 when the pile was excited with frequency of 50Hz by shaker.

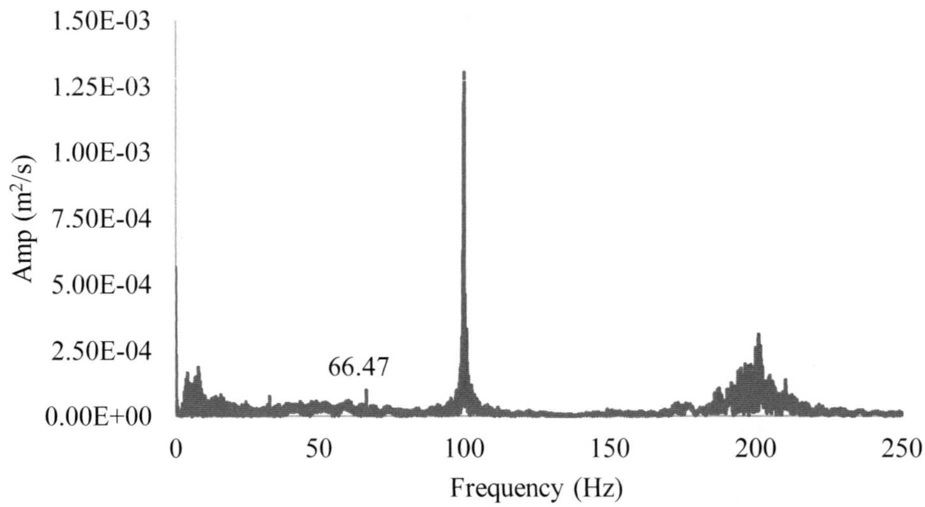

Figure 6. The FFT results from accelerometer Channel 1 when the pile was excited with frequency 100Hz by shaker.

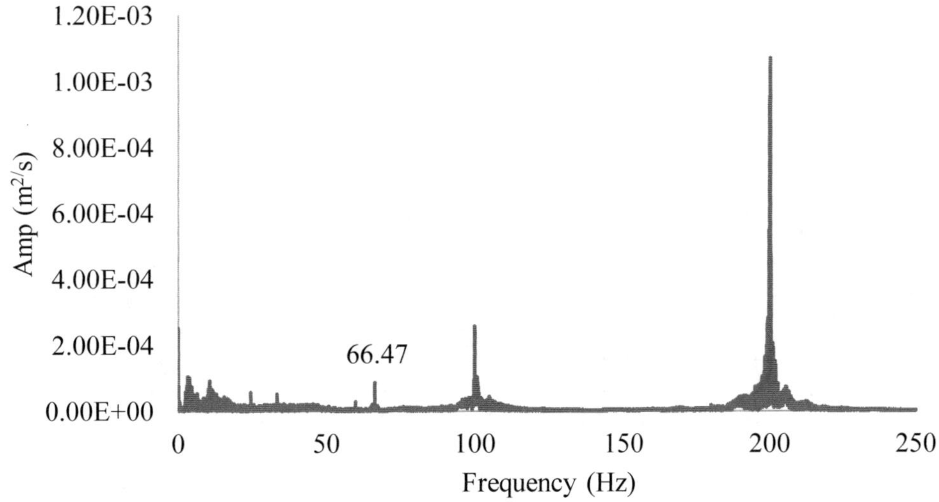

Figure 7. FFT result from accelerometer Channel 1 when the pile was excited with frequency of 200Hz by shaker.

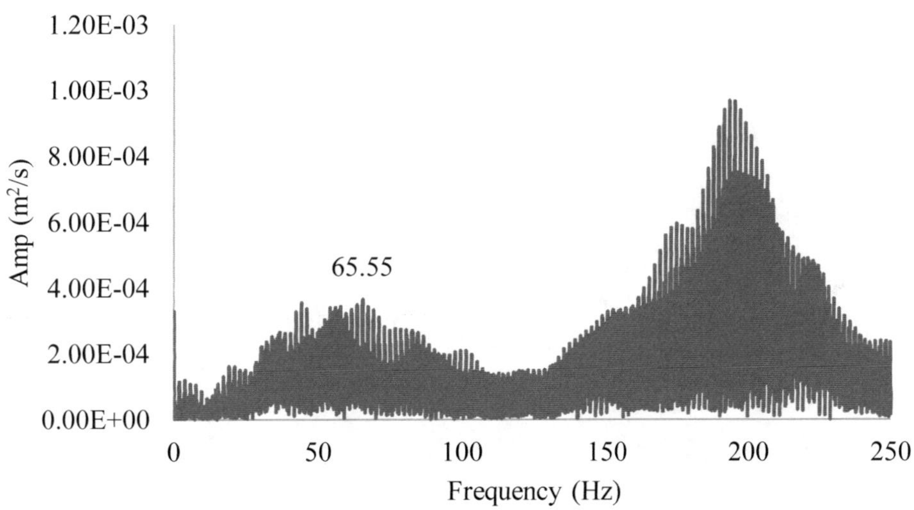

Figure 8. The FFT results from accelerometer Channel 1 when the pile was excited by impact hammer.

4. Concluding Remark

The appropriate usage of vibration testing in instrumentation and analysis benefits understanding of problems in the structure. The important information related to structural characteristics including modal parameter can be further obtained. Preliminary vibration tests have proven to be useful in the optimization of methodology and equipment to be used for detailed investigation. Moreover, vibration measurements give advantages to forecasting the extent of damage in a structure from modal parameters obtained during analysis.

IOP Conf. Series: Materials Science and Engineering **713** (2020) 012033 doi:10.1088/1757-899X/713/1/012033

5. References

[1] Ma L and Peng X 2011 The study of excitation methods in experimental modal analysis of lightly damped structures, Electric Information and Control Engineering (ICEICE) *2011 International Conference on Electric Information and Control Engineering*

[2] Kudus S A, Suzuki Y, Matsumura M and Sugiura K 2018 Damage Assessment Based on Modal Analysis of Pipe Structure *Jurnal Teknologi* **80**(5) 37 - 44

[3] Kudus S A, Suzuki Y, Matsumura M, Sugiura K 2018 Vibration-response due to thickness loss on steel plate excited by resonance frequency *IOP Conference Series: Earth and Environmental Science* vol 140 (1) (United Kingdom: IOP Publishing) pp 012123

[4] Sakhiah B A K 2018 *Study on Damage Assessment of Steel Plated Structures Based on Local Vibration Characteristics* PhD Thesis (Kyoto University)

[5] Kudus S A, Suzuki Y, Matsumura M and Sugiura K 2018 Damage Assessment Based on Sensitivity of Modal Parameter in Plated Structure *Malaysian Construction Research Journal* **24**(1) 65-82

[6] Vandepitte D V, Olbrechts T and Sas P 1997 FRF measurements errors caused by the use of inertia mass shaker *Proceedings of the 15th International Modal Analysis Conference* pp 188-194

IOP Conf. Series: Materials Science and Engineering **713** (2020) 012034 doi:10.1088/1757-899X/713/1/012034

Preliminary Study on Modified Foamed Concrete-Filled Hollow Section under Cyclic Load

Siti Amirah Azra Khairuddin[1], Norashidah Abd Rahman[2], Mohd Luqman Rossam[1] and Anis Nazirah Mustafa[1]

[1]Faculty of Civil and Environmental Engineering, Universiti Tun Hussein Onn Malaysia, 86400 Parit Raja, Batu Pahat, MALAYSIA
[2] Jamilus Research Centre Faculty of Civil and Environmental Engineering, Universiti Tun Hussein Onn Malaysia, 86400 Parit Raja, Batu Pahat, MALAYSIA
Corresponding author: nrashida@uthm.edu.my

Abstract. The use of green materials has become a hot topic for discussion in steel construction. Foamed concrete is a green and lightweight material used in concrete mix. The use of foamed concrete as a fill in steel construction has been discussed by many researchers. However, the performance of foamed concrete under cyclic load has not been evaluated. Therefore, this study employed a modified foamed concrete-filled hollow section under cyclic load to determine the performance of concrete-filled hollow section under cyclic load. The modified foamed concrete consists of foamed concrete added with rice hush ash and fibre. Different types of fibres (steel and polypropylene fibres) were utilised. Results showed that the concrete-filled hollow section with steel fibre slightly improved the strength and ductility of the hollow section.

1. Introduction

The use of concrete-filled hollow section (CFHS) as a steel structure has been widely discussed. However, the use of normal concrete increases structure dead weight and construction costs. Therefore, many researchers studied lightweight concrete [1-3]. As an example of a lightweight concrete, foamed concrete has a density of 400–1600 kg/m^3 and a strength of 10–35 MPa [4]. Hunaiti [5] assessed the performance of foamed CFHS and found that the strength of the hollow section is insufficient. Jaini et al. [6] reported that adding rice hush ash (RHA) in the foamed concrete mix increases the strength of CFHS.

The design of the structure has become increasingly critical with the recent earthquakes. Many construction works nowadays consider the effect of earthquakes to the structure. Therefore, the performance of foamed CFHS under cyclic load should be investigated. Previous studies reported that the strength of CFHS under cyclic load is decreased, which leads to high ductility [7-9].

The present study determined the strength and ductility of steel structure in accordance with the recommendation of the European Convention Construction Steelwork (ECCS). The parameters considered in the equation are yield strength (F_y) and yield displacement (e_y). The values of F_y ad e_y can be determined from the intersection of the two tangents from the F-e graph as shown in Figure 1.

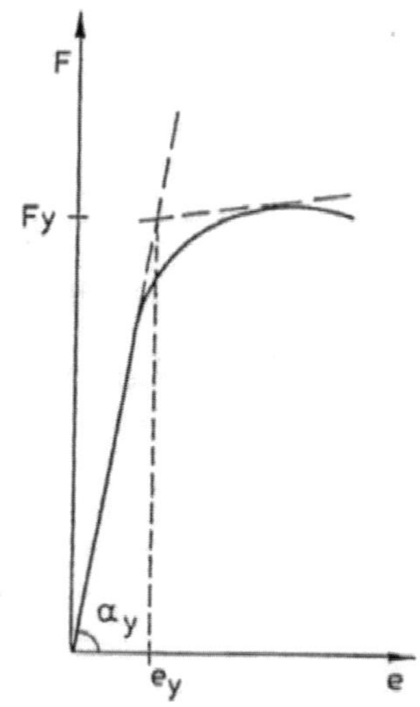

Figure 1. F-e curve [10]

The earthquake-resistance model for steel structure under continuous conditions must be established to analysis the behaviour of steel structure under cyclic load. Therefore, the graph was recorded of one cycle, interpret from the whole test. Figure 2 shows the parameters obtained for one cycle.

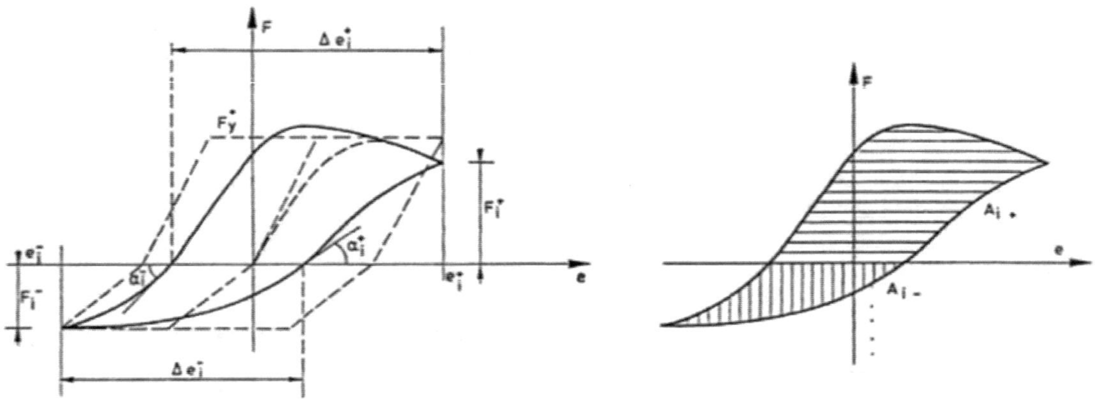

Figure 2. F-e diagram after one cycle [10]

To analyse the data, the ductility and the strength in terms of resistance ratio can be used as a recommended procedure [10]. The ductility and strength can be measured using Equations 1 to 8:

Partial ductility

$$\mu_{oi}^{+} = \frac{e_i^{+}}{e_y^{+}} \tag{1}$$

$$\mu_{oi}^{-} = \frac{e_i^{-}}{e_y^{-}} \tag{2}$$

Full ductility

$$\mu_i^{+} = \frac{\Delta e_i^{+}}{e_y^{+}} \tag{3}$$

$$\mu_i^{-} = \frac{\Delta e_i^{-}}{e_y^{-}} \tag{4}$$

Full ductility ratio

$$\psi_i^+ = \frac{\Delta e_i^+}{e_i^+ + (e_i^- - e_y^-)} \tag{5}$$

$$\psi_i^- = \frac{\Delta e_i^-}{e_i^- + (e_i^+ - e_y^+)} \tag{6}$$

Resistance Ratio

$$\varepsilon_i^+ = \frac{F_i^+}{F_y^+} \tag{7}$$

$$\varepsilon_i^- = \frac{F_i^-}{F_y^-} \tag{8}$$

2. Experimental work

2.1. Material

Nine samples of CFHS with 350 mm length and 2 mm thickness were prepared (Figure 3). The samples were divided into three groups: unfilled hollow section (HS), modified foamede concrete (CFHS–RHA–FC) and modified foamed concrete containing steel fibre (CFHS–RHA–SF). A foamed concrete mix was prepared with a density of 1600 kg/m^3. The modified foamed concrete mix contained ordinary Portland cement, sand, foaming agent and RHA to improve the strength of foamed concrete. Details of the modified foamed concrete are shown in Table 1. Nine (9) samples of cube specimen with a size of 100 mm × 100 mm × 100 mm were prepared and placed under compression test to determine the compressive strength of the modified foamed concrete. The cube samples were tested under compression load for 7, 14 and 28 days. The compression test result is summarised in Table 2.

Table 1. Material composition

Mixture	Foamed Concrete	SF Foamed Concrete
Cement–Sand Ratio (C/S)	0.5	0.5
Foamed–Cement Ratio (F/C)	0.7	0.7
Water–Cement Ratio (W/C)	0.55	0.55
RHA (%)	40	40
Steel Fibre (%)	-	0.8

Table 2. Modified foamed concrete compressive strength

Sample	Compressive strength (MPa)		
	7 days	14 days	28 days
FC	7.6	13.9	15.1
SF	10.1	15.3	16.6

Modified foamed concrete was later poured on 100 mm × 100 mm × 350 mm with 2 mm thickness short column steel hollow section (Figure 3). The strength (f_{yk}) of the steel is 349 MPa.

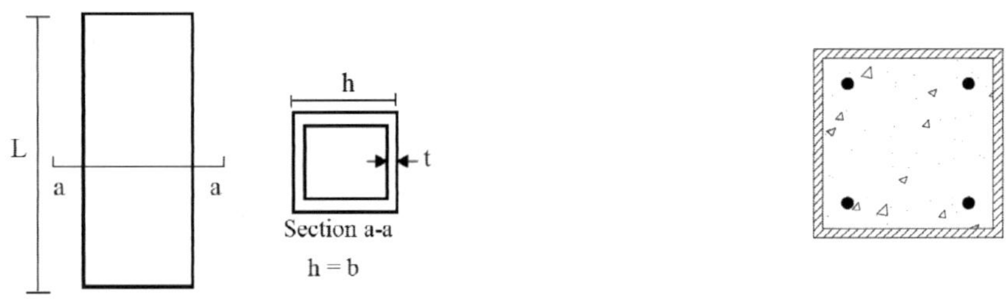

Figure 3. Illustration of a cross section

2.2. Specimen test.

The HS and CFHS were tested under cyclic load as shown in Figure 4. The displacement amplitude was applied as shown in Figure 4. In accordance with the cyclic test by ECCS [10], yield displacement (e_y) was measured from the F-e curve as discussed. The displacement e_y was taken from a previous study on CFHS using foamed concrete by Khairuddin et al. [11], where $e_y = 4.28$ mm at yield strength (f_y) = 371 kN. Given the limited load (400 kN) applied at the cyclic loading machine, all specimens applied the e_y for the foamed concrete. The effect of CFHS at the same range of amplitude was also determined. The amplitude range applied is shown in Figure 4, and the frequency for the test is 0.25 Hz.

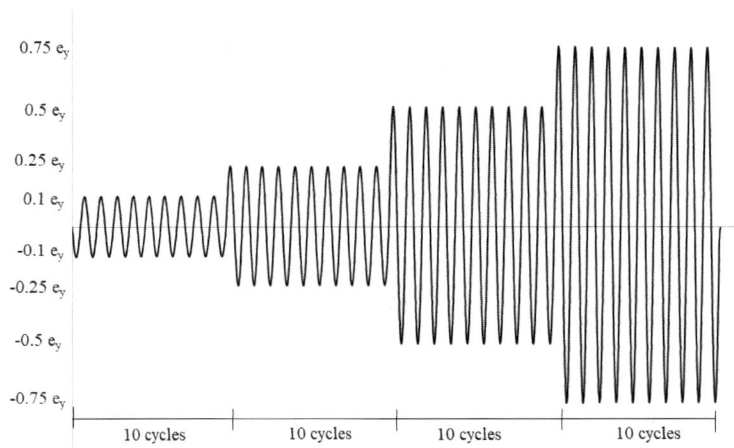

Figure 4. Cyclic amplitude

3. Results and discussion

The hysteretic loops of HS, CFHS–RHA–FC and CFHS–RHA–FC are shown in Figures 5–7. The hysteretic loops illustrate that the columns were elastic at the initial stage of the test. The tensile loading capacity was maintained after the peak for all samples. In the infilled hollow section columns, the strength dropped drastically after the compressive peak load. Under the tension side, the CFHS and the inclination insignificantly influenced the tension-resisting behaviour.

The hysteretic loop of the CFHS under compression was more stable than that of the unfilled columns. The compressive capacity gradually decreased after the peak compressive load. Equations 1–8 were used to discuss the strength and ductility of the sample.

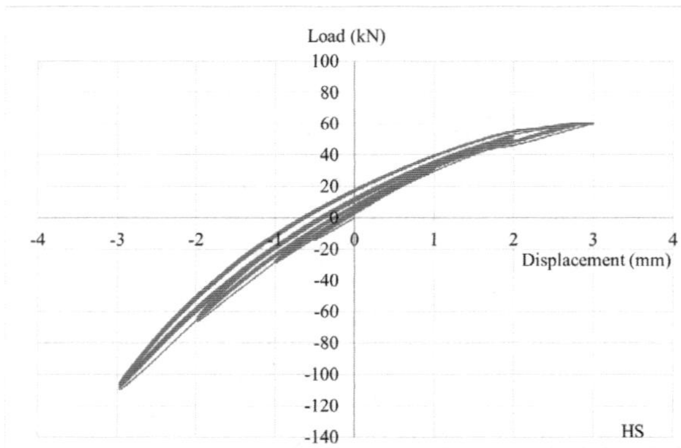

Figure 5 Hysteresis loop of hollow section (HS)

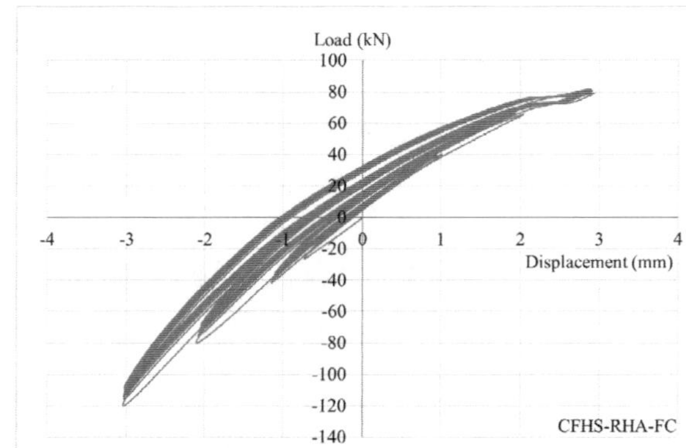

Figure 6 Hysteresis loop of CFHS–RHA–FC

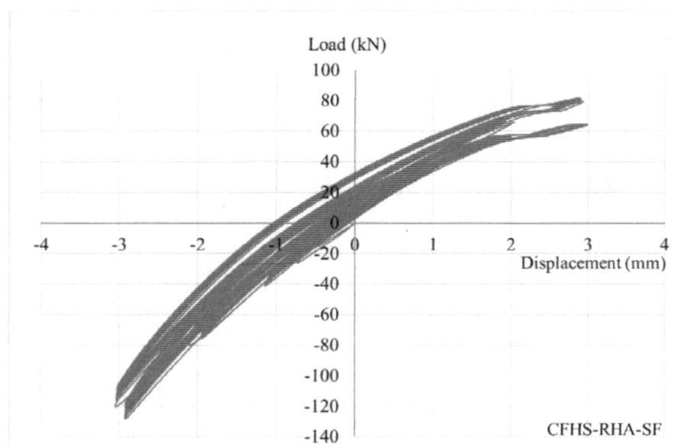

Figure 7. Hysteresis loop of CFHS–RHA–SF

One complete cycle of each sample as shown in Figure 8 at 3 mm displacement cycle was examined. The maximum tensile and compressive resistances occurred during the first cycle at each amplitude. The specimens exhibited high stiffness under compressive load. Furthermore, the modified foamed concrete exerted minimal influence on the shape of the hysteresis loop. From the graph, the resistance ratio and ductility of the sample were calculated by using Equations 1–8. Table 3 shows the resistance ratio and ductility of the sample under cyclic load. On the basis of the result of resistance ratio, the strength of the CFHS with steel fibre was smaller than that of the CFHS without steel fibre because the hysteresis loops of one cycle were small as shown in Figure 8. From the data, the resistance ratio of the CFHS without steel fibre was higher than that of the CFHS with steel fibre. However, the strength of the CFHS with steel fibre was slightly higher than that of the unfilled hollow section. The structure performance of the steel fibre was not achieved, which resulted in poor strength. This result indicates the small energy dissipation in the system is ineffective to maintain under cyclic load. Consequently, there is no failures occurred. Overall, the results indicated that the concrete infill did not greatly enhance the load per cycle but delayed the local buckling.

CFHS–RHA–SF also showed a better ductility behaviour than CFHS–RHA–FC and HS. The presence of steel fibre delayed the local buckling of the specimen. When the inward local buckling was prevented and delayed by concrete as an infill, the ductility of CFHS increased. The buckling effect was not clearly observed because the fibre can postpone the buckling effect of CFHS [11]. Usami et al. [12] concluded that the ductility of CFHS can be increased by filling the steel section with concrete, allowing the column to undergo large inelastic deformation during earthquakes.

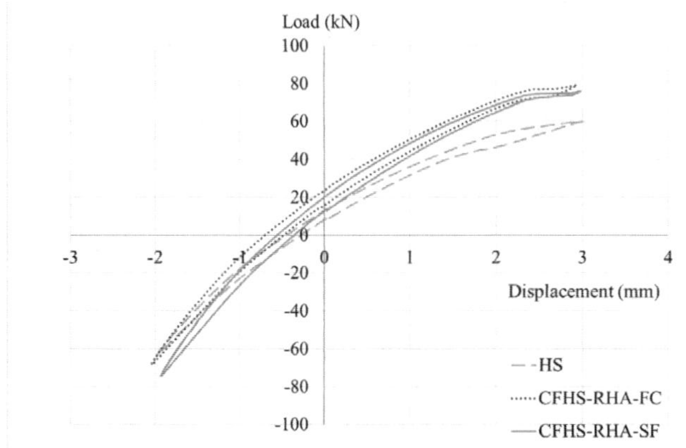

Figure 8. One cycle of hysteresis loop

Table 3. Resistance ratio and ductility of column under cyclic test.

Specimen	Resistance ratio		Partial ductility		Full ductility		Full ductility ratio	
	ε_i^+	ε_i^-	μ_{oi}^+	μ_{oi}^-	μ_i^+	μ_i^-	ψ_i^+	ψ_i^-
HS	0.157	−0.162	1.038	−0.692	1.055	−0.588	−1.614	0.899
CFHS–RHA–FC	0.213	−0.184	0.687	−0.477	0.848	−0.360	−1.074	0.456
CFHS–RHA–SF	0.167	−0.164	0.273	−0.176	0.327	−0.145	−0.362	0.160

4. Conclusion

Nine specimens of CFHS were prepared and tested under cyclic load. Results showed that the modified foamed concrete with SF as an infill demonstrated good performance because of its high compressive strength. SF also delayed the local buckling of the specimen. Thus, the high compressive strength on the modified foamed concrete delayed the local buckling behaviour and enhanced the strength of the foamed concrete with large load-carrying capacity. In addition, the CFHS with steel fibre showed better ductility than the foamed concrete and hollow section because of the inclusion of fibre in the structure. The inclusion of concrete infill with fibre not much improved the cyclic performance due to limited load applied. Therefore, Further investigation need to be conducted by applying the displacement control higher than e_y let say $2.0e_y$,$3.0e_y$.

5. References

[1] Giakoumelis, G and Lam D 2004 Axial capacity of circular concrete-filled tube columns *Journal of Constructional Steel* Research **60**(7) 1049–1068

[2] Bagherinejad K, Hosseinpour E and Hosseini S H 2015 Evaluation of Rectangular Concrete-Filled Steel-Hollow Section Beam-Columns *Journal of Asian Scientific Research* **5**(1) 46-54

[3] Antonius and Imran I 2012 Experimental Study of Con-fined Low, Medium and High-Strength Concrete Subjected to Concentric Compression *Journal of Engineering Science* **44**(3) 252–269

[4] Mugahed Amran Y H, Farzadnia N and Abang Ali A A 2015 Properties and application of foamed concrete: A review *Constr. Build. Mater.* **101** 990-1005

[5] Hunaiti, Y M 1997 Strength of composite sections with foamed and lightweight aggregate concrete *ASCE-Journal of Materials in Civil Engineering* **9**(2) 58-61

[6] Jaini Z M, Rum R H M and Boon K H 2017 Strength and fracture energy of foamed concrete incorporating rice husk ash and polypropylene mega-mesh 55. *IOP Conference Series: Materials Science and Engineering* Vol 248 (United Kingdom: IOP Publishing) pp. 12005

[7] Ge H and Usami T 1996 Cyclic Tests of Concrete-Filled Steel Box Columns *Journal of Structural Engineering* **122**(10) 1169–1177

[8] Broderick B M, Elghazouli A Y and Gonggins J 2004 Cyclic behaviour of hollow and filled axially-loaded members.*In the 13th World Conference on Earthquake Engineering* (Vancouver, Canada)

[10] Hansapinyo C, Buachart C and Chaimahawan P 2018 Tests of Inclined Concrete-Filled Steel Tubular Stub Columns under Vertical Cyclic Loading *Advances in Civil Engineering* **2018** 1–9

[11] European Convention on Construction Steel (ECCS) 1986 Recommended Testing Procedure for Assessing the Behaviour of Structural Steel Elements under Cyclic Loads *European Convention for Construction Steelwork* First Edition

[12] Khairuddin S A A, Rahman N A, Jamaluddin N, Jaini, Z M and Ali N 2017 Ultimate Strength Capacity of a Square Hollow Section Filled with Fibrous Foamed Concrete. *IOP Conference Series: Materials Science and Engineering* Vol 271 (United Kingdom 012103

[13] Usami T and Ge H B 1994 Ductility Of Concrete Filled Steel Box Columns Under Cyclic Loading *Journal of Structural Engineering ASCE* **120**(7) 2021-2040

Acknowledgments
The authors are grateful to the Ministry of Education Malaysia and Universiti Tun Hussein Malaysia for funding this research and providing the necessary facilities. This study was conducted under the grant scheme FRGS-1625.

A study on effectiveness of Building Information Modelling (BIM) on the Malaysian construction industry

S W R Kong[1], L T Lau[1], S Y Wong[1] and D T Phan[1]

[1] Department of Civil Engineering, FOSE, University of Nottingham Malaysia, Selangor, Malaysia
Corresponding author: duoc.phan@nottingham.edu.my

Abstract. The introduction of Building Information Modelling (BIM) in Malaysia has been fairly recent and is gaining popularity. Although most literature have been focused on the adoption and application of BIM in Malaysia construction industry, the effectiveness of BIM has not been investigated. This study aims to investigate the influence of BIM on the Malaysian construction industry. A qualitative approach was taken and interviews were conducted with professionals with prior knowledge in both the construction industry and BIM. Information obtained from the interviews was used to identify and categorise the determining factors. It was found that the application of BIM in Malaysian construction industry is somehow not fully effective in terms of time and cost, owing to the absence of standards in modelling and the frequent request for design changes. However, the results also show that BIM in Malaysia has the potential to be as effective as other developed countries if the major problems addressed being solved.

1. Introduction

The Architecture, Engineering and Construction (AEC) industry has been experiencing remarkable transformation, in which how building designs and information is created, portrayed and handled. Beginning with the use of two-dimensional (2D) drawings and text, it evolved into Computer-Aided Design (CAD) that enabled design process to be modelled in three-dimensional (3D) space [1]. However, CAD system only stores information such as building graphics and models that are not fully optimised for creating and handling nongraphic, as well as other information about the building. Therefore, development of an advanced processing system and software is necessary in order to utilise the benefits of information technology in the building industry. By integrating information technology into building graphics and models, Building Information Modelling (BIM) was introduced as the inevitable solution.

BIM was considered as the facility for 3D model when first introduced. The full capability of BIM, namely, digital, object-oriented and interoperable processes is far beyond what initially perceived, when used with modern communication technology. For instance, BIM has the potential to be modelled up to 6D with types of models of 2D drawings, 3D models, 4D time scheduling, 5D cost estimates and 6D facilities management [1]. The National BIM Standard-United States® (NBIMS-USTM) defined BIM as "a digital representation of physical and functional characteristics of a facility. As such, it plays the important role as a shared knowledge resource for information about a facility, which forms a reliable basis for decisions during its life cycle from inception onward" [2]. In addition, McGraw Hill Construction defined BIM as "the process of creating and using models for design, construction and/or operations of projects".

IOP Conf. Series: Materials Science and Engineering **713** (2020) 012035 doi:10.1088/1757-899X/713/1/012035

Many surveys and researches show the effort of using BIM worldwide. However, the challenges still exist since industrial players have certain reasons for not adopting BIM in their projects. For developed countries, it was reported that the lack of adequate training is the greatest challenge based on research conducted in 2008. Cost of software and required hardware upgrades were reported to be the second greatest challenge. Smart-Market Reports also showed that for the years of 2009, 2012 and 2014, the lack of demand is the top reason for not adopting BIM among non-users in Australia, New Zealand and North America. An Asian country like Singapore is also experiencing similar challenges [3].

In Malaysia, the government plays an important role in ensuring BIM being successfully adopted and implemented. In 2007, BIM was introduced by the Director of Public Works Department (PWD). The potential of BIM in improving project delivery and facilities management propelled PWD to adopt BIM in Malaysia. In the same year, a BIM committee was established to identify the appropriate BIM platforms to be implemented. As suggested by the committee, the platform should be Autodesk software, such as Revit Architectural, Revit Structural, Revit MEP, Navisworks and BIM 360 [4].

Private sectors in Malaysia have been the key driver in implementing BIM since 2009. Although government initiatives are taken to adopt BIM for building industry, some contractors on the property development and Construction Company in Malaysia still find that, "BIM is mainly a private initiative driven by the larger property developers and contractors". They also predicted this issue to go on because of the lack of government support for BIM adoption. In 2010, the construction of the National Cancer Institute (NCI) in Putrajaya became the first government project to implement BIM. The Malaysian government has also decided that BIM would be implemented in government projects by year 2016 [4]. However, evidences show that the rate of BIM adoption is still behind. A survey was conducted during a recent seminar organized by CIDB to investigate the adoption and awareness of BIM among contractors in Malaysia. Results showed that the percentage of BIM adoption was as low as 5.2%, while only 52.6% of the respondents were aware of BIM [5].

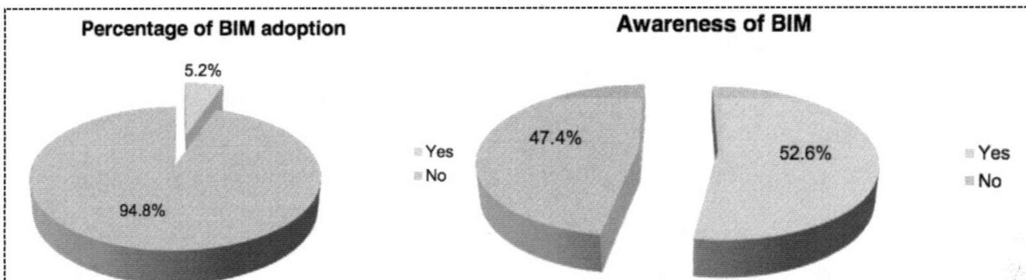

Figure 1. BIM Adoption and Awareness among Contractors in Malaysia [5]

The application of BIM can be divided into three phases, i.e. pre-construction phase, construction phase and post-construction phase. BIM is applied in the pre-construction phase by providing early collaboration, which is efficient platform for sharing information and estimate accurate cost and materials take-offs based on the 3D model. For the construction phase, BIM can be used to track cash flow and work progress in real time, as well as keeping progress on schedule, reduce rework and avoid cost overrun. Post-construction benefits from BIM are better facility management, operations and maintenance works.

This paper aims at conducting an in-depth investigation to discover whether the application of BIM is really effective in Malaysia. In addition to the interview methodology used as qualitative approach, this paper critically reviews the scale of BIM's implementation in some countries around the world. The application in the construction industry and the challenges in adopting BIM are also considered. The findings of this study would provide useful information for stakeholders in the Malaysian construction industry to identify and discover the ways of improving BIM's effectiveness. In the long term, this study could certainly encourage its adoption and contribute to the advancement of BIM technology in Malaysia.

2. Qualitative approach for assessing effectiveness of BIM

A research on BIM can be taken either a qualitative or quantitative approach. Selecting the appropriate research strategy depends on the objectives of the study and the approaches to data collection. Quantitative approach is objective in nature and involves using variables to justify a theory or hypothesis. On the contrary, qualitative approach is more subjective in nature and emphasises on opinions, views, and experiences etc., which are often verbally expressed. A qualitative approach can be further classified under two categories, which are exploratory research and attitudinal research [6].

As mentioned in previous section, the primary aim of this research is to conduct an in-depth investigation to discover the effectiveness of BIM in the Malaysian construction industry. Due to the nature of this research, a qualitative approach was taken in order to achieve the intended aim and objectives. This approach was chosen over a quantitative approach because this research is exploratory and very subjective in nature. Many literature written and research studies conducted are focused on the adoption, implementation and challenges of BIM in Malaysia. There is limited knowledge and evidences on the effectiveness of BIM in the Malaysian context. Hence, an exploratory research was chosen, i.e. interview questionnaire forms to explore and provide a better understanding in this area of study. This approach has the potential to develop new products and shed light upon poorly understood scenarios or identify possible business opportunities.

The technique used for data collection in this research is personal interview with relevant professionals currently practicing BIM in the industry (see Table 1). Interview forms can be divided in three categories, namely structured, semi-structured and unstructured [5].

Table 1. Interviewees' profiles

Interviewees'	A1	A2	A3	A4	A5	A6
Company background	Contractor	Project Management-Construction	Turn Key Builder	Turn Key Builder	C&S Consultant	M&E Consultant
Position	BIM Engineer	BIM Engineer	Manager	Senior Engineer	Executive Engineer	BIM Manager
Years of experience in the industry	4	5	14	5	9	6
Years of experience in BIM	4	5	4	1	1	4
Number of projects involved using BIM	8	5	2	2	1	12

The most suitable form of interview for this research is semi-structured interview due to its unique characteristics [7]:

- It is carried out with individuals with known prior experience
- The interview process is based on a handout stating questions about the relevant topics in the research area
- The individual's experiences in the relevant areas being studied is the main focus

The potential candidates to be interviewed were screened through personal contacts, friends and LinkedIn webpage, the world's largest professional network with more than 400 million users. After that, nine candidates were selected and letters of invitation to participate in this research were sent. However, only six candidates, namely A1 to A6, replied and agreed to participate in this research (Table 1).

The duration of each interview session was approximately 45 to 60 minutes and they were conducted at various locations, where the interviewees feel most convenient. The consent form and interview questionnaire were sent to each interviewee prior to the interview process. A total of 13 'open' and 'closed-ended' questions were used during the each interview session with the intention to allow interviewees to share their experiences and express their opinions freely. All interviewees accepted to be audio recorded for the interview session. The interview questionnaire was developed based on three determining factors of BIM effectiveness, namely time, cost and design.

2.1. Data Confidentiality

Due to the sensitivity of some information revealed by the interviewees such as comments, criticisms and personal opinions, data confidentiality needs to be upheld within the degree of consent given by each interviewee respectively. Each interviewee was requested to select the options as follows:

- Remain anonymous and not be identified with the information provided, or
- Be identified with and given credit for the information provided.

Interviewees who chose the option to be anonymous were assigned a code, e.g. A1. Interviewees who preferred to be identified and given credit were addressed by name in this study. Since all interview sessions were audio recorded, data confidentiality was upheld by ensuring all audio recordings that were well safeguarded and not circulated. Audio recordings were necessary to ensure the accurate representation of the information, which provided by interviewees. A written consent form was presented before each interview session to be signed by both the interviewee and researcher. Thereafter, each party keeps a copy of the form.

2.2. Data Credibility

In qualitative research, the sample size is often questioned in ensuring the credibility of the data obtained. Research showed that the number of interviews for phenomenology study should be in between five and twenty-five interviews. Since BIM is considerably new to the Malaysian construction industry, there are very few companies that have implemented BIM in their projects. Also, there is only a small pool of experienced BIM users in the industry and not many would be willing to participate in this research. Hence, it is a challenge to conduct more interviews. Nevertheless, the author believes that the sample size of six interviews is adequate. Moreover, a qualitative research should focus more towards analytic generalisation instead of statistical generalisation. Besides, the interviewees represent a diverse selection of industry players. This ensures that the data collected is credible because it considers different perspectives and approaches towards BIM. Among the six interviewees, two are clients, two are contractors and two are consultants representing the Civil & Structural (C&S) and Mechanical & Electrical (M&E) disciplines.

2.3. Data Analysis and Interpretation

It is a challenge to interpret collected data in qualitative approach because they consist mainly of words. Words can carry a lot of different meanings and can be interpreted differently. Hence, the key in qualitative data analysis is "to make sense of massive amounts of data, reduce the volume of information, identify significant patterns, and construct a framework for communicating the essence of what the data reveals" [8]. The method used to analyse the qualitative data in this research is as follows:

- Transcribe all recorded interview sessions
- Identify and categorise the determining factors based on information obtained from the interviews
- Formulate general findings by identifying common patterns and comparing differences in BIM practices

2.4. Limitation

It is almost impossible to completely eliminate limitations in a research. In fact, limitations can determine the extensiveness of a research and sometimes affect the conclusion that can be drawn. There are several limitations in this research, which are time allocation, sample size and reluctance to share.

The time allocated to conduct this research was limited to seven months and works have to be completed within this period. Personal interviews require the recorded sessions to be transcribed, which is time consuming. Also, time is needed to contact, schedule and interview candidates, which are heavily dependent upon factors such as the interviewee's flexibility and the distance of travel.

The sample size of data collected is small because of limited number of companies and professionals who have prior experience in BIM. Also, bigger sample size would make the obtained data harder to be analysed. Since data is collected through personal interview, it would incur higher cost due to traveling to meet the interviewees. They may be reluctant to share certain information or knowledge due to various reasons, namely, job security, commercial value, and lack of incentives.

3. Results and Discussion

3.1. Time

Theoretically, BIM is said to be effective in saving time if it helps to speed up project duration, improve the response time to design changes or even save time through automated processes. Through the interviews conducted, there is no clear evidence that BIM helps to speed up project duration. Most of the interviewees as shown in Table 1 mentioned that more time is required at the design stage in order to save time at the construction phase. Some processes save time while others consume more time instead. However, the overall project duration doesn't differ much compared to conventional methods.

In terms of project duration, A2 mentioned that "BIM can help reduce project duration only in a small margin and unless the industry is BIM-ready, the margin will be insignificant." In fact, A2 also points out that the total project duration could possibly be longer because of two main reasons. Firstly, BIM involves working with other disciplines and secondly, some of the consultants work with only know basic modelling, hence are not able to keep up with the standards. As a result, more time is needed to recheck and ensure the model is in accordance to the standards.

On another note, four out of six interviewees have similar opinions that BIM processes involves spending more time at the design stage because of a higher level of detail and consideration compared to conventional methods. A2 revealed that working on a 3D model would prolong the documentation process because more thought is put into various parameters such as elevations, height and coordination. Conventionally, they would only draw the intention of the pipe routing and not the exact pipe routing with parameters. The drawing is a formal sketch, which will be used as tender documents, and the routing is supposed to be coordinated on site. A3 also highlighted that "this is especially true for MEP works and more time is spent on running clash detection using Navisworks." Because of the longer time spent at the initial stage, A3, A2 and A6 had a common opinion that this would save time on site because of lesser design errors, lesser rework and faster installation process, hence reducing delays in the construction phase.

In the developed countries, it is experienced that BIM is widely used because of several benefits, which one of them is the ability to reduce project duration. A research conducted in Australia and New Zealand proved that having a project team with BIM expertise could reduce overall project duration. Another research conducted in US and UK reflected that the most significant benefit of using BIM is to reduce construction time [9]. However, the findings in Malaysia do not follow the trend as that in other countries. Interviewees found that BIM does not significantly reduce the overall project duration. One of the main reasons is because that BIM is still considerably new to the Malaysian construction industry and many construction companies are not BIM-proficient, especially C&S and M&E consultants. The lack of BIM expertise can lengthen many BIM processes such as drawings production, modelling, coordination and clash detections, hence affecting the overall project duration. In addition, the construction industry involves many parties working together and therefore much depends on each other to deliver a project on time. The saying "A chain is only as strong as its weakest link" is particularly applicable in this scenario where the inefficiency or weakness of just one party can cause the entire project to slow down even though the other parties are efficient. Hence, there is increasing importance on having project team members with BIM expertise.

Eastman et al. described the effectiveness of BIM in responding faster to design changes [10]. Even for BIM projects, the change in design process is inevitable but they can be managed more efficiently [11]. Findings support in the literature addressed, but also explained that design changes by clients can cause projects to suffer the delays, even with BIM. One of the causes of construction delay in Malaysia is because of too many design changes ordered by clients [12]. Shourangiz et al. [11] reported that the level of frequency of design changes for BIM projects in Malaysia is very high; 67% always, 33% sometimes and 0% never. Hence, although BIM has the flexibility and the potential to respond quicker to design changes, it should not be exploited freely. The focus should not be on how to be more flexible or respond quicker, but the root of the problem may be due to indecisive clients that impose numerous design changes throughout the construction project, sometime even up to the point of construction. The later the design change is made, the higher the impact on the project duration. The MacLeamy curve, as shown in Figure 2, illustrates the preferred design process in a construction project [13].

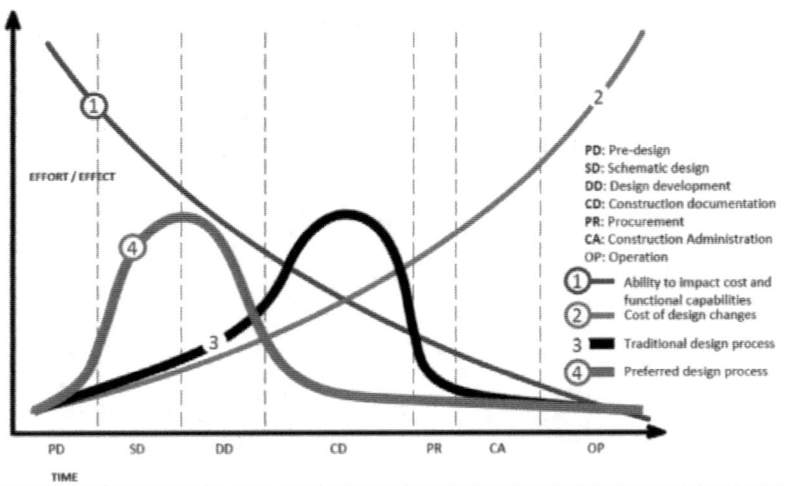

Figure 2. Frequency of Design Changes in BIM Projects [13]

The optimum scenario will be to make as many amendments to the design and draw to a conclusive design at the end of the schematic design stage. Design development stage would involve the coordination with multiple disciplines that is time consuming and also requires the highest amount of effort. Making a design change after this stage, depending on the degree of the change means that efforts made before could become redundant and unappreciated. Therefore, coordination is crucial prior to construction process. This can greatly affect the effectiveness of BIM in reducing the overall project duration if the industry does not comprehend the consequences of last minute design changes.
Companies using BIM in Malaysia have already experienced changes in terms of project timeline. Findings show that projects using BIM involves higher effort and more time in the design phase but less of both in the construction phase. The findings were in accordance to the MacLeamy Curve, as shown in Figure 4. This outcome is similar to what being practiced in other countries. This would naturally happen because BIM processes increase the level of effort and prolong the design phase. As a result of better planning and coordination, delays caused by clashes, change orders (COs) and requests for information (RFIs) can be eliminated as much as possible during the construction phase to help the project stay on schedule.

From these observations, it is clear that BIM has not been effective in reducing overall project duration in Malaysia. The current situation indicates that BIM has the potential to conserve time in some ways but at the same time being hampered by delays in other related aspects. BIM helps save time through automated calculations, drawings production, reduced design errors, lesser rework and faster installation process on site. On the other hand, BIM consumes more time on documentation, modelling, coordination and in the event of a late design change.

3.2. Cost

Cost seems to be one of the main barriers for BIM implementation in Malaysia. CREAM documented that it is a challenge for SME's in Malaysia to implement BIM because of high investment in hardware, software, staff training, employing staffs with BIM expertise, obtaining certifications and licenses as well as extra overhead cost. In fact, having BIM capabilities does not guarantee that SME's will get projects. Since the initial capital spent on implementing BIM is high, companies are uncertain on the possibility of recovering ROI [4]. These findings agree to that mentioned in the literature so it can be deduced that the same challenges can be generalised for most companies in Malaysia.

Despite the high cost, some companies are willing to take the risk because of the potential benefits they believe to gain out of BIM. Based on a research conducted in North America, key findings reveal that two-thirds of BIM users are experiencing positive ROI and 93% of BIM users believe there is potential to gain more value in the future [14]. However, this is not reflected in the findings of this study. Most of the interviewees can only say that it is a long-term investment. This can imply that they are not generating positive ROI. It can be assumed that leading companies in Malaysia using BIM generally suffer losses or only break-even at initial years but they gain experience and learn lessons along the way. Their perspective is that, investing into BIM now will reap greater benefits in the future. None of the interviewees could provide figures of their company's ROI and this is possibly because of two reasons. Firstly, companies may not have an acceptable method of calculating ROI in BIM and secondly, the information is confidential and could not be revealed at this point of time.

Findings in previous section of this report also show that the industry is generally not BIM-prepared and the Malaysian construction industry is lacking expert users in BIM. As a result, this could be one of the possible reasons why Malaysia is not experiencing positive ROI. McGraw Hill Construction [14] reported that only 38% of beginners see positive ROI with BIM compared to 87% of expert users, as tabulated in Figure 3. Hence, as more users gain experience and proficiency, it can be predicted that their ROI will improve over time.

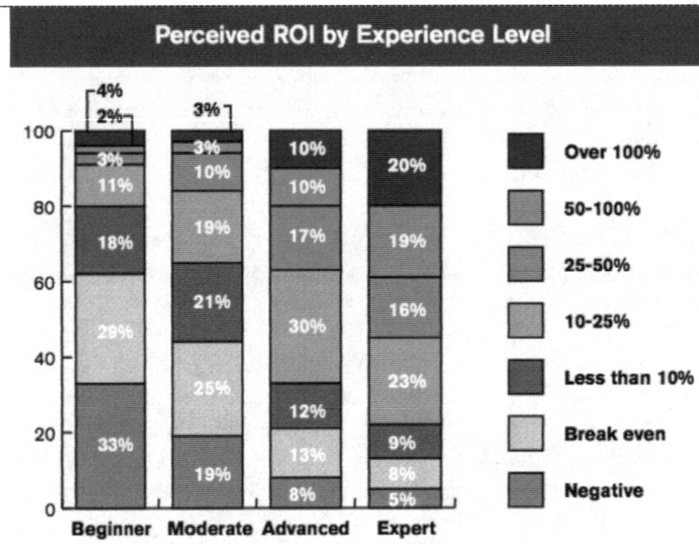

Figure 3. Perceived ROI by Experience Level [14]

Besides that, findings also show that BIM is seen as a way to stay ahead in the game and gain projects. Research reveals that some of the top business benefits of BIM are marketing new business to new clients, proposing new services with BIM and retaining repeat business with past clients [14]. As more clients discover the benefits of BIM, there will be more requests for BIM to be implemented in their future projects. Companies with BIM expertise will have an advantage and a head start in the industry. As can be seen, BIM in Malaysia has not been effective in generating ROI to recover the high implementation cost in comparison to other developed countries such as North America. Broadly speaking, BIM is currently not cost-effective in Malaysia. Since BIM is slowly gaining popularity in Malaysia, the scenario is expected to change and improve drastically in the coming decade. Because of the high initial capital needed to implement BIM, not many companies are willing to venture into BIM at this point of time.

IOP Conf. Series: Materials Science and Engineering **713** (2020) 012035 doi:10.1088/1757-899X/713/1/012035

3.3. Design

Industry players in Malaysia have benefited from some of the design benefits offered in BIM. Findings show that the ability to visualise building models in 3D helps to improve the quality of building designs. Adam Meeker, Senior Architect / Associate at FTC&H testified that the use of 3D models helps with faster and earlier decision making, improves collaboration and coordination between disciplines, and lesser clashes among systems. In the former days, the skill of visualising drawings in 2D and interpreting them comes with many years of working experience in the industry. Visualisation in 3D has assisted the younger generation of professionals, especially fresh graduates with little or even no experience in the industry to catch up and better understand the design model they are working on. McGraw Hill Construction offers supporting evidence, where clients in the UK and US acknowledges visualisation in BIM as a potent benefit because it provides a better understanding of the proposed design.

Autodesk, Inc. recorded customer success stories about how BIM tools have been utilised to help coordinate, identify and resolve design clashes before starting the construction. BIM tools have provided the suitable platform for the transition from conventional methods of overlaying 2D CAD to incorporating building models across disciplines into a single building model for coordination purposes. This process could be carried out in advance, during the design development phase. Although there have not been any recorded success stories published from Malaysia, findings proved that companies engaging BIM have been gaining design benefits in some ways. For Malaysian scenario, coordination of building designs between disciplines should be taken a step forward in future.

In addition, findings also show that there are struggles in sharing building models and integrating models from different disciplines on a single building model over a cloud server. A study showed that the reluctance of architects to share their model is because of liability reasons. As a result, contractors have opted to creating their own model for construction purposes. This is because architects and contractors may have completely different ways or standards of modelling. The current scenario in Malaysia is very much similar to the literature mentioned, in which sharing of models do not take place, every party creates their own model individually to suit their respective needs and coordination is only done over technical meetings between parties involved. The effectiveness of this working practice remains arguable because duplication of efforts exists among project teams. Cloud services hosted by Autodesk provide a platform for collaboration in the cloud and 3D models can be updated in real time by project teams. Nevertheless, the effectiveness of BIM cloud services in Malaysia is also very vague because of the working practices and lack of modelling standards.

Therefore, the key findings reveal that BIM tools generally have been effective in Malaysia. A step forward from 2D CAD to 3D modelling helped improve visualisation and coordination as well as detect clashes and design errors. Despite several challenges in design practices, BIM processes are able to make a positive impact to the industry. As more and more industry players catch up with BIM, the effectiveness of BIM in this area is expected to remarkably improve. There are still a lot of design benefits to explore in the days to come.

4. Conclusion

This study successfully investigated the effectiveness of BIM in the Malaysian construction industry by conducting a qualitative approach and through interviewing professionals. The objectives laid out in the introduction section were also well achieved. Based on the findings, BIM is effective in design but not the same in the areas of time and cost. Therefore, it could be concluded that BIM application has generally not been effective in Malaysia. However, it is proven to have good effectiveness in other countries which have implemented BIM for a longer period of time. It could be deduced that in a long term, the effectiveness of BIM in Malaysia has the potential to be on par with those countries. So, the problems addressed in the current implementation of BIM need to be solved through the cooperation of government bodies and industry players. The government can also explore or consider the option of making BIM mandatory for building projects, since this would greatly improve the adoption of BIM and its effectiveness. Trial and error, exposure and experience with BIM will equip professionals with the skills to excel in BIM.

IOP Conf. Series: Materials Science and Engineering **713** (2020) 012035 doi:10.1088/1757-899X/713/1/012035

5. References

[1] Storer D 2012 *Six Dimensional Building Information Modelling* Retrieved on March 8th, 2016 from http://www.hjhigh.com/news-and-media/market-trends/six-dimensional-building-information-modeling/

[2] National Institute of Building Sciences 2015 *National BIM Standard-United States®* Retrieved on March 4th, 2016 from https://www.nationalbimstandard.org/files/NBIMS-US_FactSheet_2015.pdf

[3] Zahrizan Z, Ali N, Haron A, Marshall P A and Hamid Z 2013 Exploring the adoption of building information modelling (BIM) in The Malaysian construction industry: A Qualitative approach. *International Journal of Research in Engineering and Technology* **02(08)** 384-395

[4] Construction Research Institute of Malaysia (CREAM) 2014 *Issues and Challenges in Implementing Building Information Modeling (BIM) for SME's in the Construction Industry.* Retrieved on April 2nd, 2016 from https://www.cidb.gov.my/cidbv4/images/pdf/announcement/BIM/bim%20seminar%20%20workshop%20for%20malaysia%20construction%20industry.pdf

[5] Construction Industry Development Board (CIDB) 2014 *Contractor's Acceptance of Building Information Modeling (BIM) Towards Improvement of Project Performance and Profitability* Retrieved on April 2nd, 2016 from https://www.cidb.gov.my/cidbv4/images/pdf/final%20bim%20for%20contra ctors2.pdf

[6] Naoum S 2013 *Dissertation research & writing for construction students*, 3rd ed. (New York: Routledge) pp 10-15

[7] Merton R and Kendall P 1946 The Focused Interview *American Journal of Sociology* **51(6)** 541-557

[8] Patton M 1990 *Qualitative evaluation and research methods* (Beverly Hills, CA: Sage) pp 169-186

[9] Yan H and Damian P 2008 Benefits and Barriers of Building Information Modelling *Proc. of 12th International Conference on Computing in Civil and Building Engineering* (Beijing: International Conference on Information Technology in Construction (INCITE 2008)) pp1-4

[10] Eastman C, Teicholz P, Sacks R and Liston K 2011. *BIM Handbook 2nd ed* (Hoboken, New Jersey: John Wiley & Sons, Inc.) pp 10-25

[11] Shourangiz E, Mohamad M, Hassanabadi M, Banihashemi S, Bakhtiari M and Torabi M 2011 Flexibility of BIM towards Design Change *Proc. of 2nd International Conference on Construction and Project Management* (Singapore: IACSIT Press) pp 79-83

[12] Hamzah N, Khoiry M, Arshad I, Badaruzzaman W and Tawil N 2012 Identification of the Causes of Construction Delay in Malaysia *International Journal of Civil and Environmental Enineering* **6(12)** 1063-1068

[13] The Division 4 Triclinium 2016 *Of the MacLeamy Curve, Efficient Design, and Expensive Money-Or Why Developers are a Breed Apart!* Retrieved on March 8th, 2016 from http://division4triclinium.blogspot.com/2013/06/of-macleamy-curve-efficient-design-and.html

[14] McGraw Hill Construction 2009 *The Business Value of BIM* Retrieved on March 8th, 2016 from https://www.trane.com/content/dam/Trane/Commercial/global/markets/Architect/ building-information modeling/SMR%20BIM%2009%20FINAL%20rev.pdf

IOP Conf. Series: Materials Science and Engineering **713** (2020) 012036 doi:10.1088/1757-899X/713/1/012036

Factors Affecting Safety Performance of Construction Projects: A Literature Review

N H Abas[1], N Yusuf[2], N A Suhaini[1], N Kariya[1], H Mohammad[2] and M F Hasmori[1]

[1] IBS Cluster, Jamilus Research Centre, Faculty of Civil and Environmental Engineering, Universiti Tun Hussein Onn Malaysia (UTHM), 86400 Batu Pahat, Johor, Malaysia
[2] Site Office Saloma Link, Level 1, Lot 2400, Jalan Raja Muda Musa, 50300 Kuala Lumpur, Malaysia
[3] Centre for Diploma Studies, Universiti Tun Hussein Onn Malaysia, Hab Pendidikan Tinggi Pagoh, KM1, Jalan Panchor, 86400 Pagoh, Muar, Johor, Malaysia.
Corresponding author: nhaslin@uthm.edu.my

Abstract. The construction industry is known as the most hazardous industry and has always been plagued with accident for a long time. Incidents or accidents that happened at construction sites has caused several drawbacks of project performance, such as delay in project completion, increase project cost, reduce productivity and create negative impressions to the organization. Therefore, it is crucial for employer to ensure safety and health at the workplace is taken care of to avoid accident. This paper presents the review of the factors affecting safety performance of construction project, focusing on project level. Among the most frequently cited factors that affect safety performance of construction project in project level are safety training, implementation of safe working environment and implementation of safe plant and equipment. The outcomes of the paper may serve as a starting point for further research in investigating the main.

1. Introduction

In many countries in the world, the construction industry plays a big role for development process which contributes towards the economic growth that generating additional demands for construction activities. Despite its significant contribution towards economic growth of the country, construction industry has also contributed to large fatality rates due to high number of accidents [1]. Construction workers who involved directly in construction activities faced a greater risk of fatality compared to the workers in other industries, as proven by accident statistics [2]. In Malaysia, the Department of Occupational Safety and Health (DOSH) [3] revealed that the highest number of occupational accidents by sector in 2017 is in construction industry, with 15 deaths from out of 70 (as in May 2017). Several studies have revealed that, there has been rapid growth in the number of accidents related to occupational safety and health, which cause injury and even death during the last five years [4-6].

Major construction projects in Malaysia are exposed to dangerous conditions which can cause accidents and consequently affect the safety performance of the construction projects. The causes of accidents in construction industry are related to unique nature of the industry, human behavior, improper site conditions, unsafe work method, equipment and procedures which affected from poor safety management [7]. These causes could result in accidents which further cause disruption of work and decrease the work rate [8]. Therefore, it is crucial for every employer to provide training and the comprehensive safety programs [7] which can improve safety performance continuously to reduce potential hazard in construction project.

Some researchers suggest that measuring contractors' own safety performance is important to ensure their organizations are aware about their safety well-being. A study by Hinze and Gambatese [9] concluded that contractor's safety performance was consistently influenced in part by a number of factors. In addition, the quality of project contractors is even more important because it has a direct bearing on the performance of the prime contractors on key elements of the work [10]. Improving OHS performance in the construction industry is crucial because it represents the excellence of the executed projects, and more importantly the protection of life for people who work in the field [11-14]. Traditionally, safety performance was measured by the goal-setting and feedback method (known as lagging indicators) such as Incident Rate (IR), Accident Rate (AR) and Experience Modification Rate (EMR). In recent years, leading indicators have been highlighted as the method to evaluate safety performance of the construction project [15]. Safety performance measurement tool has been developed in other countries, for example 'Safety Meter' in Australia. The tool is based on leading indicators that aim to involve the workforce in the method of measurement and raise site safety awareness [16].

In Malaysia, it is believed that the study on safety performance measurement is still lacking. Eventhough there exist a tool named SHASSIC developed by Construction Industry Development Board (CIDB) to measure the performance of safety and health of construction industry, the indicators or factors underpinning the tool are not comprehensive [4]. The indicators or factors in SHASSIC are more organisation-wise, such as OSH policy and OSH organisation, with the lacking in project-wise indicators, such as the implementation of audit and safety personnel. Therefore, there is a gap in existing study need to be addressed towards a more comprehensive safety performance measurement framework which takes into account factors pertinent to an organisation and its projects. As a stepping stone for the framework development process, this study seeks to identify the factors affecting safety performance in construction project, pertinent to project levels.

2. Literature Review

2.1. Construction safety management

Construction safety management is a method of controlling safety policies, procedures and practices on construction site [17]. According to Cheng et al. [18], the process involve from dynamic by small or large adjustment made to site environment to normal business. Moreover, every level of company or part of a cross-organization project should apply safety as management concept. The construction management should consider a safety culture and safety climate is important aspect should be concerned [19].

Occupational safety and health management on construction site has unique of challenges. Despite to the challenges, the incidents can be reduce through well-structured and well-funded by safety programmes and techniques applied from firms [26]. In addition, safety management should be often adjusted due to industrial needs. Hallowell and Gambatese [26] stated that most of company allocates limited resources for safety management and contractors are forced to carefully select from the available elements. Therefore, good safety procedures reflect in maximizing the safety performance of construction safety in site. According to Jaselskis et al. [27], construction safety management techniques significantly improve by following the safety related act which highlights the responsibility for construction safety on employers and increase in safety planning also in management efforts on construction industry. Improving safety performance in the construction industry is crucial because it represents not just the excellence of the executed projects, but more importantly the protection of life for people who work in the field [4].

2.2. Safety performance of construction industry

Generally, safety performance is often evaluated based on negative indicators such as accident number, days lost, injury rates and accident cost. The importance of evaluating safety performance is to gauge the effectiveness of construction firms' management in accident prevention by setting out safety objectives and targets [28]. When a project experiences high accident rates, the contractor involved can develop a poor reputation, though the size of the company does contributes to the safety performance of construction contractors [17, 23]. Therefore, both large and small contractors have an obligation to maintain their reputation in health and safety by maintaining their safety performance.

Abas [4] quoted:

"Jaselskis et al. [21] provided strategies for achieving better construction safety performance and project level, and emphasized that safety performance measurement is most effective when using both 'quantitative and qualitative' safety measurements. Whilst quantitative measures include lost time, severity rates, and Experience Modification Rating (EMR) – a method used to calculate insurance premiums for workers due to accidents; qualitative measures consist of outstanding, average and below-average performances, as determined by OHS assessors. In their study, Jaselskis et al., [21] presented numerical profiles of companies and projects with verifying levels of safety performance at the company and project level, and recommended- through their findings, that there are several factors that were significant to achieve better safety performance at both levels. For company level, such factors include upper management support, time devotes to safety issues for the company safety coordinator, number of safety inspections, meeting with the fields safety representatives and craft workers etc; whereas for project level, some important factors include increased project manager experience level, more supportive upper management attitude towards safety, reduced project team turnover, increase time devoted to safety for the project safety representative etc. [21]. This is supported by Jafri et al [24] who proposed the evaluation of safety performance to be evaluated based on the leading indicators and outcome (lagging) indicators."

Table 1 represents the several authors' perspectives on safety indicators while measuring the safety performance. Safety performance is deemed as the effectiveness of leading indicators, and the impact of those indicators to safety performance will determine the outcome [24]. By evaluating safety performance using leading indicators and observing the outcome using lagging indicators, the potential hazards can be identified at an early stage and therefore unnecessary losses in life and cost can be avoided [25]. Furthermore, current strength and weakness can be identified by continuously auditing and reviewing the management systems and operational practices in order to achieve high safety performance [25]. For instance, in Australia implementation of performance measurement by using safety meter was developed. This tool is based on leading indicators that aim to involve the workforce in the method of measurement and raise site safety awareness [16]. Thus, measuring leading indicators and get the result can help to improve safety.

Table 1 shows some of the perceived safety indicators while measuring safety performance. It is to be highlight of that the Construction Industry Development Board (CIDB)'s indicators are extracted based on government documents, not from sources of research.

Table 1. Leading and lagging indicators when measuring safety performance.

Authors	Leading indicator	Lagging indicator
Dyck D, Roithmayr T [15]	Regular safety audit conducted, good injury record keeping.	Number of accidents at work or accidents frequency rates, workers' cost compensations, number of days of absence due to accidents at work, number of occupational diseases etc.
Salminen et al, [26]	Safety audits, toolbox meeting, owner promotion, pre-task planning, drug test, number of close calls, worker observation records.	Accident cost analysis, injury/incident rates.
Jafri et al. [24]	Safety culture, behavior-based indicator, management involvement, employee empowerment, employee perception survey, safety audits, root cause analysis.	Accidents number, injury rates, days lost, accidents costs.
Lin & Mills [23]	Health and safety systems, managements responsibility, contract review, design control, work method control, purchasing, inspection and testing, corrective and prevention action,	Severity rates, lost time injury

IOP Conf. Series: Materials Science and Engineering **713** (2020) 012036 doi:10.1088/1757-899X/713/1/012036

	control on non-conformance, health and safety records, health and safety auditing, statistical techniques, training	
CIDB [27]	OSH organization, OSH policy, Risk Assessment and Risk Control (HIRARC), Hazard Identification, machinery management, training and promotion, material management, accident investigation and reporting, Emergency Response Plan (ERP), record management and performance monitoring	N/A

2.3. Factors affecting safety performance of construction project

According to Thomas et al. [22], there are two levels which affecting safety performance in construction industry which are 'project level' and 'organizational level'. The factors that affect safety performance of construction project are discussed in the next section, pertinent to these two categories.

2.3.1. Project level. The project level is known as the owner's decision to divide the work program with several inter-firm organizations [22]. Safety on construction project is related with historical factors which include economy, psychology, technical, procedure, organization and environmental work issues [14]. Ng et al. [22], in an attempt to develop the framework for safety performance evaluation, has summarized the factors that affect construction project safety performance on project level, as below (Note that the summary of factors are supported by relevant literatures):

i) *Implementation of safety inspection* - Safety inspection is the main tool to maintain safe condition on site and monitor the unsafe practices at workplace. Permana [28] affirmed that safety inspection is very essential to safety practices based on the study he did in Batam, Indonesia, which is also supported by Kai et al [29] and Tan and Razak [30].
ii) *Implementation of safe system of work* – According to Ismail [31], any safety program should be based on a policy insisting on the safety protection and safe system of work to the employees. Apart from that, serious enforcement of the written policy has to be made especially for high rise building project Khairolden et al. [32]. It is because all the contractors and workers can aware to the equipment and structures at workplace that hazardous to themselves or even public.
iii) *Implementation of safe plant & equipment* - PPE is worn by a person at work to protect him against risk to safety and health and any additional accessory designed to protect him while performing task. It is important to provide PPE to workers at construction sites in order ensure a safe and healthy condition at the workplace [33-34].
iv) *Implementation of safe working environment* - Bakri et al. [36] highlighted that providing a safe and healthy workplace as the most effective strategies to maintain the cost of construction project. Failure to adhere to safe working environment implementation could result in accident that may increase the project cost.
v) *Implementation of safety officer & supervisor* – In order to ensure better construction safety, an employer should appoint full-time safety and health officers (SHO) and site safety supervisors (SSS) with specific jobs such as conducting site safety tours and inspections [37, 38]. The appointment of both SHO and SSS is depend on the project scope that includes project value, in accordance with the legislations. This is supported by Jannadi [39] and Lee and Halpin [40] that site safety supervision and the provision of a safe environment are the factors related to the implementation of safety systems at construction site.
vi) *Safety review for safety audit* - A safety audit includes safety inspections, inspection of documents and interviews [41]. Safety audit is important to check the organization's level of compliance with OSH legislations, thus provide a safe workplace.
vii) *Safety review for site safety policy review* - Griffith and Howarth [42], defined the safety policy as a published statement reflecting the organization's vision and mission in relation to the management of health and safety matters.

viii) Emergency plan and procedures - emergency drill should be conducted once in every three months which all the workers will be given a briefing on emergency procedures [42]. Attendance of the workers at the designated assembly area should be also recorded.

ix) Safety training - According Paringga [35], education and training designed to prevent human error that may cause the accidents and to enable workers to perform a repetitive task with skill. In addition, Lai et al., [43] also stated that safety training is the most effective tool to mitigate hazards since training helps to improve worker's skills and abilities to identify hazards. Tsui and Gomez-Mejia [44] also stated that one way to encourage employee safety is to involve all employees at various times in safety training through regular sessions with supervisors, managers, and employees which should often coordinated by HR staff members. The statement also supported by other scholars such as Permana [28], Kai et al [29] and Tan and Razak [30].

Table 2 depicts the summary of the factors affecting safety performance in construction project pertinent to project level. It could be seen that safety training is the most cited factor in project level that affects safety performance of construction project. By conducting regular training on the workers for the particular project, safety performance can be improved and further, reduce the number of accident. In order to ensure the successfulness of the training program to improve safety performance, safety training must be conducted to all construction and building site workers in a project, as such to enhance awareness about safety in workplace.

Other most frequently cited factors are implementation of safe working environment, implementation of safe working plant and equipment and safety review for site safety policy. This could be done by providing means to ensure safe working environment, plant and equipment, such as through specific program that can encourage safety culture, periodical inspection of the workplace and establishing safer work practices. Both employer and employee at the workplace must comply with any legislative requirements related to safety and health.

3. Conclusions

In conclusion, through literature review, there are several identified factors that can affect the safety performance of construction project in project level. There are many advantages by implementing these safety factors, such as reduce the number of accident on the construction project, increase productivity, project complete on time, and decrease compensation cost and increase morals among employees. The good safety performance in project level can lead to better improvement of safety performance in construction industry. Future research is needed to explore the factors affecting safety performance in organization level.

Table 2. Summary of the factors affecting safety performance in construction project pertinent to organisation level.

Factors affecting safety performance in project level	Implementation of safety inspection	Implementation of safe system of work	Implementation of safe plant & equipment	Implementation of safe working environment	Implementation of safety officer & supervisor	Safety review for safety audit	Safety review for site safety policy	Emergency plan & procedures	Safety training	Accident investigation	Site safety meeting
Tsui and Gomez-Mejia [44]									/		
Mattila et al [45]				/	/		/		/		
Jannadi et al [39]											
Tam and Fung [46]									/		
Sawacha et al [47]				/			/				
Abudayeh et al[48]				/							
Hislop [49]								/			
Kartam et al [50]									/		
Griffith and Howarth [42]							/				
Jannadi and Bu-Khamsin [51]								/			
Al-Kaabi and Hadipriono [52]			/								
Reese [53]	/										
Lee and Halpin [54]					/						
Hinze and Gambatese [9]									/		
Tam et al. [55]	/	/		/							
Fang et al [56]				/							
Teo et al [58]		/							/		
Cheng et al [18]									/		
DOSH [33]			/								
Ismail [31]		/									
Bakri et al [36]				/							
Permana [28]	/								/		
Ghani et al [32]		/									
Ahmad [34]			/					/			
Yung [38]					/						
Paringga [35]			/						/		
Nikolaos [41]						/					
Lai et al [43]									/		
Kenrick [59]			/								
Tan and Razak [30]	/								/		/
Kai et al [29]	/						/		/	/	/
Frequency	5	4	5	6	4	1	4	3	12	1	2

IOP Conf. Series: Materials Science and Engineering **713** (2020) 012036 doi:10.1088/1757-899X/713/1/012036

4. References

[1] Suazo G A R and Jaselskis E J 1993 Comparison of Construction Safety Codes in United States And Honduras, *J. of Construction Eng. And Management* **119** 560-572

[2] Abas N H, Blismas N and Lingard H 2015 Knowledge-based Energy Damage Model for Evaluating Industrialized Building Systems (IBS) Occupational Health and Safety (OHS) Risk *The 3rd International Conference on Civil and Environmental Engineering for Sustainability (IConCEES 2015) (Melaka)* vol 47 (Paris: MATEC Web of Conferences) pp 1-7

[3] DOSH 2017 Occupational Accidents Statistics by Sector until May 2017. Retrieved on April 20, 2017 from http://www.dosh.gov.my

[4] Abas N H 2015 *Development of a Knowledge-based Energy Damage Model for Evaluating Industrialized Building Systems (IBS) Occupational Health and Safety (OHS) Risk* PhD Thesis (Melbourne: RMIT University)

[5] Chong H Y and Low T S 2014 Accidents in Malaysian Construction Industry: Statistical Data and Court Cases *Int. J. of Occupational Safety and Ergonomics*: *JOSE* **20**(3) 503-513

[6] Ayob A, Shaari A A, Zaki M F M and Munaaim M A C 2017 Fatal occupational injuries in the Malaysian construction sector– causes and accidental agents *IConCEES 2017* (Bristol: IOP Publishing Ltd.) 1-10

[7] Charehzehi A and Ahankoob A 2012 Enhancement of Safety Performance at Construction Site *Int. J. of Adv. In eng & Tech.* **5** 303-312

[8] Enshassi A, Mohammed S, Mayer P and Abed K 2007 Benchmarking masonry labor productivity *Int. J. of Prod. and Performance Mgmt.* **56**(4) 358-368

[9] Hinze J and Gambatese J 2003 Factors that influence safety performance of specialty contractors *J. of Cons. Eng. and Mgmt.* **129**(2) 159-164

[10] Larcher P and Sohail M 1999 *Review of safety in construction and operation for the WS&S Sector - A Literature Review: Part 1* WELL Study (United Kingdom: Loughborough University)

[11] Amartey C A 2014 *Improving Safety Performance of Ghanaian Building Contractors* Master Thesis (Kumasi: Kwame Nkrumah University of Science and Technology

[12] Awwad R, El Souki O and Jabbour M 2016 Construction safety practices and challenges in a Middle Eastern developing country *Safety Science* **83** 1-11

[13] Al-Kilani F M 2011 *Improving Safety Performance in Construction Project in Lybia (Case Study: In Tripoli)* Master Thesis (Didonegoro: Diponegoro University)

[14] Sawacha E, Naoum S and Fong D 1999 Factors affecting safety performance on construction sites *Int. J. of Proj. Mgmt.* **17** 309-315

[15] Dyck D and Roithmayr T 2004 Great safety performance: an improvement process using leading indicators *AAOHN J* **12** 511-520

[16] WorkCover 2001 *Safety Meter Positive Performance Measurement Tool* (New South Wales: WorkCover)

[17] Wilson J M J and Koehn E 2000 Safety management: problems encountered and recommended solutions *J. of Cons. Eng. and Mgmt.* **126** 77–79.

[18] Cheng E W L, Li H, Fang D P and Xie F 2004 Construction safety management: an exploratory study from department of building and real estate *Construction Innovation* **4** 224-229.

[19] Flin R, Mearns K, O'Connor P and Bryden R 2000 Measuring safety climate: identifying the common features *Safety Science* **34** 177-192

[20] Hallowell and Gambatese, 2009 Construction safety risk mitigation *J. of Cons. Eng. and Mgmt.* **135** 1316-1323

[21] Jaselskis E, Anderson S A and Russell J 1996 Strategies for achieving excellence in construction safety performance *J. of Cons. Eng. Mgmt.* **122** 61-70

[22] Ng S T, Kam P C and Skitmore R M 2005 A framework for evaluating the safety performance of construction contractors *Build. and Env..* **40** 1347-1355

[23] Lin J and Mills A 2001 Measuring the occupational health and safety performance of construction companies in Australia *Facilities* **19** 131-139

[24] Jafri H, Ali M W, Ahmad A and Kamasah M Z 2005 Effective Occupational Health and Safety Performance Measurements *Int. Conf. of Chemical and Biochecmical Eng. (Sabah)* pp 702-708

[25] Ng K, Laurlund A, Howell G and Lancos G 2012 Lean safety: Using leading indicators of safety incidents to improve construction safety *Conf. of the Int. Group for Lean Const. IGLC 20 (San Diego)* pp 1 -11

[26] Salminen S, Saari J, Saarela K L and Rasanen T 1993 Organizational-Factors Influencing Serious Occupational Accidents Scandinavian *J. of Work Environment & Health* **19** 352-357

[27] CIDB 2006 *Safety and Health Assessment System in Construction* (Kuala Lumpur: Construction Industry Development Board Malaysia)

[28] Permana 2007 *Construction Safety Practices in Batam, Indonesia* Master Thesis Malaysia: (Skudai: Universiti Teknologi Malaysia)

[29] Kai G, Hui H G, Omar M F, Tien C T, Zin A A M 2016 Accidents Preventive Practice for High-Rise Construction *The 3rd International Conference on Civil and Environmental Engineering for Sustainability (IConCEES 2015) (Melaka)* vol 47 (Paris: MATEC Web of Conferences) pp 1-6

[30] Tan C K and Razak N 2014 Case Studies on the safety Management at Construction Site *J. of Sustainability Science and Mgmt.* **9** 90-108

[31] Ismail I 2006 *Asessment of Safety Level in Performing Building Maintenance Work in Malaysia* Master Thesis (Skudai: Universiti Teknologi Malaysia)

[32] Ghani M K, Abd Hamid Z, Mohd. Zain M Z, Abdul Rahim A H, Kamar K A M, Rahman M A A 2008 *Safety in Malaysian Construction: The Challenges and Initiative* JURUTERA pp 16-19 (May 2008)

[33] DOSH 2005 *Guidelines on the Use of PPE Against Chemical Hazards* (Kuala Lunpur: Ministry of Human Resources Malaysia)

[34] Ahmad R 2008 *Best Practices in Safety Management for Conventional Civil Construction Industry in Malaysia* Master Thesis Malaysia: (Skudai: Universiti Teknologi Malaysia)

[35] Paringga L A 2010 Construction Safety in Jakarta, Indonesia. Master Thesis of Science Construction Management (Skudai: Universiti Teknologi Malaysia)

[36] Bakri A, Mohd Zin R, Misnan M S and Mohammed A H 2006 Occupational Safety And Health (OSH) Management Systems: Towards Development Of Safety And Health Culture *Procs of the 6th Asia-Pacific Struct. Eng. and Const. Conf. (APSEC 2006) (Kuala Lumpur)*pp19-28

[37] Hinze J and Rabound P 1988 Safety on large building construction projects *J. of Const. Eng. and Mgmt* **114** 286-293

[38] Yung P 2009 Institutional arrangements and construction safety in China: An empirical examination *Const. Mgmt. and Economics* **27** 439–450

[39] Jannadi M O 1996 Factors affecting the safety of the construction industry *Build. Res. and Information* **24** 108–111

[40] Lee S and Halpin D W 2003 Predictive tool for estimating accident risk *J. of Const. Eng. and Mgmt.* **129** 431–436

[41] Nikolaos 2010 The Measurement of Health and Safety Conditions at Work Theoretical Approaches, Tools and Techniques a Literature Review *Int. Res. J. of Finance and Economics* **36** 87-95.

[42] Griffith A and Howarth T 2001 *Construction Health and Safety Management* (London: Pearson Education Limited)

[43] Lai D N C, Liu M and Ling F Y Y 2011 A Comparative Study on Adopting Human Resource Practices for Safety Management on Construction Projects in the United States and Singapore *International Journal of Project Management* **29** 1018-1032

[44] Tsui A S and Gomez-Mejia L R 1988 Evaluating Human Resource Effectiveness *Human Resource Management Evolving Roles and Responsibilities* ed. Dyer L (Washington: The Bureau of National Affairs) pp 187-127

[45] Mattila M, Hyttinen M and Rantanen E 1994 Effective supervisory behaviour and safety at the building site *Int. J. of Ind. Ergonomics* **13** 85-93

[46] Tam C M and Fung I W H 1998 Effectiveness of safety management strategies on safety performance in Hong Kong *Const. Mgmt and Economics* **16** 49–55

[47] Sawacha E, Naoum S and Fong D 1999 Factors affecting safety performance on construction sites *Int. J. of Project Mgmt.* **17** 309-315
[48] Abudayyeh O, Fredericks T C, Butt S E and Shaar A 2006 An investigation of management's commitment to construction safety *Int. J. of Project Management* **24** 167-174
[49] Hislop R 2009 *Construction Site Safety* Retrieved on January 1, 2019 from https://www.noao.edu/nsf/presentations/FACILITIES/THF14c%20HISLOP%20Key%20Elements%2090416.pdf
[50] Kartam (2000) Construction Safety in Kuwait: procedures, problem, and recommendation, J. *of Safety Science* **36** 163-184
[51] Jannadi O A and Bu-Khamsin M S 2002 Safety factors considered by industrial contractors in Saudi Arabia *Build. and Env.* **37** 539-547
[52] Al-Kaabi N and Hadipriono F C 2003 Construction safety performance in the United Arab Emirates *Civ. Eng. and Env. Syst.* **20** 197-212
[53] Reese C D 2003 *Occupational and safety management: A practical approach* (Florida: CRC Press)
[54] Lee S and Halpin D W 2003 Predictive tool for estimating accident risk *J. of Const. Eng. and Mgmt* **129** 431–436.
[55] Tam C M, Zeng S X and Deng Z M 2004 Identifying elements of poor construction safety management in China *Safety Science* **42** 569-586.
[56] Fang D P, Huang X Y and Hinze J 2004 Benchmarking studies on construction safety management in China *J. of Const. Eng. and Mgmt.* **130** 424-432.
[58] Teo E and Feng Y 2009 The role of safety climate in predicting safety culture on construction sites *Archi. Science Review* **52** 5-16
[59] Kenrick V 2012 *PPE and Construction Safety in the Middle East* Retrieved on January 2, 2019 from http://ehstoday.com/ppe/construction-safety-in-the-middle-east-0131

Acknowledgement
The author would like to thank Universiti Tun Hussein Onn Malaysia (STG Vot U638) for the support and funding provided towards the completion of this research. The results presented in this paper are also part of an on-going research project (FRGS-Vot 1579).

Improvement for significant clauses in the standard form of contract for industrialized building system construction

Mohd Ashraf Mohd Fateh[1], Hanah Zakariah[2] and Shaiifa Ema Ezanee[3]

[1]School of Energy, Geoscience, Infrastructure and Society, Heriot-Watt University, 62200 Putrajaya, Malaysia
[2]Faculty of Engineering, Science & Technology, Infrastructure University Kuala Lumpur, 43000 Kajang, Selangor Darul Ehsan, Malaysia
[3]Finance Department, Perbadanan Pr1ma Malaysia, 47301 Petaling Jaya, Selangor Darul Ehsan, Malaysia
Corresponding author: mohd.ashraf.fateh@gmail.com

Abstract. Industrialized Building System (IBS) has been introduced in Malaysia since the 1960s. The government has come out with varies strategic blueprint to push IBS as a national agenda. Unfortunately, the pace of adaptation of IBS is still slow and below the government's target. Construction players are still facing various issues when adopting IBS particularly on contractual and procurement aspects; thus it contributes to the low adoption of IBS in Malaysia. Lack of provision in the Malaysia standard form of contract is one of the issues. There are six (6) significant clauses in the Malaysia standard form of contract that can be improved to make it more conducive to IBS construction approach. Therefore this research will attempt to investigate the factors of improvement for the six (6) significant clauses in the standard form of contract with IBS construction. A semi-structured interview was used in the data collection exercise. The study revealed that there are eleven (11) factors of improvement for the six (6) significant clauses in the standard form of contract with IBS construction approach. The findings will be useful in order to enhance the local standard form of contract to suit the IBS construction approach hence able to accelerate the adoption of IBS construction in Malaysia.

1. Introduction

Report by the Leng [1] highlighted that the construction sector for 2019 is 4.9% which indicate a slight increase from 2018 which is 4.5%. The growth up marginally due to the increase in new planned supply in affordable housing projects. In September 2015, the Construction Industry Development Board (CIDB) launched the Construction Industry Transformation Programme 2016-2020 (CITP). One of the CITP's objective is to transform the construction industry to become highly productive sector using IBS technology. Based on the findings by Mohd Fateh & Mohammad [2] and Kamaruddin *et al.* [3] Malaysia government has been pushing the IBS as the national agenda with numerous initiative namely CIMP in 2006, IBS roadmap in 2010 and CITP in 2015. Nevertheless, a report CIDB [4] highlighted that the adoption rate is still low and did not meet the government's target.

Most of the previous research on IBS is focus more on the technical parts which include the components/products itself and construction process. There is a lack of research on the contractual aspect according to Abd Jalil *et al.* [5] and Mohd Fateh & Mohammad [2]. Therefore, this paper will focus on the eleven (11) factors of improvement for the six (6) significant clauses in the standard form of contract with IBS construction approach. This research will give an insight of information and guidance to all the related stakeholders such as CIDB, Public Works Department (PWD), expert panels, academicians and related IBS players. The challenges in boosting the adoption of IBS projects would be looking at the enhancement of the standard form of contract for IBS construction approach. It will give a value-added impact on the adaptation of IBS construction approach. CIDB supported [6] stated that a suitable standard form of contract for IBS would invariably expedite the project success.

This paper is structured into three parts. Firstly, the review covers a broad range of literature providing an overview of the construction industry, IBS scenario in Malaysia and the standard form of contract. The second part discussed the methodology used for this research. The last part discussed findings and conclusions derived from evidence from the literature review and semi-structured interview.

2 Literature Review

2.1 The Construction Industry

The construction sector is one of Malaysia's key economic drivers and a significant enabler of the Economic Transformation Programme as highlighted by Hussain [7]. The construction industry offers goods and services. For goods, it creates a vast source of demand for metals, ceramics, cement and other building raw materials while also being a large buyer of higher-value-added equipment and machinery. With this, the local construction industry contributes 15% of the output of Malaysia's manufacturing sector. For services, the construction industry is a vital consumer of a range of services from knowledge-driven consultancy and engineering, as well as a broad base of financial services. With this, the industry accounts for 5% of the total output from the Malaysian services sector [8].

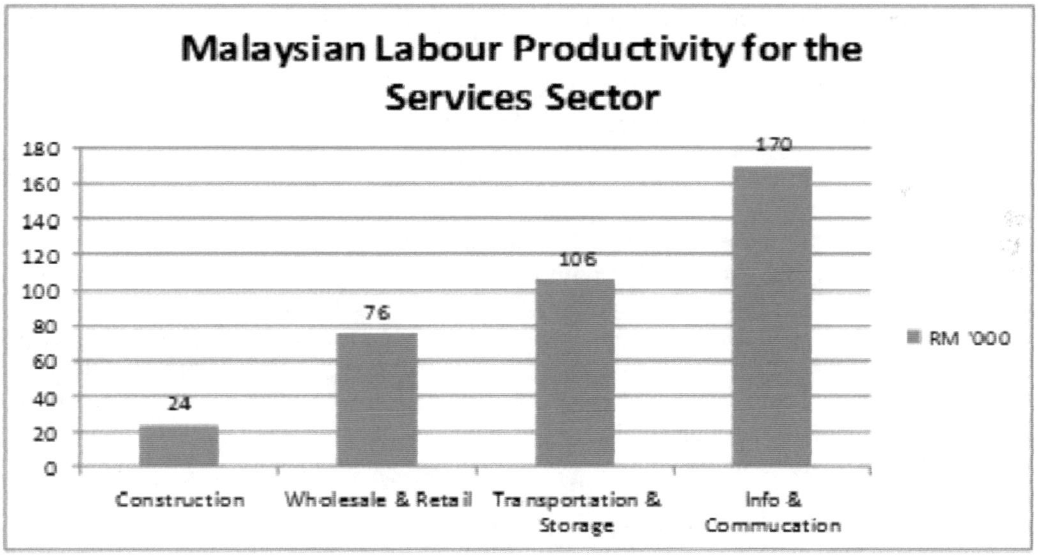

Fig. 1: Malaysian Labour Productivity for the Services Sector (Department of Statistics [9])

Nevertheless, the construction industry records low productivity levels relative to other sectors in Malaysia as reported by Department of Statistic [9], shown in figure 1. The relatively low productivity is a reflection of limited modernisation of construction methods such as IBS and the reliance on low-skilled labour.

2.2 IBS in Malaysia

CIDB [10] defined IBS as a construction technique in which components are manufactured in a controlled environment (on or offsite), transported, positioned and assembled into a structure with minimal additional site work. Previous research have proved that IBS able to offer considerable benefits in terms of cost and time certainty, attaining better construction quality and productivity [11-22]. Nevertheless, the pace of adaptation of IBS is still slow and below the government's target. Construction players are still reluctant to adopt the IBS construction approach due to uncertainty in the provision of the standard form of contract. According to Mohd Fateh *et.al* [12] and Mohamad Kamar *et.al* [23] stated that the construction players are still facing some issues and challenges in contractual aspects when adopting IBS; thus it contributes to the low adoption of IBS in Malaysia. Based on the preliminary study being done by Mohd Fateh *et.al* [12] stated that all respondents agreed that it is a necessity to enhance existing Malaysia's standard form of contract to suit IBS construction approach. Therefore, this research is to investigate the factors of improvement of the existing provision in Malaysia's standard form of contract to suit IBS construction approach, hence propel the acceleration adoption of IBS construction approach parallel with the CITP's vision.

2.3 Standard Form of Contract

The standard form of contract is a pre-printed form with a set of terms and clauses specially drafted by professional or public organisations for the use in the construction project. This was agreed by Abdullah [24] mentioned the standard form of contract contains terms and agreement (clauses) between the client and contractor which includes right, obligations, responsibilities and work to be done. The standard form of contract will be a part of the document contract for each project. The principal parties in a construction contract will be the client and the contractor. The client acts as the owner of the project while the contractor acts implementer of the project.

Although there are varies types of standard form of contract in the industry as explained in the previous section, nevertheless overall structure is the same as highlighted by Abdullah [24]. The standard of form contract consists of three (3) sections namely:

- The articles of agreement: In this section, it spells out the overall information of the contract which includes when the contract is being formed, the parties in the contract, the scope, value, location, and the period of the contract. The definitions and interpretation of words and phrases in the contract also be explained in this section. Lastly, the attestation space. It is an act of witnessing the signing of the contract where both parties will put pen to paper and 'execute' the contract.
- The condition of contract: This is the main body of the standard form of contract. In this section, it spells out in particular clauses laying down the rights and obligations of the parties in the contract. All of the clauses are presented in a numbered format for ease of references. The clauses include the usual performance required from each party, rules/procedures for conduction the contract, dealing with disputes arise and also terms of payment to the contractor.
- The appendix: this section contains the essential contract particulars such as the contract period, completion date, defect liability period and the rates of the liquidated damages. All the blanks space for the contract particulars need to be filled because some of the clauses are not applicable at all unless the appendix section is filled. For some clauses, if there are not values made than the default values apply for example if no defect liability periods inserted, then it will be 12 months.

2.4 Standard form of contract with IBS construction approach

According Jaafar and Radzi [25], the procurement system in Malaysia has not evolved dramatically since the British colonial era. Many clients in the local industry prefer to choose procurement systems considered familiar, even though the criteria and purposes or every project are different. Previous research by [12] stated that every project is unique and dynamic in terms of processes, resource allocation, risk exposure and responsibilities between all parties, therefore, there is a necessity to develop a framework for improving the significant clauses in the standard form of contract to make it more suitable for the IBS construction approach in Malaysia. Jaafar & Radzi [26] suggested that when there are changes in the method of construction, there is also a need to adopt a new procurement system and standard form of contract. Using unsuitable standard form of contract in IBS will not only affect the progress of the project but will also affect the construction team in terms of understanding and interpretation of the regulations. According to Gandu *et.al* [27] and Lutz Preuss [28], the system of procurement and standard form of contract are considered as the key to project success. Blismas & Wakefield [29] also agreed that in producing a successful IBS project, the contractual approach must be suitable. Findings by Abd Jalil *et.al* [5] highlighted that a specific procurement system and standard form of contract is required as the IBS construction approach involves unique producers and processes which are different from traditional construction. Previous researchers had concluded that the current traditional standard form of contract was not suitable for the IBS construction approach, especially in terms of payment and project coordination [5]. This was agreed by [23] who reported that the existing standard form of contract does not favour the industry players that want to adopt the IBS construction approach. The [30] also highlighted that several of the barriers that hinder the adoption of the IBS construction approach involves the lack of a procurement method and provision in the standard form of contract. Findings by [2] identified that there are six (6) clauses that are significant to the IBS construction approach.

2.4.1. Clause 1: Definition of unfixed material and goods. In this clause, it stated that all unfixed material and goods need to be on or adjacent to the site. The contractor is fully responsible for the loss or damages to the unfixed material and goods. The material and goods need to be delivered to the site following the project progress; it cannot be 'premature' delivered to avoid any unwanted circumstances (site congested, material damages or lost). When the client has made the payment, the unfixed material and goods will be the client. In the context of the IBS construction approach, it is significant due to most of the unfixed material and goods are not being delivered to the site. All of the unfixed material and goods are being stored/used off-site. Therefore, the provision in the clause is not friendly to the IBS construction approach and might need to be improved. This clause also is related and will give implication to the evaluation of interim payment to the contract which will be discussed next.

2.4.2. Clause 2: Evaluation of interim payment. In this clause, it stated that the client needs to evaluate at least once a month on the progress of work done. The value of the interim payment will be based on two (2) component; the progress of work done that fulfils the specification and the material that has been delivered on-site according to the project progress. The client needs to issue the interim payment certificate within 14 days of the evaluation of the site, and the payment needs to be done by 30 days upon the interim payment certificate issued. In the context of the IBS construction approach, both of the components to calculate the value of the interim payment are not applicable. As discussed earlier, most of the activities in IBS construction approach are being done off-site thus making the contractor cannot claim any value to the progress of works and off-site activities. This will create a very tight cash-flow for the contractor because the contractor needs to pay the manufacturer up front for the manufacturing process of the components to start. The contractor also needs to pay for other works to start such as preliminary works or piling. In other words, the contractor needs to upfront much capital without able to claim within or next month. This might propel other issues in the project such as delay and termination. Therefore, the provision in the clause is not friendly to the IBS construction approach and need to be improved.

2.4.3. Clause 3: Inspection, testing of material, goods and equipment. In this clause, it stated that the contractor needs to carry out the inspection and test as approved in the contract or elsewhere as the client required. The inspection and test of the material, goods, and equipment whether or not already incorporated in the works. The contractor from time to time also needs to provide such assistance such as instrument or machine that is required the inspection and testing. This includes supplying samples before incorporation in the works. In the context of the IBS construction approach, this provision to allow the inspection and test are done whether or not already incorporated in the works is favourable to IBS construction approach. Nevertheless, the researcher still selects this as one of the significant clauses to the IBS construction approach because the researcher wants to test and get the opinion from the industry players on these matters. Industry players might not be on the 'same page' because it might incur additional cost to conduct an inspection and test for the works that have not incorporated in the works (off-site).

2.4.4. Clause 4: Insurance/Bond. In this clause, it stated that the contractor needs to take insurance in the joint names of the client. The insurance will cover against loss, damages, lightning explosion, storm, flood and many more. The coverage protects all work executed and all unfixed material/goods that are delivered, on or adjacent the site (on-site only). In the context of the IBS construction approach, the provision in the clause only allows the material onsite to be covered by the insurance. Since most of the activities in IBS construction are offsite, the researcher believes that the clause is not favourable to IBS construction approach thus making this clause significant to IBS construction approach. Therefore, the provision in the clause is not friendly to the IBS construction approach and need to be improved.

2.4.5. Clause 5: Submission of the supervision report. In PWD 203A, no clause required the contractor to submit a supervision report to the client. On the hand for PWD DB, the contractor is required to submit on-site report activities which include works correctly done, the progress of works, test done and safety measures on site. In the context of the IBS construction approach, it is okay if there is a clause for the submission of the supervision report. Nevertheless, the scope is of the supervision report should be extended which includes the offsite activities since most of the activities for the IBS construction approach are offsite. Therefore, the clause should be added to make it more IBS construction approach.

2.4.6. Clause 6: Extension of time (relevant events. In this clause, it stated the rules, procedures and relevant events that the contractor can refer to apply the extension of time. In brief, the contractor should write to the client on the causes and calculation of the length of delay with the supporting documents. The causes of delay should be within the relevant events stated in the clauses. Nevertheless, all of the relevant events stated in the clauses in focusing on site activities only. In the context of IBS construction approach, the provision in the relevant events to focus only on site is not favourable. The researcher believes that for the IBS construction approach, the relevant events should extend to the off-site activities such as at the factory or fabrication yard where the IBS components are being manufactured.

In summary to the literature reviews, it shows that there is room for improvement for provision in the significant clauses in the existing Malaysian standard form of contract with the IBS construction approach. As of to date, there is still a lack of provisions in the Malaysia standard form of contract to suit the IBS construction approach as reported by Mohd Fateh *et. al* [12]. Abd Jalil *et. al* [5] stated that despite the varies Government's efforts in pushing IBS as national agenda (IBS roadmap, CIMP and CITP), there are not detailed IBS building guidelines or standard regulations for procurement systems or contract documents in term of tendering, design, construction and operation have been produced. Research by Abd Jalil *et. al* [5] add up that the current standard form of contract has not addressed this very pertinent issue. Therefore this research will attempt to investigate and identify the factors of improvement for the six (6) significant clauses in the standard form of contract with IBS construction.

3. Methodology

Literature review and semi-structured interview were used in the data collection exercise. For the first phase, literature review derived from relevant books, journals articles, thesis and dissertations, conference proceedings and reports were examined. For the second phase, a semi-structured interview was done to the government officials, private clients and contractors. The justification for focusing on these three (3) groups is as follows:

- As the project owner, a client (government or private) always has the final say in deciding the appropriate standard form of contract to be used in the project.
- As the project implementer, a contractor's opinion is critical to ensure that the research is not biased. A contractor also plays essential roles in the adoption of IBS construction.
- Both client and contractor are the main parties bound by the main contract using the standard form of contract that is the subject of this research.

The justification for the semi-structured interview was used because when dealing with contracts and clauses, the quantitative analysis cannot record the qualitative aspects of the research. Ideas, opinions and comments (qualitative) are needed. By doing the semi-structured interviews allows the respondents to express themselves autonomously, thus providing a type of information that is eminently qualitative [31]. Leiva *et.al* [31] also added that the data collected will have a greater richness and depth in the responses compared to quantitative data. This is because the respondent can express his/her thoughts and opinions and be able to provide more lengthy explanations. Patton [32] stated that semi-structured interviews are used to develop an idea that will attempt to understand how people think and feel about the issues. To ensure that the respondents were able to answer transparently and without bias, the researcher had already stated that all information collected would be kept confidential although the results of the research would be published. Therefore, all respondents' details will not be published and only be coded as "Respondent 1 (R1) or Respondent 2 (R2)".

Based on the guidelines provided by Miles *et. al* [33], content analysis was applied to the qualitative data (semi-structured interview) from this research. Content analysis can be defined as an observational research method that is used to systematically evaluate the symbolic contents of all forms of recorded communication Sekaran and Bougie, Kolbe and Burnet [34-35]. [34] also added that content analysis is suitable to analyse newspapers, websites and recordings of interviews.

4. Results and Discussions

4.1 Respondents' demographic

The demographic profile contains information regarding the personal and behaviour-related information of the respondents. In this study, the personal information that is collected includes the respondent's name, designation, sectors, contact details and years of involvement with IBS. Table 1 shows the frequency and percentage analysis according to the organisation sector of the respondents. The highest sectors recorded were private client and contractor with 7 (36.8%) respectively. The lowest will be the government sector with 5 (26.4%). Table 2 shows the frequency and percentage analysis according to respondents' years of experience with IBS. A majority, 11 (57.9%), of the respondents, have more than 10 years of experience. This is followed by respondents with 6 to 10 years of experience, 5 (26.3%). A small number of respondents, 3 (15.8%), have 1 to 5 years of experience. The data collected show that respondents have vast experienced in IBS construction approach. Therefore, the data collected can be considered reliable and good for the study.

Table 1. Organization sectors of the Respondents (N=19).

Sector	Frequency	Percentage (%)
Private Client	7	36.8
Contractor	7	36.8
Government	5	26.4
Total	19	100

Table 2. Respondents' Years of Experience in IBS (N=19).

Years of experience	Frequency	Percentage (%)
1 to 5 years	3	15.8
6 to 10 years	5	26.3
More than 10 years	11	57.9
Total	19	100

4.2 Clause 1: Definition of unfixed material and goods

The definition of unfixed material and goods in the standard form of contract only acknowledges that they are on-site. As discussed in the literature review, most of the activities for the IBS construction approach take place off-site. 'Acknowledge offsite material as well to be unfixed material and goods' had the highest frequency of 77 (87% positive) while 'quality stated in specifications' had a frequency of 9 (95.7% positive) and 'acknowledge only onsite material as unfixed material and goods' the lowest frequency of 3 (87% negative). Table 3 summarises the frequency of occurrence of themes and the respondents' positive and negative reactions. Extracts of typical comments made by the respondents are also presented.

- R1: "Most of the components for IBS construction will be in the factory (offsite); therefore it is reasonable to acknowledge offsite material as unfixed material and goods as well."
- R12: "... definitely need to acknowledge offsite material as well, because all of the raw material for IBS components (cement, sand and reinforcement) will be offsite (factory). The casting and curing process will be offsite. Once it is done, and then only delivered to site. When it reached the site, it will proceed directly for installation thus is not considered as unfixed material and goods already."

Table 3. Summary of content analysis for clause 1: definition of unfixed material and goods.

Rank	Themes	Frequency of occurence	% Response	
			Positive	Negative
1	Acknowledge offsite material	77	87	13
2	Quality stated in specifications	9	95.7	4.3
3	Acknowledge only onsite material	3	13	87

4.3 Clause 2: Evaluation of interim payment

Evaluation of interim payment has always been a profound issue in IBS construction. In the standard form of contract, the evaluation of interim payment was based on two (2) items, namely, the progress of work done on site and the material on site. Based on the literature findings, the mechanisms of the evaluation need to be reviewed to suit IBS construction. Based on the content analysis for this clause, it can be broken down into four (4) themes. All respondents mentioned that the mechanism of interim payment needs to be enhanced to suit IBS construction (72 occurrences, 95.7% positive). This was followed by the need to certify offsite material as interim payment as well (71 occurrences, 87% positive). The negative impacts on the evaluation of the interim payment are the difficulties of proving that the progress of works done onsite (29 occurrences, 82.6% negative) and the materials need to be onsite only (6 occurrences, 91.3% negative). Table 4 summarises the frequency of occurrence of these themes and the positive and negative reactions recorded from the respondents. Extracts of typical comments made by the respondents are also presented.

- R20: "The mechanisms of the interim payment for IBS components need to be changed. For example, for M&E works, there will be a certain amount of payment once approved for certain test/requirement at the factory. Even though that item has not been delivered to the site, the same principle can be applied to IBS components."
- R2: *"It will be much better if the evaluation of interim payment will include the off-site material as well. For IBS construction, most of the activities and progress are being done offsite. Therefore the contractor cannot claim for the progress of work done and material onsite. It will create a very tight cash flow that might lead to other problems in constructions".*

Table 4. Summary of content analysis for clause 2: evaluation of interim payment.

Rank	Themes	Frequency of occurence	% Response	
			Positive	Negative
1	Review the mechanism of interim payment.	72	95.7	4.30
2	Need to certify offsite material as well.	71	87	13
3	Proof of progress of works.	29	17.4	82.6
4	Material needs to be onsite.	6	8.70	91.3

4.4 Clause 3: Inspection, testing of materials, goods and equipment

In the standard form of contract, the inspection and testing of material, goods and equipment are only being done for the work already incorporated on site. There is a mixed pattern of reactions (positive and negative) of respondents for this clause. The highest frequency of 99 occurrences (65.2% positive) is for the respondents' stated need to allow for the inspection even though the items have not yet been incorporated on site. Some respondents stated that there is no need for inspection if an item has not yet been incorporated because all components will come with a certification of inspection (24 occurrences, 73.9% negative). Lastly, respondents highlighted that a proper procedure of inspection is needed to allow for the inspection even though the items have not yet been incorporated onsite if required (7 occurrences, 82.6% positive). Table 5 summarises the frequency of occurrence of the themes and the positive and negative reactions from the respondents. Extracts of typical comments made by the respondents are also presented.

- R4: "...need to allow for the testing being done even though the work is not incorporated yet. This can prevent any failure or error on the ground before it is installed on site. The client also will have the opportunity to inspect not only materials but also the facilities of the factory."
- R7: *"... most of the material for IBS will be received offsite; therefore it is good if want to inspect before it is incorporated in works. The client can ensure the quality (for example consistency) of the components straight in the factory. If it does not meet the standard, the manufacturer can be notified and rectify early. These steps will save time and cost of transportation."*

Table 5. Summary of content analysis for Clause 3: Inspection, testing of materials, goods and equipment.

Rank	Themes	Frequency of occurence	% Response	
			Positive	Negative
1	Allow for the inspection even though not yet incorporated onsite.	99	65.2	34.8
2	The material will come with a certification of inspection.	24	26.1	73.9
3	Proper procedure of inspection.	7	82.6	17.4

4.5 Clause 4: Insurance/Bond

Three (3) themes were identified for this clause. These were, whether to include offsite (activities/facilities) scored the highest frequency (23 occurrences, 69.6% negative) followed by to maintain the existing practice in the standard form of contract (23 occurrences, 60.9% positive) and, lastly, respondents highlighted that it is most likely if the offsite (activities/facilities) are included that the cost of the project will increase (5 occurrences, 82.6% positive). For this clause, there is a mixed response (positive and negative) from the respondents. Table 6 summarises the frequency of occurrence of the themes and the positive and negative reactions recorded from the respondents. Extracts of typical comments made by the respondents are also presented.

- R11: "It is required to include offsite as well. In case of anything happening offsite, the insurance will cover it. As we are aware, most of the IBS construction activities will be done offsite. It also will able to secure both parties."
- R16: "*... if it can be done, it is good for contractor and client. Proper clauses need to be mentioned in the standard of form to ensure that all parties understand and are aware of it. Another option is only extending the coverage of the existing clauses.*"

Table 6. Summary of Content Analysis for Clause 4: Insurance/Bond.

Rank	Themes	Frequency of occurrence	% Response	
			Positive	Negative
1	Include offsite (activities/facilities).	23	30.4	69.6
2	Existing practice is sufficient.	12	60.9	39.1
3	Increase cost.	5	82.6	17.4

4.6 Clause 5: Submission of the supervision report

The clause in the standard form of the contract stated that the submission of the supervision report is only for onsite activities. Based on the content analysis of the interview, two (2) themes were identified, namely, the scope of the report needs to be extended to cover the offsite activities (59 occurrences, 78.3% positive) and the quality of the report needs to be standardized and needs to be correctly set out in the standard form contract to prevent disputes and discrepancies later (15 occurrences, 52.2 % positive). Table 7 summarises the frequency of occurrence of the themes and the positive and negative reactions recorded from the respondents. Extracts of typical comments made by the respondents are also presented.

- R11: "The report should cover offsite activities as well as more work is done off-site. The offsite reports also will help to plan the onsite activities (in terms of delivery dates, installation and site readiness)."
- R12: "*... it is good if we can have it. So we will know what is going on in the factory. The information will be useful for planning and works execution. It also will be written evidence if in the future the contractor wants to claim for extension of time.*"

Table 7. Summary of content analysis for clause 5: submission of the supervision report.

Rank	Themes	Frequency of occurence	% Response	
			Positive	Negative
1	Cover offsite activities.	59	78.3	21.7
2	Quality of the report.	15	52.2	47.8

4.7 Clause 6: Extension of Time (Relevant Events)

Three (3) themes were derived from the responses recorded; namely, the practice of extension of time remains the responsibility of the contractor and needs also to consider the offsite activities but make some amendments on the relevant events. Most of the respondents stated that things should remain as spelt out in the standard form contract (32 occurrences, 60.9% positive). Followed by the extension of time is the sole responsibility of the contractor (27 occurrences, 78.3% negative) and lastly, need to consider offsite activities as well but make some amendments on the relevant events (13 occurrences, 56.5% negative). Table 8 summarises the frequency of occurrence of the themes and the positive and negative reactions recorded from the respondents. Extracts of typical comments made by the respondents are also presented.

- **R1:** "... extension of time needs to consider only onsite relevant events. Extending the scope of extension of time to offsite will open plenty of problems and loopholes. The contractor needs to organise and plan their works according to the IBS construction approach."
- **R3:** "*We can consider offsite as well, but the 'relevant events' in the standard form of the contract need to be amended as well. Clearer scope and limitations need to be stated to*

prevent misuse of the clauses. As a contractor, it might be a sweet deal, but as a client, it is not very practical. The contractor might use it as an excuse to claim for extension of time. Therefore the relevant events and implementation need a fine tune if want to include the offsite delay."

Table 8. Summary of content analysis for Clause 6: Extension of time (relevant events).

Rank	Themes	Frequency of occurence	% Response	
			Positive	Negative
1	Fine-tune the relevant event to suit IBS environments.	32	60.9	39.1
2	Responsibility of contractor.	27	21.7	78.3
3	Consider offsite with amendments on the relevant events.	13	43.5	56.5

5. Conclusion

From the study conducted, it shows that there is room for improvement to Malaysia's standard forms of contract in the provision for IBS construction approach. All of the improvement might be able to reduce the uncertainty and potential risk to the industry players when dealing with IBS construction approach thus accelerate the adoption in IBS construction approach. For IBS construction approach to be adopted effective and efficiently, attention needs to be given not only on the technical but also in the contractual parts. Both parts play a vital role in the adoption of the IBS construction approach. Previous findings by Mohd Fateh & Mohammad [2] show that the current standard form of contract does not provide enough provision for the IBS construction approach. Therefore, it is necessary to improve the standard form of contract to suit the IBS construction approach hence accelerate the adoption of IBS construction. The output of this research will hopefully illustrate good insights into the industry and help to accelerate the adoption of IBS construction in Malaysia as a whole.

6. References

[1] Leng T A 2018 *Economic Report 2019: Construction sector to grow at moderate pace* (Kuala Lumpur: Edge Property)

[2] Mohd Fateh M A and Mohammad M F 2017 IBS Provision in Local and International Standard Form of Contracts *J. Constr. Dev. Ctries.* **22** 1–12.

[3] Kamaruddin S S, Mohammad M F, Mahbub R and Yunus R 2013 Perception Towards Cost Implication of Mechanisation and Automation Approcah in IBS Projects in Malaysia *Procs. 29th Annual ARCOM Conf (Reading)* (UK: Association of Researchers in Construction Management) pp 213–222

[4] Construction Industry Development Board (CIDB) 2010 *IBS Road Map* no. 72 (CIDB: Kuala Lumpur) pp 1-10

[5] Abd Jalil A, Nuruddin R, Jaafar M, and Mydin A O 2015 ANew Procurement Method for Housing Projects Implementing IBS Modular System *International Conference on Advances in Civil and Environmental Engineering (Penang) (*Kuala Lumpur: Pusat IBS) pp 1–8

[6] Construction Industry Development Board (CIDB) *Construction Industry Master Plan 2006-2015* 2006 (CIMP 2006-2015) (CIDB: Kuala Lumpur) pp 100

[7] MPC 2015 *Productiviy Report* (Selangor: Malaysia Productivity Corporation) pp 25

[8] Maybank 2012 *Maybank Sector Update : Construction* (Kuala Lumpur: Maybank) pp 2-11

[9] Department of Statistics 2017 *Ecomonic Census of Contruction* Retrieved on April 25, 2015 from https://www.dosm.gov.my/v1/index.php?r=column/ctwoByCat&parent_id=76&menu_id=OEY5SWtFSVVFVUpmUXEyaHppMVhEdz09

[10] CIDB 2014 *IBS-Roadmap 2011-2015 (*Kuala Lumpur:CIDB) pp 12-30

[11] Abd Hamid Z 2011 *The Current State of Industrialised Building System (IBS) Construction in Malaysia: Proceedings of 1st IBS Roundtable Workshop* (Kuala Lumpur: Makmal Kerja Raya Malaysia)

[12] Mohd Fateh M A, Mohammad F M and Abd Shukor A S 2016 Review in formulating the standard form of contract for Industrialized Building System (IBS) construction approach in Malaysia 2016 *9th Int. Unimas Stem Eng. Conf. (ENCON 2016) Innovative Solut. Eng. Technol. Challenges* vol 87 (Paris: MATEC Web Conf.) pp 1-7

[13] Gibb A G F 2001 *Pre-assembly in Construction: A review of recent and current industry and research initiatives on pre-assembly in construction In Construction Research & Innovation* Strategy Panel CRISP Consultancy Commission 00/19 London pp 20-30

[14] The UK government Department of Trade and Industry (DTI) 2004 *Modern methods of construction in Germany – playing the off-site rule In Global Watch Mission Report* (United Kingdom: Pera Innovation Ltd.) pp 50-56

[15] Nawi M N M, Lee A, Kamar K A M and Hamid Z A 2011 A critical literature review on the concept of team integration in industrialised building System (IBS) project, *Malaysian Constr. Res. J.* **9** 1–17

[16] Mohamad Kamar K A, Mustafa A, Zuhairi A H, Mohd Nasrun M N, Ahmad Tarmidzi H and Mohd Rofdzi A 2009 Industrialised Building System (IBS): A Review of Experiences in UK and Malaysian Construction Industry *2nd Constr. Ind. Res. Achiev. Int. Conf. (CIRAIC) (Kuala Lumpur: Construction Research Institute of Malaysia)* pp 1–12

[17] Construction Industry Development Board (CIDB) 2015 *Construction Industry Master Plan 2 (CIMP 2 2016-2020)* no. 1 (Kuala Lumpur: CIDB)

[18] Mohammad M F 2013 Construction Environment: Adopting IBS Construction Approach Towards Achieving Sustainable Development *Procedia - Soc. Behav. Sci.* **85** 8–15

[19] Shukor A A S, Mohammad M F and Mahbub R 2011 Supply Chain Integration Challenges in Project Procurement: Contractor's Perspective *Proc. of the Int. Conf. on Manag. Innov. a Sustain. Built Environ. (MISBE2011) (Amsterdam)* (Holland: Delft University of Technology) pp 495-504

[20] Musa M F, Mohammad M F, Yusof M R and Mahbub R 2015 The Way Forward for Industrialised Building System (IBS) in Malaysia *Proceedings of the International Civil and Infrastructure Engineering Conference 2014 (InCIEC2014) (Kuala Lumpur)* (Singapore: Springer) pp 163–175

[21] Muhamad Halil F, Mohammad M F, Mahbub R and Abd Rahim S F 2009 Sustainable value through the implemantation of strategic partnering in Industrialised Building System (IBS) *Proc. of the Int. Symp. Dev. Econ. Commonalities Among Diversities (Penang)* pp 188–200

[22] Kamaruddin S S, Mohammad M F and Mahbub R 2013 Enhancing the Quality of Life by Adopting IBS: An Economic Perspective on Mechanisation and Automation *Procedia - Soc. Behav. Sci.* **101** 71–80

[23]` Mohamad Kamar K A, Alshawi M and Abd Hamid Z 2009 Bariers to Industrialised Building Systems (IBS): The case of Malaysia *9th International Postgraduate Research Conference (IPGRC) (Salford)* pp 1-16

[24] Abdullah N A 2019 *Basics of Construction Contract (PAM, CIDB, PWD and FIDIC Standard Forms)*(Kuala Lumpur: UiTM Press)

[25] Jaafar M and Radzi N M 2012 Building procurement in a developing country: A comparison study between public and private sectors *Int. J. Procure. Manag.* **5** 608–626

[26] Jaafar M and Radzi N M 2013 Level of satisfaction and issues with procurement systems used in the Malaysian public sector *Australasia J. Constr. Econ. Build.* **13** 50–65

[27] Gandu Y J, Ali A A and Chindo P G 2009 Bidding model for sustainable projects using the traditional procurement method *The Information Manager* **9** 36–43

[28] Preuss L 2009 Addressing sustainable development through public procurement: the case of local government *Supply Chain Managemant* **14** 213-223

[29] Blismas N and Wakefield R 2009 Drivers, Constraints and the Future of Offsite Manufacture in Australia. *Constr. Innov.* **9** 72–83

[30] IBS Centre 2007 IBS Roadmap Review (Final Report) (Kuala Lumpur:CIDB)

[31] Leiva F M, Ríos F J M and Martínez T L 2006 Assessment of interjudge reliability in the open-ended questions coding process *Qual. Quant.* **40** 519–537

[32] Patton M Q 2016 Two Decades of Developments in Qualitative Inquiry *Qual. Soc. Work* **1** 261–283

[33] Miles M B, Huberman A M and Saldana J 2013 *Qualitative Data Analysis: A Methods Sourcebook 3rd ed.* (Oaks: Sage Publications)

IOP Conf. Series: Materials Science and Engineering **713** (2020) 012037 doi:10.1088/1757-899X/713/1/012037

[34] Sekaran U and Bougie R 2009 *Research methods for business: A skill building approach 5th ed.* (United Kingdom: John Wiley & Sons)

[35] Kolbe R H and Burnet M S 1991 Content-Analysis Research: An Examination of Applications with Directives for Improving Research Reliability and Objectivity *J. Consum. Res.* **1** 243–250

IOP Conf. Series: Materials Science and Engineering **713** (2020) 012038 doi:10.1088/1757-899X/713/1/012038

The on-site waste minimization practices for construction waste

Muhammad Fikri Hasmori[1], Ahmad Faizul Md Zin[1], Sasitharan Nagapan[1], Rafikullah Deraman[1], Norhaslinda Abas[1], Riduan Yunus and Mustafa Klufallah[2]

[1] Building & Construction Department, Universiti Tun Hussein Onn Malaysia, 86400, Batu Pahat, Johor, Malaysia
[2] School of Built Environment, University of Reading Malaysia, Persiaran Graduan Kota Ilmu, Educity 79200 Iskandar Puteri Johor, Malaysia
Corresponding author: mfikri@uthm.edu.my

Abstract. Rapid development in construction industry had caused the increasing of construction waste. This situation brings many negative impacts towards the environment, cost, productivity, time, social and economy. The increasing of construction wastes is mainly due to the inefficient waste management practices in the construction projects. Hence, the effective management practices in handling construction waste should be highlighted. This study seeks to identify the waste management practices that are able to engender on-site waste minimization. As much as 54 practices had been found from previous research publications. Mapping technique was used to determine the frequency for each practice. This study have found five (5) significant management practices that need to be emphasized in order to achieve on-site waste minimization. The findings of this paper will help researchers to further investigate the significant management practices in minimizing on-site construction waste.

1. Introduction

Construction industry is very important in every developing country. The construction industry is an economic investment and its relationship with economic development is well posited [1]. Both developing and developed nations have realized and comprehended the important of construction industry in socio-economic level and sustainable development of a country. It plays a vital role towards a growth of a nation by providing essential infrastructures and physical structures for activities such as commerce, services and utilities [2]. The development of construction sector is a part of the living environment which is affecting the living conditions, social well-being and health. The industry also engenders employment opportunities and inject money into nation's economy [3]. However, despite of these contributions, rapid development of construction industry has led to the increasing of construction wastes and this situation had become one of the serious issue in Malaysia.

Generally, construction waste is defined as the unused products or materials which resulted from construction activities during pre-construction, construction and post construction phase [4]. In addition, Tam *et. al,* [5] stated that construction waste is defined as any unwanted products from the construction, renovation and demolition works. Meanwhile, Shen *et. al,* [6] mentioned that construction wastes are any building materials, concrete, steel, timber and other material which are resulted from various construction activities. It can be concluded that construction waste is the excess materials generates from the construction activities. In Peninsular Malaysia, the amount of solid wastes generated per day increased from an estimated 23,000 tonnes in 2010 to 25,000 tonnes in 2012 [7]. Currently, Solid Waste and Public Cleansing Management Corporation of Malaysia confirms that approximately 8 million tonnes of construction wastes per year generated from construction projects [8]. The increasing of construction wastes is caused by inefficient waste management practices in the construction projects [9]. All the construction wastes that produced from construction activities will cause negative impact towards surrounding environment, cost, productivity, time, social and economy [10]. In construction industry perspective, this issue will contribute to the value reduction of construction productivity and reduce the performance of overall projects.

Based on the discussion above, an effective construction waste management is the solution to overcome the issue. According to Yuan [11], construction waste management is one of the sustainable development approaches to minimise wastes and to avoid negative impacts on the environment. It is necessary to highlight construction waste management in order to provide mutually beneficial to the environment, society and the economy. Hence, this paper seeks to identify construction management practices which will cause the minimization of on-site construction wastes.

2. Construction waste management

The increasing awareness regarding environmental impacts from construction wastes has led to the development of waste management as an important approach for construction project management. Managing construction wastes is vital in order to cope with future sustainable development. The practices of waste management for construction activities comprised with the aim to protect the environment and with the recognition that wastes from construction and demolition works contribute significantly to pollution of the environment [6]. Fauziah & Agamuthu [12], defined waste management as the discipline that encompasses solid waste generation, storage, collection, transport, processing, and disposal by considering the environmental, economic, aesthetics and public concerns. In addition, the management of waste includes monitoring, collection, transport processing and waste disposal.

There are many efforts that have been carried out by the Malaysian government to minimise the generation of waste. Nevertheless, many contractors failed in implementing good waste management which led to the mismanagement of construction waste [13]. There are several approaches to construction waste management. The process of managing construction waste goes far beyond the disposal of the wastes itself. It is encompassing a strategy to effectively utilize construction resources, with the view to reduce the quantity of waste and utilizing the generated waste in the most effective manner. In Malaysia, disposing the wastes directly to landfill sites is the most common approach in managing construction wastes. This method is chosen among contractor in Malaysia because the waste materials is assumed to have a little premium value. However, the practices will no longer be applicable for a long term since construction industry had generated significant amount of wastes and there is increasing of construction wastes year by year that will further congest the already over-flowing landfills. As the evidence, it was reported that among 289 landfills which also includes dumpsites, 113 of these landfills are no longer in operation due to protest from surrounding residents or have reached their full capacity [14]. This has prompted the need for alternatives of waste prevention and the initiatives to reduce, reuse and or recycle wastes produced which are referred to as the 3R's of construction waste management.

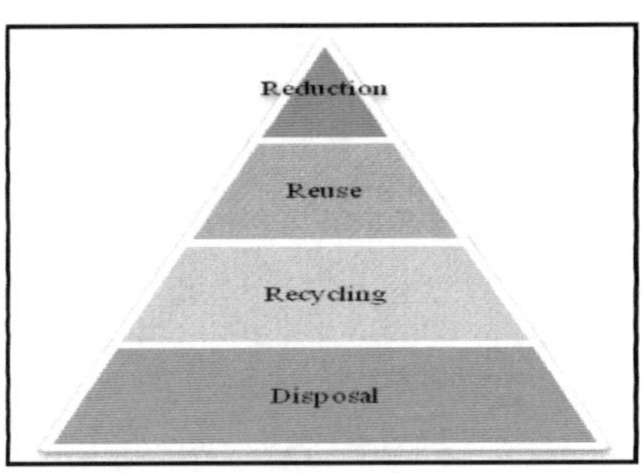

Figure 1. Waste management hierarchy [14].

Basically, construction wastes need to go through a pre-treatment process before being disposed to landfills. It should be treated according to proper waste management hierarchy as shown in Figure 1 [14]. There are 3R concepts in waste management that need to pass through before being disposal at landfills. The criteria of waste management started with waste reduction, reuse, recycling and lastly disposed to landfills. The process of construction waste management at the end will end up at landfill [12]. Hence, it is vital to effectively utilise the construction resources in order to reduce the generation of wastes.

2.1. Reduce, Reuse & Recycle Strategy

The term of waste reduction is related with the practices to reduce waste generation at it source. This ensure waste is minimized before it become much bigger problem. In construction project, it could be achieved through reducing or avoiding any activities and process that caused waste generation. As early as the planning stage, the reduce strategy can be implemented and continuously embedded throughout the construction lifecycle [15]. The reduction in construction waste is considered the most effective and efficient method to manage construction waste. But, it is an utmost important to recognize the influencing factors in waste generation before the reduce strategy can be implemented. These factors comprised of design changes, poor material handling, lack of capability among the labours, poor management planning, poor site condition, procurement of construction material and external factor (weather) [16].

Reduction of construction wastes at it sources not only able to reduce waste generation, but also reduce waste transportation costs, waste disposal and waste recycling. Even in the most sustainable construction site, waste can't be eliminated totally. The site definitely will produce certain types of waste and this waste is called as unavoidable waste. Since the waste generated is not fully avoidable, the next step should be considered is to reuse some of the construction waste materials such as using broken brick and concrete as a sub-grade of access road to the construction site [14]. In addition, [15] stated that reuse is defined as the using of construction resources which is the material, such as timber formwork, more than once either by means the same function or not. Recycle strategy is defined as a process of collecting the used, re-used and also the unused things that already considered as waste but turned into usable new item [17]. At site, these items are sorted inside the construction site or off-site and will be sent to processing centre to be turned into raw materials or re-manufactured into new or same products. The sorting of waste for recycling purposes could be done as an on-site or off-site waste management technique.

According to [18], the on-site recycling technique is the isolation of the construction waste and then is used as a raw material in other construction processes. Meanwhile, the off-site recycling technique is the separating and transporting of the construction waste to other organizations so that it can be use as the raw material. Government participation is crucial to ensure a successful recycling program in the construction industry. Recycling will not only reduce negative impact to the environment but also will ensure the availability of the materials in the future.

3. The on-site waste management practices

Mapping method of practices is adopted from [19], in order to provide preliminary idea on what past researchers had discovered regarding construction waste management practices that contributed to on-site construction waste reduction. The matrix analysis identified the most prominent practices based on the frequency of the measures which identified by past researchers around the world. Through a review of the identified literature, some site management practices for engendering construction waste minimization are identified as presented in Table 1. All of the stated site management practices for successful on-site waste management in Table 1 are most likely suitable and applicable to be implement in the construction project within Malaysia.

As much as 54 measures for successful on-site waste management are identified from 20 literature review on the research articles. The measures are divided into 5 management groups which are Human Resources, Material and equipment, Construction method, Administrative and Regulation. The grouping for construction management practice is adapted and modified from [20] research. There are 4 measures in Human Resources management group, 9 measures in material and Equipment group, 19 measures in

Construction Method group, 17 measures in Administrative group and 5 measures in Regulation management group.

4. Mapping process and analysis of the on-site waste management practices

Based on the on-site waste minimization management practices in Table 1, the highest frequency for the waste minimization practices are presented in figure 2 with accordance to the code given.

Based on the Figure 2, the most significant practices to minimize on-site construction wastes is A1 (Human resources group) practice following respectively by E1 (Regulation group), C1 (Construction Method group), D1 (Administrative group) and B1 (Material and Equipment group). In 'Human and Resources' group, 18 out of 20 researchers agreed that education and awareness is the important factor to engender waste minimization. A study by [38] found that the major barrier in the industry is the lack of awareness among local contractors, construction labor and architects about waste management techniques and approach. They perceived that conducting construction waste management usually will increase the project costs and therefore will reduce company profits. The construction practitioners are giving more focus on the cost, time, and the quality specified in the contract, less attention is giving towards waste reduction [39]. Thus, the construction industry practitioners' awareness regarding resource saving and environment protection is a vital driver for construction waste minimization. Education and awareness aspect regarding waste management at site will lead construction actors to implement effective practices in reducing generated wastes. However, the awareness of having a proper management of construction wastes is still lacking in Malaysia. Hence, there is an urgent need to improve the education and awareness of construction industry practitioners in Malaysia.

The second highest frequency of on-site waste minimization practices is E1 which is the usage of off-site products and components in construction project. In 'Regulation' group, as much as 16 researchers admitted that this practice is able to provide directly waste minimization at site. Construction Industry Development Board (CIDB) Malaysia, has been dynamically encouraging the practice of industrialized building system (IBS) and other off-site construction techniques in local construction industries since 1998 as a method to overcome construction waste problem, conserve landfill capacity and also helping to achieve the concept of sustainability [40]. Waste reduction is one of the major benefits when using this practice compared with conventional construction. According to [41], the off-site construction techniques such as prefabrication, is perceived as a solution to major causes of waste generated during both the design and construction stages. This practice also contributes to other benefits on-site such as improved quality control, provide clean and safer working place, improved the environmental performance, and reduction in construction time and labour requirements. Unfortunately, because of the higher initial cost that incur during the construction process, there is a significant percentage of construction projects do not adopt the use of off-site construction techniques in Malaysia. Hence, it is important for local authority to force a regulation regarding the application of off-site techniques in construction projects. This practice not only able to reduce construction waste generation at site, but also will lessen the burdens related with its management and disposal.

Next, the third practice for on-site waste minimization is C1, by providing waste skips for specific materials (waste segregation). In 'Construction Method (on-site practice)" group, about 13 of 20 researchers stated this practice is able to reduce waste generated at site. In order to reduce total waste generated, there is a need for effective separation of waste, by providing waste skips for specific materials. This approach is based on the understandings that recycling construction material is one of the best option to reduce negative impact on the environment which also includes in 3R concept of waste minimization. Although waste segregation itself is not a strategy for waste reduction, but it is a requisite act to ease the construction materials reuse and recycling. This practice is applied as a method to facilitate construction waste for the recycling purposed after the waste generated in construction site. It is because, the application of construction waste recycling requires sorting of generated waste into ''recyclable and non-recyclables" during the construction activities or at the recycling site. This practice will ease recycling operations and ensures accurate separation of inert and non-inert materials. By providing waste skips for waste segregation, there is likelihood of on-site reuse of the materials in waste skips or for other projects [23]. This practice also contributes in preventing waste mixture with soil [20].

As such, waste segregation provides both short and long-term benefits of on-site materials reuse and ease of waste recycling.

The highest frequency for the practice in 'Administrative' management group is the standardization of design and material. Standardization is defined as the extensive use of components, methods or processes in which there is regularity, repetition and a background of successful practice. The aim of this practice is mainly to improve the buildability of a structure along with the benefit in reducing construction waste. This study found that 12 out of 20 literature review had stated this practice is able to minimize on-site waste generation. Standardisation has the potential to dramatically reduce the current production of construction waste. This argument is supported by [22] that a substantial reduction in off-cuts construction materials could be achieved by designing room areas and ceiling heights in multiples of standard material sizes. Nowadays, the increase in complexity of a structure had caused a lot of design changes in construction projects. This situation raised due to the increased integration of components making it more likely that a change in one area of the design will require other areas to be redesigned [29]. This issue will directly contribute towards the generation of construction waste at site. Hence, administrative management play a vital role in the implementation of standardize practice for design and material in order to increase the buildability of a structure. This practice at once will contribute efficiently in reducing on-site construction waste.

Based on the 'Material and Equipment' management group, the study found that the highest frequency of on-site waste minimization is the proper material handling. There are 8 journals out of the 20 review journals stated that the proper material handling practice could reduce waste production. Generally, improper handling of construction material is one of the causes waste generation in construction project. This issue often arises because of the wrong material handling by construction workers at site which contribute towards waste production [19]. The incompetent supervisor and project manager, lead to poor workmanships and improper material handling done by construction workers. Besides that, the used of improper equipment in handling of material will cause damage and loss on the construction materials, subsequently will contribute to waste generation at site. This issue normally occurs during delivery process where the wrong equipment is used to bring down construction material at site. In order to overcome this problem, it is important to use the right piece of handling equipment which aiming to prevent waste due to breakage and loss on the construction materials. In addition, the wrong materials handling by construction workers can be overcome by assigning a good and dedication supervisor at the site.

Table 2 presented the most significant practices in reducing on-site construction waste for each management groups. These factors are prominently able to engender on-site waste reduction in construction project.

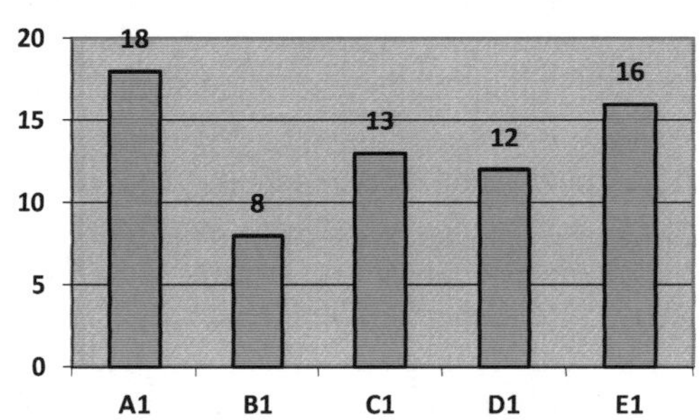

Figure 2. Frequency for the highest practices in each group based on the given code.

Table 1. Construction management practices in reducing on-site construction waste.

Group	On-site Waste minimization practices	Code	References																				Freq.
			[21]	[22]	[11]	[23]	[24]	[25]	[26]	[27]	[28]	[29]	[30]	[31]	[32]	[16]	[20]	[33]	[34]	[35]	[36]	[37]	
(A1) Human Resources	Improved education/ awareness regarding on-site waste management of workforce or/and staff	A1	*	*	*		*	*	*	*	*		*	*	*	*	*	*	*	*	*	*	18
	Appointment of waste manager on-site	A2						*									*			*			3
	Appointment of labors solely for waste disposal	A3															*			*			2
	Preventing of waste materials by labors	A4															*			*			2
(B1) Material and Equipment	Using mechanical handling to reduce damage on the material during delivery (proper handling)	B1						*				*	*	*	*	*	*			*			8
	Prevention of over ordering of material	B4		*		*		*					*			*			*	*			7
	Proper selection of materials (e.g. avoiding fragile material being used)	B2			*					*						*	*		*	*	*		7
	Use materials with a high content of recycled material	B3					*				*		*				*	*		*			6
	Prevention of double handling materials	B5				*									*	*	*			*			5
	Buy materials avoiding unnecessary packaging	B6			*		*						*						*				4
	Just in Time Deliveries (JIT) ensure less material waste	B7	*	*								*	*										4
	On-site materials shredders or compactors machine	B8		*		*	*		*														4
	Store construction materials collectively in a separate area	B9			*	*							*										3
(C1) Construction Method (on-site	Provision of waste skips for specific materials (waste segregation)	C1	*	*	*	*	*	*	*			*	*				*		*	*		*	13
	On-site reuse construction materials	C2	*		*	*		*	*			*	*		*		*	*	*		*		12
	Provide adequate and safe storage to lessen the amount of damaged material on-site.	C3	*					*	*				*	*			*			*			7

	Set-up an effective line of communication at construction site to avoid bad practice	C4			*		*			*		*				*					*	*	7
	Provide sufficient space to ease the management of C&D waste	C5				*	*		*								*			*			5
	Providing bins for collecting wastes for each subcontractor	C6				*	*										*			*			4
	Setting up temporary bins at each building zone	C7				*	*										*			*			4
	Periodic check on the use of C&D waste containers	C8				*	*								*								3
	Soil remains to be used on the same site	C9				*	*														*		3
	Storing waste at an easily accessible area	C10															*		*	*			3
	Noticing staff to reuse recycle materials	C11															*	*		*			3
	Preventing waste mixture with soil	C12				*											*			*			3
	Detect the construction activities that can admit reusable materials from the construction	C13				*	*					*											3
	Follow the project drawings / design	C14				*	*																2
	Informing methods to deal with the rest of wastes after recycling	C15															*			*			2
	Keeping the site clean to minimize loss of material and waste generation	C16											*	*									2
	Time scale of project can ensure effective waste reduction on-site	C17	*										*										2
	Use of demolition and excavation materials for landscape	C18				*																	1
	Central areas for cutting and storage	C19				*																	1
(D1) Administrative	Standardization of design and material to improve buildability	D1		*	*	*			*	*		*	*	*	*		*			*	*		12
	Adequate supervision of waste management activities with clear instruction	D2								*	*	*	*			*	*			*		*	8
	Ensure fewer design changes during construction process	D3			*	*					*	*			*	*			*			*	8

	Supply chain alliances with supplier and recycling companies	D4	*	*	*			*				*	*									*	7
	Waste auditing to monitor and record on-site waste management	D5		*		*		*							*		*			*		*	7
	Making sub-contractors responsible for waste disposal (cooperation)	D6				*			*			*					*			*		*	6
	Contract suppliers managing their product waste	D7		*			*					*	*				*			*			6
	Plan layout of construction projects properly (site planning)	D8			*				*					*	*	*					*		6
	Designing out waste will reduce material management issues on-site.	D9	*				*											*			*		4
	Recycling target to be set for every project phases	D10				*						*			*			*					4
	Involvement of contractors and product manufacturer during design stage	D11				*							*									*	3
	Positive incentive for decreasing or recycling by contractors/ subcontractors	D12															*		*	*			3
	Rules on dealing with waste-generators	D13															*			*			2
	Shortening a period of collecting waste in a site	D14															*			*			2
	Design management to prevent the over specification of materials	D15		*																	*		2
	Corrective actions to reduce waste when a change has happened	D16																				*	1
	Adequate planning to stabilize work process on-site	D17																				*	1
(E1) Regulation	Usage of offsite products and component (low waste construction technology)	E1		*	*	*	*	*		*	*	*	*	*			*	*	*	*	*	*	16
	The use of a Site Waste Management Plan (SWMP) is important.	E2	*					*	*	*		*	*		*		*	*		*	*	*	12
	Contractual clauses to penalize poor waste performance	E3		*	*							*	*				*			*			6

IOP Conf. Series: Materials Science and Engineering **713** (2020) 012038 doi:10.1088/1757-899X/713/1/012038

	Additional tender where waste initiatives are to be implemented	E4		*												*			*		*	4
	Issuing waste management levels from owners/clients to contractors	E5														*			*		*	3

Table 2. The highest frequency of on-site waste minimization practices based on management groups.

Management group	**On-site waste minimization practices**
Human Resources	Improved education/ awareness regarding on-site waste management of workforce or/and staff
Material and Equipment	Using mechanical handling to reduce damage on the material during delivery (proper handling)
Construction Method	Provision of waste skips for specific materials (waste segregation)
Administrative	Standardisation of design and material to improve buildability
Regulation	Usage of offsite products and component (low waste construction technology)

5. Conclusion
This study has been carried out with the aim to identify construction management practices which will engender the minimization of on-site construction wastes. As much as 20 research papers had been reviewed and the result of most significant practices for each management group has been clarified through mapping method. This study found that improving awareness and education of workforces regarding construction waste management is very important to achieve on-site waste minimization. In addition, the practices of the usage offsite products and component (low waste construction technology), provision of waste skips for specific materials (waste segregation), standardization of design and material, and proper handling of construction materials also important in order to encourage on-site construction waste reduction. This study provided soft measures that could be applied in construction site management practices, irrespective of the construction techniques.

6. References

[1] Olanrewaju A L and Abdul-Aziz A R 2015 *An Overview of the Construction Industry. In Building Maintenance Processes and Practices* (Singapore: Springer) pp 9-32

[2] Papargyropoulou E, Preece C, Padfield R and Abdullah A A 2011 Sustainable construction waste management in Malaysia: A contractor's perspective *Proc. of the Int. Conf. on Manag. Innov. a Sustain. Built Environ. (MISBE2011)(Amsterdam)* (Holland: Delft University of Technology) pp 1-10

[3] Wibowo A 2009 The contribution of the construction industry to the economy of Indonesia: A systemic approach *Proc. of the Int. Symp. Const. in Developing Economies: Commonalities Among Diversities (Penang)* pp 279-287

[4] Rahim M H I A, Kasim N, Mohamed I, Zaina R, Sarpin N and Saikah M 2017 Construction waste generation in Malaysia construction industry: illegal dumping activities *GCoMSE2017 (Johor Bahru)* vol 271 (Bristol: IOP Publishing) pp 1-8

[5] Tam V W Y, Tam C M, Zeng S X and Ng W C Y 2007 Towards adoption of prefabrication in construction *Int. J. of Build. Science and Its Applications* **42** 3642-3654.

[6] Shen L Y, Tam V W, Tam C M and Drew D 2004 Mapping approach for examining waste management on construction sites *J. of Const. Eng. and Mngmt.* **130** 472-481

[7] Eusuf, M A, Ibrahim M and Islam R 2012 The construction and demolition wastes in Klang Valley, Malaysia *J. of the Malaysian Institute of Planner* **X** 99-124
[8] Saadi N, Ismail Z and Alias Z 2016 A review of construction waste management and initiatives in Malaysia *J. of Sustainability Science and Management* **11** 101-114
[9] Marzouk M and Azab S 2014 Environmental and economic impact assessment of construction and demolition waste disposal using system dynamics *Resources, Conservation and Recycling* **82** 41-49
[11] Yuan H 2013 A SWOT analysis of successful construction waste management *J. of Cleaner Production* **39** 1-8
[12] Fauziah S H and Agamuthu P 2012 Trends in sustainable landfilling in Malaysia, a developing country *Waste Management and Research* **30** 656-663
[13] Tey J S, Goh K C, Kek S L and Goh H H 2013 Current practice of waste management system in Malaysia: Towards sustainable waste management *1st FPTP Postgraduate Seminar 'Towards Sustainable Management' (Batu Pahat)* (Batu Pahat: Universiti Tun Hussein Onn Malaysia) pp 1-5
[14] Nagapan S, Rahman I A and Asmi A 2012 Construction waste management: Malaysian perspective *Int. Conf. on Civil and Env. Eng. Sustainability (IConCEES) (Johor Bahru)* vol 2 (Batu Pahat: Penerbit UTHM) pp 299-309
[15] Esa M R 2017 *Moving towards sustainable construction in Malaysia: a holistic model for construction and demolition (C&D) waste management* Master Degree Thesis (Brisbance: University of Queensland)
[16] Nagapan S, Rahman I A and Asmi A 2012b Factors contributing to physical and non-physical waste generation in construction industry *Int. J. of Adv in Applied Sciences* **1** 1-10
[17] Sreenivasan J, Govindan M, Chinnasami M and Kadiresu I 2012 Solid waste management in Malaysia: a move towards sustainability *Waste management-An integrated vision ed.* L F M Rebellon (London: Intech) 55-70.
[18] Franchetti M J and Apul D 2012 *Carbon footprint analysis: concepts, methods, implementation, and case studies* (Florida: CRC Press)
[19] Nagapan S, Ismail A R and Ade A 2011 A review of construction waste cause factors *Proc. of the Asian Conference of Real Estate: Sustainable Growth Managing Challenges (ACRE) (Johor)* pp 1-8
[20] Jingkuang L and Yousong W 2011 Establishment and Application of Performance Assessment Model of Waste Management in Architectural Engineering Projects in China *Syst. Eng. Procedia* **1** 147–155
[21] Marinelli M, Dolan M, Spillane J and Konanahalli A 2014 Material Waste in the Northern Ireland Construction Industry: On-Site Management, Causes and Methods of Prevention *Proc. of 30th Annual ARCOM Conference (Portsmouth)* (United Kingdom: Association of Researchers in Construction Management) pp 113–122
[22] Dainty A R J and Brooke R J 2004 Towards Improved Construction Waste Minimisation: A Need for Improved Supply Chain Integration? *Struct. Surv.* **22** 20–29
[23] Ajayi S O, Oyedele L O, Bilal M, Akinade O O, Alaka H A and Owolabi H A 2017 Critical Management Practices Influencing On-site Waste Minimization in Construction Projects *Waste Management* **59** 330–339
[24] Saez P V, del Río Merino M, González A S A and Porras-Amores C 2013 Best Practice Measures Assessment for Construction and Demolition Waste Management in Building Constructions *Resour. Conserv. Recycl.* **75** 52–62
[25] Al-Hajj A and Iskandarani T 2012 Reducing Waste Generation on the UAE Construction Sites *7th International Conference on Innovation in Architecture, Engineering and Construction (São Paulo)* pp 1-13
[26] Gangolells M, Casals M, Forcada N and Macarulla M 2014 Analysis of the implementation of effective waste management practices in construction projects and sites *Resources, conservation and recycling* **93** 99-111

[27] Udawatta N, Zuo J, Chiveralls K and Zillante G 2015 Improving waste management in construction projects: An Australian study *Resources, Conservation and Recycling* **101** 73-83
[28] Wang J, Li Z and Tam V W Y 2014 Critical Factors in Effective Construction Waste Minimization at The Design Stage: A Shenzhen Case Study, China *Resour. Conserv. Recycl.* **82** 1–7
[29] Oyedele L O, Regan M, Meding J V, Ahmed A, Ebohon O J and Elnokaly A 2013 Reducing waste to landfill in the UK: identifying impediments and critical solutions *World J. Sci. Technol. Sustain. Dev.* **10** 131–142
[30] Marinelli, M, Dolan M, Spillane, J P and Konanahalli, A., 2014. Material waste in the Northern Ireland construction industry: On-site management causes and methods of prevention.
[31] Hassan S H, Ahzahar N, Fauzi M A and Eman J 2012 Waste management issues in the northern region of Malaysia *Procedia-Social and Behavioral Sciences* **42** 175-181
[32] Yeheyis M, Hewage K, Alam M S, Eskicioglu C and Sadiq R 2013 An overview of construction and demolition waste management in Canada: a lifecycle analysis approach to sustainability *Clean Technologies and Environmental Policy* **15** 81-91
[33] Lu W and Yuan H 2011 A framework for understanding waste management studies in construction *Waste Management* **31** 1252-1260
[34] Begum R A, Siwar C, Pereira J J and Jaafar A H 2007 Implementation of waste management and minimisation in the construction industry of Malaysia *Resour.. Conserv. Recycl.* **51** 190–202
[35] Cha H S, Kim J and Han J Y 2009 Identifying and Assessing Influence Factors on Improving Waste Management Performance for Building Construction Projects *J. Constr. Eng. Management* **135** 647–656
[36] de Magalhães R F, Danilevicz Â D M F and Saurin T A 2017 Reducing construction waste: A study of urban infrastructure projects *Waste Management* **67** 265-277
[37] Domingo N, Osmani M and Price A D 2009 Construction Waste Minimisation in the UK Healthcare Industry *Procs. of the 25th Annual ARCOM Conference (Nottingham)* vol 2 (United Kingdom: Association of Researchers in Construction Management) pp 1021–1030
[38] Jain M 2012 Economic Aspects of Construction Waste Materials in terms of cost savings–A case of Indian construction Industry *Int. J. of Scientific and Research Publications* **2** 1-7
[39] Lu W and Yuan H 2010 Exploring critical success factors for waste management in construction projects of China, *Resources conservation and recycling* **55** 201-208
[40] Azman M N A, Kamar K A M and Nawi M N M 2013 Industrialised Building System in Reducing Waste of Construction Industry *J. of Science and Technical Education* **2** 96-103
[41] Jaillon L, Poon C S and Chiang Y H 2009 Quantifying the waste reduction potential of using prefabrication in building construction in Hong Kong *Waste management* **29** 309-320

Acknowledgment
This research was supported by Universiti Tun Hussein Onn Malaysia under the *Geran Penyelidikan Pasca Siswazah* (GPPS) - U783.

IOP Conf. Series: Materials Science and Engineering **713** (2020) 012039 doi:10.1088/1757-899X/713/1/012039

A Review on The Critical Influencing Factor for Life Expectancy (LE) and The Life Expectancy (LE) for Malaysian Housing

K Norfarahayu[1], M Hairuddin[2], A Nor Haslinda[1] and Y Siti Khalijah[1]

[1] Depart. of Building Eng. and Construction, Faculty of Civil and Environmental Eng., Univ. Tun Hussein Onn Malaysia, Parit Raja, 86400 Batu Pahat, Johor, Malaysia
[2] Centre for Dip. Studies, Univ. Tun Hussein Onn Malaysia, Pagoh Higher Edu. Hub, KM1, Jalan Panchor, 86400 Pagoh, Muar, Johor, Malaysia.
Corresponding author: hairuddinm@uthm.edu.my

Abstract. In 2013, there are two death cases reported due to the broken railing at the Projek Perumahan Rakyat (PPR) flats in the capital city of Malaysia. These cases occurred due to rusting of the balcony railing, which has been neglected in maintenance. These were the evidence of awareness lacking on the need for maintenance, which frequently only be realised at the latter stage of deterioration on house components. The situation aggravated when the house owners do not have any technical knowledge about defects and have zero building construction background, which later on resulting in less to none periodic assessment throughout the life expectancy (LE) of house components. Nevertheless, the absence of Malaysian study on the life expectancy of house components is deemed the centre of the predicament. Therefore, the paper aims to review the critical influencing factor for life expectancy (LE) and the Life Expectancy (LE) for Malaysian housing. Thus, a qualitative undertaking is conducted through multi-layered thematic analysis. As a result, there were eight influencing factors of life expectancy (LE) namely; weather and climate condition; design; maintenance and replacement; the intensity of use; quality of materials and products; technological obsolescence; workmanship; and insects damaged. Afterwards, within 114 number of generic house components, the findings indicated that there were differences of LE for house components; averaging from 14 years to more than 100 years. The particular differences also depend on the influencing factors of LE. These findings were then become the basis for the subsequent research endeavour.

1. Background of research

In a construction project, there are several phases involved, namely, planning, development, and maintenance. The maintenance phase is undoubtedly an extended period, which falls under the scrutiny of the project owner during Defect Liability Period (DLP) (usually 18 months to 24 months) and legitimate house buyers afterwards [1,2]. Maintenance is the processes and services required to undertake to preserve, protect, enhance, and care of the building [3]. The previous study has been concluded that building maintenance practices are lack of a comprehensive maintenance management framework that guides the decision-making processes. Besides that, building maintenance is also able to increase the performance and life span of the building, and it also helps in term of investments' value [4]. However, according to Alshehri, Motawa and Ogunlana [5], one of the problems in maintenance is inadequate financial [5]. That statement is in coherence with Che-Ghani [6] which indicate that the cost for maintaining a house is relatively high due to poor maintenance practices [6]. Thus, to avoid a massive flow of money, building maintenance needs to be addressed at the early stage after the completion of the building. Only then it will ensure the optimum performance of the building throughout their lifespan.

In term of house owners, most of them are not aware of the needs of maintenance until it is too late, i.e. when the deteriorated components were severely damaged and exposed. As a result, failure to replace or repair the deteriorated components at the earliest time possible will cause injury to people and further damage to the property. Among others, this was transcended to the tragedies in 2013 on two recorded death cases involving infants falling from 14th and 6th floor of Projek Perumahan Rakyat (PPR) flats in Petaling Jaya – due to a defect of railing [7,8]. Therefore, realising the importance of having a pre-rectification action at early stages, this study seeks to establish the term of references for LE factors and their average LE towards house components for the improvement of housing maintenance in Malaysia.

Nonetheless, due to the lengthy nature of the study, the particular paper only reports on the preliminary study on identification of influencing LE factors for generic house components, and also the LE for each components through literature review. Hence the objectives, the study focuses on three (3) categories of house components, which are structural, architectural, and electrical – which can be assessed using sense perceptions such as vision, hear, touch, and smell [1]. Limitation of sense perceptions is to compensate the nature of house buyer that lack a technical background. Meanwhile, buildings category A (landed houses) which comprised of detached houses, semi-detached houses, terrace houses, and cluster houses were selected as a study parameter [9].

In general, the paper has contributed towards bridging the knowledge gap on factors influencing LE for generic house components and LE for each component in Malaysia. Apart, through the initial qualitative processes, a clear understanding of LE is developed. Which unveiled that the average LE is varied according to the identified LE factors, especially weather and climate condition of the respective country.

2. Life Expectancy (LE)

House components' life expectancy (LE) is defined as the lifetime or the average number of years that users could expect for the lifespan of the house components, which generally based on certain factors. Some researchers also determined LE as the statistical range of years (e.g. 10-20 years, etc.) of the remaining life of the components before they break, damage, need further maintenance or need replacement [10]. Besides, they also stated that the age of 10 to 12 years is the common starting point of the defects' visibility throughout the building's lifetime [10].

A study conducted towards LE was carried out in 2006 by the National Association of Home Builders (NAHB) from United States of America (US). The research has emphasised that the level of maintenance is significantly affected by the LE of house components in the US. Meanwhile, other factors that influence the LE of house components are; changing of styles and technology, exposure to weather and climate condition, the intensity of use, maintenance and replacement, qualities of products and materials, and installation's quality (i.e. workmanship) [11].

On the other hand, while associating LE with construction procedures and practices, Bakri and Mydin [12] and Mohamad and Annuar [13] stated that, the house floor must be designed at a reasonable height above the ground level to prevent water from creeping into the house. On construction, the earth fillings should be appropriately watered and rammed, and good quality of soil should be used below the floor. Apart, concrete bases from dry bricks or metals, with or without sand and cement/lime should be provided to prevent floor settling. Besides, in term of floor materials, it should also be in good quality, low maintenance, and with minimal environmental effect [12,13].

3. Methodology

The paper serves as an exploratory study by exploring and understanding the central phenomenon of the research through qualitative data. The use of exploratory study is a well-established approach to understand the central phenomenon of a study from multiple sources at various time points [14]. The objectives of this study are to identify the factors influencing the LE of generic house components and the LE of each component through preliminary literature review. Therefore, data collection processes have been divided into two phases which are; (i) the identification of generic house components in

Malaysia (Phase I), (ii) and identification of factors influencing the LE of generic house components and LE of each component (Phase II).

3.1. Phase I: Identification of generic house components in Malaysia

A literature review was carried out to identify the generic house components in Malaysia towards numerous published materials (e.g. journals, magazines, government reports, proper standards and guidelines, unpublished thesis and website document). Through 'Multi-layered Thematic Process' (MLTP), house components were collected and analysed based on the basis of a completed house after construction. These were further vetted on components which can be assessed using sense perceptions (i.e. vision, hear, touch, and smell) and as well as its compatibility with the Malaysian housing environment (see Figure 1).

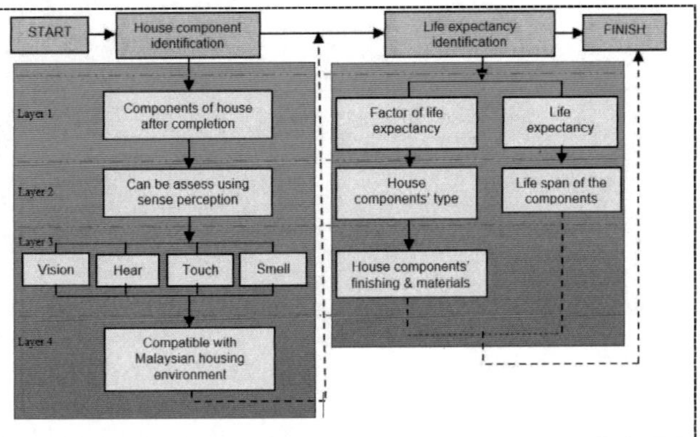

Figure 1. Multi-layered Thematic Process (MLTP) [1]

3.2. Phase II: Identification of factors influencing the life expectancy (LE) of generic house components and LE of each component

Subsequently, by referring to the checklist of generic house components in Phase I, a critical literature review was carried out to identify the factors influencing the LE of the components and the LE of each them. Data collection for both aspects were also embedded within the MLTP processes in order to present meaningful outcomes. For factors' identification, layers taken were the factors, the type of house components, and their finishing and associated materials. Meanwhile, LE identification will just screen the information on the life span of the particular components (see Figure 1).

4. Result and discussion

4.1. Phase I: Identification of generic house components in Malaysia

The identification of generic house component in Malaysia was carried out through reviewing established literature such as standards, guidelines, books, published articles, and product magazines and brochures around the globe.

The initial results found that there are 114 generic house components including their sub-components, materials, and finishing. Furthermore, according to the previous studies conducted by Kariya *et. al.* [15] and Mohammad *et al.* [16], the generic components were manually analysed and subdivided into several clusters including macro-components, micro-components, and materials. As a result, 10 main house components (e.g. column & beam, etc.), 17 macro-components (e.g. stair, etc.), 24 micro-components (e.g. panel, etc.), and 16 finishes (e.g. concrete, etc.) were identified as generic house components in Malaysia. The following table 1 shows the findings for generic house components in Malaysia and their references. As an example, the staircase is one of the generic house components which can be assessed through vision (e.g. through eyes-on cracking issue) and touch (e.g. through the hand-on the evenness of tile finishes). Their macro-components are stair (e.g. riser), handrail, baluster, and landing. Meanwhile, their finishes are carpet, tiles, and laminated flooring. However, the table only shows the main house components, macro components, and their finishes.

Table 1. Generic house components in Malaysia

House Component	Macro component	Finishes	Sources
Column & beam	Concrete Wood		Published literatures: [9,11,15-18]
Staircase	Stair (e.g. Riser, tread) Handrail Baluster Landing	Carpet Tiles (i.e. marble, granite, etc.) Laminate flooring	Published literatures: [5,11,17-25] Catalogues: *Wira Sky Steel Sdn Bhd, Supercool Creative Culture Sdn Bhd, COD Marketing Sdn Bhd, Weng Heng Stainless Steel Sdn Bhd, Wira Wrought Iron & Stainless Steel Sdn. Bhd., Solid-Glass & Metal Industries Sdn Bhd.*
Door	Folding door French door Sliding door	Glazing Vinyl Metal Wood Fibreglass	Published literatures: [5,9,11-12,15,17-20,22,25] Catalogues: *Amax Lifestyle Product Sdn Bhd, Reliance Home Sdn Bhd, Space Kete Sdn Bhd, THC Metal Engineering Sdn Bhd, Tropicale Window & door Sdn Bhd, COD Marketing Sdn Bhd, Wira Wrought Iron & Stainless Steel Sdn. Bhd., Honglee Group (M) Sdn Bhd, Solid-Glass & Metal Industries Sdn Bhd, Vin Art Building Materials Sdn Bhd.*
Window	Casement window Top hung window Fixed panel window	Glazing Vinyl Metal Wood	Published literatures: [5,9,11,16-20,22,23] Catalogues: *Amax Lifestyle Product Sdn Bhd, Breezway Window & Door Marketing Sdn Bhd, Reliance Home Sdn Bhd, Space Kete Sdn Bhd, Tropicale Window & door Sdn Bhd, Wira Wrought Iron & Stainless Steel Sdn. Bhd., Solid-Glass & Metal Industries Sdn Bhd.*
Floor	Wood Concrete	Tiles (i.e. marble, granite, porcelain, etc.) Laminated flooring Resilient (i.e. Rubber, plastic, PVC, Linoleum, Vinyl) Carpet	Published literatures: [5,9,11,13,17-20,22-23,25-26] Catalogues: *Orion Contract Sdn Bhd, Lumber Mart Sdn Bhd, Netwood, Sugawa Lonwood Sdn Bhd, Swoods Sdn Bhd.*
Wall	Internal wall External wall	Facing brick Plaster Paint Tiles (i.e. marble, granite, porcelain, etc.) Wallpaper Wood wall panel	Published literature: [5,9,19-26] Catalogues: *Orion Contract Sdn Bhd, Decasa Marbles Sdn Bhd.*
Ceiling	Suspended ceiling	Plaster / skim coat	Published literatures: [5,9,17-20,23-24,26]
Roofing	Flat roof Pitched roof Skylight & awning	Concrete Metal PVC Asbestos Wood	Published literatures: [5,9,11-12,16-20,22,24-25,27] Catalogues: *Wira Sky Steel Sdn Bhd, De Skylight Roofing Sdn Bhd, Delux Structural Works Sdn Bhd, Monier Malaysia Sdn Bhd, Supercool Creative Culture Sdn Bhd, COD Marketing Sdn*

			Bhd, Delux Stainless steel Sdn Bhd, Solid-Glass & Metal Industries Sdn Bhd, Vin Art Building Materials Sdn Bhd.
House fittings & fixtures	Toilet & bathroom Kitchen Electrical		Published literatures: [5, 9, 11, 17-20, 22] Catalogues: *Gold Rich Stoneworks Sdn Bhd, Bathcare Marketing Sdn Bhd, Duravit & Hansgrohe, Ga Hing Trading Sdn Bhd, Kosa Malaysia Sdn Bhd, Lg Global (VTO Bath), Molecq Sdn Bhd, Solid-Glass & Metal Industries Sdn Bhd, The Ultimate Bath Gallery.*
Ironmongery	Hinges Door closer Lock Handle	Metal	Published literatures: [5,12,17-21,23,28]

4.2. Phase II: Identification of factors influencing the LE of generic house components and LE of each component

Consequently, the identification of factors influencing LE of generic house component and LE of each component was also carried out through reviewing established materials based on the previous checklist developed in Phase I.

Based on the data collection process for influencing LE factors, 16 established sources were analysed. The findings shows that, there are eight (8) common factors, namely; (1) weather and climate condition, (2) design, (3) maintenance and replacement, (4) intensity of use, (5) quality of materials and products, (6) technological obsolescence, (7) workmanship, and (8) insects damage. Table 2 below shows those factors which were tied to house components and its sources.

Table 2. Generic house components in Malaysia

Generic house components	Factors influencing the LE		Sources
Column & beam	▪ Design ▪ Insect damaged ▪ Maintenance & replacement ▪ Quality of materials & products	▪ Technological obsolescence ▪ Weather & climate condition ▪ Workmanship	[5, 11, 13, 15, 17-18, 20, 22-23, 25, 28-29]
Staircase	▪ Design ▪ Usage & user activities ▪ Insect damaged ▪ Level of maintenance	▪ Quality of materials & products ▪ Technological obsolescence ▪ Weather and climate condition ▪ Workmanship	[5,11,13,15,17-18,20,22-23,25,28-29]
Door	▪ Design ▪ Intensity of use ▪ Insect damaged ▪ Level of maintenance	▪ Quality of materials & products ▪ Technological obsolescence ▪ Weather and climate condition ▪ Workmanship	[5,11,12,15,17-18,20,22-23,25,28-29]
Window	▪ Design ▪ Intensity of use ▪ Insect damaged ▪ Level of maintenance	▪ Quality of materials & products ▪ Technological obsolescence ▪ Weather and climate condition ▪ Workmanship	[5,11,15,17-18,20,22,25,28-29]

Floor	▪ Design ▪ Level of maintenance ▪ Quality of materials & products	▪ Usage & user activities ▪ Weather and climate condition ▪ Workmanship	[5,11,13,17-18,20,22-23,25, 26,28-29]
Wall	▪ Design ▪ Intensity of use ▪ Insect damaged ▪ Level of maintenance	▪ Quality of materials & products ▪ Technological obsolescence ▪ Weather and climate condition ▪ Workmanship	[5,11,13,15,17-18,20,22-24,28-29]
Ceiling	▪ Design ▪ Intensity of use ▪ Insect damaged ▪ Level of maintenance	▪ Quality of materials & products ▪ Technological obsolescence ▪ Weather and climate condition ▪ Workmanship	[5, 11, 15, 17-18, 20, 23, 28-29]
Roofing	▪ Design ▪ Intensity of use ▪ Insect damaged ▪ Level of maintenance	▪ Quality of materials & products ▪ Technological obsolescence ▪ Weather and climate condition ▪ Workmanship	[5, 11-12, 15, 17-18, 20, 22, 24-25, 29]
House fittings & fixtures	▪ Design ▪ Intensity of use ▪ Insect damaged ▪ Level of maintenance	▪ Quality of materials & products ▪ Technological obsolescence ▪ Weather and climate condition ▪ Workmanship	[5,11,15,17-18,20,22,28-29]
Common finishing for all components	▪ Design ▪ Level of maintenance ▪ Usage & user activities	▪ Quality of materials and products ▪ Weather and climate condition ▪ Workmanship	[5,12,13,15,17,18,20,22,28-30]
Ironmongery	▪ Climate condition ▪ Design ▪ Level of maintenance	▪ Usage & user activities ▪ Weather & climate condition ▪ Workmanship	[5,11-12,15,17-18,20,22,25,28,29]

Meanwhile, the identification of LE for each generic house components was carried out similarly through MLTP process of five (5) established standards and guidelines comprised of Adena Certified Inspections (US) [20], Economics Group of NAHB (US) [11], International Associations of Certified Home Inspectors (US) [34], The Cascade Team Real Estate (US) [29], and British Columbia (Canada) [35]. Table 3 below shows the minimum, maximum, and average years of the LE towards house components and their materials.

Based on Table 3, it can be seen that generic house components which have minimum LE of below than ten years are vinyl material for roofing, toilet and bathrooms fitting and fixtures, as well as house finishing. Whereas, brick/concrete material for wall, column and beam has a lifetime of more than 100 years. In a nutshell, it can be concluded that the overall generic house components have different LE towards a different type of materials, which happened to be depending on the certain influencing factors.

Table 3. Life Expectancy (LE) of generic house components based on several sources

House components		Summary LE (years)		
		Min. LE	**Max. LE**	**Average**
Column & beam	Concrete	10	100+	70
	Timber	10	100+	34
Staircase	Concrete	-	100+	100+
	Timber	10	50	31
	Metal (e.g. copper, stainless steel, iron, etc.)	15	100	33
	Glazing	10	40	23
Door	Glazing	10+	20	17
	Vinyl (e.g. PVC, uPVC) – for shower	20	50+	29
	Metal (e.g. aluminium, stainless steel, etc.)	15	50+	28
	Fibreglass	10+	20	17
	Timber	15	100	44
Window	Glazing	10+	20	22
	Vinyl (e.g. PVC, uPVC) – for shower	20	40	28
	Metal (e.g. aluminium, stainless steel, etc.)	15	50	27
	Timber	15	50	29
Floor	Concrete	30	100+	72
	Timber	15	100+	45
Wall	Brick/concrete	-	100+	100+
	Glazing	10+	20	17
	Timber	10	100+	43
Ceiling	Gypsum board	30	75	58
	Timber	20	50	36
	Vinyl (e.g. PVC, uPVC)	20	30	25
	Metal (e.g. aluminium, etc.)	15	80	34
	Fibreglass	10+	20	14
Roofing	Concrete	20	100+	73
	Metal (e.g. aluminium, zinc, etc.)	15	100+	41
	Vinyl (e.g. PVC, uPVC)	8	25	17
	Tiles (e.g. clay, etc.)	20	100+	61
	Timber	15	30	27
House fittings & fixtures	Toilet & bathrooms	5	100+	25
	Kitchen	14	30	19
	Electrical	8	40	43
Common finishing for all house components	Indoor/Outdoor	5	100+	34
Ironmongery	Metal (e.g. stainless steel, aluminium, etc.)	15	50	28

Thus, from the overall findings, a list of the suggested range of LE which compatible to be used has been developed. Is was being stratified into six (6) scales which are; (1) 0-10 years, (2) 11-20 years, (3) 21-30 years, (4) 31-40 years, (5) 41-50 years, and (6) 51 years and above [36]. The interval between scales are in 10 years, given the summary of LE from literature review indicated five years as a minimum value and 100 years as a maximum value. Thus, to compensate for the wide differences in range and to offer a meaningful representation, hence the interval. Last but not least, figure 2 shows the conceptual framework for the literature review results.

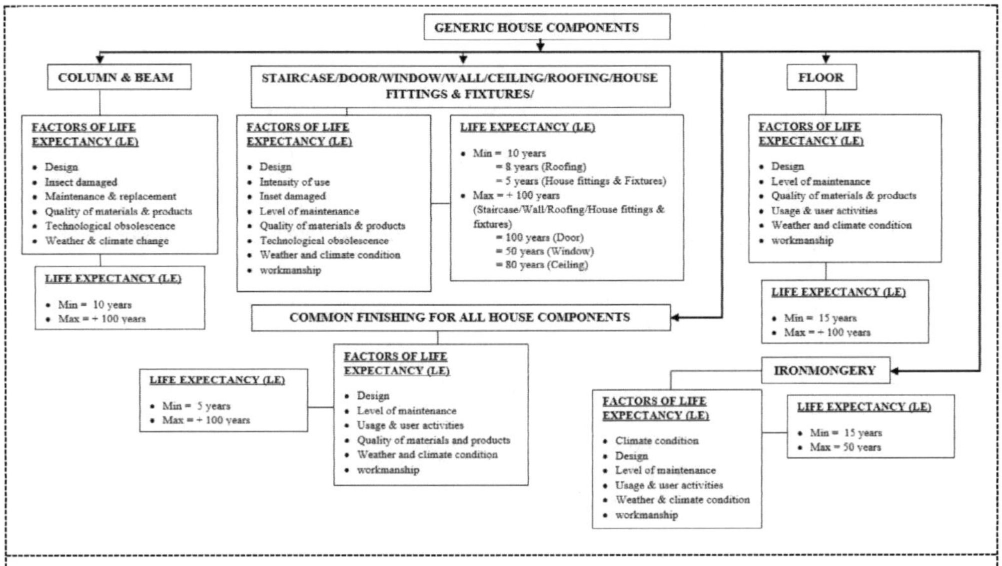

Figure 2. Conceptual framework.

5. Conclusion

The paper set out to identify the factors influencing the LE of generic house components and their LE through literature review processes employing the qualitative MLTP. The findings found that in general there are various LE's factors including weather and climate condition, design, maintenance and replacement, the intensity of use, quality of materials and products, technological obsolescence, workmanship, and insects damaged. One of the findings that emerged from this study is that every component has different LE depending on the influencing factors towards the LE. Apart from enhancing the related body of knowledge, the findings will serve as a basis for subsequent exploration of research on building maintenance's realm.

Nevertheless, noted that the findings of the paper are subjected to a limitation in term of the type of house involved. The findings were not suitable to be applied for high-rise building, which comprises of mechanical elements such as lift, HVAC, and others since they are not being covered in the study. Therefore, as a recommendation, further research on overall building maintenance, including high-rise residential and the business building was applauded.

6. References

[1] Kariya N, Abas N H, Mohammad H and Yaman S K 2017 Conceptualising the Life Expectancy of House Components in Malaysia *Int. Symposium on Civil and Environmental Engineering 2016* (*Melaka*) vol 103 (Paris: MATEC Web Conf.) 03017

[2] Mohammad H, Yaman S K, Hassan F and Ismail Z 2016 Determining the Technical Competencies of Construction Managers in the Malaysia's Construction Industry *The 3rd International Conference on Civil and Environmental Engineering for Sustainability* (Melaka) vol 47 (Paris: MATEC Web Conf.) pp 1053–1059

[3] Lateef O A, Khamidi M F and Idrus 2010 A Building Maintenance Management in a Malaysian University Campus: A Case Study *Australasian J. of Constr. Economics and Build.* **10** 76–89

[4] Akasah Z A, Abdul R M A and Zuraidi S N F 2011 Maintenance Management Success Factors for Heritage Building: A Framework *12th Int. Conf. on Structural Repairs and Maintenance of Heritage Architecture (Italy)* vol 118 (UK: WIT Trans. on the Built Environ.) pp 653–658

[5] Alshehri A, Motawa I and Ogunlana S 2015 The Common Problems Facing the Building Maintenance Departments *Int. J. of Innovation, Management and Technology* **6** 234-237

[6] Che-Ghani N Z, Myeda N E and Ali A S 2016 Operation and Maintenance Cost for Stratified Buildings: a Critical Review *The 4th Int. Building Control Conf. 2016 (Kuala Lumpur)* vol 66 (Paris: MATEC Web Conf.)

[7] Ch'ng B 2015 *Emergency RM1.8mil Allocation to Repair Rusty Railings at Kota Damansara PPR Flats* The Star Online, Malaysia (29 January 2015)

[8] Ch'ng B 2015 *Five-year-old Falls to His Death from Sixth Floor* The Star Online, Malaysia (29 January 2015)

[9] Construction Industry Development Board Malaysia 2014 *CIS 7:2014 Quality assessment system for building construction works* (Kuala Lumpur: CIDB)

[10] Lee K Q 1990 Building Maintenance & Modernisation Worldwide *Proc. of the Int. Symposium on Property Maintenance Management and Modernisation (singapore)* (Singapore: Longman) pp 138–145

[11] Economics Group of NAHB 2007 *Study of Life Expectancy of Home Components* (United States of America: NAHB)

[12] Omar Bakri N N and Othuman Mydin M A 2014 General Building Defects: Causes, Symptoms and Remedial Work *Eur. J. of Technology and Design* **3** 4–17

[13] Syed <ohamad S B H and Wan Annuar W F H 2011 Engineering Approach System to Assess Defect and Deterioration of Building Structures *Int. Seminar on the Application of Science & Mathematics (Kuala Lumpur)* (Johor: UTHM)

[14] Creswell J W 2012 *Educational research: Planning, conducting, and evaluating quantitative and qualitative research* (United States of America: Person Education)

[15] Kariya N, Yaakob Z, Mohd Sairi M N, Mohammad H, Yaman S K and Abas N H 2016 Investigation of Generic House Components and Their Practical Ways to be Assessed by House Buyers during Defect Liability Period (DLP) in Malaysia *Int. J. Eng. Trans. A-Basics* **29** 1354-1363

[16] Yaman S K, Abdullah A H, Mohammad H and Hassan F 2015 Technical Competency of Construction Manager in Malaysian Construction Industry *Appl. Mech. Mater.* **773–774** 1053–1059

[17] Ashworth A 1996 Estimating the life expectancies of building components in life-cycle costing calculations *Struct. Surv.* **4** 4–8

[18] Home Builders Association of Northwest Indiana 2015 *How Long Will It Last? The Life Expectancy of Your Home's Components* Retrieved on January 19, 2016 from https://www.nwitimes.com/lifestyles/home-and-garden/hba-how-long-will-it-last-the-life-expectancy-of/article_2f4e9fdd-f70c-50d2-8395-ca300d3dc463.html

[19] Che-Ani A I, Zaharim A, Mohd Zain M F, Mohd Tawil N and Surat M 2009 Timber Defects In Building: A Study of Telapak Naning, Malacca, Malaysia, *WSEAS Trans. Environ. Dev.* **5** no. 1 109–118

[20] Larson B 2016 *Life Expectancy of a Home's Components and Systems* Retrieved on January 21, 2016 from https://www.adenacertifiedinspections.com/home-components-info/component-lifespan

[21] Abdul Rahman S and Agus Salim N A 2013 A Study on Defects Affecting Maintenance *The 5th Undergraduate Maintenance and Facilities Maintenance Conference*

[22] Ahzahar N, Karim N A, Hassan S H and Eman J 2011 A Study of Contribution Factors to Building Failures and Defects in Construction Industry *Proc. Eng.* **20** 249–255

[23] Chew M Y L, Tan S S and Kang K H 2004 Building Maintainability-Review of State of the Art *J. Archi. Eng.* **10** 80–87

[24] Syed Mohamad S B H, Akasah Z A and Rahman A 2014 Factors Contributing to Building Maintenance Performance of Heritage Buildings *Proc. of the Int. Civil and Infrastructure Engineering Conference 2013* (Singapore: Springer Science + Business Media) pp. 978–981

[25] Rotimi F E, Tookey J and Rotimi J O 2015 Evaluating Defect Reporting in New Residential Buildings in New Zealand *Buildings* **5** 39–55

[26] Ahmad Zamri N, Wan Alwi W A and Mat Hassan H S 2012 Kualiti Bahan Binaan Projek Perumahan; Kajian Kes Perumahan Iskandar Perdana, Seri Iskandar *1st Int. Conference on Innovation and Technology for Sustainable Built Environment 2012* (Perak) pp 825–833

[27] Wahab S N A and Hamid M Y 2011 A Review Factors Affecting Building Defects of Structural Steel Construction. Case Study : Student Accommodation in UiTM Perak *Procedia Eng.* **20** 174–179

[28] Bagdiya N V and Wadalkar S 2015 Review Paper on Construction Defects, *IOSR J. Mech. Civ. Eng.* **12** 88–91

[29] The Cascade Team Real Estate 2012 *Remodeling Cost & Life Expectancy Guide* (United States of America: Washington D.C.)

[30] The Building and Construction Authority Singapore 2014 *The BCA Construction Quality Assessment System (CONQUAS)* (Singapore: BCA)

[31] Suffian A 2013 Some Common Maintenance Problems and Building Defects: Our Experiences *Procedia Eng.* **54** 101–108

[32] Buys F and Roux M 2013 Causes of Defects in the South African Housing Construction Industry: Perceptions of Built-environment Stakeholders *J. Phys. Dev. Sci.* **20** 78–99

[33] Othuman Mydin M A, Agus Salim N A, Tawil S W and Ulang N M 2014 Assessment of Significant Causes to School Building Defects *Emerging Technology for Sustainable Development Congress (ETSDC 2014) (Malaysia)* vol 3 (E3S *Web of Conf.)* pp 1–7

[34] InterNACHI International Association of Certified Home Inspectors 2006 *InterNACHI'S Estimated Life Expectancy Chart for Florida Homes* Retrieved on December 04, 2015 from http://www.nachi.org/florida-life-expectancy.htm.

[35] Homeowner Protection Office 2014 *Residential Construction for New Home Covered by Home Warranty Insurancein British Columbia* (British Columbia: Homeowner Protection Office)

[36] Joseph F B and Marianne B 2011 Shedding new light on auditor switching: this article is based on a study supported by the IMA[R] foundation for applied research (FAR) (United States: Institute of Management Accountants) pp 49–53

Acknowledgments
The authors would like to convey sincere gratitude to Universiti Tun Hussein Onn Malaysia (UTHM) for sponsoring this research through their in-house research grant (STG U638).

IOP Conf. Series: Materials Science and Engineering **713** (2020) 012040 doi:10.1088/1757-899X/713/1/012040

The influence of the stakeholders' perspective for the performance measurement in local authorities

N Rahmat[1] and A H Nawawi[2]

[1]Center of Postgraduate studies, Faculty of Architecture, Planning and Surveying, Universiti Teknologi MARA, Shah Alam, Selangor, Malaysia
[2]Faculty of Architecture, Planning and Surveying, Universiti Teknologi MARA, Shah Alam, Selangor, Malaysia
Corresponding author: norashikinrahmat84@gmail.com

Abstract. This research constructs a performance evaluation of the influence of stakeholders' perspective on the performance measurement in the local authorities. The stakeholders' perspective has been proven as one of the main factors that influence higher rating on the performance measurement in many public organizations including the local authorities. However, very little attention has been paid towards the attributes of the stakeholders' perspective that affecting the performance measurement in the local authorities. This research focused on the elements that contribute on the stakeholders' perspective from the Balanced Scorecard's perspective and Multi-Criteria Decision Making tool namely Analytic Network Process (ANP) to acquire the priorities of the attributes for the stakeholders' perspective. The attributes were used as the performance indicator for performance measurement in the local authorities. There were eight (8) attributes that have been measured in order to acquire the priorities of the attributes in affecting the performance measurement under the stakeholders' perspective (customer, customer retention, new customer acquisition, customer profitability, community support, employees, management and service providers). From this result, it will provide the management of the local authorities to improve the performance of the local authorities which are meaningful for managerial and improvement purpose in the service delivery towards the public.

1. Introduction

Performance measurement had been discussed by many researchers that has led to the importance of determining the best performance measures. However, performance measurement in local authorities in Malaysia is still in infancy. Public organization such as local authorities are facing pressures for change which require the local authorities to be more efficient in order to provide better services to the public. According to Myeda *et al.* [1] the literature in performance measurement system in Malaysia especially in facilities management is still limited even though performance measurement is one of important element in increasing the productivity of employees in the local authorities as well as increasing the level of customer satisfaction. In Malaysia, there are 149 local authorities including city council, municipal, and district council [2].

Performance measurement in facilities management is very important especially in the local authorities. Therefore, a study has been made to obtain reliable performance measures that can improve the performance of facilities management in the Malaysian local authorities. There are need to measure the performance of facilities management in the local authorities due to the need of obtaining the level of past performance, the level of current performance, the level of the local authorities wants to achieve, the approach in order to achieve the level of required performance and how the level of set target has been achieved [3]. Furthermore, according to Rahmat and Nawawi [4], there are many issues occurred in the Malaysian local authorities as it provides services and facilities to the public such as operational issues, transparency, poor services, not proactive in improving public services, cleanliness issue, provision of public facilities, the efficiency of service deliveries, poor financial planning, accountability, poor management and environment and safety issues.

Therefore, the purpose of this study is to obtain a set of reliable performance measures using Balanced Scorecard (BSC) as the base of the model. In this paper, only the sub-criteria for the stakeholders' dimension will be discussed. Saaty [5] had argued that the need of identifying and understanding the level of expectations of the stakeholders in the local authorities are important in order to deliver more effective services.

Facilities management involves various stakeholders including the internal and external stakeholders and processes that need a holistic performance measures. Stakeholders have been studied in many research studies including performance measurement [6 - 9], strategic performance management [7] and decision making [10]. However, these studies have not identified the priority of performance measurement criteria under the stakeholders' dimension in order to achieve higher performance level. Therefore, it is essential to identify the effective performance measures for the stakeholders in the local authorities. In this study, the performance measures for stakeholders' perspective for facilities management in the local authorities has been measured using Analytic Network Process (ANP) developed by Saaty [11].

The ANP helps in solving the decision problems among the interdependent criteria within the structure. In the ANP, the structure of the criteria will be not in the hierarchy structure that will show in linear flow but it will appear as a network structure with different cycles to connect its clusters of criteria and the loops that connect the cluster [12]. In the ANP, the pairwise comparisons are used same as in the AHP. The 1-9 scales are used to obtain the level of importance of the criteria and measurement indicators chosen for the BSC. Table 1 shows the fundamental scale of using the absolute numbers developed by Saaty [11].

Table 1. Fundamental scale of ANP.

Intensity of importance	Definition	Explanation
1	Equal importance	Two objectives that contribute equally towards the objective chosen
2	Weak	
3	Moderate importance	Based on the experience and judgment, the preferences of criteria is slightly favour on one activity over another
4	Moderate plus	
5	Strong importance	Based on the experience and judgment, the preferences of criteria is strongly favour on one activity over another
6	Strong plus	
7	Very strong or demonstrated importance	One criteria is preferred very strongly over other criteria (very dominance)
8	Very, very strong	
9	Extreme importance	The evidence favouring one criteria over another is of the highest possible order of affirmation

Reciprocals of above	If activity i has one of the above nonzero numbers assigned to it when compared with activity j, then j has the reciprocal value when compared with i	A reasonable assumption
Rationals	Ratios arising from the scale	If consistency were to be forced by obtaining n numerical values to span the matrix

The scale was used to pair-wise comparison in the stakeholders' dimension. The respondents were asked to answer the questionnaire based on the scale in order to gain a supermatrix values.

2. The stakeholders' performance criteria

The key dimensions of the service in the local authorities are performance measures for stakeholders' dimension which are summarized in table 2. The selected performance measures were chosen after a thorough literature search and consultation with local authorities' representatives. The selected performance indicators had to meet the following criteria, most of which are standard requirements. First, indicators needed to be easy to interpret, to maximize the probability that they would be used by local authorities' management, officers, and members.

Table 2. Key performance measures for stakeholders' dimension.

Code	Criteria	Explanation
S1	Customer satisfaction	Measures on how well the local authorities is doing according to the satisfaction level of the users
S2	Customer retention	Measurement of customers' loyalty in local authorities
S3	New customer acquisition	Measurement of the number of new customers that are using the local authorities services
S4	Customer profitability	Newly acquired customers for more profitable tax collection
S5	Community support	Measurement of how well the service and the facility are accepted by the community
S6	Employees	Measurement of how the employees in local authorities are perceived to be more effective in delivering services
S7	Management	Measurement of how the management in the local authorities is perceived to be more efficient in managing local authorities
S8	Service Providers	Measurement of how the outsourcing service providers are perceived to be more efficient and effective in providing services

3. ANP methodology

Although many studies about performance measurement have been applied in ANP, considering facilities management needs to have a method beyond the hierarchical relations. The hierarchy network can demonstrate the linear top-down connection and interdependence relations in each criteria and sub criteria. Therefore, the ANP is able to find the best strategy. In order to obtain the performance measures in the stakeholders' criterion, an ANP methodology was carried out.

3.1 Respondents

For the purpose of this study, prioritisation of indicators for performance measures in facilities management for Malaysian local authorities which the respondents consist of officers from four departments in the local authorities mainly from the valuation and property department, engineering department, human resource department, and financial department. The respondents were chosen from officer and assistant officer which directly involved in managing facilities in the local authorities and making decision for facilities management. The reason on the selection of respondents from these departments was based on the high influences in FM as well as the selection of performance measures for the BSC in FM due to their experience, knowledge and working skills. Furthermore, the data gathering process will be more reliable due to the involvement of direct person in charge in the local authorities.

The respondents consist of 105 officers and assistant officers from 44 local authorities; 10 city councils and 34 municipalities. The District councils are excluded from the population sample as their functions and purposes are geared specifically towards the infrastructure facility and public utility rather than services at city councils and municipal [13].

3.2 Process of ANP approach
The structure of analyse the BSC using ANP approach are shown as below [14, 15]:

Step 1. Design a questionnaire survey for collecting responses from the respondents. The questionnaire used in this research involved pairwise comparison of measures on a nine-point scale based on the AHP method with one criteria was extremely more important than the other and one point will be awarded if the two elements were equally important.

Step 2. In the next step involves the pairwise comparison in the pairwise comparison matrix. The pairwise comparison considers the impact of each element on others. The pairwise comparison was obtained by computing the matrices using Super Decisions software.

Step 3. Calculated the eigenvector of priorities for testing the consistency ratio (CR). According to [11], if the CR was less than 0.1, it can be considering as acceptable value.

Step 4. The other steps are forming super matrix and calculating the weights. Its elements are derived from pairwise comparison. The super matrix consisting of N clusters is presented as follows.

Step 5. After calculating the weights of the measures, the priority of the performance criteria can be selected based on the results. The aggregation of the individual judgements will be carried out by means of the geometric mean [16].

4. Findings
After a set of questionnaire has been set up, the questionnaire was distributed among the officers and assistant officers in the local authorities. The respondents were asked to pair-wise the criteria of the stakeholder's dimension according to the importance of each criterion. After a pair-wise was made, a pair-wise comparison has been made. Table 3 shows the pairwise comparison of stakeholders' dimension.

Table 3. Pairwise comparison matrix of the performance criteria with respect to the Goal.

Stakeholders dimension	S1	S2	S3	S4	S5	S6	S7	S8	CR
S1	1	1/4	1/6	1/4	1/8	1/4	1/5	1/5	
S2	4	1	1/2	1/6	1/5	1/6	1/2	1/3	
S3	6	2	1	4	3	5	5	4	
S4	4	6	1/4	1	1/5	1/6	1/5	6	0.09
S5	8	5	1/3	5	1	3	4	5	
S6	5	6	1/5	6	1/3	1	1/7	4	
S7	5	2	1/5	5	1/4	7	1	7	
S8	5	3	1/4	1/6	1/5	1/4	1/7	1	

During the evaluating process, the issue of consistency of the pair-wise comparisons may occur. Therefore, to check this consistency, the consistency ratio (CR) was calculated as equations (1) and (2):

$$CR = \frac{CI}{RI} \quad (1)$$

whereby:

$$CI = \frac{\lambda\ max - n}{n-1} \quad (2)$$

in which *CI* is the Consistency Index, *RI* is the Random Index and *n* is the size of the matrix. A CR value less than 0.10 can be acceptable [11,17]. In this analysis, the result has a CR value of 0.09 that can be acceptable.

After a pair-wise comparison of elements on the same layer with respect of other elements in the same category were conducted, the geometric mean values were computed for synthesizing the results. The geometric mean values had been conducted in order to obtain the combines comparisons of the results among the respondents.

Table 4 shows the priorities on the stakeholders' measurement criteria based on the supermatrix.

Table 4. Priorities on the stakeholders' measurement criteria.

Name	Priorities	Ranking
S1 Customer satisfaction	0.010661	2
S2 Customer retention	0.005072	6
S3 New customer acquisition	0.002425	8
S4 Customer profitability	0.006043	5
S5 Community support	0.00855	3
S6 Employees	0.004399	7
S7 Management	0.020805	1
S8 service providers	0.006216	4

From table 4, it shows that Management is ranked in the first rank among the stakeholders' measurement criteria. Most of the respondents agree that management gives more opportunity for improvement in services and facilities in the local authorities. Followed by customer satisfaction in the second rank and community support in the third rank. From this table, it shows that decision from the management in the local authorities based on the policy from the government can heavily affected the whole operation of the local authorities especially in FM. Customer satisfaction also plays an important role in the performance measures as the local authorities provides services and public facilities to the users. While community support can affect the level of performance in the local authorities if the public can support the provision of the facilities by ensuring the facilities are in a good condition while using it.

5. Discussion

From the results, it indicates that Management plays an important role under the stakeholders' dimension. This due to as a stakeholder, they must be able to make decision as well as able to influence others including the organization and the employees especially in managing over resources, technical skill and be able to access to the people who can finish the required tasks. The top management has

IOP Conf. Series: Materials Science and Engineering **713** (2020) 012040 doi:10.1088/1757-899X/713/1/012040

legitimate power because it can control resources, cost control and managing employees. Furthermore, the management are supported by the federal government.

Followed by customer satisfaction in the second rank indicates that customer also plays an important role in determining the performance of the local services in delivering services providing public facilities. If the customer is not satisfied with the service delivery by the local government, the performance of the local authorities will have to upgrade in order to be more effective in delivering services.

In the third rank is community support which requires the community to contribute to the local authorities by paying the taxes in order for the local authorities to provide more efficient service delivery and better facilities to the customer. This is important as the taxes paid by the customer determine the financial management by the local authorities to be more stable and cost control can be managed.

6. Conclusion

Performance measures for stakeholders' dimension of facilities management for Malaysian local authorities is one of the most important processes that must systematically considered from the decision maker especially in the local authorities. For this reason, performance measures for the stakeholders' dimension is evaluated using the integrated of BSC and ANP method. This paper identified the problem of selecting performance measures for facilities management in Malaysian local authorities especially in the stakeholders' dimension. An ANP model was subsequently designed based on the BSC perspectives to incorporate the judgements of respondents using the identified measures. This performance measures were applied by 105 respondents from the local authorities to select the most priority and suitable performance measures under the stakeholders' perspective.

From the results, management is the most priority measure for performance measures under the stakeholders' perspective followed by customer satisfaction and community support in the third rank. In this paper, performance measures were considered as a multi-criteria decision problem and an integrated BSC and ANP was used. An overview of the performance measurement criteria of stakeholders' dimension for facilities management in Malaysian local authorities also has been obtained. By using the ANP for the stakeholders' performance criteria, the priorities of each criterion can be obtained and therefore the performance in the local authorities can be improved.

Another important finding is that the proposed criteria is more reflecting with the importance of each criterion that affected the stakeholders' dimension for facilities management in the local authorities. This paper can be extended to other dimensions in the BSC for improvement of facilities management in the local authorities.

7. References

[1] Myeda N E, Zaid S M, Sulaiman R and Mahyuddin N 2014 The implementation of performance measurement system (PMS): Malaysian facilities management (FM) industry *Building Surveying, Facilities Management and Eng. Conf. (BSFMEC 2014)* vol. 15 (Paris: MATEC Web of Conferences) 01014

[2] Portal OSC (One Stop Center) 2019 *Number by State and Local Authorities Status* Retrieved on April, 20 2019 from http://www.epbt.gov.my/osc/

[3] Lebas M J 1995 Performance measurement and performance management. *International Journal of Production Economics* **41** 23-35

[4] Rahmat N and Nawawi A H 2017 Facilities management in Malaysian local authorities: identifying current issues *The Social Sciences* **12**(5) 755-761

[5] Saaty L T 1999 Fundamentals of the analytic network process *The International Symposium on the Analytic Hierarchy Process* (Kobe, Japan) pp 12–14

[6] Wisniewski M and Steward D 2004 Performance measurement for stakeholders *International Journal of Public Sector Management* **17(3)** 222-233

[7] Kloot L and Martin J 2000 Strategic performance management: a balanced approach to performance management issues in local government *Management Accounting Research* **11** 231-251

[8] Hildebrand R 2007 *Measuring and managing performance in local government: a literature review* Local Government Institute Working Paper Series May 2007 (Victoria: School of Public Administration University of Victoria) pp 1-68

[9] Harrison J A, Rouse P and Viliers C J D 2012 Accountability and performance measurement: a stakeholder perspective *JCC The Business and Economic Research Journal* **5**(2) 243-258

[10] Gomes R and Gomes L 2008 Who is supposed to be regarded as a stakeholder for public organizations in developing countries? *Public Management Review* **10**(2) 263-275

[11] Saaty L T 2005a Analytic hierarchy process *Encyclopedia of Biostatistics 1* (New York: John Wiley & Sons, Ltd)

[12] State Council Meeting for Local Government (MNKT) 60th On 3rd Of June 2008 Retrieved on April 21, 2019 from http://jkt.kpkt.gov.my/en/SUK%26PBT/PBT/KategoriPBT

[13] Saaty L T 2005b *Theory and applications of the analytic network process* (Pittsburgh: RWS Publications)

[14] Lin C T 2018 A decision-making model for evaluating and selecting suppliers for the sustainable operation and development of enterprises in the aerospace industry *Sustainability* **10** 735

[15] Saaty T L 2001 Decision making with dependence and feedback: the analytic network process (Pittsburgh: RWS Publications) pp 37

[16] Saaty L T 2013 Analytic network process *Encyclopedia of Operations Research and Management Science* (United States: Springer US) pp 64-72

[17] Saaty L T 1980 *The analytic hierarchy* process (New York: McGraw-Hill)

IOP Conf. Series: Materials Science and Engineering **713** (2020) 012041 doi:10.1088/1757-899X/713/1/012041

Project Manager's Skills for improving the performance of complex projects in Kuwait Construction Industry: A Review

Faisal Alshammari[1], Khairulzan Yahya[1], and Zaiton Binti Haron[1]

[1] Department of Structure and Materials, School of Civil Engineering, Faculty of Engineering, 81310 UTM Johor Bahru, Johor, Malaysia.
Corresponding author: Faisal.albedawi@gmail.com

Abstract. The role of project manager is one of the crucial factors in achieving project success. Project managers handling large and complex construction projects have to deal with several factors to ensure coordination in achieving project delivery based on time, cost, and quality. However, recently, several large and complex projects in Kuwait s are facing several delays and overrun. According to previous research, the main factors were communication, teamwork & teambuilding, planning & coordination, problem solving, and interpersonal skills. The aim of this paper is to provide a review on project manager's skills framework (PMSF) that are required in improving complex construction projects in Kuwait. Extensive literature reviews on the project managers skills in the management of the project were identified. It was found that skills such as teamwork, effective communication with staff and contractors, and effective resource management, effective planning and training, risk management are some of the important factor. The findings of this research contribute to exiting literature on developing a project managers skills framework (PMSF) that can be used by academic professional, business practitioners, and policy makers in identifying the issues facing construction projects and providing successful outcomes. Project managers involved in complex projects need to have strong skills to handle their team and every project participant. Project manager has the highest importance and critical to the success of the project. The literature review has indicated several skills that are important for project managers. The type of project determines the type and extend of skills that is required in successful project outcomes.

1. Introduction

Complex projects face many challenges during implementing its activities. The critical key for ensuring the success of complex projects is having a skilled project manager [1]. Some construction projects are technically complex such as when using a new technology that are not understandable by the project members [2]. In addition, this type could be specified when there is no qualified manager that could manage the new technology involvement although it is recognized to the organization [1]. Project complexity could be produced through two main features; which are uncertainty and structural complexity. Uncertainty could be produced through the uncertainty in utilised construction methods or either in the required goals [3].

In addition, the complex projects can be classified from another perspective. Amable et al. [4] identified it as complex projects that will demand managers with special technical and managerial skills that could make them able to show a high-quality performance level. Shane et al. [5] demonstrated that in complex project each organisation could face the challenge of continuous changes to cope with the newly technology innovations that could enhance the organisation performance [5]. Complex project will require a highly committed and hard worker manager with strong planning abilities that could enable such managers to adapt with the nature of complex projects [1].

Meanwhile, Kuwait is one of the Gulf countries which is witnessing a great development in the field of construction industries. In view of the great importance of the construction industry in this country, it is necessary to ensure that construction projects are being completed on time and within the required cost. The construction projects in Kuwait face a delay in delivery and thus exceed the planned budget (timeskuwait.com, 2019). The Kuwait Accounting Office has revealed that there was a delayed in construction projects with hundreds of millions due to errors in management, planning and design. In addition, the actual implementation rate of construction projects reached 67.6%, with a decreased by 32.4%, which necessarily contributed to the delay in completing the construction projects. Previous research by Al-Almutairi [7] related delay of project with project managers' skills.

There is so far very little knowledge regarding Kuwaiti project managers' skills and abilities in managing complex construction projects. The researcher also noticed that there is only a few studies conducted in Arab countries and especially in the gulf region, and with regard to the differences between countries cultures, the availability of skilled and talented managers in each country; all of this make it difficult to generalise these studies findings on other countries. The only vast majority of prior relevant researches, regarding the project managers' skills, were often conducted in Europe and American construction industry. Therefore, this study investigates the challenges faced by complex projects in Kuwait construction industry and determine the potential skills that must be owned by project managers to manage complex projects successfully.

2. Construction Project Complexity

The complexity of construction projects is constantly increasing due to the newly technological and operational initiatives. Construction projects that are recognized as technically complex could be recognized in several forms, such as when using a new technology that are not understandable by the project members [2]. In addition, this type could be specified when there is no qualified manager that could manage the new technology involvement although it is recognized to the organization [1]. The project complexity could be produced through two main features; which are uncertainty and structural complexity. Uncertainty could be produced through the uncertainty in utilised construction methods or either in the required goals [3].

Complex projects from another perspective, where he clarified that these projects could be identified as the projects that will demand managers with special technical and managerial skills that could make them able to show a high-quality performance level [4]. It is also argued that each organisation could face the challenge of continuous changes to cope with the newly technology innovations that could enhance the organisation performance [2]. This will require a highly committed and hard worker manager with strong planning abilities that could enable such managers to adapt with the nature of complex projects [1].

The literature points out that there are various factors that contribute to the success of complex projects. However, one of the common and most important factors is the role of PM. Complex projects require highly skilled and knowledgeable individuals to head the project. There should be a strong link between skills of the PM and the elements of the projects to ensure successful outcomes.

2.1 Challenges faced by Complex Projects

Kuwaiti construction projects are facing several problems and issues that cause projects' delays, cost overruns, and poor quality. Several researchers had investigated the reasons and the reasons and causes of delays showing up the most common causes and the most influential ones that lead to financial loss and project workflow crashes. Al-Tabatabi [8] conducted a survey in which he ranked the causes of delay in governmental construction projects utilizing 53 delay questionnaires. According to a relative importance index, the top ranked ones are slow decision-making process, payment and financial procedure, lack of planning and utilization of CPM, lack of design coordination and poor subcontractors' scheduling.

Koushki, AL-Rashid and Kartam [9] reported the three main causes of delays in construction projects, which causes cost overruns and complexity of projects were: owners' financial problems and his lack of experience in the construction business as well as the changing orders. In accordance to cost overruns, the three key causes were recognized as material-related issues, owners' financial issues and contractor-related problems. They recommended several measures in order to mitigate the delay and complexity of construction residential projects would including the allocation of adequate money and time at the design phase, availability of satisfactory funds, and choosing of a reliable contractor and competent consultant to execute the project.

Moreover, Al Zubadi and Al Otaibi [10] identified five critical factors to delays and overrun of Kuwait's building and infrastructure projects are: delay in in approval and preparation in variation orders as well as the delays in government permits and approval , additional work requested by client beyond the original scope and outline , variations and alterations on engineering conditions on the contract document and the decrement in the productivity of labours due to the climatic conditions. A different study by Soliman [11] survey the delay causes in Kuwaiti construction industry utilizing of 30 questionnaires. It was noticed that the most important delay factors were the conflicts occurred between the consultant and contractor, owner financial problems, delaying from owner in payments of contractor and contractor financial problems.

Additionally, Al Marri, Ibrahem and Nassar [12] found Kuwaiti highway and infrastructure projects utilizing a thirty questionnaire divided into five sections according to project time stages. They obtained that the most important causes of delay in infrastructure projects in Kuwait were at all stages of the project and not only in the construction stage. They concluded also that most important causes were: multiplicity of agencies in order to gain approvals, coordination problems between the different ministries that are responsible for project's implementation, delay in receiving the project contractor document by the consultant and the increased cost of materials and projects.

Al-Almutairi [7] added that delay in construction projects represents one of the most important factors that is significantly affecting Kuwaiti construction projects in a negative way and required a solution and a well management from their project managers. He also stated that time and cost overruns in Kuwait have largely been because of issues of attempting to utilize technically advanced solutions in construction while there is a huge lack in Kuwaiti project managers' skills and experiences. He also demonstrates that the lack of qualified Kuwaiti project managers that are unable to manage simple projects, not just complex projects, and the non-existent of specific Kuwaiti Project Manager's Skills Framework (PMSF), explains why the Kuwaiti construction industry is always faced with financial, delaying and insoluble conflict problems between its parties. This can be both time consuming and costly which may develop into great risks that affect the achievement and management of different Kuwaiti construction projects. Khalid [13] also supported that delays can be reduced or prevented by an increased pre-project planning and through successful project management as they are one of the most critical success factors of the construction project accomplishment.

3. Project Manager

Various authors have emphasised that the complex demanding, and the dynamic requirements of the complex projects call for project managers with high capabilities who can manage the construction process from the early stage to the final stage of the construction project successfully [3, 4, 14]. Remington and Crawford [15] and Mouchi et al. [1] pointed out that complex projects require a project manager who holds critical specifications that differ from the specifications of the project manager of simple projects. The complexity of complex projects requires the manager to be able to develop new initiatives and new ideas that are not consumed in order to ensure the success of the project. On the other hand, Pollack and Remington [14] stressed that the successful complex project manager must be able to transform the complex areas of the project into easy elements that are easily understood and realized by employees. This means that the project managers must have the necessary management skills that will help them succeed in the project.

Leung et al. [16] stated that there is a set of characteristics and attributes that should be owned by the project manager to ensure the success of the project, where earning these qualities and characteristics is being through exposing to a wide range of complex projects. One of the main skills that each project manager must have is the ability to deal with various kinds of individuals in good way in different situations throughout the whole life cycle of projects. As the main responsibility of the project outcomes comes to the project managers; they must have the required skills that could enable them to influence the work members in effective way to cover the project objectives [1].

The discussion here indicates that there is a difference between PM that handle complex project and simple projects. The characteristics and attributes of PMs in complex projects relates to the outcome of complex projects. This indicates the need to study the PM skills and project elements to identify and design a framework.

3.1 Project Manager Skills

Patanakul and Milosevic [17] added that each project manager must create strong relations with the team members to meet the project objectives, in addition to increasing organisation profits with keeping in mind customers' expectations. Fisher [18] clarified that there are six main skills must be owned by each project manager to make them succeed in their managerial practices. One of those skills is the manager cooperation, sincerity and his /her ability to understand various behavioural characteristics of the team members. Adding effective leadership style is another skill that must be owned by project managers. The third proficiency is managers ability to influence and motive work team toward covering the organization objectives. This reflects the ability of project manager to respect each work member and accept individuals on their situation and without any attempt to change them by force. Managers must also have the ability to recognize the conflict sources, influences and should own the managerial skills to solve conflicts. The final ability that must be owned by project manager is the awareness toward the cultural differences of the work members.

Furthermore, Paglis and Green [19] clarified that self-confidence is another factor that could influence the ability of project managers to succeed. Self-confidence could provide managers with the ability to take the right decisions in the correct time. Self-confidence could also enable managers to perform complex responsibilities and solve expected problems smoothly [1]. Additionally, project managers feeling toward their performance has an enormous effect on the project success, and this managerial skill play a vital role in succeeding construction projects [20].

3.2 Project Manager Elements

Shane et al. [2] clarified different elements for project complexity, those elements were classified in different categories and each category comprised of several elements related to it. Some of those elements are: cost elements (material cost, cost analysis and estimation, payment restrictions) , schedule elements (unclear sequence of the construction process, resource availability, uncertainty in process and materials used in structure, risk analysis), technical elements (design methods and specifications, scope of the project, delays, technology usage and details required to complete the project clearly), political issues and legislative process, size and location of the project.

Mouchi et al. [1] have also investigated a number of complexity elements from the viewpoints of five senior managers of complex projects in five different organizations including: unclear project scope, type of project participants, technical problems of the projects, complex requirement of customer, uncertainty in scope and aspects of projects, uncommon feature and unfamiliar construction elements, political issues around clients, unusual structure design, incomplete and unknown elements in design, size and location of the project, lack of details, high level of risks and resources unavailability.

Dao et al. [21] mentioned that construction project complexity refers to the interactivity between project elements and the number of project complex elements that make it difficult to be managed and achieved. So, he represented thirty-seven different attributes and elements that were truly representative of construction project complexity. Some of those mentioned elements are: stakeholder management (complexity of customer requirements, inspection and approvals from regulatory agencies and entities, etc.), Governance (number of project's partners, number of times a change order on project has occurred, etc.), Fiscal planning (difficulties in paying to labours, number of funding phases, etc.), Quality (quality of materials utilized in project), Legal (number of permits to be required, difficulty in obtaining design approvals, etc.) , Design and Technology (difficulty and integrity in structural design, unfamiliar and uncommon construction elements, Uncertainty in the utilised construction methods, lack of details in design, company unfamiliar with technologies involved in construction phase), Location and size of projects (execution hard site, include many risks, huge size of the project), Scope (undefined and uncertain project scope), Project Resources (quality of labours, quality of materials, sequence of the projects due to the availability of sources).

3.3 Project manager potential skills for managing complex construction projects

Several researches have proposed the project manager to the most successful skills that must be possessed in order to achieve and manage their projects successfully. Other than that, Project Management Body of Knowledge (PMBOK) guide was also available, issued by the Project Management Institute in order to be an assistant and guidance for project managers in their complicated management journey. This manual is certified for application within the United States and is also certified by the American National Standards Institute (ANSI), which is highly valued and significant in the project management's world. This also explains why many institutions and individuals interested in project management and contracting sector are interested in the PMP Certificate and in having a project managers with a PMP certificate, in which he is following the management strategies of this Guide and characterized by its professional managerial skills , in order to maximize the chances of success of the projects, and also to ensure the appropriateness and successfulness of mangers in managing complex projects in which large investments have been pumped annually to plan and implement them [2].

This PMBOK Guide defines three characteristics of an effective project manager which are personal, performance and knowledge. Performance indicates what the project manager can actually accomplish and achieve. Is the project moving forward? Moreover, Knowledge specifically refers to the project managers' possessed knowledge about project management. Such as, can the project manager develop a project management plan and write a project schedule? However, Personal refers to the project manager's personality, leadership and attitude. It can be presented as: is the project manager relating effectively to all the stakeholders with whom they must interact [22].

Furthermore, PMBOK Guide describes eight interpersonal and managerial skills that must be possessed by project managers in order to manage projects successfully and effectively. Such skills are Leadership, Motivation, Communication, Team building, Influencing, Political and cultural awareness, Decision making and Negotiation [23]. It would be interested to note that a significant number of these mentioned skills are straightforwardly related with the capacity of manager to impact others to do what is important for the project to convey an item, administration, or result on time, on target and on budget. The capacity to impact and influence on others is particularly imperative since project managers frequently don't have formal authority over the general individuals with whom they are working [23]. Also, PMBOK Guide the most important skills for project manager which qualified him/her to manage complex construction project efficiently are Self-Awareness, Leadership, Self-Confidence, Quick Thinking, Flexibility, Integrity and Honesty, Intelligence, Imaginative and Creativity, Persistence and Decision making.

Additionally, several researches had proposed the most common skills that must be owned by project managers (project managers' skills, PMS) in order to manage complex projects successfully is shown in Table 1. It can be seen that the skills are in line with the PMBOK Guide the most important skills.

Table 1. Summarized required complex construction project managers' skills obtained from literatures.

No.	Skills	Previous researches
Category: Communication		
1	Owning high communication skills	1, 17, 20, 23, 29, 30, 31
2	Good communication skills with different stakeholders.	1, 17, 20, 23, 29, 30, 31
3	The ability to negotiate	1, 17, 20, 23, 29, 30, 31
4	The ability to give constructive feedback	1, 17, 20, 23, 29, 30, 31
Category: Teamwork & Team building		
5	The ability to create strong relations with the team members	1, 17, 20, 23, 24, 29, 30, 31
6	The ability to understand various behavioural characteristics of the team members	1, 17, 20, 23, 24, 29, 30, 31
7	The ability to influence and motive teamwork	18, 23, 24, 30
8	Supporting the team	18, 23, 24, 30
Category: Planning & Coordination		
9	Be able to develop new initiatives and new ideas	1, 17, 18, 20, 31
10	The ability to keep up with changes	1, 24
11	The ability to plan effectively	1, 24
12	The ability to strategically planning	1, 24
Category: Problem Solving		
13	Be able to transform the complex areas of the project into easy elements	18, 24
14	The ability to recognize the conflict sources	18, 24
15	The ability to solve conflicts	18, 24
16	The ability to deal with risk easily	1, 24
Category: Interpersonal Skills		
17	The ability to deal with various cultures in good way and in different situations	18, 24
18	Owning self-confidence	1, 17, 20, 32
19	The ability to focus	1, 24
20	Owning a clear project vision	1, 32
21	The ability to organize the work	1, 24
22	Keeping Flexible	1, 17, 20
23	Owning high technical skills	1, 24, 31

There are altogether 23 skills that project managers must have to successfully manage complex projects have been mentioned in several studies, some of them has been mentioned in specific studies and not mentioned in other studies. This is due to the fact that most of those studies have investigated the skills needed by the project manager for successfully managing construction projects in general regardless of their types and complexity, so they weren't directed specifically for complex construction projects which is the major target of this study. However, it was noticeable that the studies of [24] and [1] were the closest studies to this study because they mentioned and examined the majority of skills matching the skills chosen in our study as long as these two studies were specifically designed to investigate the skills required by managers for complex projects only and not for all projects generally as the situation was for the other studies. In general, these 23 skills come under the categories of teamwork, effective communication with staff and contractors, and effective resource management, effective planning and training, risk management are some of the important factor.

3.3.1 Communication. Communication can be through various forms such as oral/verbal, written, and comprehensive communication skills. Oral communication skills are the ability to communicate efficiently with other orally, and the ability to make effective presentation. Written communication skills mean the ability to communicate efficiently by writing documents, letters and reports. Comprehensive skills are the ability to understand both the stated and implied meaning of others [25].

3.3.2 Teamwork & Team building. Teamwork and team building skills are important skills for a project leader. These skills include team structure, design skills, identifying the ability of different project team members, delegation skills and integration skills [26].

3.3.3 Planning & Coordination. Planning and coordination include being able to build harmonious relationship in order to achieve project goals and dealing with conflicts from both inside and outside [27]. The project leader must know how to plan, control and coordinate the various stages and move it in an objective-oriented direction.

3.3.4 Problem Solving. Problem solving skills include problem-finding, analyzing and solving skills, which are the ability to discover problems in a timely manner, problem-analyzing skills, judging and decision-making skills, as well as problem solving and controlling skills [25].
3.3.5 Interpersonal Skills. Interpersonal skills include being able to deal with people of different backgrounds, which means the skill of developing relationships with different kinds of people when needed [27]. Interpersonal skills also involve persuading motivating and incentive skills. Persuading skills refer to persuading and influencing others to support you in realizing the objectives of the project [28]. Motivating and incentive skills refer to carrying out special strategies to motivate team members to work hard by identifying their feelings, needs and expectations [27, 28].

4. Kuwait

It is noticed that the Gulf Cooperation Council (GCC) is witnessing a remarkable development in the construction industry due to many reasons which include the existing economic conditions, the increased population growth and prosperity of various sectors. Kuwait is one of the Gulf countries which is witnessing a great development in the field of construction industries. The facts have pointed out to the significance important of the construction industry at Kuwait. Further, decreasing the oil prices in Kuwait will contribute in continuing the expanding of Kuwait's projects market and thus will affect the government finances. According to MEED Projects, Kuwait has awarded KD 1.5 billion ($4.8 billion) worth of contracts in first quarter of 2016, where this has effect on the total size of the local projects market (private and public, active and planned) to reach KD 76.6 billion, i.e. an increase of 2.8% year-to-date.

The previous figures emphasised on the importance of improving the performance of complex projects in Kuwait regarding the significant contributions of this sector in the Kuwaiti economy. In addition, the previous studies indicated that complex projects will demand mangers with high managerial abilities and skills that could enable them to manage such projects successfully, including; strong communication, managerial and personal skills and intensive managerial knowledge. As a result of the importance of this aspect and the seriousness of its impact on the performance of complex construction projects in Kuwait; this study aims to develop a framework that could be utilized to enhance the complex projects overall performance, especially in Kuwait, through improving project managers skills and abilities.

5. Conclusion

This study describes that the challenges faced by complex projects in Kuwait construction industry. The biggest challenge faced by the complex project are the of delay and overrun that lead to financial loss and project workflow crashes. To overcome it, a technical solution and a well management from their project managers are required. The latter requires the project manager to have skills as suggested by PMBOK Guide and several previous literature reviews. The potential skills for project manager such as shown in Table 1 consist of 24 elements. It was found that skills such as teamwork, effective communication with staff and contractors, and effective resource management, effective planning and training, and risk management are some of the important factors. The findings of this research will contribute to exiting literature on developing a project managers skills framework (PMSF) that can be used by academic professional, business practitioners, and policy makers in identifying the issues facing construction projects and providing successful outcomes.

6. References

[1] Mouchi G, Rotimi J O and Ramachandra T 2011 The skill sets required for managing complex construction projects *Business Education & Administration* **3** 89-100

[2] Shane J S, Gransberg D and Strong K C 2014a *Project management strategies for complex projects. Strategic Highway Research Program (SHRP 2)* (Washington DC: Transportation Research Board of the National Academies)

[3] Williams T M 2013 *Managing and Modelling Complex Projects* (Netherlands: Springer)

[4] Aimable S 2015 *Effects of Risk Management Methods on Project Performance in Rwandan Construction Industry: A Case Study of the Multi-Storey Buildings Construction Project of RSSB* Master Theses (Kigali: Jomo Kenyatta University of Agriculture and Technology)

[5] Shane J, Strong K, Gransberg D, Ahn J, Allan N, Brisk D, Hunt J, del Puerto C L, Owens J, Scheepbouwer E, Scott S, Tighe S and Touran A 2014b *Project Management Strategies for Complex Projects Case Study Report* Strategic Highway Research Program 2 (Washington DC: Transportation Research Board of the National Academies) pp 1-334

[6] Times Kuwait Report *Project Delays Hurting Kuwait Development* Retrieved on September 11, 2019 from from ttps://www.timeskuwait.com/news/project-delays-hurting-kuwait-development/

[7] Almutairi N 2016 Causes of delays on Construction Projects in Kuwait according to opinion of engineers working in Kuwait *Int. J. of Eng. Res. and Application* **6** 84-96

[8] Al-Tabtabai H 2002 Causes for Delays in Construction Projects in Kuwait, *Eng. J. of Qatar University* **15** 19-37

[9] Koushki P A, Al-Rashid K and Kartam N 2004 Delays and cost increases in the construction of private residential projects in Kuwait *Construction Management and Economics* **23** 285-294

[10] Al Zubadi H and Al Otaibi S 2008 An Empirical Approach for Identifying Critical Time-Overrun Risk Factors in Kuwait's Construction Projects *J. of Economic and Administrative Sciences* **24** 35-53

[11] Soliman E M 2010 Delay causes in Kuwait Construction Projects. Structural Engineering Dept., Faculty of Engineering. Alexandria University *The Seventh Alexandria International Conference on Structural and Geotechnical Engineering (AICSGE) (Alexandria)* pp 57-67.

[12] Al Marri M, Ibrahem M and Nassar G 2012 Time delays in highways construction projects in Kuwait, *J. of American Science* **8** 194-197

[13] Khair K, Mohamed Z, Mohammad R, Farouk H and Ahmed M E 2018 A management framework to reduce delays in road construction projects in Sudan *Arabian J. of Science and Engineering* **43** 1925–1940

[14] Pollack J and Remington K 2012 *Tools for Complex Projects* (Aldershot: Gower Publishing Ltd.)

[15] Remington K and Crawford L 2004 Illusions of control: philosophical foundations for project management IRNOP *VI Conference (Turku)* (Turku: Abo Akademi University Press) pp 563-577

[16] Leung M Y, Chan Y S and Yu J 2009 Integrated model for the stressors and stresses of construction project managers in Hong Kong *J. of Construction Engineering and Management* **135** 126-134

[17] Patanakul P and Milosevic D 2009 The effectiveness in managing a group of multiple projects: Factors of influence and measurement criteria *Int. J. of Project Management* **27** 216-233

[18] Fisher E 2011 What practitioners consider to be the skills and behaviours of an effective people project manager *Int. J. of Proj. Management* **29** 994-1002

[19] Paglis L L and Green S G 2002 Leadership selfefficacy and managers' motivation for leading change *J. of Organizational Behavior: The Int. J. of Industrial, Occupational and Organizational Psychology and Behavior* **23** 215-235

[20] Turner J R and Muller R 2005 The project manager's leadership style as a success factor on projects: A literature review *Proj. Management J.* **36** 49-61

[21] Dao B, Kermanshachi S, Shane J, Anderson S and Hare E 2017 Exploring and assessing project complexity *J. of Construction Engineering and Management* **143** 1-10

[22] Reich B H and Wee S Y 2006 Searching for knowledge in the PMBOK guide *Proj. Management J.* **37** 11–26

[23] Rose K H 2013 A guide to the project management body of knowledge (PMBOK® Guide) – 5th Ed. *Proj. Management J.* **44** e1-e1

[24] Sense A and Kiridena S 2014 Building workforce competencies through complex projects *Workforce Development: Perspectives and Issues* eds. R Harris and T Short (Singapore: Springer Publishing) pp 153-171

[25] Awan M H, Ahmed K and Zulqarnain W 2015 Impact of project manager's soft leadership skills on project success *J. of Poverty, Investment and Development* **8** 27-46

[26] Deepa S S and Seth M 2013 Do soft skills matter? – Implications for educators based on recruiters' perspective *IUP Journal of Soft Skills* **7** 7-20

[27] Brenton B and Levin D 2012 The softer side of innovation: the people *J. of Product Innovation Management* **29** 364-366

[28] Cornelius D 2012 *Conference Focuses on Leadership and Project Management as the Crossroads of Project Success* (Pennsylvania: PM Network)

[29] Grisales R and Lopez N 2011 *The Link Between Project Management Leadership and Project Success* Master Thesis (Karlskrona: Blekinge Intitute of Technology)

[30] Zakaria I B, Mohamed M R, Ahzahar N and Hashim S Z 2015 A Study on Leadership Skills of Project Manager for a Successful Construction Project *Int. Academic Research J. of Social Science* **1** 89-94

[31] Khamaksorn A 2016 Project Management Knowledge and Skills for the Construction Industry *2016 International Conference on Civil, Architecture and Sustainable Development (CASD-2016). (London)* (London: IICBE) pp 93-97

[32] Zhang L and Fan W 2013 Improving performance of construction projects, *Engineering, Construction and Architectural Management* **20** 195-207

Vibration Criteria Analysis on Floor due to Human Walking

T N T Chik[1], P C Xian[1] and N A Yusoff[2]

[1] Faculty of Civil and Environmental Engineering, Universiti Tun Hussein Onn Malaysia, 86400, Parit Raja, Batu Pahat, Johor, Malaysia
[2] Research Center for Soft Soil, Universiti Tun Hussein Onn Malaysia, 86400, Parit Raja, Batu Pahat, Johor, Malaysia
Corresponding author: thayati@uthm.edu.my

Abstract. The development of concrete material technology and modern structural design pattern in Malaysia may not consider the effect of vibration from seismic and other sources. Floor vibration may affect the serviceability of the structural and also provide uncomfortable environment to occupants. The main purpose of this study is to investigate the dynamic analysis on floor at the first floor of Faculty of Technology Management and Business (FPTP) building. The analyses are compared with the vibration criteria guidelines by Gordon (1991) to identify the vibration sensitivity of the floor. The investigation on floor vibration due to people walking is using Finite Element Modelling (FEM) in ANSYS software in order to perform modal analysis and transient analysis. Modal analysis shows the result of the natural frequency of the floor and different mode shapes react with subjected various frequency. The time history of the vibration signal will be performed by transient analysis. Furthermore, the results are transferred to MATLAB software (e.g. VSATs) to assess the vibration criteria curve. The results of the vibration plot show the middle of floor panel and laboratory room are categorized under ISO level. Therefore, it can be concluded that this room is not suitable to accommodate the sensitive equipment particularly at the middle of the floor.

1. Introduction

Floor vibration generally makes people anxious and creates panic of structural collapse, even though there is only small displacements and stresses that are actually produced [1]. FPTP building is located in Universiti Tun Hussein Onn Malaysia (UTHM) campus received a complaint regarding the floor vibration in survey and construction laboratory. Therefore, the purpose of conducting this study is to analyse the amplitude of the vibration and sensitive area due to human walking vibration. Dynamic analysis is used to find the physical structures behaviour dynamically subjected to the forces. Modal analysis is the investigation of the dynamic testing of the structures under vibration excitation. Modal analysis can verify the parameter of natural frequency, natural mode, damping factor, modal scaling and etc. [2]. Two major parameters concerned in this study which are natural frequencies and mode shapes in designing a structure for dynamic loading condition. While, transient analysis is used to verify the dynamic response of the structure under the action of any general time-dependent loads. It can produce the time-varying displacement, strains, stresses and forces in a structure as it response to any combination of static, transient and harmonic motion [3].

First floor of FPTP building are investigated due to the dynamic response in serviceability condition by using ANSYS software. This software is used to run modal and transient analysis. The results are transferred to MATLAB software to perform the vibration curve and vibration plot of the floor. Moreover, the vibration curve and vibration plot results are compared with the vibration criteria curve by Gordon [4] in order to determine the vibration sensitivity of floor. Vibration criterion curves were enhanced in the 1980's, it is a design standard to assist a wide range of tools and instruments used by microelectronics, medical and biopharmaceutical industries [4]. Vibration is produced when subjected to the significant of dynamic, it might cause the building structure damage and human comfort when the amplitude is high. This study can be a reference sources due to vibration sensitivity on low consideration of vibration safety factor in the design building guidelines.

2. Literature review

The development of the concrete material technology and modern structural design may focus on cost, material and beautiful. Modern structures adopt high strength and lightweight construction materials resulting in decreased structural mass and damping properties. Vibration system involve the transfer of potential energy to kinetic energy and kinetic energy to potential energy. Energy in a cycle of vibration dissipated should replace with external sources to maintain the state of steady vibration [5].

2.1 Types of vibration

2.1.1 Free vibration

Free vibration is the natural response of a structure to some influence or displacement to produce oscillation where the total energy stays the same over the time. This is theoretical idea because in reality the energy will be dissipated to the surrounding over the time amplitude decays away to zero, this dissipation of energy is called damping. The damping effect will depend on the characteristics of the materials and also the condition of vibration.

2.1.2 Forced vibration

Forced vibration is the reaction of a structure with forces that cause the structure vibrate at the frequency of the excitation. When present the external force in the system correspond with one of the natural frequency of the system will produce resonance amplitude in the system. Therefore, the amplitude of the force is corresponding to the vibration level.

2.2 Sources of vibration

Floor vibration is produced by dynamic loads applied either directly or indirectly to the floor system producing motion sufficient to be perceptible by the occupants. The temperature and humidity effect will affect the frequency occur from sources [6]. When the temperature increase will deteriorate the modulus of elasticity of concrete significantly causing the natural frequency decrease. Natural frequency also decrease with humidity increase because the specific absorbs more water in high humidity environment and increase mass.

2.2.1 External forces

Floor vibration that caused by the external sources includes ambient vibration at the site, nearby road and rail traffic, construction activities and machinery running nearby buildings. For a tall building, wind load also consider as sources to contribute vibration to floor structure. With rapid urbanization, high-density development with housing, industrial and commercial areas are organised caused the building being constructed close to the road and highways. Traffic induced ground vibration could be examined as continuous in nature and concluded that can damage the structure if peak particle velocity exceeds 5mm/s and structural damage may take place if exceed 10 mm/s [7].

2.2.2 Internal forces

The source of internal force may include human activities, elevator and conveyance system and service machines. This research will only consider the floor vibration due to people walking. According to research done by Murray, Allen and Ungar concluded a repeated force can be represented by a combination of sinusoidal force whose frequencies, f are multiples or harmonics of the basic frequency of the force repetition [8]. Walking vibration action on long-span floor has shown that the higher modes of vibration need not be appraised because they die out quickly and do not cause discomfort [9].

2.3 Arup prediction method
According to Arup, a floor with fundamental frequencies above 10Hz can use an empirical formula, $L_{eff} = 42f^{1.43}/f_n^{1.3}$ (where f is pacing rate and f_n is floor frequency). Figure 1 shows the damping is largely irrelevant when the maximum velocity occurs the instant the impulse is applied and the decay is redundant [10].

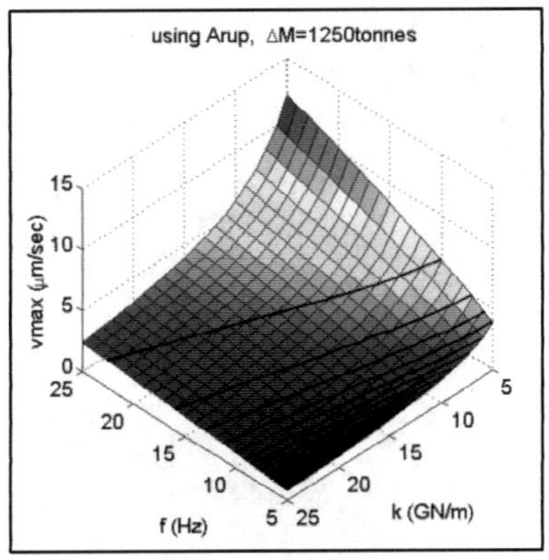

Figure 1. Response using Arup formula [10]

2.4 Dynamic analysis
Dynamic analysis is a simple extension of static analysis by adding the inertia force from the loads or displacement applied. In real structures, computer model is used to simulate the behaviour of the structure because structures potentially have an infinite number of node (joint) displacement [11]. There is various structural dynamic analysis can perform by using modal analysis and transient analysis to analyse the floor vibration.

2.4.1 Modal analysis
Modal analysis is used to verify the vibration characteristics of a structure or machine component to perform the natural frequencies and mode shapes. Design a structure for dynamic loading condition, natural frequencies and mode shapes are the parameters. The parameter depends on the material properties (mass, damping and stiffness) and boundary conditions of the structure [12]. Silva et al. (2003) have evaluated the dynamic performance of composite slab with various load frequencies induced by humans which are slow walk, standard walk, fast walk, standard run and fast run [13].

The reciprocal of the period is the natural frequency of an object that oscillation of the mass continuously. The period is a time interval to complete a cycle of oscillation and it can be derived as,

$$f_n = \frac{1}{2\pi}\sqrt{\frac{k}{m}} \tag{1}$$

where the relation of natural frequency to static deflection is shown by diagonal-dashed line with substitute the stiffness, $k = mg/\delta_{st}$ [14].

2.4.2 Transient analysis
Transient analysis is used to calculate the dynamic response of a structure under the forces of any general time-dependent loads. Transient analysis can find out the time-varying displacements, strains, stresses and forces in a structure as it responds to any combination of static, transient and harmonic loads [15]. Pavic et al. [16] has investigated the time-domain analysis established on the application of a single or several pulse representing footfall impact. It analyse with applied a single impulse and a series of impulses loads at either antinodes of the modes or test point. Time history is a linear or nonlinear evaluation of dynamic structural response under loading which may vary according to specified time

function. The basic elements equation of the dynamic system is $F(t) = ma+cv+kx$, where m, c and k refer to as mass, viscous damping coefficient and stiffness [17].

2.5 ANSYS software

ANSYS software is based on finite element algorithmic the powerful numerical value simulation software, it is exploited by the Swanson Analysis System Inc. The characteristic of ANSYS software is convenience, comprehensiveness, facility, supernal performance and agility plot function, meshing grid function, multi-selective iterative solution function and powerful postprocessor function. Pavic et. al [16] have used ANSYS software to analysis the walking induced vibration at three storey blocks which have both office and laboratory space on each of the floors. Figure 2 shows the entire floor surface was performed in 3D finite element model.

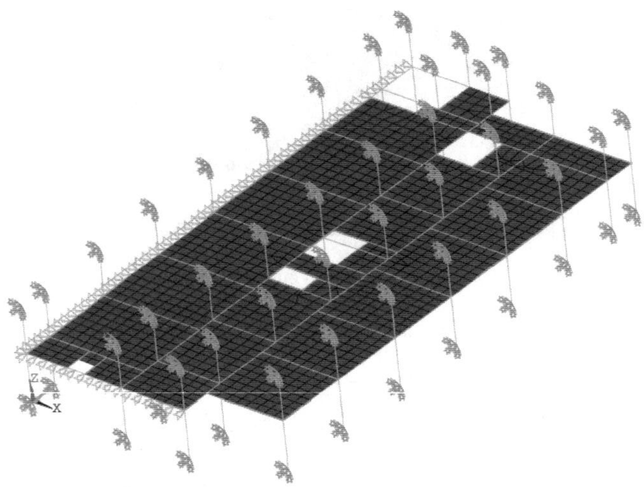

Figure 2. 3D pre-test finite element model [16]

2.6 MATLAB software

Vibration Serviceability Assessment Tools (VSATs) is a MATLAB based GUI-driven software developed in Vibration Engineering Section with the aim to analyse and assess the vibration serviceability limit state of large slender civil engineering structure. Moreover, ModalV function is used to perform root-mean-square velocity spectra by input the result of transient analysis from ANSYS in one-third octave band generic vibration criteria [18].

2.7 Generic vibration criteria

The vibration criterion (VC) curves were developed by Gordon and his colleague in 1980, but due to time passed the curves were extended to reflect the needs of technology and tool developments, primarily in the microelectronics industry. There were five types of VC curve, which are curve VC-A to VC-E that shows particular equipment category the measured one-third octave band velocity spectrum shown in Figure 3 [4].

3. Materials and methods

The structural plan of FPTP building was analysed to obtain the required information such as parameter and geometry by using ANSYS structural analysis software. It is essential to solve dynamic analysis with using batch mode. In modal analysis, the first step is build the model with the information from structural plan and apply load by using data from previous researcher. Then, the analysis was ran and produced the natural frequency and mode shapes of the structure. Next, transient analysis was performed with mode superposition method. It sums factored mode shapes (eigenvectors) from a modal analysis to calculate the building response.

The next stage is ModalV analysis which is in house developed in MATLAB software was ran to perform vibration curve from ANSYS result. The results obtained is one-third octave RMS velocity spectra will compare to the vibration criteria curve to determine the vibration sensitivity level of floor. Then, Vibration Serviceability Tools (VSATs) analysis with import and specify analysis was performed to produce the floor vibration response using loaded modal parameters. The vibration response was generated by plotting the response due to vibration criteria guidelines.

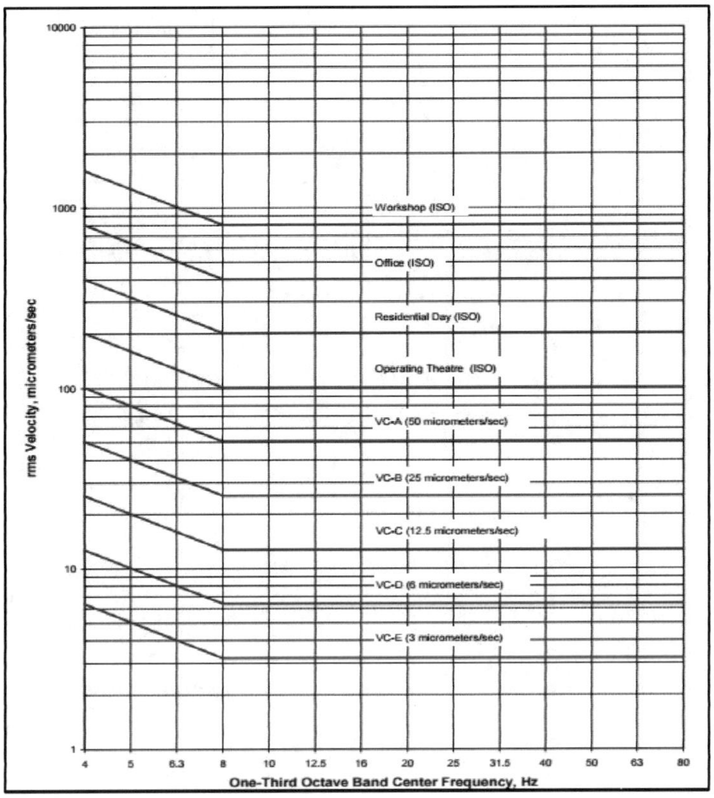

Figure 3. Generic VC curve for vibration-sensitive equipment [4].

Table 1. Application and interpretation of the generic vibration criterion (VC) curves [4].

Criterion curve	Amplitude[1] μm/s (μin/s)	Details size[2] μm	Description of use
Workshop (ISO)	800 (32 000)	N/A	Distinctly perceptible vibration. Appropriate to workshops and nonsensitive areas.
Office (ISO)	400 (8 000)	N/A	Perceptible vibration. Appropriate to offices and nonsensitive areas.
Residential day (ISO)	200 (8 000)	75	Barely perceptible vibration. Appropriate to sleep areas in most instances. Usually adequate for computer equipment, hospital recovery rooms, semiconductor probe test equipment, and microscopes less than 40x.
Operating theatre (ISO)	100 (4 000)	25	Vibration not perceptible. Suitable in most instances for surgical suites,

			microscopes to 100x and for the other equipment of low sensitivity.
VC-A	50 (2 000)	8	Adequate in most instances for optical microscopes to 40x, microbalances, optical balances, proximity and projection aligners, etc.
VC-B	25 (1 000)	3	Appropriate for inspection and lithography equipment (including steppers) to 3 μm line widths.
VC-C	12.5 (500)	1-3	Appropriate standard for optical microscopes to 100x, lithography and inspection equipment (including moderately sensitive electron microscopes) to 1 μm details size, TFT-LCD stepper/scanner processes.
VC-D	6.25 (250)	0.1-0.3	Suitable in most instances for demanding equipment, including many electron microscopes (SEMs and TEMs) and E-Beam systems.
VC-E	3.12 (125)	< 0.1	A challenging criterion to achieve. Assumed to be adequate for the most demanding of sensitive systems including long path, laser-based, small target systems, E-Beam lithography systems working at nonometer scales, and other systems requiring extraordinary dynamic stability.
VC-F	1.56 (62.5)	N/A	Appropriate for extremely quite research spaces; generally difficult to achieve in most instances, especially cleanrooms. Not recommended for use as a design criterion, only for evaluation.
VC-G	0.78 (31.3)	N/A	Appropriate for extremely quite research spaces; generally difficult to achieve in most instances, especially cleanrooms. Not recommended for use as a design criterion, only for evaluation.

4. Results and discussions

The model of the FPTP Building by Floor, Room and Panel have been modelled by ANSYS software. ANSYS software is used to carry out modal analysis to perform the natural frequency and mode shape of building. In addition, transient analysis perform the time history (acceleration versus time graph) of the response from the results of previous researcher due to human walking vibration. Furthermore, all of the data will transfer to MATLAB perform the vibration curve and vibration plot with compared to the generic vibration criteria.

4.1 Modal analysis

The first tenth mode shape for the first floor, laboratory room and corner panel is shown in Figure 4. The column of the modelling is consider fixed for single floor, room and panel, therefore the displacement of the column is not consider to the horizontal direction and only consider the vertical direction of the beam and slab displacement. The maximum vertical displacement of the slab for first

IOP Conf. Series: Materials Science and Engineering **713** (2020) 012046 doi:10.1088/1757-899X/713/1/012046

floor, laboratory room and panel were mode 7 with 8.8 mm displacement, mode 4 with 13.7 mm and mode 8 with 13.6 mm respectively. The result shows an increment trend of natural frequency while the mode is increased.

Mode Shapes		
First floor	Laboratory room	Corner panel
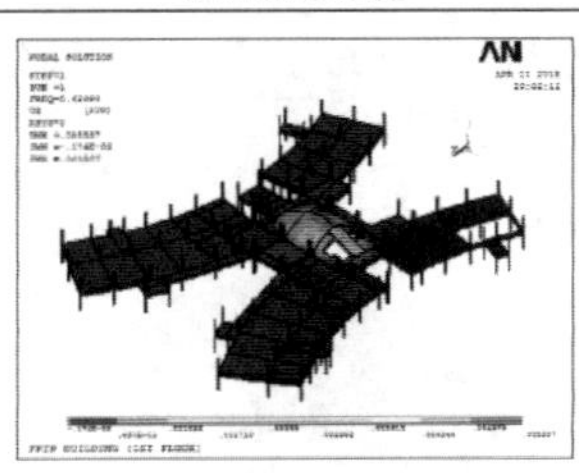	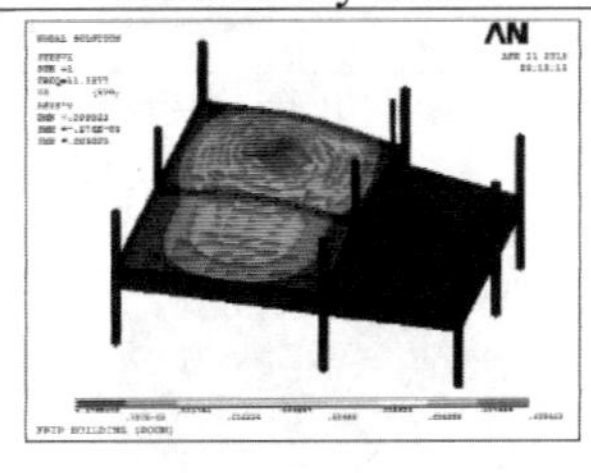	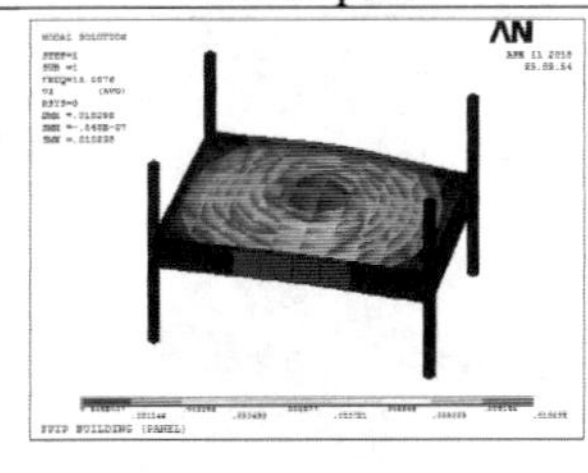
Mode 1, 5.62Hz	Mode 1, 11.83Hz	Mode 1, 13.09Hz
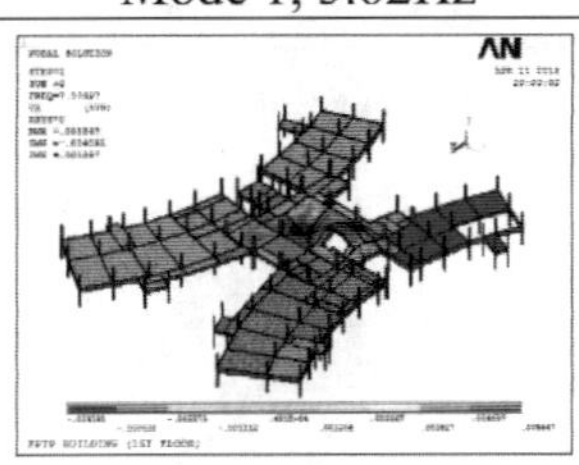	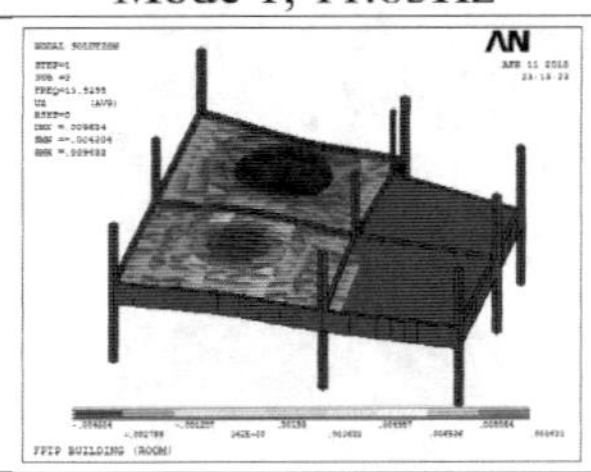	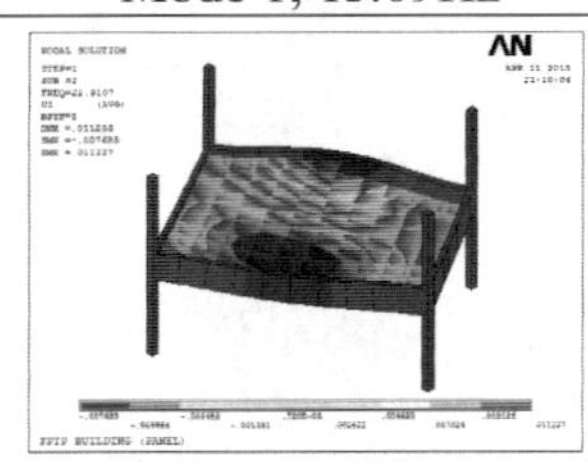
Mode 2, 7.50Hz	Mode 2, 13.53Hz	Mode 2, 22.91Hz
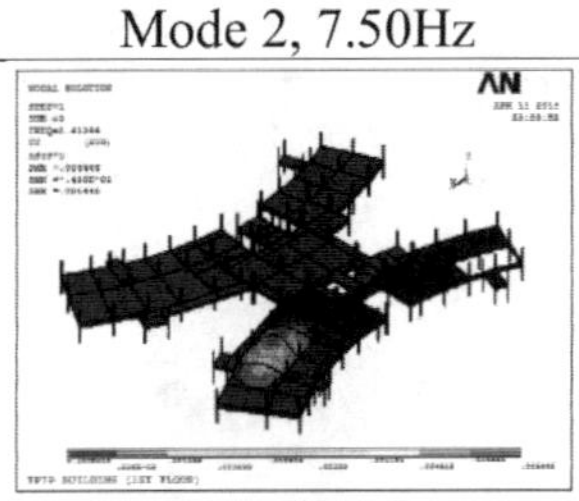	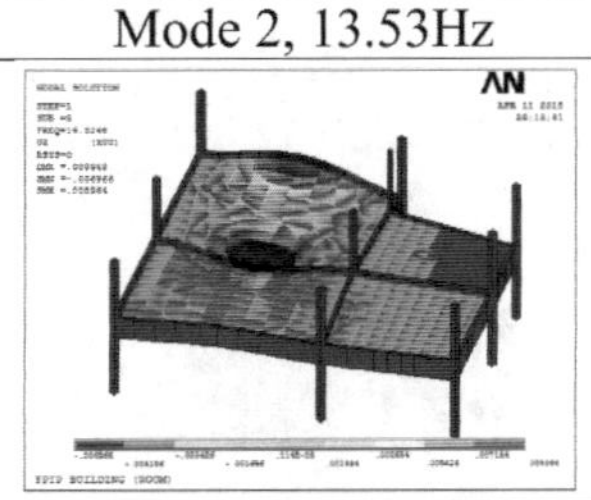	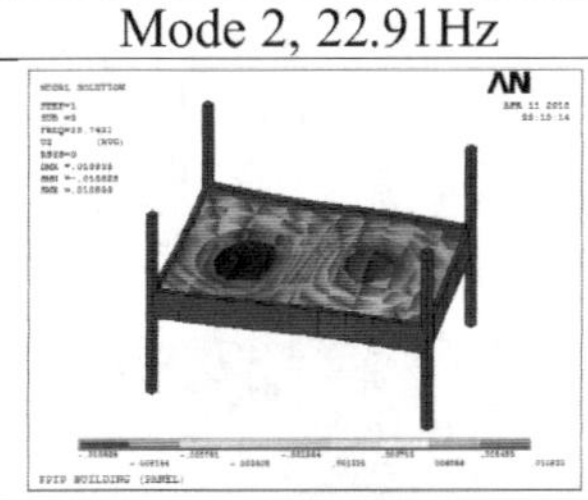
Mode 3, 8.42Hz	Mode 3, 19.52Hz	Mode 3, 25.74Hz
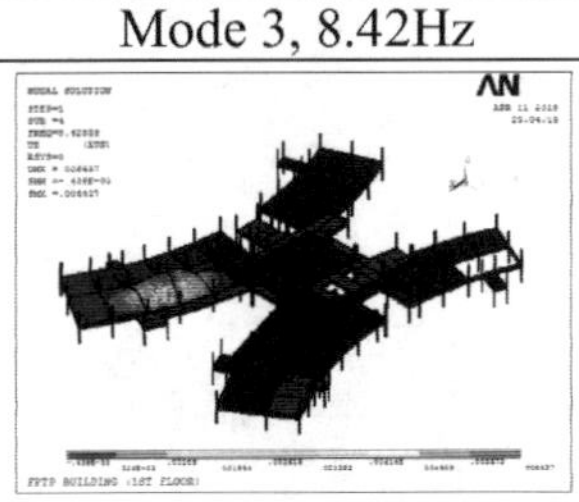	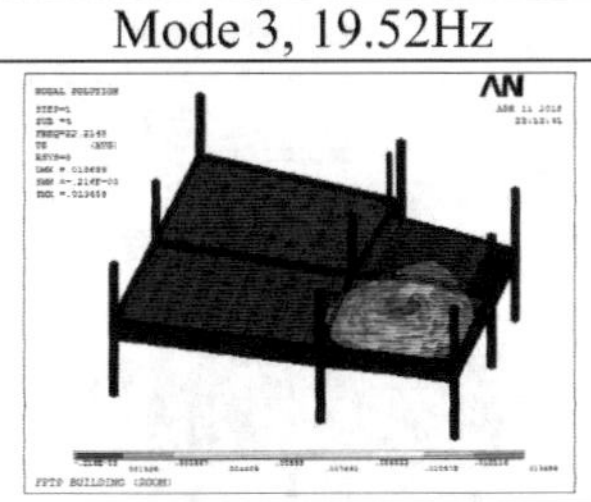	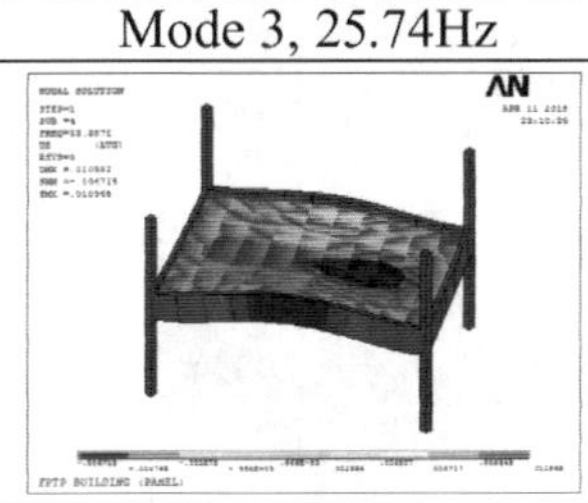
Mode 4, 8.43Hz	Mode 4, 22.21Hz	Mode 4, 33.36Hz

Figure 4. Mode shapes of FPTP building structure

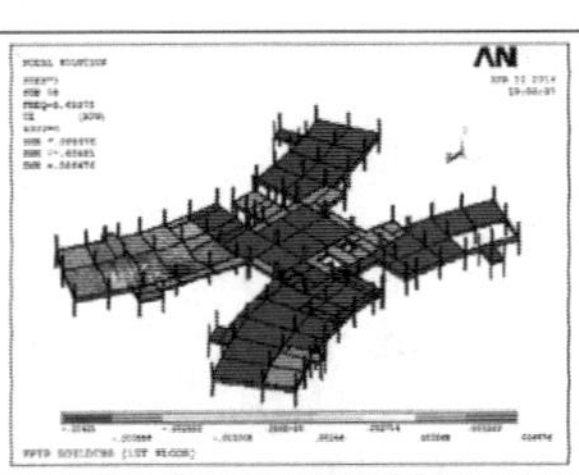	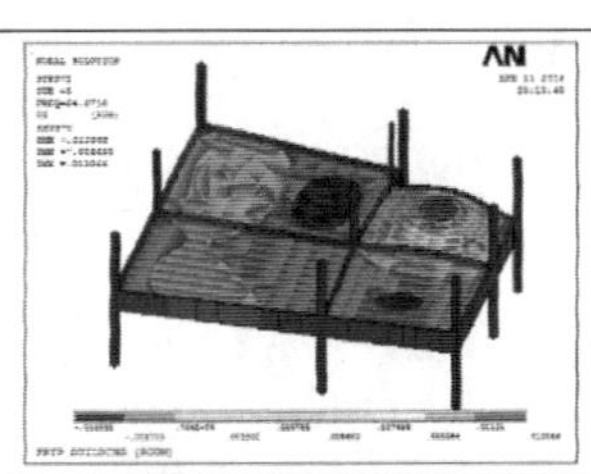	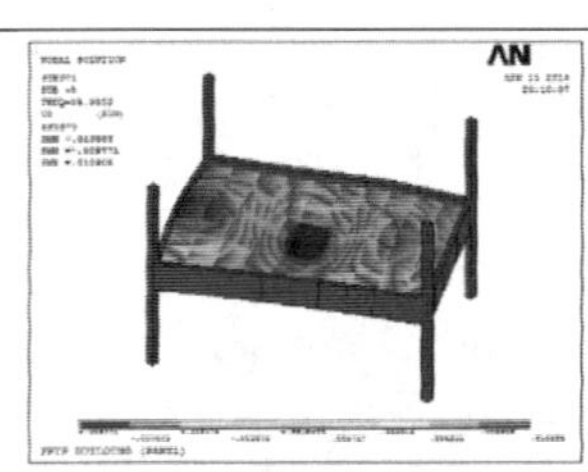
Mode 5, 9.69Hz	Mode 5, 24.68Hz	Mode 5, 39.91Hz

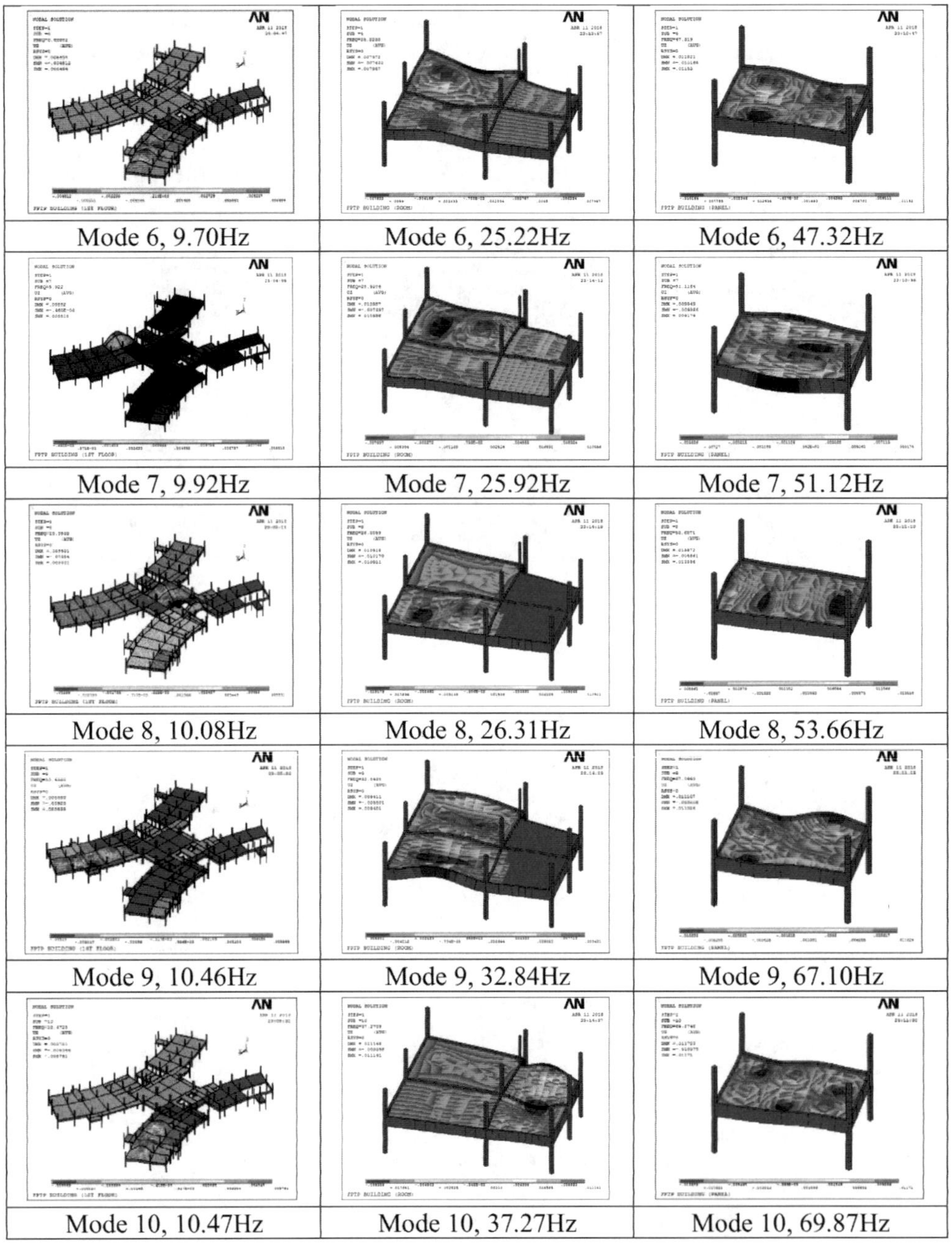

Mode 6, 9.70Hz | Mode 6, 25.22Hz | Mode 6, 47.32Hz
Mode 7, 9.92Hz | Mode 7, 25.92Hz | Mode 7, 51.12Hz
Mode 8, 10.08Hz | Mode 8, 26.31Hz | Mode 8, 53.66Hz
Mode 9, 10.46Hz | Mode 9, 32.84Hz | Mode 9, 67.10Hz
Mode 10, 10.47Hz | Mode 10, 37.27Hz | Mode 10, 69.87Hz

Figure 4. Mode shapes of FPTP building structure (continued)

4.2 Transient analysis

Transient analysis was ran as the input for the problem laboratory room and the panel. The result will transfer to MATLAB to run ModalV function to generate the vibration curve. Moreover, the vibration curve is not perform due to the result input from ANSYS. This is because the acceleration reaction of the floor in laboratory room and panel was very small, therefore the vibration curve is not perform or near to zero. Figure 5 shows the results of transient analysis by using MATLAB. It illustrated the acceleration time graph for Floor, Room and Panel structure. The acceleration wave became more stringent due to smallest structure modelled.

Acceleration time graph	
Floor	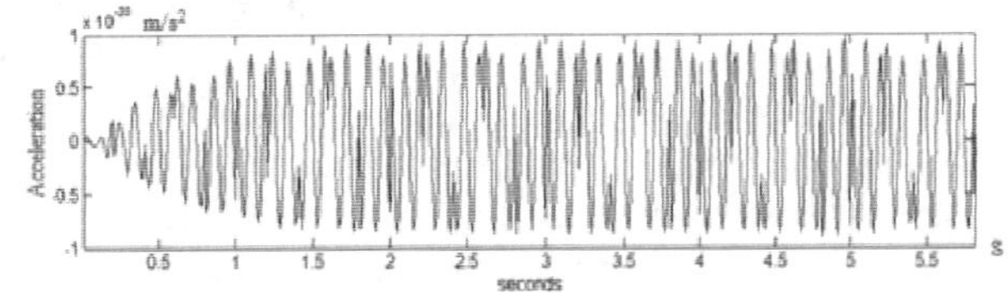
Room	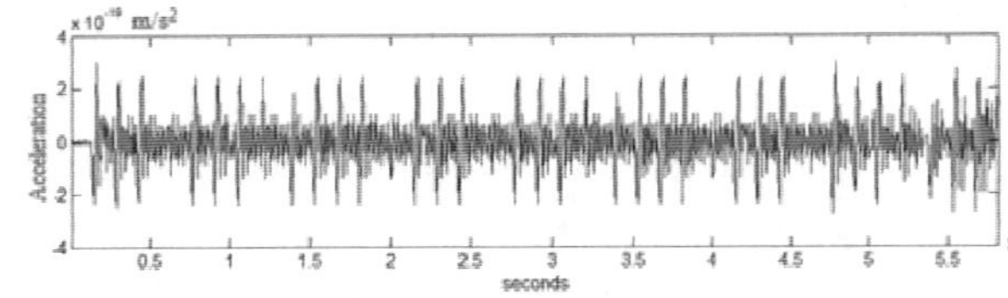
Panel	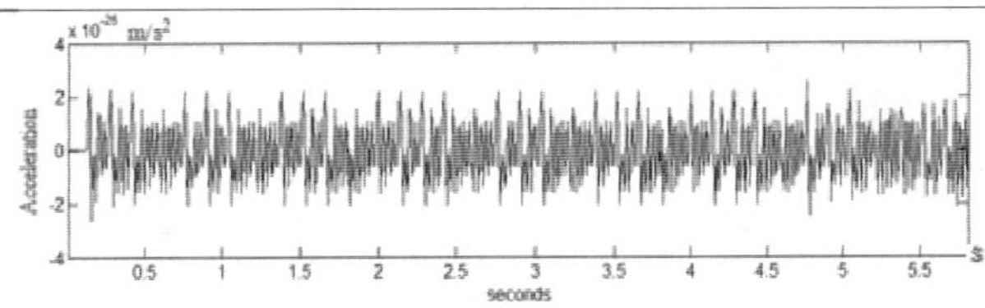

Figure 5. Transient analysis from ANSYS transfer to MATLAB.

4.3 Vibration criteria analysis

VSATs is used to calculate the response of the floor vibration by using the data input by ANSYS and also Arup equation to determine the vibration criteria. The highest VC level was ISO that suitable to workshops and non-sensitive areas while the lowest level was VC-E that provide to sensitive systems required extraordinary dynamic stability. Figure 6 shows the two method of VC respond at the middle of slab in laboratory room and panel was in the range over the ISO level when referred to generic vibration criteria. It is proven that the slab is prone to vibration and not suitable to place sensitive equipment on that particular location as described in Table 1.

Structure	Vibration testing data	Arup method
Room	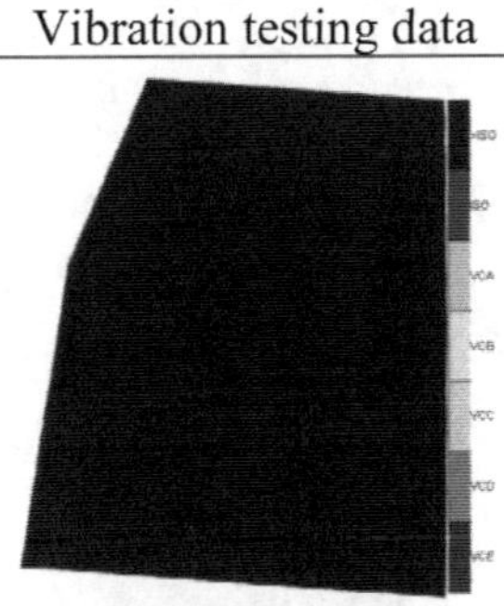	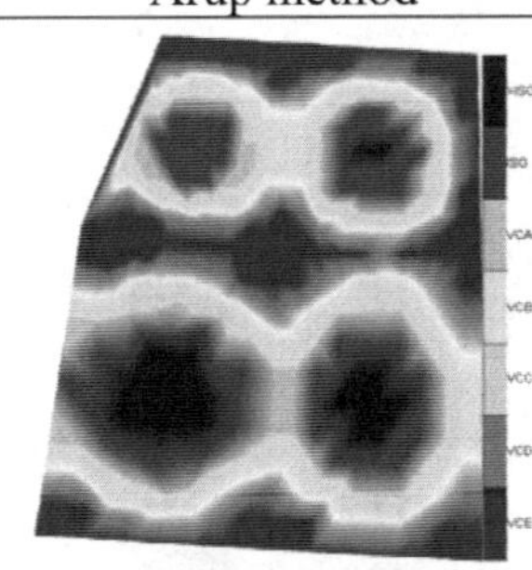
Panel	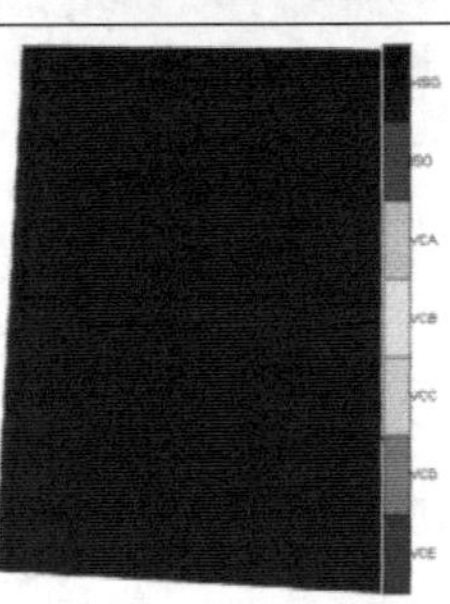	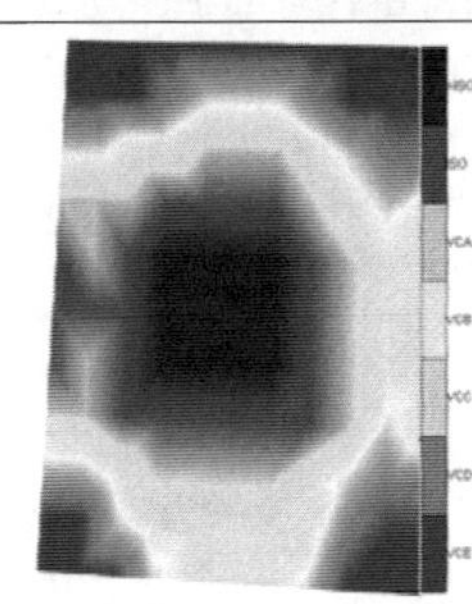

Figure 6. Vibration plot for the laboratory room and panel.

5. Conclusions

As a conclusion, the first floor of the FPTP building was modelled by floor, room and panel to perform the natural frequency and mode shapes of the model by using ANSYS software. The sensitivity of the floor was identified after the process of transient analysis and transfer to MATLAB for further analysis to compare with the generic vibration criteria. From the result obtained, it shows the laboratory is not suitable to allocate sensitive equipment particularly at the middle of floor. It is because the ISO level is achieved at that location which is perceptible to any vibrations. The suitable location is on top of beam which showing at the VC-A to VC-E level based on Arup method. It shows the certain location is at VC-E level which is around the column area. Therefore, this is the most suitable place if the owner of the room planned to accommodate the sensitive equipment.

6. References

[1] Allen D E and Pernica G 1998 *Control of Floor Vibration* (Ottawa: Institute for Research in Construction) pp 1-4.
[2] Qu Z Q 2004 Model order reduction techniques *Model Order Reduction Techniques: With Applications in Finite Element Analysis* (London: Springer London) pp 31-46
[3] ANSYS I 1998 *Structural Analysis Guide* Retrieved on Mac 15, 2018 from http://www. ansys.stuba.sk/html/guide_55/g-str/GSTRToc.htm
[4] Gordon C G 1991 Generic vibration criteria for vibration sensitive equipment *Proc. Int. Society for Optical Engineering* **1619** 71-85
[5] Struik D J 1948 *Fundamentals of Vibration In History of Mathematics* (Dover Publications)
[6] Xia Y Hao H Zanardo G and Deeks A 2006 Long term vibration monitoring of an RC slab : Temperature and humidity effect *Engineering Structures* **28** 441-452
[7] Whiffen A and Leonard D 1971 *A Survey of Traffic Induced Vibrations* (Berkshire: Road Research Laboratory)
[8] Murray T M Allen D E and Ungar E E 2003 *Floor Vibrations Floor Vibrations Due to Human Activity* (Chicago: American Institute of Steel Construction)
[9] Allen D E and Rainer J H 1976 Vibration criteria for long-span floor *Canadian J. of Civil Engineering* 3(2): 165-173
[10] Brownjohn J M W 2006 Dynamic performance of high frequency floors *Proc. of IMACXXIV* St. Louis pp 1-8
[11] Wilson E L 2002 *Three-Dimensional Static and Dynamic Analysis of Structures* (Berkeley: Computer and Structures)
[12] Guillaume P 2006 Modal analysis *Vrije Universiteit Brussel* **117(10)** 2585-2588
[13] Silva J G S D, Vellasco P C G D S, Andrade S A L, De Soeiro F J D C P and Werneck R N 2003 An evaluation of the dynamical performance of composite slabs *Computers and Structures* **81** 1905-1913
[14] Blake R E 2002 Basic Vibration Theory *Harris' Shock and Vibration Handbook* (New York: McGraw-Hill) pp 1-32
[15] Wang E and Nelson T 2002 *Structural Dynamic Capabilities of ANSYS* (Int. ANSYS Conf.)
[16] Pavic A, Reynolds P, Prichard S and Lovell M 2002 Evaluation of mathematical models for predicting walking-induced vibrations of high-frequency floors *Int. J. of Structural Stability and Dynamics* **3(1)** 107-130
[17] Wright C 2012 *Introduction to Structural Impact* (Fairfax: PDH Center) pp 12
[18] Pan T, You X and Lim C L 2008 Evaluation of floor vibration in a biotechnology laboratory caused by human walking *J. of Performance of Constructed Facilities* **June** 122-130

Acknowledgements

This research work is supported by the TIER 1 Grant code H259 at Universiti Tun Hussien Onn Malaysia (UTHM), Batu Pahat, Johor.

IOP Conf. Series: Materials Science and Engineering **713** (2020) 012047 doi:10.1088/1757-899X/713/1/012047

Enhancing the Knowledge and Proficiency for Interior Designers in Malaysia through the Implementation Building Information Modelling

A B Abd Hamid[1], M R Embi[2], M Z Mohd Taib[1] and A H N Abdul Razak[3]

[1]Centre of Interior Architecture Studies, Faculty of Architecture, Planning and Surveying, Universiti Teknologi MARA, 42300 Bandar Puncak Alam, Selangor Malaysia
[2]Department of Architecture, Faculty of Built Environment and Surveying, Universiti Teknologi Malaysia, 81310 Skudai, Johor Malaysia
[3]Department of Interior Design, Faculty of Architecture, Planning and Surveying, Universiti Teknologi MARA, 32610 Seri Iskandar, Perak, Malaysia
Corresponding author: abah7591@uitm.edu.my

Abstract. Building Information Modelling (BIM) is the process to develop and enhance the process of work to ensure the accuracy in the scope of work and efficiency in the documentation are less error during the construction project. Having the knowledge and proficiency for interior designers on the required tasks and scope of work for interior designers are critical as a tools to manage interior project. This study is to emphasize the significance of knowledge and skill among interior designers in Malaysia while implement BIM in the interior projects. This research was conducted through preliminary study by choosing interior design firms in Klang Valley. Sixteen (16) interior design firms were choose randomly in this survey. The survey were conduct through questionnaire survey and discussion with the interior design expertise. The data was analysed the elements of barrier and key factors the enhancement of knowledge and proficiency for interior designers in Malaysia.

1. Introduction

Nowadays, in the construction industry, it is more challenging in term of technology, process, budgets, documentations and management. The construction players in the building process are facing this challenging to delivere flourishing project, despite tight budgets, and accelerated schedules which transpire to the Architecture, Engineering and Construction (AEC) industry [1, 2]. This phases, AEC industry has required to adopt another system to reduce project cost, diminish project delivery time, increase productivity and quality, and abolish waste as well as for AEC industry to improve in construction process [3]. The progress development can be become proper if the right approaches implement in construction and Building Information Modelling (BIM) has potential and purpose for construction players adopt in building process [4].

In the universal construction industry, the most significance innovation tool is Building Information Modelling (BIM) technology. According to Chen and Qu [5], BIM is a momentous approach technology to apply and maintain the integrated digital for all building development in different phases of the complexity construction building and abstract of data repository [6]. The integrated among design process and BIM were more organize include all the progress working in term of design, planning and documentation.

Today, many countries in the world using innovation tools and new technologies in the construction industry for enhanced the skills and knowledge in their expertise. For example, in construction it has been proven by using existence tool such as two-dimensional AutoCAD has been raising the efficiency and accuracy of project in the design phases. For that purpose, construction players have been created a building information modelling (BIM) as a latest tool to be present innovative and constructive by improvement several tools and techniques to increase the performance in the construction industry [7]. The effectiveness of building information modelling implementation in the projects is a one of the paradigm for architect, engineering and construction (AEC), project managers, facility managers and contractors-subcontractors to develop from a traditional approach to modern and innovative approach in the process of design project. More integration technology can enhance by implementation BIM in the construction approach. Lately, BIM have growing rapidly in several countries such as Australia, United State, Finland, Denmark, Middle East, and India – industry players required submit the project document in BIM format [8]. However, the adoption BIM in Malaysian construction industry are still freshly [9, 10].

2. Literature review

Interior design is a multi-faceted profession with creative and technical solutions are applied within a structure to achieve a built interior environment. With the complying the solutions of element are functional, enhance the quality of life and culture of the occupants, and are aesthetically attractive [11]. Interior design task is including a scope of services performed by a professional design practitioner, qualified by means of education, experience, and examination, to protect and enhance the life, health, safety and welfare of the public. Professional interior design encompasses programming, space planning, construction, detailing, and finish selection as well as considerations of interior decor. As in building design, the process of interior design defines client and user requirements, develops conceptual approaches, and creates and documents solutions.

In the construction industry, interior design also has a role in construction industry development. For that, knowledge and performance skill should be considered to ensure the development progress in the construction were smooth in progress. The consideration for interior design industry to implement Building Information Modelling (BIM) should be consider to avoid from some element of risk such as cost overruns, project delays and building not performing as expected. According to Pryke [12], by using BIM can have provided greater certainty to ensures the key project milestone are encountered and delivered as expected while the deliver the building at the right time, cost and quality. Quality performance can have defined as the long term the effectiveness permanent [13] to improve the productivity and profitability of contractors for client's satisfaction.

2.1. Introduction of BIM

There are many definitions of BIM that defines from designers, contractors and operators. BIM is known as a digital about the physical and function of characteristic the building by sharing the knowledge resource and forming about the building [14]. BIM is defined as a modelling technology and associated set of processes to produce, communicate and analyse digital information for construction life-cycle in the construction projects [15]. BIM is not any single act or process and it is not creating a 3 dimensional (3D) model but it is design for isolate and utilised computer-based fabrication to aware the needs in the process. it is contained the information and data of design, construction, logistics, operation, maintenance, budgets, schedule and more.

In figure 1, is shows the life-cycle in BIM principles there are process, people, policy and technology. This is as a guide for construction industry to implement BIM in their projects as paradigm in construction technology.

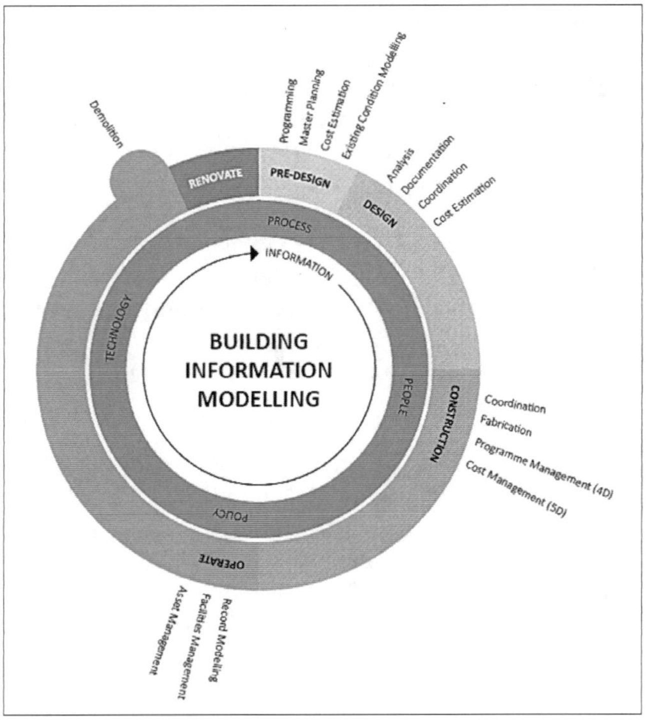

Figure 1. The principles of BIM in life-cycle projects.

In table 1, shows the characteristics of BIM implementation among the stakeholders in the construction industry. Each stakeholder was identified the role and function to enhanced the quality of works and documentation.

Table 1. Major stakeholders and characteristics of BIM implementation [15].

Stakeholders	Characteristics of BIM Implementation
Developer	• To improve the time, deliver and quality of project. • To enhance the project information and design interest in industry. • To reduce the cost of project implementation
Academic	• To assist teaching and advance knowledge by using technology. • To provide the graduate with BIM knowledge related research.
Design consultant	• To improve the collaborative design process and productivity. • To optimize the analysis implementation and potential use of information in BIM model.
Facilities Management	• To acquire accurate data and information. • To reduce the cost of operation during facilities management and time.
Contractor & Quantity Surveyor	• To produce accurate automated Bill of Quantities (BQ). • To better understand the information to construct the building. • To have better administer the contract and project documentation.
Vendor & BIM Consultant	• To provide platform and technical advice about collaboration. • Train the industry and user

2.2. Interior design

Interior designers need to practice to educate itself by following the process in scope of work to ensure the design phases and construction process will be coordinate accordingly the task and stage process as a professional practice. As mentioned [16], interior designers must satisfy a certain level of skills and knowledge with familiar with scope of work as a tool to ensure the projects delivery successful. To produce knowledgeable interior designers, the improvement in several issues must be rise to ensure interior designers capable using BIM technology in the design stage. Starting from highest education level it can be enhance the knowledge and potential talent to grab the opportunity of learning in BIM technology. Interior design industry can be competing with others construction industry such as architecture, engineering, construction, and facilities management in term of knowledge, skill and reliable to using BIM.

Todays, interior design has improvised the technology and the understanding of digital technology in the design process and divided at the following phases: Programming, Conceptual Design, Design Development, Contract Document and Evaluation [17]. The design process has improved [18] by adding several phases: Analysis, Development, Implementation and Evaluation as the different step of the interior design process.

3. Research methodology

This study involved questionnaire survey to extract the relevant data and information from the respondents. The issues about implementation BIM among interior design industry were justify through literature review and previous research in construction industry. This research was done by preliminary study by interior design firms are as respondents in Klang Valley. In this study, the interior design firms registered with Lembaga Akitek Malaysia (LAM) were chosen. This survey was distributed by face to face (appointment) and email (by request) of firms. This survey was discovery the engagement interior design industry implement BIM technology in term of knowledge and skill.

This research was conduct with the literature study to review the challenge factors, barriers element, solution and benefits of BIM implementation for interior design industry in Malaysia context. Beginning with the preliminary study and following with the questionnaire, discussion, analysis and finding to find out the problems and issues as shown in figure 2.

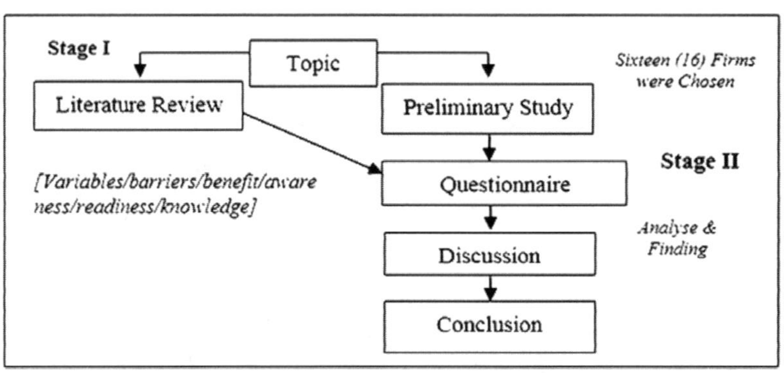

Figure 2. Research Methodology Process.

This methodology has two (2) stages which is Stage I and Stage II. It is started with identify the research topic through literature review and previous study. Then all the variables and key determine the factors were conduct via preliminary study by questionnaire survey which is sixteen (16) interior design firms were chosen randomly in Stage II. From the collecting data, analyse data and finding were justify in discussion and conclusion.

4. Results and analysis

This study was conduct through preliminary study to find out the engagement of interior design firms implement BIM technology as a tools in the design process. The survey is gather from sixteen (16) interior design firms that randomly selected in Klang Valley. The survey was conduct through face to face by appointment with respondents to collect data and discussion regarding the research topic. This survey is to find out the current situation for interior design industry implement BIM in their profession as a tool to develop on the design process and documentation.

IOP Conf. Series: Materials Science and Engineering **713** (2020) 012047 doi:10.1088/1757-899X/713/1/012047

4.1. Demographic of respondents

This study, it is sixteen (16) respondents was chosen to justify the current situation about implementation BIM technology for interior design industry through the questionnaire survey. In table 2 shows the frequency of demographic of interior design firms in the data collection. Through the data the responses of year of firm established for interior design which is the data analysed six (6) respondents represented 38% of the total respondents were established in year 2006 to 2010 while five (5) respondents represented 31% were established in year 2011 to 2015. It sis following with three (3) respondents represented 19% established in year 2001 to 2005. However, in year 1990 to1995 and year 1996 to 2000 each representing one (1) respondent represented 6% only.

In the categories of firm, the respondents mostly are interior design consultant with represented fifty-six percent (56%) followed design and build with represented twenty-five percent (25%) and contractor with represented nineteen percent (19%). Most of the interior design industry are from interior design consultant because they comfort to this type of business and can generate more return of investment (ROI).

Table 2. Demographics of Respondents.

	Characteristic	Frequency	Statistics (%)
Year of Established	1990-1995	1	6%
	1996-2000	1	6%
	2001-2005	3	19%
	2006-2010	6	38%
	2011 - 2015	5	31%
Categories of Firm	ID Consultant	9	56%
	Design and Build	4	25%
	Contractor	3	19%
	Others	0	0%
BIM Usage	Less than 1 Year	3	19%
	2 to 5 Years	7	44%
	6 to 9 Years	4	25%
	More than 10 Years	2	12%

4.2. Knowledge level and experienced implement BIM

In figure 3 shows the responses of level of knowledge instead of experience of BIM, justify that four (4) respondents know about BIM while twelve (12) respondents do not know or have never heard of BIM. Most of the interior design industry they are not aware about BIM benefit in their design process and documentation. From the graph in figure 3, the knowledge is very crucial which is among of the interior design firms are not implement BIM.

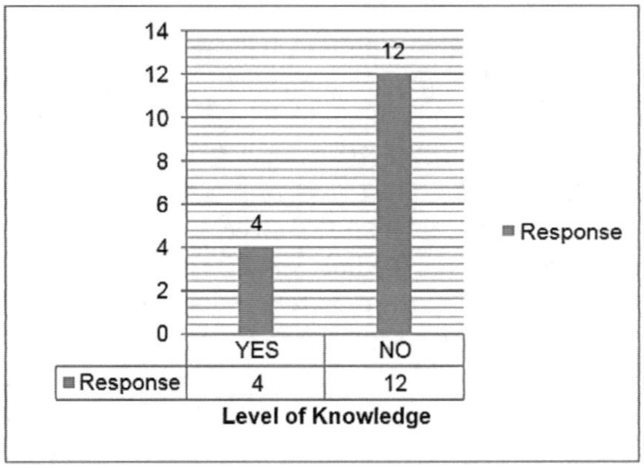

Figure 3. The Knowledge Level among Interior Design Firms.

4.3. Significant of BIM in the interior design

In table 3 shows the significance of BIM in interior design firms which is thirteen (13) respondents represented eighty-one percent (81%) agreed that BIM is very importance but three (3) respondents represented nineteen percent (19%) not agreed the significance to implement BIM in the interior design industry. The reason why they are not agreed to used BIM are they are more comfortable to used traditional method such as sketches and 2D drawing like AutoCAD. The high cost also take consideration to implement BIM in the interior design firms and in term of knowledge among interior designs are very low and the level of awareness also very poor.

Table 3. The significant of BIM implementation in interior design.

Response	Frequency	Percentage
Yes	13	81%
No	3	19%
Total	16	100%

4.4. Software usage

Table 4 shows the responses of software usage for interior design. Accordingly, table 4, 1 respondent representing 50% of the total respondents were using Revit while 1 respondent representing 50% were using ArchiCAD as their tool. From the data, the numbers of interior design firms using BIM software is very crucial with only two (2) respondents answered they are using BIM technology. In the BIM technology, software knowledge and skill are the things very important for construction industry to expose more about BIM especially the beneficial and advantages while using BIM.

Table 4. BIM software for designers.

Software Usage	Frequency	Percentage
Revit	1	50%
ArchiCAD	1	50%
AutoCAD	0	0%
Others	0	0%
Total	2	100%

IOP Conf. Series: Materials Science and Engineering **713** (2020) 012047 doi:10.1088/1757-899X/713/1/012047

5. Conclusion

Interior design is still struggle in their skill and abilities in term of public and perceptions [19]. The associated of knowledge and current industry technology are very critical for interior design industry to perceived in the professionalism to value the project especially interior design project. The three-dimensional design based are very effective while implement BIM technology in the design process. BIM technology produced and provided the integration between environment and technology for the designers from the conceptual to construction drawing, visualization and documentation in the construction projects. Through the development of information of BIM technology, it is can achieve to the high level of perfection to pursuit in the design process.

With BIM technology, interior design industry can improve the quality of work and can reduce the cost on the project construction. The ability to using BIM would be developing the professionalism of interior design industry with others construction industry in the same level. The capability to using BIM software can help interior design industry lacking on the knowledge and it is secure in term of cost, time and design development in the project. Thus, they are still lacking on the skill and knowledge. The process to ensure interior design to used BIM are need to has initiative and enforcement to ensure they will have used BIM in the future.

6. References

[1] RCS 2014 *Engineering design and project risk: do more of the first to take on less of the second (RCS-Rapid Construction Solution)* (Baku, Azerbaijan: MacKay Group)

[2] Man and Machine 2014 *Collaborate effectively and deliver better design, construction and operational models (white paper)* (UK: Man and Machine white papers/Autodesk)

[3] Azhar S, Nadeem A, Mok J Y and A Leung B H 2008a Building Information Modelling (BIM): A new paradigm for visual interactive modelling and simulation for construction projects. *First Int. Conf. on Construction in Developing Countries (ICCIDC-I)* (Karachi, Pakistan: Advancing and Integrating Construction Education, Research and Practice) pp 435-446

[4] Azhar S, Hein M and Sketo B 2008b Building Information Modelling (BIM): Benefits, risks and challenges *Proc. of the 43rd ASC National Annual Conf.* (Auburn, Alabama: The Associated Schools of Construction ASC) pp 1-11

[5] Chen L and Qu H 2011 Evolutions for "Economics and Legislative Factors Influence the Design Team and Contractor throughout a Building Project from Inception to Completion" *Journal of System and Management Sciences* **1**(1) 94–108

[6] Gu N and London K 2010 Understanding and facilitating BIM adoption the AEC industry *Automation in Construction* **19**(8) 988-999

[7] Baiden B, Price A and Dainty A 2006 The extent of team integration within construction projects *International Journal Project Management* **24** (1) 13–23

[8] Won J and Lee G 2010 Identifying the consideration factors for successful BIM projects critical success factors in BIM adoption Construction *Proc. of The Int. Conf. on Computing in Civil and Building Eng. (The University of Nottingham)* (Nottingham: Nottingham University Press) pp 1–6

[9] Zakaria Z, Mohamed Ali N, Tarmizi Haron A, Marshall-Ponting A J and Abd Hamid Z 2013 Exploring the adoption of Building Information Modelling (BIM) in the Malaysian construction industry: A qualitative approach *Int. J. Res. Eng. Technol.* **2** (8) 384–395

[10] Ahmad Latiffi A, Mohd S and Brahim, J 2014 Application of building information (BIM) in the Malaysian construction industry: A story of the first government project *Appl. Mech. Mater.* **773** 943–948

[11] National Council for Interior Design Qualification (NCIDQ) 2004 *Definition of interior design* (Virginia: Council for Interior Design Qualification Inc)

[12] Pryke A 2016 *BIM: Risk Management During Project's Lifecycle* (Canada: Royal Institute of Chartered Surveyors (RICS))

[13] Idrus A and Sodangi M 2010 Framework for Evaluating Quality Performance of Contractors in Nigeria *International Journal of Civil & Environmental Engineering IJCEE-IJENS* **10** (1) 31–36
[14] New Zealand BIM Handbook 2014 *A Guide to Enabling BOM on Building Projects* (New Zealand: Ministry of Business, Innovation and Employment) pp 4
[15] Construction Industry Development Board (CIDB) 2013 *Building Information Modelling (BIM) Roadmap for Malaysia's Construction Industry Workshop Report (Series 2)* (Kuala Lumpur: Construction Industry Development Board Malaysia) pp 4
[16] Ahmad Noorhani N M 2012 *Interior Design Project Manager Competency* Unpublished PhD draft thesis (Shah Alam: Universiti Teknologi MARA, Malaysia)
[17] Kang M 2009 A Computer Database of Design Methodological Tool Pattern for Interior Designer *ITDL* **4**(3) 45–61
[18] Dodsworth S 2009 *The Fundamentals of Interior Design* (Switzerland: AVA Publishing)
[19] Clemons S and Waxman L K 2007 Student perceptions: debunking television's portrayal of interior design *Journal of Interior Design* **32**(2) pp v–xi

Acknowledgments
The authors would like to acknowledge the Universiti Teknologi MARA (UiTM) and Ministry of High Education (MOHE) for supporting the financial under the awarded Bestari Grant (600-IRMI/MyRA 5/3/BESTARI (003/2017). The authors also would like to thankful to interior design firms and to who assist along the research progress for providing comprehensive and important information for contributing part of this research.

IOP Conf. Series: Materials Science and Engineering **713** (2020) 012048 doi:10.1088/1757-899X/713/1/012048

A Review of Green Roof: Definition, History, Evolution and Functions

F Abass [1], L H Ismail [1], I A Wahab [1] and A A Elgadi [1]

[1] Department of Architecture, Faculty of Civil and Environmental Engineering, University Tun Hussein Onn Malaysia, Parit Raja, Batu Pahat, Johor, Malaysia
Corresponding author: izudin@uthm.edu.my

Abstract. In cities, buildings contribute large unused panels, especially on their wall and roofs. These external surfaces generally made from materials that listed as a reflector to the heat such as concrete, glass, and bricks. Covering the external surfaces with green vegetation could be a solution to many environmental, economic and social problems, especially in the urban areas. A review was directed of studies that reported and measured on the relationship between characteristics of vegetation and their influence on building thermal performance, air and water pollution, and building energy-efficient. A numerous of published studies were investigated for their: (a) definition based on roof typologies and combinations; (b) historical evolution periods; (c) observation the green system functions. These analyses cover to afford scholars with a review of conventional techniques to understand one of the passive cooling strategies adopted in warm and cold climates. The techniques provide a significant contribution to the field of sustainable architecture, also to improve understanding of the progressive developments of a green roof. Despite other new designs of green system, this conventional and simple green roof has remained to contribute significantly to provide indoor thermal comfort.

1. Introduction

Buildings construction is always moving together with economic development. By 2030, the number of megacities with more than 10 million populations is estimated to be 43 [1]. The growth of building sectors is directly reported to be related with an increase of the greenhouse gas by 3% between 2000 and 2010, alongside with the increase of energy consumption caused by human activities; where nearly 40 % of total global energy consumption is consuming by building sector [2].

Due to the vulnerability of living area, it will be essential to implementing mitigation techniques across private and government sectors and countries, especially on sectors that substantially be contingent on fossil fuel use [3]. As building roof surfaces covers of 20–25% of the urban areas, they can effectively be used to lessen the surface and air temperature of the urban areas. A green roof is a horizontal living system, which helps to mitigate several environmental problems [4] [5].

The terms "green roof, living roof, eco-roof, vegetated roof, and rooftop garden" [6] [7] were used to define two types of green roof types extensive and intensive roof, with naturalistic or self-established vegetation. Jim [8], define the term green roof to refer to the human-made establishment product on the roof of a house, including erecting a structural framework with appropriate mechanical strength. Another definition mentioned by Yu *et al.* [9], is a building roof which is entirely or partially covered with vegetation and growth medium. It can be a sloped roof surface or a flat designed to sustenance vegetation besides working as a fully functioning roof.

IOP Conf. Series: Materials Science and Engineering **713** (2020) 012048 doi:10.1088/1757-899X/713/1/012048

A green roof consist of many components such as plants, furnish substrate to source nutrients, water system to help the root to grow, and a drainage layer to shed remaining water. Generally, a green roof offers an appropriate environment to sustain vegetal growth. Table 1 illustrates the classification and comparison of three types of a green roof, which broadly installed on building's roof base on required functions and cost. The comparison varies on the structure system, types of plants, prevalence, and the installation cost. It should note that the final cost might vary among countries and green roof installer between these green roof types.

Table 1. Types of green roofs according to classification [10] [11].

Green system	Comparison
Single course extensive	Thickness: 7.6 cm – 10 cm
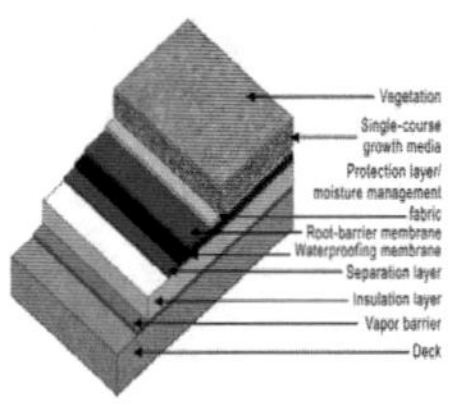	Drainage layer: No separate drainage layer
	Vegetation layer: Ornamental and Succulents plants
	Media type: Coarse media
	Irrigation layer: Not required
	Prevalence: Used in an area with sufficient precipitation
	Weight: 60 – 150 kg/m^2
	Cost: 543 – 690 RM/ m^2
Semi-intensive	Thickness: 15 cm – 31 cm
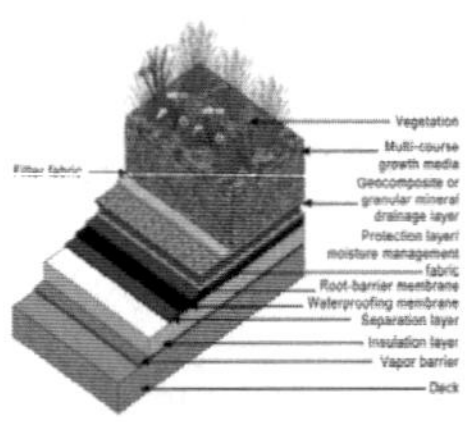	Drainage layer: Separate drainage layer
	Vegetation layer: Ornamental, meadow species, turf grass and woody perennial
	Media type: Multi-course media
	Irrigation layer: Required if meadow grass used
	Prevalence: Common and provide more plants choice
	Weight: 25% above or below 150 kg/m^2
	Cost: 25% above or below 690 RM/ m^2
Intensive	Thickness: Over 31 cm
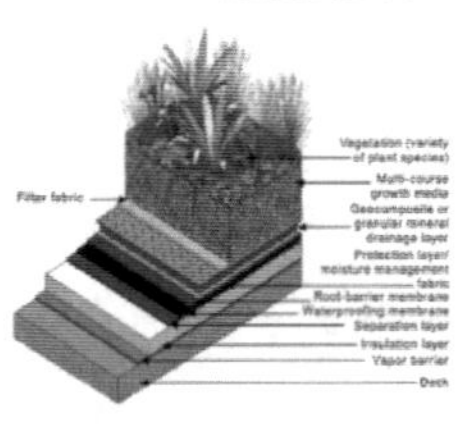	Drainage layer: Separate drainage system
	Vegetation layer: Ground-level plants
	Media type: Intensive growth media layer
	Irrigation layer: Required
	Prevalence: Less common due to it is structure and maintenance factor
	Weight: 200 – 500 kg/m^2
	Cost: 2259 RM/ m^2

2. History evolution of green roofs

2.1. Green roof in the 20th century

In the early modern era, different continents kept the idea of the green roof alive; this concept was broadly adopted in various regions and cultures. In the mid of 1880s, the new technology brought the idea of a living roof on the top of the concrete roof, the first model of this roof appeared in the World Expo in Paris in 1867. The model has illustrated a green roof with waterproofing and drainage system, which consider the first design of an extensive green roof [12].

During the 20th century, the originator of modern architecture (Le Corbusier, Alvar Aalto, and Frank Lloyd), start to implement the green roof and walls in their design to merge the natural with construction. Their famous designs are a clear sign of this concept (Villa Shodhan, Villa Mairea, and Millard House). Figure 1 presents famous buildings designed with a green roof between the 1920s and 1930s.

IOP Conf. Series: Materials Science and Engineering **713** (2020) 012048 doi:10.1088/1757-899X/713/1/012048

In the late of 20th century, the appearance of industrial era tracked back the concept of green roof in Germany, after the innovation of mixed gravel and sand with tar to produced non- inflammable waterproof by H. Koch. With the help of nature, the new materials became the base system to herbaceous plants that grow on building's roof, later in the 1960s, both sand and gravel layers were replacement with simple drainage system and new design of lightweight green roof [8]. The innovation and development of roof technology made German the first country in the world adopted the principle of green roof in the building followed by North Europe and North American and Finally a few countries in Asia.

Figure1. (A) Green gardens at the top of Rockefeller Centre in New York, (B) Roof garden at Villa Mairea designed by Alvar Aalto, (C) Green roofs at Monastery of La Tourette designed by Le Corbusier's.

2.2. Green roof in a hot climate

The definition of a hot climate in this section refers to the Mediterranean, tropical, steppe, and hot climate. These climates adopted green roofs as construction elements or upper ground garden in both vernacular and monumental architecture in Asia, Africa, Eurasia, America, and Australia. The first appearance of the green roof was in Ziggurat of Ancient Mesopotamia [14], from the fourth millennium until 600 BC. The green roof located in the courtyard temples, shrubs and trees were planted at terraces formed by a gran-stepped pyramid. Whereas, the most famous type of green roof is which constructed at about 500 B.C. is "The Hanging Garden of Babylon" that known as the first botanic garden in the world, whereas various types of plants do not exist in the community were cultivate.

In the Mediterranean era, the existence of the green roof was found in Pompeian buildings were the patch of green in the heart of the city. The appearance of the green garden in the shape of Atrium not only found in public spaces and but on the roof of dwellings building. Villa of the Mysteries (Villa Dei Misteri) is evidence of intensive roof garden in Pompeii, according to Osmundson [14], the entrance of the house is clear evidence of the roof garden, while the façade is a kind of hanging garden.

The previous paragraphs presented a brief of some examples of green roof that existence in Monumental architecture, whereas used to mitigate the hot climatic condition, as a recreational area, or to express the social status of the owner. Not far away from the concept of green roof of Monumental architecture, the appearance of a dry roof in vernacular architecture due to the lack of water in some areas, the use of bamboo, grass, leaves, and reeds as construction element were found in different regions, times and civilization [13]. In table 2, a comparison of different type of green roof that exist in vernacular architecture in other regions and civilizations with hot climate is highlighted.

IOP Conf. Series: Materials Science and Engineering **713** (2020) 012048 doi:10.1088/1757-899X/713/1/012048

Table 2. Examples of a thatched roof found in vernacular architecture in a hot climate.

Figure	Description
	Dome-shaped dwelling made from dry reeds and covered by dry grass in Zulu Land [15]. Dwellings designed with round or geometrical shapes were covered with dry grass on the top of clay, stone houses or grass houses. The thatched roofs used since ancientness time until now, to cope with the hot climatic condition.
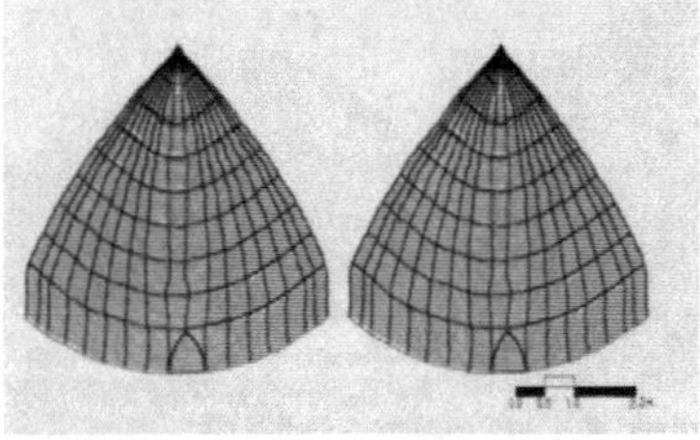	In some Europe countries, the thatched roof used commonly to keep out the heat faraway from the indoor. The designs of the house are different from that area to another. As showing in figure two, grass hunts were used by the farmers in Greece as a temporary shelter and storing area [13].
	Dry turf used by Native Americans as construction materials even before the Spain invasion. The grass was used the steep cover roof, and to protect the palm leaves from outdoor weather. Later, the (Choza house see the figure), developed with the design of the square house after Spain invasion or by the Africans migrants to round hunts.
	In Asia, dry plants used as a raw material to construct their settlements in China, India, Japan, and Indonesia. Turf roof used by villagers to cover their straight, curved, or even pyramids roof shape. The figure shows pyramid shape thatched hut in Bay. Thatched roof and walls were considered a technique to mitigate the climate condition.

2.3. Green roof in cold climate

In a cold environment, there is plenty of green roofs existed; the lack of water usually is not an issue in this climate. Therefore, the green roofs are used as a type of insulation material, to lower the heat losses from inside to outside. The use of dry grass in construction materials in Europe has appeared from prehistoric time; it has used in central Europe in Germany, Poland, France, some Scandinavian countries, Ireland, and Britain. The existing of green roof shown in burial construction, while the burial location under a hill covered by grass and plants and has a bath way to chambers. The burial's layout could be small, large, single, twin or triple. These simple layouts help the prehistoric nation to draw their footprint on the landscape design.

The man-mad burials also used by many ethnic groups, such as Vikings, Celts, Pagan Saxons and ethics groups of central Europe. The time estimated to be between 1500 to 800 B.C. for Scandinavia groups, and 1400 to 1200 B.C. for the ethics groups in the South of France [16], and continued to Christian periods. As adaption to the green roof that shows in burials since 1500BC, the reed roof mainly used in holy places, like a church in Christian's periods and mosques in Africa. An English traveler, John Barrow recorded a church built in 1796 called Thingvellir church applied the concept of a thatched roof. Besides, that Sifrastadir church (built-in 1842) in Northern Iceland and Vidmyri church was also designed with green living roof [13].

Thatched roof in vernacular architecture in cold climate is not the only elements that used was to cover buildings decks, but also the intensive roof. The weather condition it is possible of the grass to grow easily. Both techniques of the roof (hatching and living roof) were widely famous from prehistory ear until the 19th century in many areas around in central Europe, France, British Isle, and Russia. Excavations evidence revealed that between 3600BC until 2600 BC the implementation of the thatched roof on stonewalls, strawbale or brick walls was applied widely. This technique proved that the green roof and brown roof widespread through different types of building from churches, temporary sheltering industries, to houses. Even when people shifted to agriculture, and villages started to appear on 3900-3600 BC, the concept of green vegetation remains on huts and building [14]. In Table 3, some examples of the application of green roof in vernacular architecture in cold climate are presented.

Table 3. Examples of living/ green roof found in vernacular architecture in a cold climate.

Figure	Description
	Turf wall and turf pitched roof were essential elements used during the Viking era to protect the building from a low temperature during winter. This technique dated from 800 AD until the late 9th century. The figure presents a living roof in the Mallhaugen Open Air Museum in Lillehammer, Norway.
	The concept of turf roof and walls found in North America, L' Epaves Bay, Newfoundland, as well as L'Anse. According to [16]. Turf roof covering the ground and the top of winter house or oval family settlement appeared in Yup'ik Eskimo in Alaska. The attached figure shows large vertical dwellings cover with the turf at Yukon River, Alaska.
	This figure illustrates the living grass on the roof of a dwelling in Canada. During the 19th century, the Russian migrants passed their green roof technique to Canada; also the turf walls and roof shown in miners dwellings at Thompson river valley, in Canada whereas dry twigs and turfs used as construction materials, to insulate indoor from the cold climate.

3. Adaption of a green roof to overcome environmental problems

The idea behind green roofs in the metropolitan for environmental benefits looks to be more accepted in the late of 20th century. Individuals become more conscious of environmental issues. The implementation of the green roof in buildings in some cities presented as an approach to developing the urban environment. The benefits of horizontal green systems under the three segments: environmental, social, and economic benefits are presented in table 4.

Table 4. The benefits of the implementation of a green roof in an urban area.

Benefit	Findings
Environmental	i. Natural filtration: green roof works as natural air filtration to absorb dust and clean the air [17]. ii. Plants have photosynthesis that consumes CO_2 emission and release oxygen; this makes the air fresh and reduces CO_2 dioxide emission [18] iii. The green roof could be trapped up to 4% of heavy metal city dust iv. Acoustics: The ability of green roof on controlling and reduce sound reflection [19]. The reduction of noise level could reach 8–10 DB. v. Stormwater management: Geen roof acts as porous surfaces and control storm waters [20]. vi. Green roof or pavement could reduce 4 °C less than the original roof [21].
Economic	i. Energy: green roof has a benefit to reduce energy consumption through shading, insulation, evapotranspiration, and increase in thermal mass [22]. ii. Due to the capability of horizontal plants to lessen temperature, they are an appropriate technique for dropping cooling energy request and improving the energy efficiency of buildings [23]. iii. A green roof is lessening contraction and growth of building material due to temperature, besides works as a shield from acid rain and UV ray [24]. iv. Thermal comfort: Plants have severe effects of systems on reducing urban heat island which is a significant problem in cities and urban areas, due to it is capacity to absorb shortwave radiation and cooling the atmosphere [25]. v. Green roof nominated as a tool to mitigate urban heat island, the ambient temperature lower from 0.3 °C to 3 °C when used green roof vi. Green roof increases the life span of the building roof by protecting the roof from thermal, UV radiation fluctuations, and stress diurnal [22].
Social	i. The green roof creates places for recreation and rest [26]. ii. Green roof proved that contact with nature has a psychological impact and increases human health and wellbeing [27]. iii. Plants reduced stress and lower obesity achieved by proximity to green areas. iv. Plants: Have a positive impact on the people working or living around them. A report indicates that employee productivity in building with a sign of green plants is higher than in buildings that are less living environmental. v. Plants in urban areas and compound by buildings attract people more than plants in gardens [28]. vi. Roofs can offer very vital environments for expert rare or imperiled species [12]

4. Discussion

In this paper, it was shown that greening the roof for improving life and social activities in cities are not merely a scientific vision of the 20th century. The concept of the implementation of dry or living plants and grass on building roof and walls appeared in different climates as a construction element, or as a mitigation technique to lower the high temperature in a hot climate or to insulate building envelopes from cold weather. In some cultures, either the green roof showed the social status of dwelling owners,

used for socializing or enjoying the outdoor terrace garden, even the concept in some culture could be for bidden to more impoverished people. Whatever the use of a green roof to symbolize a social status or as to mitigate the climatic conditions, thatched roofs were found in permanent and temporary buildings since the prehistory era until the beginning of last century.
Since the early 20th century, visionaries have been trying to find a solution to improve the lives of urban dwellers. Despite their approach and social beliefs, vegetation is still a central element for the well-being of the inhabitants. At the beginning of the 20th century, the concern was more about people and society rather than the environment itself.

This paper concludes the evaluation of green roof started from its definition, system components, historical evolution, and the impact of green roof performance. The paper is a part of a complete study that focuses on the green system in building envelope. To present a level of understanding the effectiveness of these systems on the building environments as insulation materials in a hot, humid climate and passive techniques that is a help to reduce the energy consumption and increase thermal envelope performance.

5. References

[1] Urbanization Statistics UN 2018 *Revision of World Urbanization Prospects* (New York: UN DESA)
[2] World Energy Council 2013 World Energy Resources Survey 2013 Retrived on Febraury 24, 2019 from https://www.worldenergy.org/publications/2013/world-energy-resources-2013-survey.
[3] Toe D H C 2013 *Application of Passive Cooling Techniques to Improve Indoor Thermal Comfort of Modern Urban Houses in Hot-Humid Climate of Malaysia* Phd Thesis (Hiroshima: Hiroshima University)
[4] Arabi R, Shahidan M F, Kamal M S M, Bin Jaafar M F Z and Rakhshandehroo M 2015 Mitigating Urban Heat Island Through Green Roofs *Curr. World Environ.* **10** Special-Issue1 918–927
[5] Poptani H 2014 Extensive Green Roofs : Potential for Thermal and Energy benefits in buildings in central India *30th Int. PLEA Conf.* (Ahmedabad: CEPT University) pp 1–8
[6] Francis R A and Lorimer J 2011 Urban reconciliation ecology: The potential of living roofs and walls *J. Environ. Manage.* **92**(6) 1429–1437
[7] Ahmed S F, Ahasan T, Rasul M G, Khan M M K and Azad A K 2014 Performance Evaluation of Hybrid Green Roof System in a Subtropical Climate Using Fluent *J. Power Energy Eng.* **2**(4) 113–119
[8] Jim C Y 2017 Green roof evolution through exemplars: Germinal prototypes to modern variants, *Sustain. Cities Soc.* **35** August 69–82
[9] Yu WD, Cheng S T, Miao C M and Perng G Y 2017 Green innovation of green roof technology - a case study *Materwiss. Werksttech* **48**(5) 420–429
[10] GSA.US 2011 *The Benefits and Challenges of Green Roofs on Public and Commercial Buildings* A Report of the United States General Services Administration (United States: General Services Administration)
[11] Hossain MA, Shams S, Amin M, Reza MS and Chowdhury TU 2019 Perception and Barriers to Implementation of Intensive and Extensive Green Roofs in Dhaka, Bangladesh *Buildings* **9**(4) 79
[12] Dunnett N N and Kingsbury 2008 *Planting Green Roofs and Living Walls* 2nd illustrated (Portland, Ore: Timber Press)
[13] Alexandri E 2005 *Investigation into Mitigation the Heat Island Effect through Green Roofs and Green Walls* PhD Thesis (Cardiff: Cardiff University) pp 94 -141
[14] Osmundson T1999 *Roof Gardens: History, Design, and Construction* (New York: W.W. Norton)
[15] Oliver P1997 *Encyclopedia of Vernacular Architecture of the World: Cultures and habitats.* (Cambridge: Cambridge University Press)
[16] Donnelly M C1992 *Architecture in the Scandinavian Countries* (Cambridge, Massachusetts, UK: The MIT Press)

[17] Liu C, Li Y and Li J 2017 Geographic information system-based assessment of mitigating flash-flood disaster from green roof systems *Comput. Environ. Urban Syst* **64** 321–331

[18] Korol E and Shushunova N 2018 *Modular Green Roofs in Urban Ecospace* Landscape Architecture - The Sense of Places, Models and Applications (London: IntechOpen Limited) chapter 7 pp 127 – 140

[19] Azkorra Z, Perez G, Coma, J., Cabeza L F, Bures S, Alvaro J E, Erkoreka A and Urrestarazu M 2015 Evaluation of green walls as a passive acoustic insulation system for buildings *Appl. Acoust.* **89** 46–56

[20] Chen X, Huang P, Zhou Z and Gao C 2015 A review of green roof performance towards management of roof runoff *J. Appl. Ecol.* **26**(8) 2581–90

[21] Peron F, De Maria M M, Spinazzè F and Mazzali U 2015 An analysis of the urban heat island of Venice mainland *Sustain. Cities Soc.* **19** 300–309

[22] Feng H and Kasun H 2018 Economic Benefits and Costs of Green Streets *Nature-Based Strategies for Urban and Building Sustainability* 307–318

[23] Coma J, Pérez G, de Gracia A, Burés S, Urrestarazu M and Cabeza L F 2017 Vertical greenery systems for energy savings in buildings: A comparative study between green walls and green facades *Build. Environ.* **111** 228–237

[24] Liang E 2014 *What is Living Wall ?* (Singapore: Institute of Technical Education)

[25] Santamouris M 2014 Cooling the cities – A review of reflective and green roof mitigation technologies to fight heat island and improve comfort in urban environments *Sol. Energy* **103** 682–703

[26] Susca T, Gaffin S R and Dell'Osso G R 2011 Positive effects of vegetation: Urban heat island and green roofs *Environ. Pollut.* **159**(8–9) 2119–2126

[27] Gidlöf-Gunnarsson A and Öhrström E 2007 Noise and well-being in urban residential environments: The potential role of perceived availability to nearby green areas *Landsc. Urban Plan* **83**(2–3) 115–126

[28] Blanc P 2012 *The Vertical Garden: In Nature and the City* (New York: W. W. Norton & Company)

Acknwledgements

I would like to thank Assoc. Prof. Ts. Dr. Lokman Hakim bin Ismail, Dr. Izudinshah bin Abd Wahab, and Dr. Ahmed Elgadi for their discussions, guidance, and sharing of experience which enrich this study.

IOP Conf. Series: Materials Science and Engineering **713** (2020) 012049 doi:10.1088/1757-899X/713/1/012049

Traffic Noise Assessment among Residential Environment in Batu Pahat, Johore, Malaysia

V C Segaran[1], Y G Tong[1], N H Abas[1], B David Daniel[2], S Nagapan[1] and R Kelundapyan[1]

[1] Jamilus Research Centre, Faculty of Civil and Environmental Engineering, Universiti Tun Hussein Onn Malaysia, Batu Pahat, 86400, Malaysia
[2] Micropollutant Research Centre, Faculty of Civil and Environmental Engineering, Universiti Tun Hussein Onn Malaysia, Batu Pahat, 86400, Malaysia
Corresponding author: ygtong@uthm.edu.my

Abstract. Noise pollution arises as a severe problem to the community and ground transportation has highly impacted noise environment of the neighbourhood. In the present study, six stations were selected from three different locations in order to assess the traffic noise impact among residential area along Jalan Kluang, Batu Pahat. The objectives of this study focused on two main stages: (1) determination of noise indices; and (2) assessment of traffic noise impact. The noise indices such as equivalent continuous noise level (LAeq), noise level exceeded 10% (L_{10}), 90% (L_{90}) and the maximum noise level (LAFmax) of peak hours measurement time were determined using sound level meter (SLM) at all six stations. Besides, traffic noise index (TNI) and noise pollution level (LNP) were calculated based on the measured noise indices to expose the traffic noise impact. Overall, the result of the present study showed that residential area faces traffic noise disturbance partially whereby 50% of the stations exceeded the DOE standard which was 55dBA. Nevertheless, the residents living nearby Jalan Kluang are commonly facing traffic noise nuisance, but stations located further inside residential areas tend to have more residential background noises apart from traffic influence.

1. Introduction

The rising number of ground transportation due to the development of cities and population daily demands has directly been associated to urban noise pollution in terms of elevated traffic noise levels [1]. The traffic noise mainly occurs from the frictional effect between road surface and tires which causes high negative impact to the urban communities and the surrounding environment [2-3]. Dominance of traffic noise level in the surrounding environment affects the sleeping, resting, studying and communication among population and together with traffic noise annoyance, they constitute threatening impacts to human health in long term [4].

2. Traffic Noise Limits

Noise pollution has become a serious problem for humans and has been officially announced as the second threatening pollution in the world by World Health Organization (WHO). Noise pollution inducement from transportation varies into several sources such as traffic noise, aircraft noise, marine noise and railway noise. Traffic noise can be defined as an unwanted sound generated from vehicles while operating on roadways. Traffic noise is known as the most pervasive and serious among transportation form of noise sources [5]. According to the European Environment Agency (EEA) in year

2014, urban noise has been categorized as major environmental health threats in overall Europe [6]. From the report, road traffic is the significant source of environmental noise with an estimated 125 million people affected by noise level greater than 55 decibels dBA which is about 40% from the entire population living among Europe.

In Asia, developing countries such as China, India and Vietnam are facing serious problem of traffic noise pollution in their cities [7]. Malaysia is one of the developing countries included in this problem due to the development pace that constantly occurs with the increasing number of road transportation networks in supporting the development process. Noise pollution awareness is not a new scenario as it has received the attention of the authorities in Malaysia as early as 1979 [8]. The Star Online (2016) stated that 132 noise pollution complaints have been reported to the Department of Environment, Malaysia in 2015. Most of the complaints were reported due to the noise from the commercial and construction sites. Other noise pollution complaints were reported regarding the noise that came from industrial and transportation system. Although traffic noise annoyance is not the major source of nuisance, the expert has highlighted that road traffic noise is the most pervasive noise pollution over the world [9].

Based on the environmental control limits set by the Malaysian Department of Environment (2007) at suburban and residential areas, the maximum permissible noise level outside the low density residential areas should not exceed 55 (dBA) in order to protect the community from this urban noise. However, the noise limit varies according to the sensitivity of area, type of location and different regulations set by the countries. Table 1 shows noise limit level among the selected countries which are classified into daytime and night-time noise level which are compared to WHO noise limit [10-12]. Hence, specific evaluation of traffic noise level will help to understand the actual condition of noise pollution in residential environment. Furthermore, the obtained results contribute in identifying enough precautions to control traffic noise annoyance in residential areas along Jalan Kluang Batu Pahat.

Table 1. Noise level standard by WHO and selected countries [10-12].

Noise Level Limit	**Noise level, L_{eq} dBA**	
	Daytime	**Night-time**
WHO	55	45
Malaysia (**DOE** Low Density)	55	50
Germany (Noise level guidelines)	45	35
Australia (Recommended outdoor background noise level)	45	35
Japan (Environmental quality standards)	45	35
Korea (Environmental quality goal)	50	45
Philippines (Environmental quality noise standards)	50	40
Iran		
(Residential area)	55	45
(Commercial area)	65	55
(Industrial area)	75	65

Thus, the present study was conducted to evaluate traffic noise indices and to assess the traffic noise impact towards nearby residents of Jalan Kluang. The assessment of traffic noise pollution at these areas was also compared with the noise limit and control that has been set by Malaysian Department of Environment.

IOP Conf. Series: Materials Science and Engineering **713** (2020) 012049 doi:10.1088/1757-899X/713/1/012049

3. Methodology

The main objectives of the present study were divided in two main stages: (1) determination of noise indices; and (2) assessment of traffic noise impacts. The details of each stage are described in the following sections.

3.1. Study area

In recent years, Batu Pahat has become a rapidly growing township due to several establishments of colleges, institutions and universities. In the recent year of 2015, about 41632 number of vehicles were recorded using Jalan Kluang daily according to data from Transport Statistic Malaysia [13]. The noise assessment of this study was specifically conducted on the residential areas which are directly located along Jalan Kluang Batu Pahat. A total of three main residential locations were chosen in this study to evaluate the traffic noise effects to the population surrounding. Two respective stations were chosen for each location to effectively identify the traffic noise level and the details are shown in Table 2 and Figure 1. All the observed stations were chosen by considering similar basic criteria as according to DOE guidelines which are being at least 75 meters away from intersections, traffic signal and other noise source such as schools, stadium, mosque and playground.

Table 2. Measurement locations.

Location	Stations	Coordinates
Taman Lundang	L1	1.8571° N 103.1000° E
	L2	1.8477° N 103.0717° E
Taman Arked	L3	1.8664° N 103.1168° E
	L4	1.863216°N 102.955564°E
Taman Gading	L5	1.863159°N 102.958450°E
	L6	1.859886°N 102.949631°E

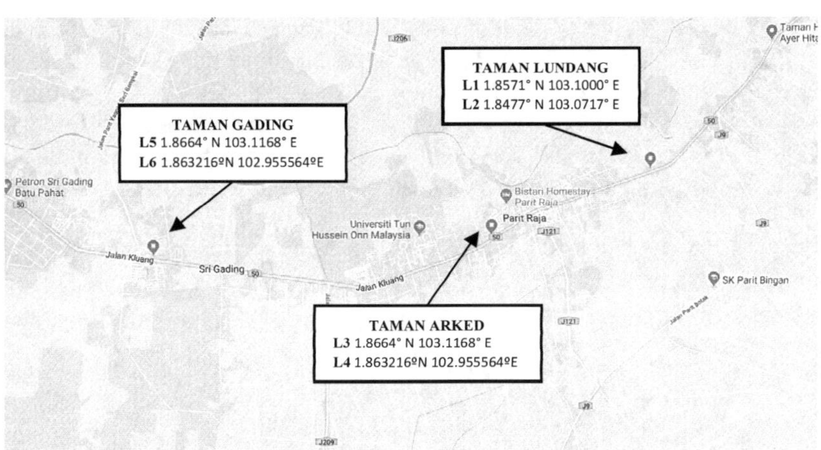

Figure 1. Map of study areas.

IOP Conf. Series: Materials Science and Engineering **713** (2020) 012049 doi:10.1088/1757-899X/713/1/012049

3.2. Determination of noise indices

The field measurement in this study was conducted to evaluate the noise indices using sound level meter of Bruel & Kjaer 2250 band class 0 which complies to the guidelines set by Malaysian Department of Environment [12]. Sound level meter (SLM) was configured with 1/3 octave band and A-weighted units to capture sound measurement in the present study. SLM was approximately placed on level ground about 1.5 meters high and 70 degrees to noise source with the aid of s tripod. The microphone of SLM was protected by windshield cover and the stations were located by avoiding any reflective sources nearby as shown in Figure 2. The traffic noise assessments were conducted during weekdays. Noise surveys were taken during the peaks hours which consist of morning peak hours (7-9 am) and evening peak hours (5-7 pm).

(a) (b) (c)

Figure 2. On-site traffic noise measurement locations for (a) Taman Lundang, L1, (b) Taman Arked, L3, and (c) Taman Gading, L6.

3.3 Assessment of traffic noise

The noise data were recorded in SLM and analysed using logging software BZ5503-Measurement Partner Suite to generate the equivalent continuous noise level (LAeq) for respective period, percentile of noise exceedance, maximum and minimum sound level. The field measured data were analysed and converted to traffic noise index (TNI) and noise pollution level (LNP) to identify the noise pollution experienced by the residents in the targeted residential areas [14-15]. The analysed noise indices such as LAeq, noise level exceeded 10% (L_{10}) and noise level exceeded 90% (L_{90}) were used to derive the TNI and LNP values. The traffic noise index (TNI) has an input of L_{10} and L_{90} as un-linear noise variability to make the respective allowance and it was derived as shown in Equation 1 [16]. On the other hand, combination of noise indices LAeq, L_{10}, and L_{90} derived the noise pollution level (LNP) as shown in Equation 2. In overall, the TNI and LNP values should not be more than 74dBA and 88dBA respectively as to be a good indicator to evaluate healthy noise environment for residential areas.

$$TNI = 4\,(L_{10} - L_{90}) + L_{90} - 30 \quad (1)$$

$$LNP = LAeq + (L_{10} - L_{90}) \quad (2)$$

4. Results and discussion

Field measurements on different time interval were observed to evaluate the seriousness of the traffic noise pollution in Batu Pahat. Overall, the measured equivalent continuous noise level (LAeq) was analysed using 2-hour period as shown in Table 3. In addition, noise parameters such as noise level exceeded 10% (L_{10}), 90% (L_{90}), maximum noise level (LAFmax), traffic noise index (TNI) and noise pollution level (LNP) were calculated to determine the traffic noise pollution rate among the observed residential areas. Analysed noise indices and parameters were presented in chart for better illustration as in Figure 3 and 4.

Overall, the field measurement result of LAeq was satisfactory whereby 50% of the observed stations were under the recommended DOE standard which is 55dBA. Each location consists of two different stations. The setup for the first station was chosen to be nearer to the road than the second station which provided the variability of distance to the noise source. The variation in distance to the main road or the targeted noise source plays a major role in determining the noise level. Thus, the nearer stations of L1, L3 and L5 recorded higher LAeq than the other stations. Apart from the distance variation to noise source, the locations layout and quantity of traffic vehicle service played more important roles in determining the noise level. For instance, location such as Taman Arked and Taman Gading recorded higher LAeq than Taman Lundang. Although the distance to noise source in all three locations were similar, the quantity of traffic vehicles at Taman Arked and Taman Gading were slightly higher. The respective two locations had similar capacity of road users since the locations were nearer to Bandar Penggeram Batu Pahat. The highest equivalent noise level was obtained in station L3 and L5 which is 57.1dBA both respectively. Both of those respective stations have similarities in term of the location's layout and the capacity of road users.

The field measurement obtained the maximum noise level (LAFmax) to identify the characteristic of noise produced in the respective residential areas. The results recorded about 71.6dBA as the lowest and 92.2dBA as the highest for LAFmax. The lowest LAFmax value obtained in this study is higher than 60dBA which like an outside environment of residential area possibly cause discomfort but only physically related effects to humans occurs during 110dBA or higher according to WHO guidelines [17]. The LAFmax has very short term presence, yet the higher noise has increased the impacts towards humans in psychology and physiology ranging from discomfort up to mortality [18]. The LAFmax is totally the case depending variable factor in each residential area. The objective is to obtain those parameters in order to identify the ongoing noise circulation in a residential area. Each station has independent level of LAFmax based on the exposure and ongoing daily events in the respective location. For instance, residential layout consists of housing area, shop lots, free ground, public utility area and playgrounds. Each different area has independent noise emission which is totally uncontrollable. This scenario shows that the arising background noise level is strongly associated with some of the resident behaviours such as honking from the vehicles, speeding of the vehicles, producing intense noise emission from heavy engine of their vehicles and other modifications of the vehicles. Apart from daily human activity, their nature noise disruption also occurs in each of the observed area such as dog barking, bird and crickets chirping which highly impacted the background noise level during the field measurement.

From table 3, it can be noticed that the traffic noise index (TNI) values and noise pollution level (LNP) were analysed to measure the physiological and psychological impact to residents living at the residential area along Jalan Kluang Batu Pahat. According to the result in table 3, it shows that all the assessed residential location recorded lesser TNI values than 74dBA. Thus, this proves that these sampling locations were not affected by traffic noise. The noise pollution levels (LNP) at the respective 3 residential areas were under 88dBA even though the LAFmax and certain LAeq values exceeded the DOE standards. Since Jalan Kluang is the main road that connects Ayer Hitam, Parit Raja and Batu Pahat, the drivers tend to speed their vehicles especially toward Parit Raja where the road is straight and wide. More detailed studies in the future are necessary to identify the extent of factors that affect the level of noise along Jalan Kluang. Besides that, other qualitative factors that may affect communication, sleepless disturbance, teaching and living environment of these areas also need to be taken into consideration as well.

Table 3. Summary of traffic noise assessment.

Location	*Station*	*Time*	*LAeq*	*LAFmax*	L_{10}	L_{90}	*TNI*	*LNP*
Taman Lundang	L1	AM	52.4	83	53.2	42.9	54.1	62.7
		PM	55	83	55.8	44.5	59.7	66.3
	L2	AM	51.9	76.5	53.4	45.4	47.4	59.9
		PM	51.1	83.85	51.1	42.7	46.3	59.5
Taman Arked	L3	AM	57.1	83.8	58.5	49.1	56.7	66.5
		PM	56.1	81.2	57.8	49.3	53.3	64.6
	L4	AM	51.3	76.1	52.1	42.6	50.6	60.8
		PM	47.5	71.6	49.1	43.5	35.9	53.1
Taman Gading	L5	AM	57.1	92.2	56.7	48.2	52.2	65.6
		PM	54.4	75.4	56.9	47.9	53.9	63.4
	L6	AM	52.2	82.6	54.1	44.1	54.1	62.2
		PM	55.1	78.1	57.3	47.7	56.1	64.7

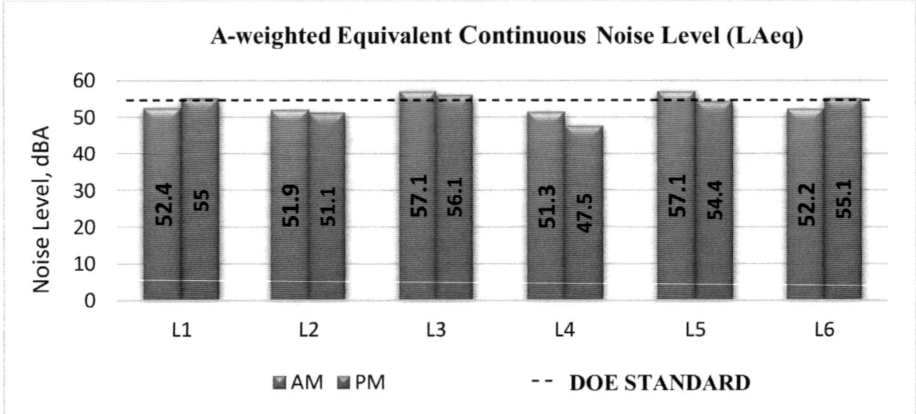

Figure 3. A-weighted equivalent continuous noise level (LAeq) for 2 hours (AM) Morning peak hours; (PM) Evening peak hours.

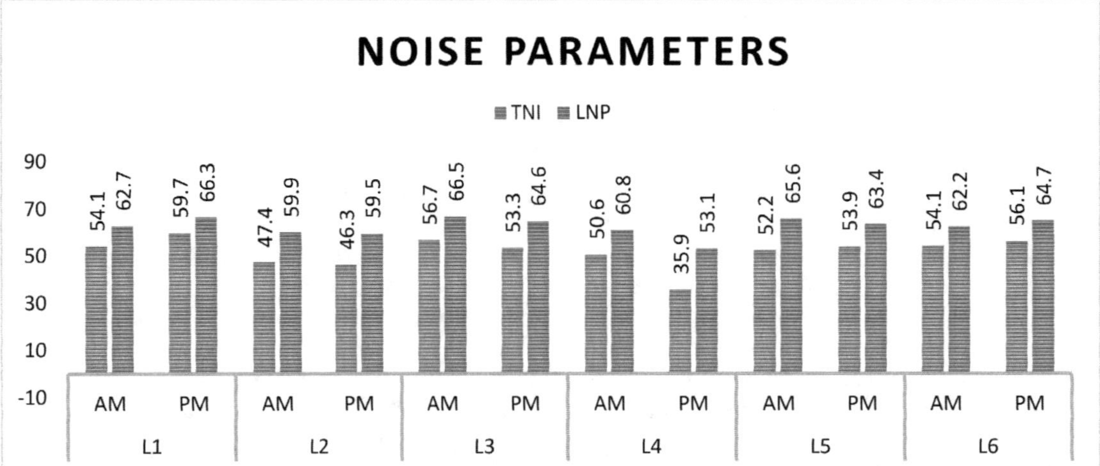

Figure 4. Traffic Noise Index (TNI) values and Noise Pollution Level (LNP) for (AM) Morning peak hours; (PM) Evening peak hours.

5. Conclusion and recommendation

In the present study, field measurements were carried out at 6 locations to assess the traffic noise impact at 3 different residential areas along Jalan Kluang Batu Pahat. From the results, two stations exceeded the DOE limits were found to be L3 and L5 over six stations by clocking the largest LAeq at about 57.1dBA. The largest LAFmax value acquired was 92.2dBA, which has a greater tendency for people living with discomfort alongside Jalan Kluang, Batu Pahat. The stations which exceeded the recommended standard have helped the researcher to understand the sound characteristics and real life scenarios throughout the field measurements. The largest traffic noise index (TNI) and noise pollution rate (LNP) were achieved at 59.7dBA, 66.5dBA respectively, which remains below the recommended boundaries. In a nutshell, resident houses near Jalan Kluang commonly face more traffic noise nuisance, while the houses further inside residential areas faces higher background noise from natural factors. The field measurement has exposed that background noise in residential area has to be considered severely for further researchers to specify the traffic noise related field of study.

Further research on traffic noise impact can be carried out more precisely which focuses on the targeted traffic nuisances to reduce the disruption of mentioned uncontrollable variables in study area. More detailed studies in the future are necessary to identify the extent to which factors affect the level of traffic noise along this busy road together with traffic characteristics data such as speed and volume of vehicles. Besides, other factors that may affect communication, sleepless disturbance, teaching and learning environment of these areas can also be included in the investigation.

6. References

[1] Abdullah M E, Shamsudin M K, Karim N, Bahrudin I A and Shah S M R 2009 A Study on Noise Level Produced by Road Traffic in Putrajaya using Sound Plan Road Traffic Noise Software *Proc. of MUCEET2009 Malaysian Technical Universities Conf. on Engineering and Technology (Kuantan, Pahang)*

[2] Zulkepli H I, Hazel M and Richard K 2000 Noise Pollution at School Environment Located in Residential Area *Malaysian Journal of Civil Engineering* **12** (2) 47-62

[3] Hayden R E 2005 Roadside Noise from the Interaction of a Rolling Tire with the Road Surface *The J. Acoust. Soc. Am* **50**(1A) 113

[4] Ouis D 2001 Annoyance from road traffic noise: A review *Journal of Environmental Psychology* **21**(1) 101–120

[5] Mehdi M R, Kim M, Seong J C and Arsalan M H 2011 Spatio-temporal Patterns of Road Traffic Noise Pollution in Karachi, Pakistan *Environment International* **37**(1) 97–104

[6] European Environment Agency (EEA) 2014 *Noise in Europe in 2014* (Luxembourg: European Union) pp 6-55

[7] Ma G, Tian Y, Ju T and Ren Z 2006 Assessment of Traffic Noise Pollution from 1989 to 2003 in Lanzhou City *Environ. Mont. Assess* **123** 413-430

[8] Rahim L A, Hashim M and Nayan N 2011 Road Traffic Noise Pollution and its Management in Tanjong Malim, Perak *Journal of Techno-Social* **3**(2) 1-12

[9] Christina C 2016 *Drowning out the noise.* Retrieved on April 29, 2019 from https://www.thestar.com.my/news/nation/2016/07/31/drowning-out-the-noise

[10] Yuen F 2014 A vision of the environmental and occupational noise pollution in Malaysia *Noise and Health* **16**(73) 427

[11] Mirhossaini S H and Pourzamani H R 2008 Evaluation and Analysis of the Environmental Noise of Arak, Iran *Journal of Applied Sciences* **8**(7) 1333-1336

[12] Department of Environment, Malaysia 2007 *The Planning Guidelines for Environmental Noise Limits and Control. Book 1 of 3* (Putrajaya: Ministry of Natural Resources and Environment) pp 10-19

[13] Transport Statistics Malaysia 2016 *Statistik Pengangkutan Jalan 2015* (Kuala Lumpur: Ministry of Transport, Malaysia)

[14] Jacques L, Joël L and Chrystèle P 2013 ENNAH - *European Network on Noise and Health.* (Luxembourg: European Union) pp 45-47

[15] Hashim M, Misran H F, Saleh Y, Nayan N and Che Ngah S Y 2014 Analisis Bunyi Bising Trafik Persekitaran Sekolah di Bandar Batu *Malaysia Geografi* **2**(2) 66–79
[16] Marathe P D 2012 *Traffic noise pollution IJED* **9**(1) 63–68
[17] Berglund B, Lindvall T and Schwela D 2000 New WHO guidelines for community noise *Noise & Vibration Worldwide* **31**(4) 21–29
[18] Yuen F 2014 A vision of the environmental and occupational noise pollution in Malaysia *Noise and Health* **16**(73) 427

Acknowledgments
Authors would like to thank UTHM for providing TIER 1 Research Fund (Grant Vot: U841) and Research Fund GPPS (U941) to support this study. Also, thanks to Faculty of Civil and Environmental Engineering, UTHM for supporting and providing facilities to accomplish the study.

IOP Conf. Series: Materials Science and Engineering **713** (2020) 012050 doi:10.1088/1757-899X/713/1/012050

Finite Element Modeling of Innovative Shallow Raft Foundation with Granular Pile Anchor System for Expansive Clays

Ehab Hamad Sfoog [1], Alvin John Lim Meng Siang[1*], Wael M. Albadri[2], Nahla Naji[3], Sim Sy Yi[4], Nickholas Anting Anak Guntor[5]

[1] Research Center for Soft Soils, Department of Infrastructure and Geomatic Engineering, Faculty of Civil and Environmental Engineering, Universiti Tun Hussein Onn Malaysia, MALAYSIA
[2]Faculty of Civil Engineering, University Technology MARA, UiTM. MALAYSIA.
[3]Faculty of Engineering, Anbar University, Anbar, Iraq
[4]Faculty of Engineering Technology, University Tun Hussein Onn Malaysia, MALAYSIA
[5]Department of Structural and Materials Engineering, Faculty of Civil and Environmental Engineering, Universiti Tun Hussein Onn Malaysia, MALAYSIA
Corresponding author: alvin@uthm.edu.my

Abstract. Granular pile anchor foundations (GPAF) are considered a significant promising foundation system to alleviate the serious effects of changes in the volume of expansive soils that occur throughout shrinkage and expansion. In this paper, 3D finite element analyses are presented by applying PLAXIS software, which is carried out on a typical double-story building built over a GPAF system in expansive soil. An investigation on GPAF system is presented in terms of its resistance ability to the forces caused by the soil movement as a result of variant moisture and the effect of the resistance on the superstructure induced by the straining actions. The results indicate the significance of the GPAF system in restricting the soil movement with high efficiency, which results in a noticeable improvement in the building structural responses in terms of uplift forces, heave and induced deformations.

1. Introduction

Expansive soils are considered as clays that could shrink and expand with variation in the moisture content, and it can be found in many dry and semi dry regions around the globe. For instance, 20% of the surface soil in Australia can be considered as reactive [1]. There are some elements that affect the shrinking and expansion of clay soils, which include cation, exchange capacities, the type and volume of clay soil minerals and the existence of initial water content and moisture. The challenges associated with reactive soils in terms of shrinking/swelling ground movements are uprising to the geotechnical community due to the resultant effects of such soils in terms of the distress caused to the structures of the lightweight foundations, the cracking in retaining walls, linings, canal beds, and pavements [2-3].

There are still huge financial losses in many regions of the world, in spite of having various foundations systems for controlling the movements associated with reactive soils. Statistics indicated by the American Society of Civil Engineers show that approximately one quarter of all US houses have suffered some damage cause by reactive soils, besides, the financial losses of the property owners are more than caused by a combined natural disasters such as hurricanes, floods, tornados, and earthquakes [4]. In Australia, lightweight buildings experience an early damage caused by the reactive soils in spite of the strict Australian regulatory requirements [5]. There are several solutions in the literature implemented by researchers on reactive soils, such as pile foundations, replacing the whole layer of reactive soil [6], stabilizing the soil by using additives [7-8] or by implementing special foundations such as friction piers and drilled [3]. The granular pile anchor foundation (GPAF) is a recent proposed promising foundation solution and it is has shown noticeably a great potential in reactive soils with high efficiency in reducing the damages caused by such clays, hence it is an innovative solution. Phanikumar and Ramachandra Rao [9] firstly proposed the GPAF system under heave conditions for reactive soils then was followed later by other researchers through filed trials and laboratory work [10-11].

In order to determine the controlling parameters of the GPAF system, further investigation on the efficiency of such technique was needed using the finite element method (FEM) as it was firstly examined by Shahin and Ismail [12]. In spite of the significance of the GPAF technic in restricting the movement of the soil as mentioned in the literature, however, it is not applied in practice yet mainly due to the limited field trials. In this paper, a numerical 3 dimensional (3D) analysis for a foundation structure of a typical double-storey, four- bay structure in reactive soil to examine the performance of the GPAF during shrinkage and expansion. The efficiency of the GPAF technique in controlling the ground movements associated by soil shrinkage/heave is examined and also the association of this on the superstructure internal forces is further investigated.

2. Concept of GPAF system

The concept of GPAF system is illustrated in Figure 1 which is considered as a hybrid solution that indicates the support of a shallow foundation on a granular pile in which derives its resistance from the interface between the nearby reactive soil and the granular pile. Figure 1 shows the GPAF system which contains of the pile of granular material set up into the reactive zone and continues to the saturated layer to increase the uplifting and heave resistance ability of GPAS. A steel anchor is built above the granular pile and it is connected to the pile through a concrete footing that is casted rigidly to be able to transfer the load between the pile and the footing. In this system, along the boundary of granular pile, the uplift resistance is deployed as skin friction at the interface region with the surrounding soil.

Furthermore, a rigid base plate that is connected to the pile anchor is considered as a medium for interface where the pile anchor force is transmitted to. Based on this system design arrangement, the granular pile can efficiently reinforce the ground (for example the loose sand soft clay) and resist the expansive soil induced uplift forces effectively. The GPAF system uplift resistance is a function of various factors such as the self-weight of the pile footing, the granular pile surface area, normal stress caused during the soil expansion and the interface shear strength. As can be seen in [13], there is an increase about 6.6 % in the shallow shear strength of the expansive soil around the pile compared to the free field zone clay as a result of the normal stress induced from the expansion.

IOP Conf. Series: Materials Science and Engineering **713** (2020) 012050 doi:10.1088/1757-899X/713/1/012050

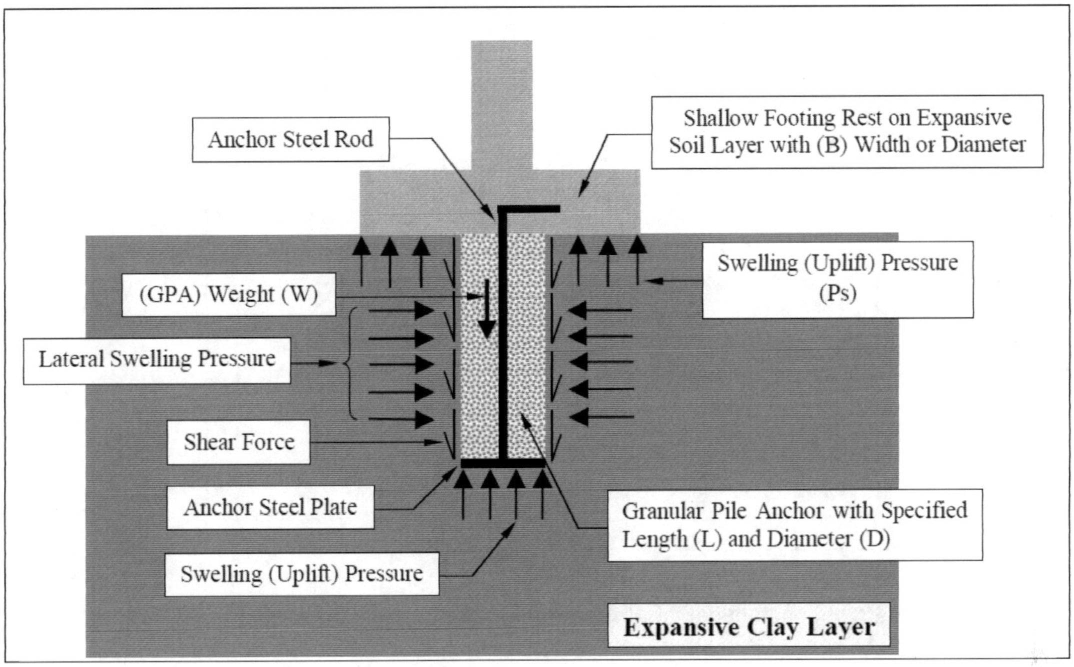

Figure 1. Typical GPAF system

3. Double-storey building numerical analysis based on GPAF system.

In order to examine the performance efficiency of the GPAF system in real practice, three models were developed; one is for free swelling of the expansive clay, second model was for building rest on raft foundation with load 30kN/m^2 and third model is for building resting on raft foundation which rest on group granular pile anchor. Models were analysed numerically by using PLAXIS 3D foundation software for finite element [14-16].

3.1. Problem identification

The dimensions of the proposed double-storey building is 3 m height for each storey which accumulates 6 m height for the double-storey and it is 11.20m × 11.2 m (W x L) and 2.4 m × 2.4 m dimensions of each bay. In each storey, an assumed thickness of 160 m for the ceiling slab. The dimensions of the beams supporting the slabs are 300 mm deep and 300 mm wide. These slabs rest on square columns with dimensions of 300 mm × 300 mm. One main factor is considered for calculating the structural component dead load of the frame building which is the weight of the material unit in that component, besides assuming a distributed live load of 9 kPa applied on the top of the slabs. The specifications of the concrete materials were considered as a concrete with unit weight of 24 kN/ m^3, Poisson's ratio of 0.2 and an elastic modulus of 37 GPa. The raft foundation distribution load was 30kN/m^2.

The GPAF system consists of raft foundation with dimensions 11.20m ×11.20 m x 0.25m, is supported by 25 angular piles with 8 m length and a diameter of 0.8 m with a measured distance of 2.4 m between pile and other. Figure 2 displays the organization of the group granular piles within the pad footing. The enhancement of the stability of the system and the rotational stiffness is implemented by using a group of piles instead of using a single pile. Figure 3 shows 3D raft footing with distribution GPAF.

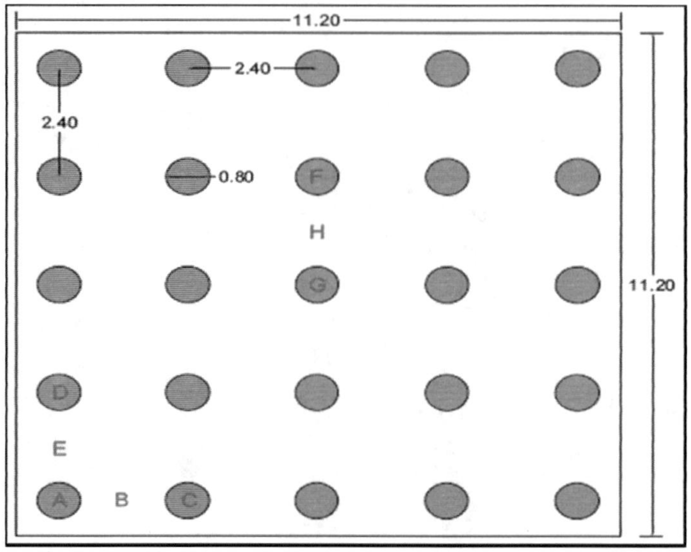

Figure 2. Arrangement of the group granular piles dimensions

The ideal ground profile has an expansive clay layer of 4 m and an average thickness covering the non-expansive clay (saturated clay) with 16 m. It is worth mentioning that the 4 m layer is the responsible for causing the heave and shrinkage event while the stable zone (saturated clay) is considered stable because there is no moisture variation over time. In order to enhance the analysis accuracy, the model was refined around the granular piles and the footing and in order to reduce the effect of the boundary, these boundaries are placed far from the area of interest. The footings of the concrete were placed about 0.0 m on the surface of the ground and a place element of Mindlin's with 0.25 m thickness were used to model the concrete footings.

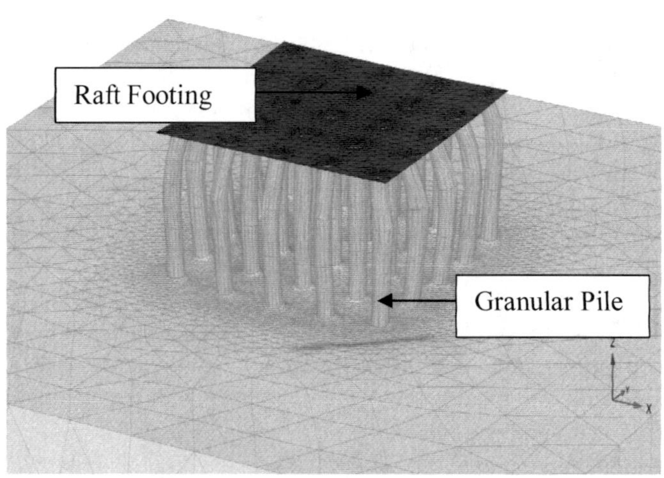

Figure 3. FEM 3D model of Raft and GPAF

3.2 The models and the parameters of the Soil

The model used for reactive clay was the hardening soil constitutive model (HS) with an assumed undrained manner behavior during expansion. Mohr-Coulomb (MC) model was used to model the granular pile. The HS model is considered as a non-linear plastic elastic formulation that embraces multiple loci as plastic shear strain function and as a cap to permit volumetric hardening [14]. The hyperbolic formula represents the relationship of the non-linear stress strain, with an initial loading that is governed by a secant deformation modulus (E50) at a material strength of 50 %. An elastic unloading and loading were assumed within the current yield surface and the Mohr-Coulomb failure criterion governs its failure. According to the following formula, the evolvement of E50 and Eur occurs with the minor effective stress, σ\3 as shown below:

$$E_{50} = E_{50}^{ref} \left(\frac{c.\cos\emptyset - \sigma 3 sin\emptyset}{c.cos\emptyset + p^{ref}.sin\emptyset} \right)^{m} \quad (1)$$

Where: c is the soil effective cohesion, Ø represents the friction angle, m represents the exponential factor that controls the dependency of the stiffness on stress and p^{ref} is the reference stress corresponding to E_{50}^{ref}. Table 1 summarizes the parameters of the expansive clay evolving after the wetting/drying and during expansion/shrinking event. It is worth mentioning that, in reality, the strength of expansive soils decreases during expansion and increases during shrinking (due to suction) which was not modeled in this study.

However, the rate of volume change of the expansive clays depends on the magnitude of overburden pressure and location from the source of moisture. In the current study, a leaking occasion of an underground water capability existed underneath the column center was assumed to heave of 6.6% over the thickness of the affected area underneath the central footing, as shown in Figure 3b. Both heave and shrinkage were modeled by applying equal volumetric strains over the layer thickness of modeled soil. The heave was applied independently starting from the stage after application of the dead and live loads.

Table 1. The properties of the soil for analyzing the finite element.

Parameter	Expansive Clay	Stable Clay	Granular Pile
γusat(KN/ m3)	16.33	16.33	17
γsat(KN/ m3)	19	19	20
C' (KN/ m2)	30	30	3
ϕ'	22	22	42
Volumetric strain%(swelling)	6.6	-	-
Volumetric shrinkage%	1.7	-	-
E_{50}^{ref}	3500	3500	$28x10^3$
E_{oed}^{ref}	3000	3000	$84x10^3$
Vur	0.2	0.2	0.25

4. Results and discussion

The performance efficiency of the GPAF system over a reactive soil to enhance a double-storey building was investigated based on three models namely; soil without loading, soil without GPAS (building self-weight) and building with GPAS. The comparison among the three models in terms of heave and uplift force is shown in Figure 4. The comparison was made for the top raft at central plan and at edge. It is obvious that the constructed building with GPAS foundation system enhances substantially the expansive soil in terms of heave and uplift force.

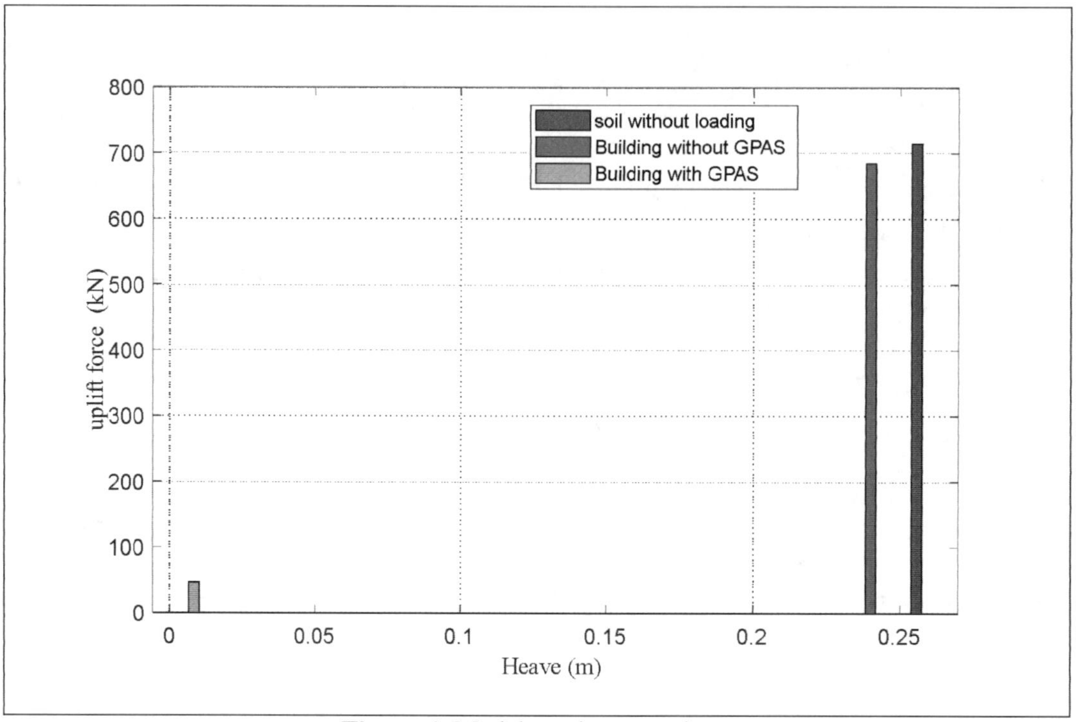

Figure 4. Model result comparison

Table 2 presents the average heave with three case model to compare the results of the induced deformations. The heave was 0.26 m in plain soil without constructed house and GPAS foundation. The heave was reduced by 8 % (0.24 m) and further reduced by 98.6 % (0.004 m) when the GPAS was introduced. Furthermore, the uplift force improvement is summarised in Table 3. The uplift force for plain soil was found to be 715 kN. On the other hand, significant improvement was found in building without GPAS and building with GPAS 4 % (685 kN) and 96 % (47) respectively. Hence, it is obvious the applicability of the GPAS system to arrest the heave and uplift force resulted in expansive soil. This improvement is induced by the frictional forces alongside the GPA and soil interface.

Table 2. Averge heave three case of models

Type of building	Soil without loading (free heave)	Building without GPAS	Building with GPAS
Heave(m)	0.26	0.24	0.004
Degree of improvement	0%	8%	98.6%

Table 3. Average Uplift forces three case of models

Type of building	Soil without loading (free heave)	Building without GPAS	Building with GPAS
uplift force kN	715	685	47
Degree of improvement	0%	4%	96%

5. Conclusion

The results analysis of the 3D FEM of the GPAF system were presented as a viable solution for lightweight structures over expensive soils. Besides, the response of the superstructure double-storey building with four-bay frame to the ground have/uplift pressure caused by reactive soils was investigated. The presented results show the significance of the GPAF system in reducing the heave and uplift forces caused by expensive soil, hence it is considered an efficient system. The degree of improvement for the heave when applying GPAS was 98.6% while the degree of improvement for uplift force was found to be 96 %. This shows the potential for the GPAS system as a simple and effective foundation system to mitigate the expansive soil problems in terms of heave and uplift forces due the frictional properties alongside the interface between the GPA and the expansive soil.

6. References

[1] Phanikumar B. R.. and Muthukumar M. 2015 Swelling behaviour of GPA-reinforced expansive clay beds subjected to swell-shrink cycles *J. Geomechanics and Geoengineering* **10** 261–270

[2] Chen, F. H. 2012 *Foundations on expansive soils* (Elsevier) pp 260-271

[3] Ibrahim Saad F, Aljornay A N and Aladly A I 2014 Heave behaviour of granular pile anchor foundation (GPA-foundation) system inexpansive soil *J. of Civil Eng.and Urbanism* **4** 213–222

[4] Zumrawi Magdi M.E. 2015 Construction Problems of Light Structures Founded Expansive Soils in Sudan *International Journal of Science and Research* **4** 896-902

[5] Kranthikumar A., Sawant V. A., Kumar P. and Shukla S. K. 2016 Numerical and experimental investigations of granular anchor piles in loose sandy soil subjected to uplift loading *International Journal of Geomechanics* **17** 04016059

[6] Nusier O K, Alawneh A S and Abdullatit B M 2009 Small-scale micropiles to control heave on expansive clays *proceedings of the Institution of Civil Engineers-Ground Improvement* **162** 27-35

[7] Phanikumar B R 1997 *A study of swelling characteristics of and granular pile anchor foundation system in expansive soils* PhD Thesis (Hyderabad: JN Technological University) pp 100-120

[8] Ismail M and Shahin M 2013 Numerical modeling of granular pile-anchor foundations (GPAF) in reactive soils *International Journal of Geotechnical Engineering* **6** 149–155

[9] Sharma A and Sharma R K 2019 An experimental study on uplift behaviour of granular anchor pile in stabilized expansive soil *International Journal of Geotechnical Engineering* 1–14

[10] Phanikumar B R., Rao A S and Suresh K 2008 Field behaviour of granular pile-anchors in expansive soils *Proceedings of the Institution of Civil Engineers–Ground Improvement* **161** 199-206

[11] Rao A S, Phanikumar B R, Babu R D and Suresh K 2007 Pullout Behavior of Granular Pile-Anchors in Expansive Clay Beds In Situ *Journal of Geotechnical and Geoenvironmental Engineering* **133** 531-538

[12] Ismail M and Shahin M 2011 Finite element analyses of granular pile anchors as a foundation option for reactive soils *Pro. Intl. Conf. on Advances in Geotechnical Engineering* 1047-1052

[13] Phanikumar B, Sharma R, Rao A S and Madhav M R 2004 Granular pile anchor foundation (GPAF) system for improving the engineering behaviour of expansive clay beds *Geotechnical Testing Journal* **27** 279-287

[14] Brinkgreve R B and Broere W 2008 *PLAXIS 3D foundation manual version 2* (Delft University of Technology and PLAXIS) pp.123-144

[15] Madun A, Tajuddin S A A, Abdullah M E, Abidin M H Z, Sani S, Siang A J L M and Yusof M F 2016 Convertion Shear Wave Velocity to Standard Penetration Resistance *IOP Conf. Series: Materials Science and Engineering* vol 136 (United Kingdom: IOP Publishing) 1 – 7

[16] Ganasan R, Lim A J M S and Wijeyesekera D C 2016 Physical and software modelling for challenging soil structure interaction *ARPN Journal of Engineering and Applied Sciences* **11**(6) 3668-3676

IOP Conf. Series: Materials Science and Engineering **713** (2020) 012042 doi:10.1088/1757-899X/713/1/012042

Benefits For Public Healthcare Buildings towards Net Zero Energy Buildings (NZEBs): Initial Reviews

M Sulzakimin[1], M A N Masrom [1], R Hazli[1], A A Adaji[1], Seow TW[1] and A R Mohd Hilmi Izwan[2]

[1] Department of Construction Management FPTP, University Tun Hussein Malaysia,
[2] Faculty of Engineering & Science, Department of Surveying, UTAR, Malaysia
Corresponding author : zakimin@uthm.edu.my

Abstract. NZEBs concept has received incrementing attention especially since European Union Parliament are progressively moving towards regulation in which all new buildings to be "nearly Zero-Energy" Buildings by 2020. Even though this concept delivers promising benefits, previous studies found that the benefit of NZEB is still in vague to the Malaysian construction industry, mostly for healthcare buildings. The NZEBs concept implementation is critical in healthcare building as it seen as a key part of the needed transition towards sustainable development. This paper aims to define benefits of NZEBs practices through initial investigation among architects. Several existing energies works of literature and pilot studies by using semi-structured interviews were conducted. The findings divulge that, although the term 'NZEBs design strategies' is not being used largely across construction industry in Malaysia, some sustainable practices related to the design stage of construction has been implemented. The findings are also promising in growing awareness, practices and implementation of NZEBs design strategies by the practitioner in Malaysia. It is foreseen that the paper will provide a straightforward knowledge for future research in NZEBs design strategies practices for healthcare buildings construction in Malaysia.

1. Introduction

Net-zero energy buildings (NZEBs) is a concept that uses renewable energy technology to produce a mix of renewable electricity and other renewable heat resources like biomass and solar to generate as much energy as they consume [1]. Similarly, [2] emphasized that NZEBs is conceptually a building with a lower of energy demand and are balanced by the on-site generation of electricity, or other energy carriers, from renewable sources. Moreover, [3] and [4] highlighted that net-zero energy buildings (NZEBs) are typically a grid-connected building with very high energy performance. On the other hand, NZEBs are buildings with the concept of cost-optimal in which ensuring the minimum cost of the estimated economic life cycle of the building, by taking the package of efficiency measures.

Despite today's buildings energy consumption demand increasing significantly, NZEBs concept practices enable to develop a network of clean domestic energy assets. The timeline for the implementation of NZEBs according to the Energy Performance of Buildings Directive (EPBD) recast is illustrated in Figure 1.

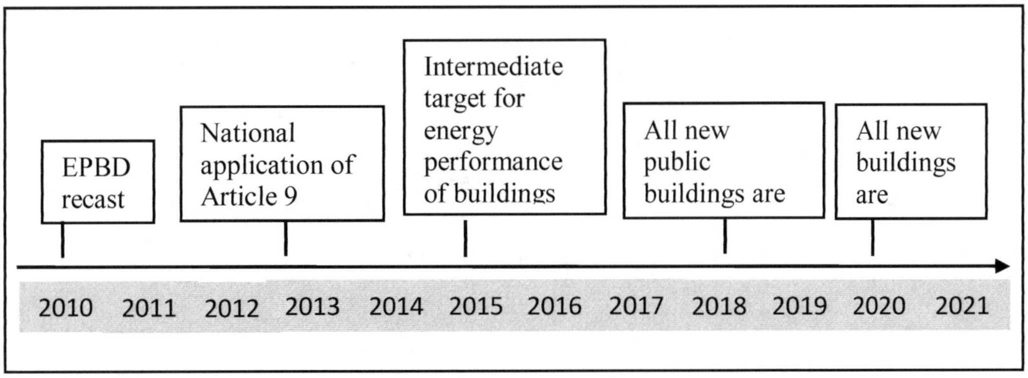

Figure 1. Timeline for NZEBs implementation according to the EPBD recast [5]

As highlighted in Figure 1, NZEBs has gained much attention in 2010 through EPBD recast. Based on the recast, European Union Parliament had agreed to set the targets for all members of Parliament to regulate a regulation in their counties that all new buildings are NZEBs in 2020. According to [6], the healthcare buildings have high electricity consumption as the result of the extensive use of various medical equipment, mechanical ventilation of building, lightings, and sterilization. Besides that, some of the countries have shown that healthcare buildings are considered as one of the most energy-intensive commercial-sector buildings. For instance, healthcare buildings in the United States are considered as one of the most intensive users of total energy among the building types where the high electricity consumptions are due to the continuity of services and the utilizations variety of complex applications in energy-intensive equipment such as medical imaging equipment [7].

In the context of Malaysia's construction industry, there is an upward trend of demand in energy consumption in building sector as the energy consumption in the commercial and residential building had an increase of 2.7% and accounts for up to 14.4% in 2013 as compared to 11.7% in 1993 [8]. Additionally, [9], analysed the energy use, energy savings and corresponding emission reductions for the energy using equipment in a public hospital in Malaysia. The authors reported that around 19,311 MWh of energy consumption for the hospital in 2008. Besides that, for the end-use energy breakdown for the hospital, the authors found that lighting uses a major fraction of total energy consumption comprises of 36% and followed by medical equipment which accounts for 34%. The findings also showed that the energy intensity of the hospital is 234 kWh/m.

2. Malaysia Healthcare Project Developments

In general, the total health expenditure for Malaysia's national spending shows an increasing trend from RM8,303 million in 1997 to RM44,748 million in 2013 [10]. From this trend, it is indicating that the Malaysia's health expenditure is growing exponentially over sixteen years. Even though there is a breakdown of several types of expenditure, but most of this expenditure is spent for gross and total output in which services provided by hospitals of the total expenditure as illustrated in Figure 2.

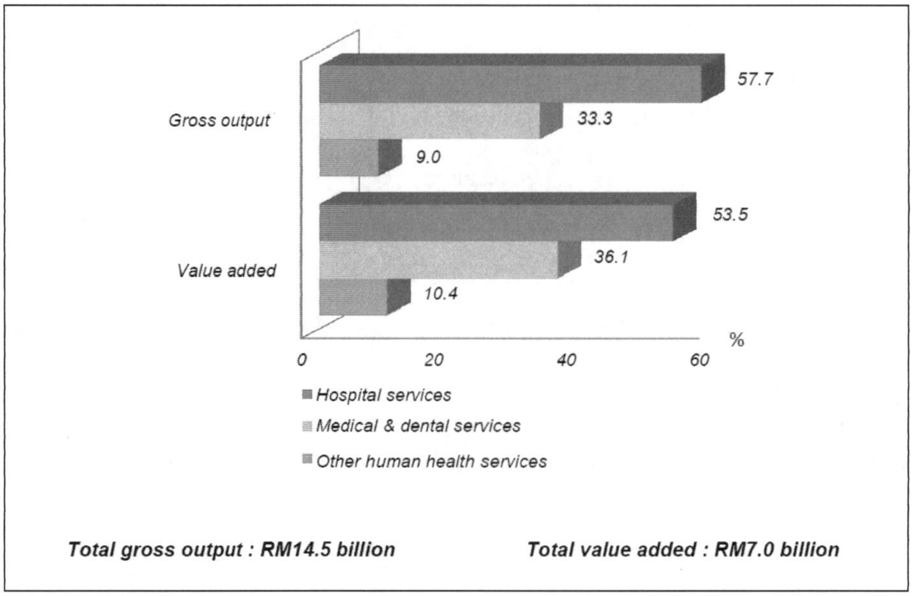

Figure 2. Expenditure by functions of health services, 2014[10]

Besides, based on the Tenth Malaysia Plan (10MP – 2011-2015), there are four strategies that need to be implemented in order to support one of the trust in 10MP which is to achieve the quality of life of an advanced nation. The four strategies involved are: 1. Establish a comprehensive healthcare system & recreational infrastructure, 2. Encourage health awareness & healthy lifestyle activities, 3. Empower the community to plan or implement individual wellness programme (responsible for own health), 4. Transform the health sector to increase the efficiency and effectiveness of the delivery system to ensure universal access. Based on those strategies, three key result areas (KRA) were identified which comprising health sector transformation towards a more efficient and effective health system in ensuring universal access to healthcare, health awareness and healthy lifestyle and empowerment of individual and community to be responsible for their health [11]. Therefore, The Country Health Plan: 10th Malaysia Plan 2011-2015 has highlights several physical healthcare development projects that will need to be implemented in order to achieve KRA [11].

As a matter of facts, since 2007 until 2014, the development progress of healthcare buildings projects in Malaysia is consistent every year although, for the past four years, the allocation budget for the healthcare project development is decreasing in 2014 as compared to 2011. However, as compared to 2010, there is an increase of 2 new hospitals in 2014. Moreover, up to 2014, there are 133 government hospitals and 9 special medical institutions which comprise of 35,318 beds and 4,942 beds respectively [12]. Overall Bed Occupancy Rate (BOR) for MoH hospitals and Institutions in 2014 was 71.79%. Besides that. in the Tenth Malaysia Plan (10MP – 2011-2015), the government has approved an allocation of RM 21.98 billion in 2014 to be allocated for MoH which consists of RM 20.49 billion for the Operating Budget and RM1.49 billion for the Development Budget.

The initial allocation for the development budget that has been approved by the government for MoH in 2014 was RM 1.662 billion which includes constructing new hospitals, upgrading existing hospitals, facilities and others [12]. Based on 71 new MoH programs, it is involved in 67 physical projects and 4 non-physical projects [12]. Moreover, for the physical projects, Public Works Department (JKR) will handle 53 projects while JKR Sabah and JKR Sarawak will handle 4 projects and 3 projects respectively. The remaining 4 projects will be handled by the Engineering Services Division (ESD) of MoH [12]. Therefore, it is important to consider the healthcare buildings sustainability in the long term. Some of the challenges including low building performance and high energy consumption for Malaysian healthcare buildings need to be addressed in order to ensure the sustainability of healthcare buildings.

3. Net Zero Energy Buildings (ZEBs)

Study by [13], stated that four main primary definitions of NZEBs which are net zero site energy, net zero source energy, net zero energy costs, and net zero emissions as illustrated in Figure 3. According to the four definitions, it pointed out that the NZEBs is a grid connected building where the annual energy generation from a renewable source is equally with energy consumption during the term of one year. Similarly, [2] emphasized that NZEBs is conceptually a building with a lower of energy demand and are balanced by the on-site generation of electricity, or other energy carriers, from renewable sources. Moreover, the authors also mentioned that the term NZEBs refer to buildings that are connected to the energy infrastructure. Furthermore, another study by Kurnitzki *et al.,*[3] and Karsten *et al.,*[4] highlighted that NZEBs are typically a grid-connected building with a very high energy performance.

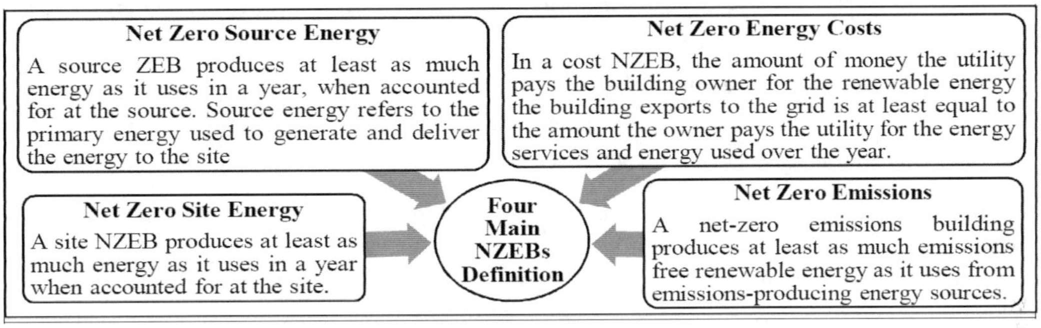

Figure 3. Four main definitions of NZEBs [9]

However in 2009, The Massachusetts Zero Net Energy Buildings Taskforce which comprised of more than 70 leaders has worked together to establish key solutions that will guide to the worldwide adoption of NZEBs and deep energy reduction retrofits throughout the Commonwealth by 2030. The taskforce defined NZEBs as a building that is capable to generate energy onsite with a quantity that is equal to or greater than the total quantity of energy consumed onsite within a period of a year, and the building uses clean renewable resources [14]. The continuous improvement on the NZEBs definition can be seen through the study of [2] which to introduces a harmonised framework for describing the important elements of NZEBs in a series of criteria. Based on their study, the authors have listed several criteria that should be questioned and addressed in order to strengthen the overall NZEB "goodness of design" as well as to give a sound NZEB definitions in a formal, systematic and comprehensive way [2]. In 2012, a consistent framework for setting NZEB definitions has been developed through the study of [15] in which evaluation of the criteria in the definition framework and selection of the related options becomes a methodology to set NZEB definitions in a systematic way. According to the study, the authors stated that there is the difference between the terms of net-zero energy buildings (NZEBs) and zero energy buildings (ZEBs) as the terms of ZEBs is broader and it could include autonomous buildings. The wording 'Net' in NZEBs indicates that the energy obtained from and provided back to the energy grids nominally within a period of a year is a balance[15]. Moreover, the study showed that the core concept of NZEBs is where there is a balance between weighted supply and weighted demand. In other words, the sum of all generated energy is equal to the sum of all delivered energy.

4. Research methodology

The method discussed here is based on the perspectives of the knowledge of experienced architects that involved in the sustainable related project and it is appropriately could be used for this study onwards. Hence, the objective of this section is to explain the methods that were used for this paper comprised of research design and interview data collection procedures. This paper reviewed an existing literature relating to NZEBs in order to gain an understanding as well as insight into previous studies. To ensure the quality of this paper, a literature review is important to provide insight into ways in which the researcher can limit the scope to a needed area of inquiry [16]. Besides that, the

IOP Conf. Series: Materials Science and Engineering **713** (2020) 012042 doi:10.1088/1757-899X/713/1/012042

semi-structured interviews were conducted with architects in order to explore their understanding on NZEBs as well as to explore the sustainable practices in the context of Malaysian construction industry particularly on the healthcare building subject. Authors used the concept of saturation in order to determine the suitable sample size for the interviews. This decision was based upon recommendations by [17] to avoid ending up with repetitive and unessential information which does not contribute to the research. Besides that, the amounts of information obtainable from interviews do not necessarily increase just because more data was obtained [17] and the authors further claimed that researchers generally use saturation as a guiding principle during their data collection towards determining appropriate sample sizes. Therefore, the authors decided to choose 3 interviewees to be interviewed for this paper. The interviewees were assigned codes as R1, R2, and R3. Besides that, to ensure the quality of the interviews, all respondents required to be active in Malaysian construction industry and had direct involvement in the buildings projects located within Malaysia. The key profiles for the interviewees are shown in Table 1.

Table 1. Key profiles for interviewees

Item	R1	R2	R3
Position	Architect	Architect	Architect
Organization	Consultant	Consultant	Developer / Client
Experience	More than 5 years working on sustainability aspects	More than 5 years working on green building	More than 5 years working on sustainable development agenda

As for the interviews, the objective of this paper involved the architect's opinions on design and practices of NZEBs based on the Malaysian construction industry context as well as to identify the benefits of NZEBs

5. Findings and disscusion

In order to identify the benefits of NZEBs, the practices that support this concept are crucial to being identified. Based on [18] there is a growing interest in NZEBs as many NZEBs demonstration projects showed that the goal of this concept is achievable. However, the need to identify NZEBs approaches is crucial in order to enable broad replication of NZEBs, especially in a healthcare building. Furthermore, all interviewees indicate that there are various of ways to achieve sustainability in construction since construction sustainability is too wide and different architect will have its own different of sustainable practices, but then all practices will be based on the same principles which are to increase the dependent on renewable sources as well as to protect the environment (R1, R2, R3). Similarly, according to [19] in the study on close inspection of the relevant design strategies and relative performance indicators of the eight case studies, the authors stated that there is no standard approach for designing NZEBs. Nevertheless, [20] highlighted that it is important for architects or engineers to have specific design guidelines and strategies in order to popularize NZEBs.

Even though there are still no specific design strategies in achieving NZEBs design, but based on [21] there are several common design elements and some consensus in designing NZEBs. The design elements that support NZEBs is important to be identified since the adaptation of NZEBs in buildings will ensuring the high performance in buildings can be achieved as well as to support the use of renewable energy for buildings.Generally, according to [22] and [21], there are three main steps of a design approach that can be applied in designing NZEBs which are the first design approach is a passive approach that focused on reducing energy demand. The second design approach is the use energy efficiency system, and the third design approach is renewable energy system which is needed to generate renewable energy as well as to offset in large measure the energy demand. The combination of this three-design approach is also crucial to be taken in designing healthcare buildings since these approaches able to succeed in reaching the desired energy performance as well as NZEBs target.

Passive approaches are one of the key element when designing NZEBs as this approach will affect directly to the energy needs for the buildings mechanical and electrical systems, and also indirectly affect the renewable energy generation [19]. Similarly, [23] also highlighted that the passive approach plays a significant part in the NZEBs design. Besides that, the improvement of energy consumption of a building can be achieved through passive design strategies as buildings are constructed that react to the environment, hence it is possible for a building to achieve high environmental quality [24]. Additionally, [20] stated that architects have an important role in controlling aspects of a passive design when designing NZEBs. Similarly, some of the interviewees have highlighted that an architect is responsible to ensure some elements of passive approach should be carried out in order to support the sustainability of construction

One of the most important aspects of passive design strategies is maximizing the amount of natural daylight into a building and reduce the amount of energy requires for artificial lighting in a building [25]. This strategy is important for designing buildings, especially healthcare buildings since these types of buildings is operating continuously 24 hours a day and required a huge amount of artificial light to provide services to patients. Besides that, it is important to control passive design strategies in a building as this approach will influence the energy performance of a building such as natural lighting, heat gain, shading, and envelope conduction. [20][26]. In addition, Aleinei *et al.*, [19] stated that the passive design strategies in buildings should be appropriately orientating building towards solar heating maximization when buildings are dealing with heating challenges.

Furthermore, according to [27] when designing NZEBs, façades have a strong impact on heating, cooling and artificial light in the building needs as well as on daylight. Thus, it is important to realise that the passive heating solutions are essential to be studied along with passive cooling solutions in order to prevent overheating in a building [19]. Therefore, it is important to consider passive approach while designing NZEBs as it will enable maximisation of natural daylight and the amount need of artificial light in the building needs will be reduced, hence it will create a pleasant environment. In regards to second design approaches which is the utilisations of energy efficiency system, it is important to realise that the pathway towards NZEBs is reducing energy demand by means of energy efficiency measures [2]. Besides that, energy efficiency has a significant part to act as a parameter that indicates the reduction level of energy consumption in carrying out a related task [28]. Generally, energy efficiency can be improved by several design strategies including airtightness to avoid infiltration and mechanical ventilation systems with heat recovery to provide air conditioning and indoor air quality (IAQ) [29]. However, the selection of high-efficiency technologies is also crucial in the improvement of energy efficiency in a building. One of the interviewees has indicated the importance of energy efficiency system in healthcare buildings

As an architect, we have emphasized the importance of energy efficiency measure for this healthcare building since this measure is much easier to be applied rather than the passive approaches and these measures do not relates to any hygienic condition requirements. (R1)

There are numerous energy efficiency technologies that can be applied in buildings. For instance, the use of low power lighting, energy efficient electrical equipment such as washing machines and dishwashers with a warm water connection are strategies in planning an equated energy balance [30]. Similarly, established by [31], healthcare building's energy demand can be significantly reduced by considering energy efficiency as a factor in buying medical equipment, besides the proper use of electricity.Moreover, in relation to the availability of energy efficiency technologies in Malaysia, there are no problems to obtain the technologies since Malaysia is one of developing countries (R2, R3). However, the cost of this technologies is quite high and for that reason, most of the developers are not fully utilised the application of energy efficiency technologies in buildings (R3). Similarly, based on Vogel *et al.*, [32] which investigated barriers that prevent energy efficiency adoption in multifamily buildings in Sweden and authors proposed a novel categorization framework for barriers to energy efficiency in buildings. The results showed that barriers related to contextual level are the most significant to prevent energy efficiency adoption. The contextual level is characterized by the rules and regulations that influence technological design and market development. In addition, as stated by interviewees R2,

The local authorities will reject the proposal of purchasing some energy efficiency technologies form the architect if the cost is too high. (R2)

Thus, based on the findings and the arguments, energy efficiency measure in a building is vital towards designing a healthcare that achieved NZEBs since energy reduction, as well as energy savings in buildings, can be achieved. Moreover, by improving the energy efficiency of the various incorporated building systems, it will help to reduce building´s energy demand [15]. It shows that NZEBs can provide a reduction in both primary energy as well as CO^2 emissions.

For the third design approaches which are renewable energy system, the necessity of renewable energy systems is vital towards NZEBs since these systems are required to reduce and as well as to offset the thermal energy need in buildings. Moreover, designing NZEBs is not only achieved by reducing the energy consumption of the building with passive design methods or the application of energy efficient system in buildings, but NZEBs should also be designed with the balances energy requirements between active energy production techniques and renewable technologies in order to achieve the NZEBs objective [33]. Besides that, the utilization of renewable energy systems in buildings can be an important solutions against global warming and greenhouse gasses [34]. Furthermore, most of the countries have already promoted the application of renewable energy systems in buildings through enforcement, regulation, target, incentive, and fund provided in the countries [35]. However, in the context of Malaysia, the interviewees agreed that there is a lack of enforcement and target set by the government even though there is some government efforts to promote renewable energy systems in Malaysia (R1, R2, R3). One of the interviewees has highlighted that,

The government has provided some incentives to promote the green concept in construction such as tax exemption, however, the enforcement is not executed strictly and thus, the utilisation of renewable technologies is not widely used. (R2)

Phuangpornpitak and Tia [36] stated that the integration of renewable energy in a smart grid system and they concluded that renewable energy system is one of incomparable option to generate energy since these systems able to provide a clean energy resource. However, they suggested that some issues such as the design, sizing, and the suitability of the system in terms cost for energy generation need to be addressed in order to ensure that it can be used thoroughly and commercially. The summary of NZEBs practices and benefits are shown in Table 2.

Table 2. Key profiles for interviewees

NZEBs practices	Literature	Key points from interviewees	NZEBs benefits
Passive approaches	Ferrara et al. [37]; Ascione et al. [38]; Becchio et al. [39]; Rodriguez-Ubinas et al. [40]; Barbolini et al. [29]	"..take into consideration on passive elements before prepare architecture plan." (R1) "..undeniably passive design elements plays significant part in architecture." (R3)	Reducing energy demand Reducing primary energy and CO^2 emissions
Energy efficiency technologies	Gaitani et al. [41]; Dalla et al. [42]; Buonomano et al. [43]; Ascione et al. [44]	"As an architect, we can propose to developer some energy efficiency system for buildings." (R1) "If we don't have any cost constraints, we can widely use energy efficiency in buildings to save electricity." (R2)	Reducing primary energy and CO^2 emissions
Renewable energy systems	Kurnitski et al. [3]; Becchio et al. [45]; Ajla Aksamija [25]	"…even though the movement is still slow, but we are still put our efforts to increase the utilisation of renewable energy technologies in buildings" (R2)	Increasirenewable energy share Reducing primary energy and CO^2 emissions

6. Conclusion

The NZEBs concept is achieve significantly since it is a realistic solution for the easing of CO^2 emissions. The acceptance of this concept into Malaysian healthcare buildings is vital because of the energy consumption for this buildings is high and this concept able to support the government initiative in Construction Industry Transformation Programme (CITP) 2016-2020 which one of strategic thrust that seeks to transform the industry is environmental sustainability. However, NZEBs concept in Malaysian construction industry is still needs to be explored and understood. But, the previous studies on NZEBs able to give an understanding into design and practice of this concept and thus, it gives the understanding to be explored based on Malaysia context. The findings showed that by the adaptation of NZEBs concept in Malaysia healthcare buildings, benefits that can be gained through this concept are it helps to reduce energy demand for healthcare buildings, reduce primary energy and CO^2 emissions for healthcare buildings, and increase the renewable energy share for healthcare buildings. However, the authors recommend that by increasing the number of sample size, it will produce any significantly different findings. Besides that, it is also recommended that this paper need to further investigate in order to get more comprehensive findings as well as reflect more on the Malaysia context. Extra note, this paper could provides some sustainable design and practices insight as basis for further investigation for public healthcare building.

7. References

[1] Antonin V D B, Bernhard V M, Lou R and Markus O 2014 *Role of building automation related to renewable energy in NZEB's* (Report No. BUINL13782) Kanaalweg: Ecofys Netherland

[2] Sartori I, Napolitano A, Marszal A J, Pless S, Torcellini P and Voss K 2010 Criteria for definition of net zero energy buildings *In Proc., of EuroSun 2010: Int., Conf., on Solar Heating, Cooling and Buildings* (Graz, Austria: EuroSun) pp 1-8

[3] Kurnitzki J, Allard F, Braham D, Goeders G, Heiselberg P, Jagemar L, Koso-nen R, Lebrun J, Mazzarella L, Railio J, Seppänen O, Schmidt M and Virta M 2011 How to define nearly zero energy buildings *REHVA J.* **48**(3) 6–12

[4] Karsten V, Eike M and Markus L 2011 From low-energy to net zero-energy buildings: status and perspectives *J. of Green Build.* **6**(1) 46-57

[5] D'Agostino D, Zangheri P, Cuniberti B, Paci D and Bertoldi P 2016 *Synthesis Report on the National Plans for Nearly Zero Energy Buildings (NZEBs)*

[6] Morgenstern P, Li M. Raslan R, Ruyssevelt P and Wright A 2016 Benchmarking acute hospitals: Composite electricity targets based on departmental consumption intensities?. *Energy and Buildings* **118** 277-290

[7] U.S. Energy Information Administration 2016 March 18 *EIA - Independent Statistics and Analysis* Retrieved July 25, 2017, from https://www.eia.gov/consumption/commercial/reports/2012/energyusage/

[8] Suruhanjaya Tenaga 2015 *Malaysia Energy Statistics Handbook 2015* (Malaysia, Suruhanjaya Tenaga Putrajaya) Pp. 1-94

[9] Saidur R, Hasanuzzaman M, Mahlia T M I, Rahim N A and Mohammed H A 2011 Chillers energy consumption, energy savings and emission analysis in an institutional buildings. *Energy* **36**(8) 5233-5238

[10] Ministry of Health Malaysia 2015 *Malaysia National Health Account, Health Expenditure Report 1997-2013* (Putrajaya: Kementerian Kesihatan Malaysia)

[11] Ministry of Health Malaysia 2014 *Malaysia National Health Account, Health Expenditure Report 1997-2013* (Putrajaya: Kementerian Kesihatan Malaysia)

[12] Mohamed A, Hasan A and Sirén K 2014 Fulfillment of net-zero energy building (NZEB) with four metrics in a single family house with different heating alternatives *Applied Energy* **114** 385-399

[13] Paul T, Shanti P and Michael D 2006 Zero energy buildings: a critical look at the definition *Preprint No. NREL/CP-550-39833*

[14] Massachusetts Zero Net Energy Buildings Task Force 2009 Getting to zero: *final report of the massachusetts zero energy buildings task force* (Boston, USA: Mass.gov) **25** 46
[15] Sartori I, Assunta N and Karsten V 2012 Net zero energy buildings: a consistent definition framework *Ener. and Build.* **48** 220-32
[16] Creswell J W 2013 *Research design: Qualitative, quantitative, and mixed methods approaches* (California: Sage publications) pp 31
[17] Mason M 2010 Sample size and saturation in PhD studies using qualitative interviews *In Forum qualitative Sozialforschung/Forum: qualitative social research* (Deutschland) **11**(3) pp19
[18] Leach M, Pless S and Torcellini P 2014 *Cost Control Best Practices for Net Zero Energy Building Projects: Preprint (No. NREL/CP-5500-61365)* (National Renewable Energy Laboratory (NREL), Golden, CO)
[19] Aelenei D, Aelenei L, Musall E, Cubi E, Ayoub J and Belleri A 2013 Design strategies for non-residential zero-energy buildings: lessons learned from Task40/Annex 52: towards net zero-energy solar buildings. In: *CLIMA 2013 - 11th REHVA World Congress & 8th International Conference on IAQVEC, Book of Proceedings (*Prague, Czech Republic) pp 1-10.
[20] Habash G, Chapotchkine D, Fisher P, Rancourt A, Habash R and Norris W 2014 Sustainable Design of a Nearly Zero Energy Building Facilitated by a Smart Microgrid *Journal of Renewable Energy* **2014** 1-14
[21] Lu Y, Wang S and Shan K 2015 Design optimization and optimal control of grid-connected and standalone nearly/net zero energy buildings *Applied Energy* **155** 463-477
[22] Aelenei L, Ayoub J and Aelenei D 2012 Net Zero Energy Residential Building Case Studies *Energy Conservation in Buildings & Community Systems* **56** 10-12
[23] Scognamiglio A, Garde F and Røstvik H N 2014 How net zero energy buildings and cities might look like? New challenges for passive design and renewables design *Energy Procedia* **61** 1163-1166
[24] Hootman T 2012 *Net zero energy design: a guide for commercial architecture* (Hoboken, NJ: John Wiley) p 186
[25] Aksamija A 2015 Regenerative Design of Existing Buildings for Net-Zero Energy Use *Procedia Engineering* **118** 72-80
[26] Omrany H and Marsono A K 2015 Optimization of building energy performance through passive design strategies *British Journal of Applied Science & Technology* **13**(6) 1-16
[27] Thalfeldt M, Pikas E, Kurnitski J and Voll H 2013 Facade design principles for nearly zero energy buildings in a cold climate *Ener. and Build.* **67** 309-21.
[28] Parameshwaran R, Kalaiselvam S, Harikrishnan S and Elayaperumal A 2012 Sustainable thermal energy storage technologies for buildings: a review *Renewable and Sustainable Energy Reviews* **16**(5) 2394-2433
[29] Barbolini F, Cappellacci P and Guardigli L 2017 A Design Strategy to Reach nZEB Standards Integrating Energy Efficiency Measures and Passive Energy Use *Energy Procedia* **111** 205-214
[30] Musall E and Voss K 2012 The passive house concept as suitable basis towards net zero energy buildings *Proc.* in *Conf.*: *Passive House Conference* (Hannover: Passiv Haus Institut) pp 1-6
[31] Maassen W 2017 *(Nearly) zero energy healthcare buildings* (Report No. I&BBE3112R001F0.1) (Netherland, Haskoning DHV Nederland B.V)
[32] Vogel J A, Lundqvist P and Arias J 2015 Categorizing barriers to energy efficiency in buildings *Energy Procedia* **75** 2839-2845
[33] Kolokotsa D, Rovas D, Kosmatopoulos E and Kalaitzakis K 2011 A roadmap towards intelligent net zero- and positive-energy buildings *Solar Ener.* **85**(12) 3067-84
[34] Ebenehi I Y, Sulzakimin M, Sarpin N, Masrom M A N, Zainal R and Mohd Azmi M A 2017 The management of building fire safety towards the sustainability of Malaysian public universities *IOP Conference Series: Materials Science and Engineering*, 271,12-25
[35] Winkel T, Rathmann M, Ragwitz M, Steinhilber S, Winkler J, Resch G, Panzer C, Busch S and Konstantinaviciute I 2011 *Renewable energy policy country profiles* (Rep. No. EIE/08/517/SI2.529243) (Intelligent Energy Europe)

[36] Phuangpornpitaka N and Tiab S 2013 Opportunities and challenges of integrating renewable energy in smart grid system *Ener Proc.* **34** 282–90

[37] Ferrara M, Fabrizio E, Virgone J and Filippi M 2015 Appraising the effect of the primary systems on the cost optimal design of nZEB: A case study in two different climates *Energy Procedia* **78** 2028-2033

[38] Ascione F, De Masi R F, de Rossi F, Ruggiero S and Vanoli G P 2016 Optimization of building envelope design for nZEBs in Mediterranean climate: Performance analysis of residential case study *Applied Energy* **183** 938-957

[39] Becchio C, Dabbene P, Fabrizio E, Monetti V and Filippi M 2015 Cost optimality assessment of a single family house: Building and technical systems solutions for the nZEB target *Energy and Buildings* **90** 173-187

[40] Rodriguez-Ubinas E, Montero C, Porteros M, Vega S, Navarro I, Castillo-Cagigal M and Gutiérrez A 2014 Passive design strategies and performance of Net Energy Plus Houses *Energy and buildings* **83** 10-22

[41] Gaitani N, Cases L, Mastrapostoli E and Eliopoulou E 2015 Paving the way to nearly zero energy schools in Mediterranean region-ZEMedS project *Energy Procedia* **78** 3348-3353

[42] Dalla Mora T, Cappelletti F, Peron F, Romagnoni P and Bauman F 2015 "Retrofit of an historical building toward NZEB" *Energy Procedia* **78** 1359-1364

[43] Buonomano A, Montanaro U, Palombo A and Vicidomini M 2015 NZEBs in Mediterranean climates: energy design and optimization for a non-residential building *Energy Procedia* **82** 458-464

[44] Ascione F, Bianco N, De Rossi F, De Masi R F and Vanoli G P 2016 Concept, Design and Energy Performance of a Net Zero-Energy Building in Mediterranean Climate *Procedia Engineering* **169** 26-37

[45] Becchio C, Bottero M C, Corgnati S P and Ghiglione C 2015 nZEB design: challenging between energy and economic targets *Energy Procedia* **78** 2070–2075

Acknowledgement

The author would like to thank Universiti Tun Hussein Onn Malaysia, Research Grant TIER 1, VOT H104, Research Management Centre Office UTHM, Department of Construction Management, Faculty of Technology Management and Business for supporting this research

IOP Conf. Series: Materials Science and Engineering **713** (2020) 012043 doi:10.1088/1757-899X/713/1/012043

Examining Risk as Guideline in Design Stage for Green Retrofits Projects: A Review

M S N Lee, M A N Masrom, S Mohamed , KC Goh, N Sarpin and N Manap

Faculty of Technology and Business Management, Universiti Tun Hussein Onn Malaysia, Parit Raja, 84000 Johor, Malaysia
Corresponding author:zakimin@uthm.edu.my

Abstract. Green buildings have generally been adopted as strategy towards sustainable development to reduce greenhouse gas. The implementation of green concept recently has widely spread over the year. However, the positive impact to the environment is still low due to heavily focus on new construction project as compare with retrofitting the existing building. A retrofit practice is not only regaining overall existing building appearance, but the improvement towards energy efficiency is remarkably significant at the expense of huge challenges and risk. As a consequence, retrofit practitioner is confronting an assortment of risks at the early stage of the project. Given that, green retrofit concept is emerging research area to Malaysia construction industry which motivates to determining the risk involve in design stage. This paper aims to review risk in design stage for green retrofit project for commercial buildings through gathering comprehensive literature review. As a result, two main category risks in design stage (technical/quality risk and financial management risk) is determined and extracted to identify risk. These risks are then quantified through analysis to enhance design performance within cost-benefit. The proposed conceptual framework model is expected to assist retrofit practitioner to identify risk and obtain feasible retrofit strategy.

1. Introduction

Green buildings concept is primarily adopted as one of the effective strategies to protect the environmental in many countries [1]. Since construction industry contributes to significant impact to the environment, the introduction to green concept is evolving year by year. However, the greenhouse gas emission is still perceived although there is various practice of green concept available to use by construction sector. It is profoundly that the green concept is heavily focus on new construction project while to greening an existing building is relatively at low level especially in Malaysia [2]. The lifecycle of the existing building is starting to deficiency in operation phase and resulting large energy usage and poor performance of the building [3]. Therefore, green retrofit concept to the existing building is the best possible method to gain the sustainable practices by improving energy and environmental performance, reduce water use, improve the thermal comfort, and overall reduce the noise level [4]. Although green retrofit concept to the existing building improves energy efficiency and overall performance of the building, the principle to gain significant energy efficiency is facing huge uncertainty as the nature of workflow is totally different as compared to the new construction project. Thus retrofitting required systematically design strategy to obtain feasible energy efficiency. However, previous researcher has been focusing on post-retrofit but poor attention to the design risk which lead to the energy performance gap [5].While, [6] stated that risk can influence the project cost, time & quality. Besides that, while government of Malaysia is committed towards promoting energy efficiency through building retrofit, there is limited study on the risk surround in design stage that will influence the final energy efficiency result. In order to fill the knowledge gap, this paper aims to review on risk involve in design stage for green retrofit project for commercial buildings in Malaysia. To accomplish this aim, the following question is addressed: What are the types of risk surround in design stage? What are the potential risks in design stage before proceeding into risk analysis? Therefore, a conceptual framework model is developed to assist stakeholders to understand the embedded risk in design risk that lead to energy performance gap.

2. Research Background

While green retrofit concepts are mostly implementing for developed countries, the study under developing countries is still on-going by most construction industry [7]. In Malaysia, the trend of building retrofit are slow-moving as there are various of uncertainties such as cost, expertise, and design strategies characteristic [8-9]. To date, there are only 24 retrofit project registered under Green Building Index (GBI) which shows the slow trend in the development of existing green building [10]. The design risk in retrofit is crucial to examine proactively and give huge impact to the environment, economic and social perspective [11-12].

A thorough understanding in design risk is ultimately would save project budget while maximizing the energy efficiency in the existing building. According to [13] the retrofit project is facing huge uncertainty compared to new construction project as the design information for existing building regarding specification, duration and cost is seems to be inaccurate. Similarly, the study by [4] show that the risks can be significantly higher if the project team such as architect, engineer and energy service company (ESCO) does not have enough capability in designing under green retrofit project. As consequences, poor retrofit design can lead to weaken energy performance throughout development of the energy model [5]. According to [12] the risk towards the development of green buildings has been identified seven categories which related to time, cost, quality, organization, policy, safety and environment.

Therefore, the previous literature review is primary focusing on the technical or quality risk in retrofit design. Next, the second part is further explored regarding financial management risk in retrofit design. Both literature reviews was conducted to examine the risk in design stage to obtain optimum retrofit design. The completion of the literature review is done by obtained through valid open-access sources database through Google scholar and Scopus journal database. The main keywords to find the related articles are 'green retrofit risk', 'sustainable refurbishment risk', 'renovation risk' and 'design risk'. These keyword is used to narrow down the finding and the term retrofit and refurbishment is interchangeable. As a result, 30 articles were selected that highlight variables which contain significant findings to support the proposed conceptual framework model.

3. Risk in design stage: Technical / quality risks

Generally, the design in retrofitting involves technical and quality risk which consists of four high impact risks that directly influence the decision making in pre-construction stage. According to [14-15] the designer related to green building is essential to understand the principle of green technologies as how it integrate with building system. Similarly, study by [16] revealed the lack of knowledge by related stakeholder such as architect and engineer in green retrofitting design will eventually affecting energy-saving measures which consequently interrupt the decision making.

Through literature review by [17] highlights that lack of design experience will not only related to energy efficiency, but overall safety aspect of retrofit building. This can be seen from previous study which involve multi story Citigroup Center building retrofit safety as engineer in design is miscalculate the wind flow which is supposedly to use welded joint instead of bolt joint and end up structural failures. This is in line with the report by [18] concluded that lack of experience in design team participation will increase the risk of failure in retrofit project. Similarly, the studies by [19-20] indicate that the design strategies in retrofit can provide variation of energy savings result. From the final result in the studies revealed there are significant differences between low impact retrofit and deep retrofit impact in term of energy saving. A simple air leakage to improve moisture in design strategy for retrofit increase the chances of mold and potential of sick building syndrome [21]. There are tendency for retrofitted building performance is below expectation due to uncertainty in design strategy since the theoretical energy saving estimation based on material manufacturing references is not standardized by body of professional accreditation [22]. Lack of standardization in design strategy for energy efficiency will resulting high risk of building performance [23]. On the contrary, based on research analysis by [24], the probabilities of occurrence for design experience risk towards green retrofit project in Sri Lanka is high but does not giving much impact.

Furthermore, the low performance team is seeing reluctant to increase the risks during design stage where it will influence the overall successful of green retrofit project. The contributions of several different stakeholders with varieties set of knowledge and skills will determine on how it will bring significant impact to overall retrofit project objectives [25]. The limited numbers of competent designer will contribute to the design failure in retrofit thus giving an impact to the project cost in the future [26]. Similarly, [27] pointed out that competent project team in design stage such as architect and engineer for retrofitting is necessary to form in based on experience. In addition, competent project team member is needed by team leader to ensure the retrofit project possibly will improve by combination of wide range of expertise throughout existing building in Singapore [28]. Therefore, to avoid possible defective design risk, it is crucial for design team to understand what client need as stated in project brief as well as establishing effective communication and collaboration in each project members to improve decision-making process [29,30].

Moreover, maximizing the energy efficiency in retrofit project require innovation which is related to the design knowledge and skills to draw a preliminary idea within project budget. According to [31,32] highlighted that the material innovation is interrelationship with the cost as each of the selection in retrofit option resulting different outcome to the energy and cost. The study concludes that replacing appropriate material from aluminium to plastic in windows section resulting same energy efficiency but with less cost. However, the availability of green technologies is presenting high impact in design stage as a material might not available and importing from other sources would increase the cost [33,34]. As pointed out by [35], the performance material of green technology is unknown and as a result, it might effect the green requirement due to insufficient minimum standards set by local regulation [36].

Therefore, this shows that the combination of each risk in technical / quality will ultimately bring a huge influence towards the project objective in order to achieve energy efficiency. Before commencing design phase, it is essential for decision makers to follow a line of investigation to every aspect in design stage to ensure feasible design can be achieved. Table 1 illustrates the summary each of the risk category in technical/quality risk to examine the risk that has been identified from the previous recent study on retrofit.

Table 1. Summary of identified risk in technical/quality risk

Author	Risk category	Risk identified
[15], [16], [17], [18], [19], [20], [21], [22]	Design experience risk	• Uncertainty in energy saving. • Lack of knowledge by consultant.
[24], [25], [26], [27], [28], [29]	Team performance risk	• Lack of knowledge and skill in retrofit. • Increase the unnecessary cost. • Lack of competencies. • Lack of communication and coordination.
[3], [14], [30], [31], [32], [33]	Material innovation risk	• Material is not available in local market. • Uncertainty in cost. • Uncertainty in material performance. • Low standard material

4. Risk in design stage: Financial management risks

The combination of optimize design in green retrofit is greatly improving energy efficiency on existing buildings but with the consideration in cost for each of retrofit option will influence the project budget. According to [37] pointed that retrofit design need a reliable financial management so that the project team especially for consultant to determine the right choice of retrofit option. A financial management in construction is related on how project team control the cost with considering of optimum quality and on time within budget allocate by the investor [38]. However, proper planning in financial management is vital as it may influence the design restriction over project budget [6].

Nevertheless, financial management risk in design stage surrounded by two important factors that contributing to the cost variance for retrofit project. In general, cost estimation risk in design stage involves uncertainty in term of cost-benefit over retrofit measures in existing building. In other word, the cost can be varied according to the design variation and since the objective to retrofit the building is to achieving energy efficiency in existing building, the risk is tends to increase the project budget due to failure in design [14, 35]). According to [37] highlighted that inaccurate cost estimation in pre-construction stage will lead to dispute after project is complete. Similarly, a cost estimating during planning stage is crucial for project team as it appear to be significant risk that contributing to the project budget [38].

For instance, a study by [23] revealed adopting green strategies in early stage for green retrofit project is essential to avoid design change that might increase project budget. The information exchange during design stage in green retrofit would consequently influence the quality of the project, thus the final result of the project will facing uncertainty in energy saving [15]. The alternative solution in design stage is evaluate based on economic perspective by project team since each alternative have an impact to the cost, lifecycle and value to the project [39]. However, the uncertainties in cost estimation are also influence by the current condition of the building as it will be too costly due to poor quality built on previous construction [40]. The cost involved in retrofit strategy is ultimately crucial to be planned in early stage to limiting the project cost beyond available budget [33]. Besides that, the cultural heritage building is limiting the retrofit option and alternative solution is needed to ensure the appearance of the building is protected [41]. In addition, the consideration of the country climates in design strategy is giving a great impact to the overall energy consumption since it contribute to the energy consumption such as warm or cold weather [42]. Therefore, the design teams are require to considering the characteristic of the building to ensure the design is accurate while avoiding risk of overrun in cost.

Apart from that, quality risk is appearing to raise an uncertainty toward design stage in retrofitting that reflects to the financial management in pre-construction stage. Since retrofitting for commercial building is normally deal with tenant operation, or even complicated if the project running in business hour, the consideration of these barrier need to be addressed to reduce tenant satisfaction while maintaining project time [43]. Besides that, using high quality product in designing retrofit project is often significantly improving energy efficiency in a long term. In addition, based on [23] revealed the quality risk is giving high impact and most likely influence to future cost in green retrofit project in Sri Lanka. Similarly, according to [4] pointed out the retrofit project will be no difference compared to conventional building project if the quality of energy saving does not meet the requirement in Green Mark Scheme in Singapore. For that reason, benchmarking tool for energy saving comparison is essential to compare each retrofit strategy to minimize design risk, hence increase the overall quality [44].

However, [45] argued through case study in LEEDBOM project where the simulation prediction of energy saving does not meet with the actual situation. Since the actual energy saving does not achieve as what in energy simulation tool plan, a quality in design stage is doubted and it is important for project team to evaluate before proceed to final decision in each retrofit strategies. In addition, throughout case study by [46] suggested that the consideration of inaccuracy factor in energy model is important to increase the confidence and enhance the quality of the design. Therefore, quality risk in design stage is significantly important to address by project team due to uncertainty in cost of the project. In overall, the identification of financial management risk is important to perceive optimum cost-benefit retrofit design. Table 2 illustrates the summary of risk that will be identified according to the previous recent study that vital to the construction industry to assess and control the risk.

Table 2. Summary of identified risk in financial management risk

Authors	Risk category	Risk identified
[23], [41], [32], [15], [39], [38], [40]	Cost estimation risk	• Probability of design change. • Probability of information exchange. • Probability of cost overrun. • Various cost according to climate. • Poor current state of the building. • Limitation of design strategy for heritage building.
[31], [42], [43], [44], [45]	Quality risk	• Tenant operation. • Energy rating requirement. • Inaccurate of energy model.

5. Conceptual framework model of risk in design stage

A definition of conceptual framework can be relate to the network that linked between each concept and grounded theory that presenting comprehensive information [46]. The combination from previous model developed by [48-49,6] can be adopted to present risk in design stage for green retrofit project to achieve target energy performance. The proposed conceptual framework was developed based on previous study as illustrates in figure 1. In overall, the early energy performance assessment can assessed through risk analysis either qualitative or quantitative technique to evaluate the risk. The qualitative analysis method defined as simple rating scale for risk but the evaluation result is vary according to the countries while quantitative analysis method is sophisticated technique which require time consuming [6]. Both analysis can be determined the feasible retrofit measure by quantifying risk to obtain optimum design while maintain cost-benefit of overall project [48-49].

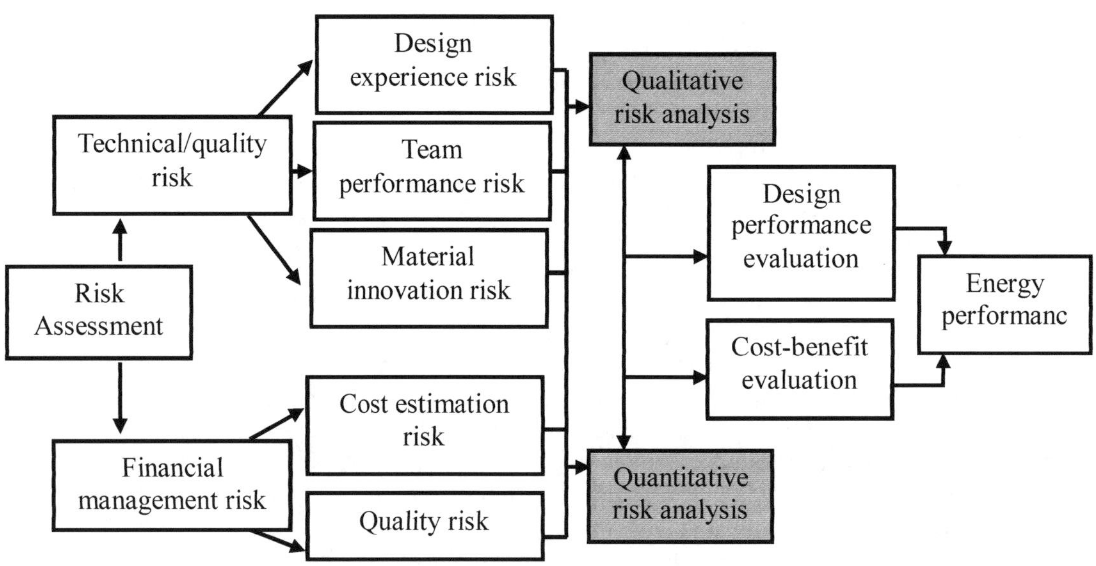

Figure 1: Proposed conceptual framework model of risk in design stage

6. Conclusion

Green retrofit project is facing large uncertainties and risks as early in pre-construction stage. The risk surrounding retrofit project will lead to undetermined energy efficiency performance on existing building thus resulting high cost and low quality in a long term. In general, it is crucial to determine design risk in green retrofit as it possibly influencing cost, time and quality of the project. By considering the design risk in green retrofit project, it is expected that the risk that has been identified will enhance the design strategy to achieve optimum energy efficiency. However, this study is limited to the internal risk in design stage and future study would expand to further investigate for external risk in design stage for green retrofit project. Nevertheless, due to huge challenges and risk in retrofit project in Malaysia, this study will assist construction industry to control the risk by taking an example from previous study on developed countries. Retrofit practitioner can modify each risk in design stage through proposed framework model to quantify which risk that need control to obtain optimum energy performance model. Throughout identifying risks towards pre-construction stage, it will therefore assist retrofit practitioners to response and minimizing potential failure or error in design strategy for retrofit project.

7.0 References

[1] Yu S, Tu Y and Luo C 2011 *Green retrofitting costs and benefits: A new research agenda* Institute of real estate studyies report *Ires* (Singapore: National University of Singapore) pp 1-16

[2] Adeyemi A, Martin D and Kasim R 2014 Improvement of existing buildings for sustainability as againts maintenance and rebuild *Int. Real Estate Research Symposium (Putrajaya)*

[3] Husin S, Zaki N and Husain M 2019 Implementing sustainability in existing building through retrofitting measures *Int. J. of Civ. Eng. and Tech.* **10** 1450-1471

[4] Hwang B G, Zhao X, See Y L and Zhong Y 2015 Addressing risks in green retrofit projects: The case of Singapore *Pro. Man. J.* **46** 76-89

[5] Mohareb E, Hashemi A, Shahrestani M and Sunikka-Blank M 2017 Retrofit planning for the performance gap: Results of a workshop on addressing energy, health and comfort needs in a protected building *Energies* **10** 50-64

[6] Zou P X W, Alam M, Sanjayan J, Wilson J, Stewart R, Sahin O and Ellis-Jones D 2016 Managing risks in complex building retrofit projects for energy and water efficiency *Int. Con. on In. Prod. and Cons Conf (Australia)* pp 1-16

[7] Darko A and Chan A P C 2016 Critical analysis of green building research trend in construction journals. *Hab. Int.* **57** 53-63

[8] Khairi M, Jaapar, A, and Yahya, Z. 2017 The application, benefits and challenges of retrofitting the existing buildings *IOP Conf. Ser.: Mater. Sci. Eng. (Melaka)* (Malaysia: IOP publishing) pp 1-8

[9] Masrom M A N, Rahim M H I A, Ann S C, Mohamed S and Goh K. 2017 A preliminary exploration of the barriers of sustainable refurbishment for commercial building projects in Malaysia *Proc. Eng* **180** 1363-1371

[10] Green Building Index 2019 *GBI Executive Summary* Retrieved on June 2019 from http://new.greenbuildingindex.org/organisation/summary

[11] Ernest K, Ankomah E N, Tengan C and Asamoah R O 2016 Challenges to retrofitting and adaptation of existing building within the major central business district in Ghana *J.of Cons.Pro. Man.and Inn.* **6** 1460-1476

[12] Yang R J, and Zou P X W 2014 Stakeholder-associated risks and their interactions in complex green building projects: A social network model *Buil. and Env* **13** 22-30.

[13] Karim S B . K, Kamal K S, Wahab L A and Hanid M 2007 Risk assessment and refurbishment: The case of Ipoh railway station Perak, Malaysia *Man. in Cons. and Res. Ass. (MICRA)*, **2** 28-29

[14] Durmus-pedini A, and Ashuri B 2010 An overview of the benefits and risk factors of going green in existing buildings *Int.J. of Fac. Man.* **1** 1-15

[15] Ebenehi I Y, Mohamed S, Sarpin N, Masrom M A N, Zainal R and Mohd Azmi M A 2017 The Management Of Building Fire Safety Towards The Sustainability Of Malaysian Public Universities *IOP Conference Series: Material Science and Engineering* IOP Publishing

[16] Bu S, Shen G, Anumba C, Wong A and Liang X 2015 Literature review of green retrofit design for commercial buildings with BIM implication *Sm. and Sust. B. Env.* **4** 188-214

[17] Al-Kodmany K 2014 Green retrofitting skyscrapers: A review *Buildings* **4** 683-710

[18] Lockwood C 2009 *Building Retro.* (United States of America: Urban Building) pp 46-57
[19] Aksamija A 2017 Impact of retrofitting energy-efficient design strategies on energy use of existing commercial buildings: comparative study of low-impact and deep retrofit strategies *J. of Gr. Buil* **12** 70-88
[20] Martinez A and Choi J H 2018 Analysis of energy impacts of facade-inclusive retrofit strategies, compared to system-only retrofits using regression models. *En. and Buil.* **158** 261-267
[21] Lubeck A and Conlin F 2010 Efficiency and comfort through deep energy retrofits: balancing energy and moisture management *Gr. Buil.* **5(3)** 3-15
[22] Jankovic L 2019 Lessons learnt from design, off-site construction and performance analysis of deep energy retrofit of residential buildings *En.and Buil.* **186** 319-338
[23] Eliopoulou E and Mantziou E 2017 Architectural energy retrofit: An alternative building's deep energy retrofit strategy *En. and Buil.* **150** 239-252
[24] Ranawaka I and Mallawaarachchi H 2018 A risk-responsive framework for green retrofit projects in Sri Lanka *Bui. Env. Pro. and Ass. Man.* **7** 1-12
[25] Sanders M D, Parrish K and Mathew P. 2012 *Ready to retrofit : the process of project team selection, building benchmarking, and financing commercial building energy retrofit projects*. Lawrence Berkeley National Laboratory technical report no. 5893E (California: Lawrence Berkeley National Laboratory) pp 1-54
[26] Wilkinson S J, Lamond J, Proverbs D, Sharman L, Heller A and Manion J 2015 Technical considerations in green roof retrofit for stormwater attenuation in the central business district *Struc. Surv.* **33** 36-51
[27] Bertone E, Stewart R A, Sahin O, Alam M, Zou P X W, Buntine C and Marshall C 2018 Guidelines, barriers and strategies for energy and water retrofits of public buildings *J. of Clean. Prod.* **174** 1064-1078
[28] Low S P, Gao S and Tay W L 2014 Comparative study of project management and critical success factors of greening new and existing buildings in Singapore *Str. Sur.* **32** 413-433
[29] Liang X, Peng Y, and Shen G Q 2016 A game theory based analysis of decision making for green retrofit under different occupancy types *J.Clean. Prod.* **137** 1300-1312
[30] Zou P X W, Zhang G and Wang J 2014 Identifying key risks in construction projects : Life cycle and stakeholder perspectives *Int.J. of Cons. Man.* **14** 1-14
[31] Liu Y, Liu T, Ye S and Liu Y 2018 Cost-benefit analysis for energy efficiency retrofit of existing buildings: A case study in China *J. of Clean. Prod.* **177** 493-506
[32] Hwang B-G, Shan M, Phua H and Chi S 2017 An exploratory analysis of risks in green residential building construction projects: The case of Singapore *Sust* **9(7)** 1116-1125
[33] Zuhaib S, Manton R, Hajdukiewicz M, Keane M M and Goggins J 2017 Attitudes and approaches of Irish retrofit industry professionals towards achieving nearly zero-energy buildings *Int. J. of Buil. Path. and Adap.* **35** 16-40
[34] Hwang B-G and Tan J S 2010 Green building project management: obstacles and solutions for sustainable development *J. of Sust. Dev.* **(10)** 21-26
[35] Taal M, Bulatov I, Klemeš J and Stehlík P 2003 Cost estimation and energy price forecasts for economic evaluation of retrofit projects *App. The. Eng.* **23** 1819-1835
[36] Cunningham T 2015 *Cost control during the pre-contract stage of a building project: An introduction.* School of Surveying and Construction Management report. (Dublin: Dublin Institute of Technology) pp 1-26
[37] Cha H and Lee D 2018 Determining value at risk for estimating renovation building projects by application of probability-based fuzzy set theory *J. of As. Arch. and Buil. Eng.* **17** 63-70
[38] Ojo G K and Odediran S J 2015 Significance of construction cost estimating risks in Nigeria *Int. J. of Civ. Engi. and Cons. Sci.* **2** 1-8
[39] Thuvander L, Femenías P, Mjörnell K and Meiling P 2012 Unveiling the process of sustainable renovation *Sust.* **4** 1188-121
[40] Rughooputh R, Cheeneebash J and Mungroo M I 2014 The impacts, benefits and cost of sustainable renovation to existing buildings in Mauritius and setting up of possible

strategies-two case studies *Env. Man. and Sust. Dev.* **3** 109-115

[41] Gao W and Zhang P 2011 *Sustainable renovation projects of residential buildings* Master Thesis (Chalmers: University of Technology) pp-80-90

[42] Mohd-Rahim F A, Pirotti A, Keshavarzsaleh A, Zainon N and Zakaria N 2017 Green construction project: A critical review of retrofitting awarded green buildings in Malaysia *J. of Des.and Bui. Env.* **17** 11-26

[43] Alam M, Phung V M, Zou P X W, and Sanjayan J 2017 Risk identification and assessment for construction and commissioning stages of building energy retrofit projects *Proc. Int. Con. of Adv. of Cons.Man. and R.Est.(Australia)* pp 1-8

[44] Jones B, and Bogus S M 2010 Decision process for energy efficient building retrofits: the owner's perspective *J. of Gr. Buil.* **3** 131-146

[45] Sun X, Gou Z, Lu Y, and Tao Y 2018 Strengths and weaknesses of existing building green retrofits : case study of a LEED EBOM gold project *En.* **11** 1-18

[46] Bozorgi A, and Jones J R 2013 Improving energy retrofit decisions by including uncertainty in the energy modeling process *Arch.Res. Conf. (United States of America)* (Charlotte: University of North Carolina) 415-423

[47] Jabareen Y 2009 Building a conceptual framework: philosophy, definitions and procedure *Int.J. of Qual.Met.* **3** 49-62

[48] Ma Z, Cooper P, Daly D and Ledo L 2012 Existing building retrofits: Methodology and state-of-the-art *En. and Buil.* **55** 889-902

[49] Nielsen A N, Jensen R L, Larsen T S and Nissen S B 2016 Early stage decision support for ustainable building renovation: A review *Buil. and Env* **103** 165-181

Acknowledgement

The author would like to thank Universiti Tun Hussein Onn Malaysia, Research Grant TIER 1, VOT H104 and GPPS, VOT H042, Research Management Centre Office UTHM, Department of Construction Management, and Faculty of Technology Management and Business for supporting this research.

IOP Conf. Series: Materials Science and Engineering **713** (2020) 012044 doi:10.1088/1757-899X/713/1/012044

Assessing the Level of Community Involvement in Post-Disaster Housing Reconstruction and Current Green Design Practices towards Net-Zero Energy Buildings

Sulzakimin M[1], Adaji A A[1,2] , Roy Hazli A[1], Mohammad Syabilee N L[1], Roshartini O[1], Indera Syahrul M R, Rozlin Z[1] and Ebenehi I Y[1,3]

[1] Department of Construction Management, Faculty of Technology Management and Business, Universiti Tun Hussein Onn, Malaysia, 86400 Parit Raja, Batu Pahat, Johor, Malaysia
[2] Department of Building Technology, Kogi State Polytechnic Lokoja, Nigeria
[3] Department of Building Technology, Federal Polytechnic Bauchi, Nigeria
Corresponding author: zakimin@uthm.edu.my

Abstract. The post-disaster housing reconstruction (PDHR) is a process discrete from the construction of housing when circumstances are undisturbed because sensitive measures are put in place following disasters occurrence. Most of the time, the successfulness of emergency relief efforts are usually recorded but the same cannot be said of PDHR projects because they often fail to meet sustainability factors. These days, the topic of sustainable building has become a leading concern in the academic, construction industries, and governmental agencies. Sustainability in the building sector has shifted emphasis towards a new concept called net-zero energy buildings (NZEBs). Data collection was done through a self-administration of structured questionnaires to 257 flood victims involved in the reconstruction projects. Findings indicated that community involvement in resource mobilisation and reconstruction process was insignificant due to misplacement of reconstruction priorities. This is capable of determining the failure of the project as beneficiaries' requirements were not considered, and the construction industry is still unaware of NZEBs. This study is vital as stakeholders beginning with flood victims will have a more resilient housing, the government will be updated with robust strategies to solving resourcing and reconstruction issues, and donors will have value for their money.

1. Introduction

In recent years, the issue of the high level of energy consumption and carbon emission is the basis for the introduction of net-zero energy buildings (NZEBs) concept, and this concept has received great attention by most countries. By having an increasing concern in environmental issues, European Union Parliament are gradually working towards a regulation that encourage the improvement of energy consumption. As well as to reduce the level of carbon emission through the recast of the EU Directive on Energy Performance of Buildings (EPBD) in which all new buildings to be "nearly Zero-Energy" Buildings by 2020 Recast [1].

NZEBs concept according to the study of Satori et. al. [2], is a concept where there is a balance between weighted supply and weighted demand. The aim is to ensure that the sum of all generated energy and all delivered energy is equal in the construction project. Construction projects are depended on substantial improvements on resource management in recent times with diverse techniques, systems and paradigms discovered for various criteria of planning, purchasing, scheduling, optimisation and apportionment of resources. Nevertheless, these findings which are principally concerned with single and non-single projects in the environment of conventional construction appear to unsatisfactorily withstand the preconditions of post-disaster housing reconstruction projects Chang et. al. [3]. From the angle of disaster situations, post-disaster environments are characterised by more complexity, volatility and dynamism in comparison with normal situations.

Reconstruction resources required for the execution of the development or project are almost difficult to obtain when the comparison is made with the conventional construction projects. Nazara et. al. [4] expressed that in the event of large scale disaster, it is likely to observe that the largest portion of local production facilities and supply systems in production industries are damaged. And the construction market seemingly becomes disordered, disputed and highly unfriendly. In a similar vein, Barenstein [5] reported that large scale reconstruction programmes regularly place great demand on natural resources for building materials, and specifically in resettlement or relocation patterns. It is in light of this observation that several known authors suggested that the combination of the authors' observation with sensitive elements can truncate the intended purpose of reconstruction after a disaster. Combination with transportation and energy supply disruption Cho et. al. [6], and pre-existing sequential problems of the local industry Singh et. al. [7], could tangibly worsen the intricacy in obtaining building proficiency and materials Jayasuriya et. al. [8]. Therefore manifesting in project failure and maintainability issues such as project abandonment, quality deficiencies, cost overruns and delay in completion date Boen [9], Steinberg [10]. However, these resourcing issues surfacing in the wake of a disaster can be curbed through proactivity.

Recognising housing reconstruction as an essential element of post-disaster recovery initiatives in developing countries, and thus, the need arises to understand what makes it effective and what does not. According to Chang et. al. [3], reconstruction planners and building practitioners are accordingly needed to unceasingly adapt themselves to the emerging resource circumstances after a disaster. Giving adequate attention to key factors or components in resource mobilisation regarding housing reconstruction after a disaster will be an appropriate strive in the right direction as calls for such undertaking have been made in recent times. The acknowledgement of Chang et. al. [3] revealed that regardless of the relationship existing between resource mobilisation and reconstruction performance, insufficient efforts have been made research wise on resourcing interconnected issues after a disaster or even proposal methods to enhance the management of resources for reconstruction projects.

This brings to bear the imperativeness to study the resource mobilisation strategy adopted in post-disaster housing reconstruction projects in relation to community level of involvement with the view of proffering adequate measures to minimize the unpleasant scene faced by flood victims in the study area. Thereby providing strategies to achieve a positive result oriented environment in future disaster recovery interventions.

To make this a reality, following the 2012 flooding in Kogi State-Nigeria, a survey was conducted on the flood affected community in 2018 to identify the level of community involvement in resource mobilisation and reconstruction process in the PDHR projects embarked upon by the Kogi State government in 2013. Although this study draws on the perceptions and experience of the beneficiaries in the housing reconstruction intervention done in the study area (Lokoja, Nigeria).

2. Resource Mobilisation for Post Disaster Housing Reconstruction

Post-disaster housing reconstruction that is not appropriately planned and instigated has the potentials to create more exposures in the disaster-stricken community. This stresses a warning that the importance of resource mobilisation for post-disaster housing reconstruction cannot be overemphasised.

As stated in the introductory section of this study, some possible consequences of inappropriate resource management in reconstruction projects motivated scholars to further elucidate on the prominent roles it plays in the cycle of disaster reconstruction and recovery process.

Several researchers have recognised resource mobilisation as a driving force needed for a successful reconstruction. Manavazhi et. al. [11] emphasised that material suspensions, especially long- lead items, are notable factors responsible for delays in construction and can initiate other relevant matters relating to resources also. Furthermore, it was identified that the unavailability of adequate plants or equipment reduces the proficiency of physical construction Yeo et. al. [12].

In the post-disaster reconstruction conditions, the mobilisation of resources is influenced by certain factors. In the submission of Singh [13], the five factors influencing resource mobilisation in reconstruction are the ranking order of works or tasks, the capacity to pool resources, the prime period of procurement, the prevailing contractual rapports and transportation in and out of the region affected by the disaster. In a related development, Singh et. al. [7] added that the mobilisation of resources is determined by the governance policies and strategies laid down by the decision makers or professionals responsible for the reconstruction.

Furthermore, the integration of renewable energy system into green buildings is crucial in reducing the demand for a non-renewable energy source. Reconciliation of materials with local desires for rebuilding houses is paramount. Barakat [14] and O'Brien et. al. [15] said the consideration of economic and environmental repercussions of several types of materials used for buildings is significant. Several humanitarian organisations including Non-governmental organisations, International non-governmental organisations like International Federation of Red Cross [16] and agencies of the United Nations such as UNISDR [17], UNISDR [18] stressed on the worth of mobilisation of resource in reconstruction projects of post- Indian Ocean tsunami. There exist a range of role facsimiles that proved that practitioners of post-disaster reconstruction gained by engaging pre-emptive resourcing schemes and planning Mitchel [19] and taking cognisance of potentials weaknesses and bottlenecks while sourcing resources Singh et. al. [7]. In spite of this, the added significance of integrating resourcing actions and policies into the post-disaster reconstruction context or field has not been proven, perhaps for three prominent motives:

i. Complications in transforming proficiencies and lessons learnt from disaster-impacted nations into other definite contexts and actions.

ii. Restricted understanding together with the nonexistence of tools in connecting mobilisation of resources with native (community or local) economic, social and environmental context.

iii. Construction professionals fail to see resource mobilisation for post-disaster reconstruction as a fundamental point in disaster recovery but often consider it as procurement obligation.

Considering the salient role of resource mobilisation in housing reconstruction and recovery intervention, this study deems it fit to explore the issues responsible for the unpleasant conditions of the flood victims on the general process of the intervention. This is systematically achieved by following the steps explained under research method of this study.

3. Study Area

Lokoja is the capital of Kogi State since 1991, the town and its suburbs constitute Kogi Local Government Area (LGA). The town and its suburb share a land boundary with 3 LGAs (Ajaokuta, Adavi and Koton-Karfi) respectively. It lies on latitude 7049' N and longitude 6044' E at an altitude of 45-125m, on the western bank of the Niger River, close to its confluence with the Benue River. The annual rainfall is about 1150 mm which usually begins in March and reaches its peak from June to September, while the dry season begins at about November Richard et. al. [20]. Being the State capital and LGA headquarter, the majority of residents are civil servants at various levels of government (Federal, State and Local).

4. Research Method

A quantitative approach was adopted for this study. The survey tool used was a structured questionnaire that was designed based on the factors derived from the literature. To adequately improve the questionnaire, both pre-test and pilot study was conducted with 3 consulting experts and 7 professionals in the built environment, and a small sample of the flood victims respectively involved in post-disaster housing reconstruction. It indicates that the respondents understand the questionnaire contents they are given, and there is face validity. The questionnaires were administered by self to the respondents who are the beneficiaries of the housing reconstruction development for field survey. The respondents were asked to rank their level of involvement in resource mobilisation and reconstruction process of the housing reconstruction from their perspective on a five-point Likert scale from 1 to 5, where 1 symbolises 'very Less' and 5 represents 'very high'. Statistical analysis was used to determine the key parameters in resource mobilisation and reconstruction process as it affects the respondents. One sample t-test of the mean was conducted using SPSS, based on the sample's ratings. This was to check if the factors identified within the questionnaire were important in affecting community level of involvement in post-disaster reconstruction situations. By using SPSS descriptive statistics, a ranking of the parameters as perceived by respondents was carried out to identify major factors which significantly affected resource mobilisation of reconstruction projects after the flood as relates to community involvement.

5. Research Findings and Discussion

The questionnaire was administered and retrieved in one week time as in Chacon [21]. The 400 sets of questionnaires were administered to flood victims in the study area. A total number of 301 questionnaires constituting 75% were retrieved from the respondents. Some forty-four (44) questionnaires were discarded due to incomplete response. Therefore, a total of 257 constituting 64% response rate were used in the analyses after removing incomplete cases, and data screening for outliers (table 1).

Table 1. Questionnaire administration.

Questionnaires	Respondents	
	Number	Percent
Administered	400	-
Collected	301	75%
Screened	257	64%

5.1. To identify the level of community involvement in resource mobilisation.

A mean ranking was conducted to assess the level of community involvement in resource mobilisation in the PDHR, as observed by the respondents in the study area and presented in table 2 below. It was observed that resource mobilisation through public financial assistance with 3.32 mean was ranked 1st as the most commonly used among the mobilisation methods, followed by resource mobilisation through international donors with 3.15 mean which was ranked 2nd. Next was gradual disbursement of bulky sums of reconstruction funding with 3.11 mean and ranked 3rd. The resource mobilisation through monetary donations from the private sector has 2.89 mean and was ranked 4th, while resource mobilisation through households' funding resources was ranked 5th. The ranking for Management of available reconstruction materials was 6th with a mean of 2.34. However, initiation of construction materials and skills that are suitable for the environment has a mean of 2.25 and ranked 7th. Supervision and coordination of dispersal of reconstruction resources were ranked 8th with a mean of 2.17, then choosing of skilled workers whether they suit local conditions ranked 9th with a mean of 2.11, where the risk assessment involving the gathering of data and the preparation of the loss estimates has 1.94 mean and ranked 10th. The least mobilisation method identified was interaction and decision making between many groups and institution, including households with 1.80 mean and ranked 11th in table 2. This is an indication that community affected were not involved in the resource mobilisation strategy for the housing reconstruction that was meant specifically for them. This result has faulted the recommendations of previous researches. The first decision to be embraced in post-disaster housing reconstruction is to fully comprehend the local context of the affected community (beneficiaries) through proper or effective or holistic needs assessment and surveys to provide correct assistance to satisfy the beneficiaries Lloyd-Jones [22], Khasalamwa [23], Mannakkara et. al. [24]. It is required that the reconstruction and recovery guidelines should then be designed based on local conditions to shape, support and preserve the culture and heritage of the affected people (beneficiaries). Other supporting researches of non-involvement of beneficiaries in housing reconstruction projects after a disaster are eminent. According to Davidson et. al. [25] and Andrew et. al. [26], it is difficult to integrate the community in the preliminaries and management of the PDHR interventions. Similarly, the findings reconcile with previous studies that community engagement in post-disaster reconstruction including resources mobilisation was not observed in the initial stages of the British Red Cross Society (BRCS) Maldives post-tsunami recovery program Lawther [27].

Table 2. Level of community involvement in resource mobilisation in PDHR.

SN	Variables	Mean	Std. Deviation	Rank
1	Resource mobilisation through public financial assistance	3.32	1.454	1
2	Resource mobilisation through international donors	3.15	1.497	2
3	Gradual disbursement of bulky sums of reconstruction funding	3.11	1.357	3
4	Resource mobilisation through monetary donations from the private sector	2.89	1.231	4
5	Resource mobilisation through households' personal funding resources	2.37	1.116	5
6	Management of available reconstruction materials	2.34	1.050	6
7	Initiation of construction materials and skills that are suitable for the environment	2.25	.992	7
8	Supervision and coordination of dispersal of reconstruction resources	2.17	.829	8
9	Choosing of skilled workers suits local conditions	2.11	.865	9
10	Risk assessment involving the gathering of data and the preparation of the loss estimates	1.94	.905	10
11	Interaction and decision making between many groups and institution including households	1.80	.994	11

5.2. To identify the level of community involvement in the reconstruction process.

A mean ranking was conducted on the general reconstruction process used for the PDHR as observed by the respondents in the study area. With regards to table 3, it was observed that Technology-driven housing reconstruction approaches with a lower mean of 2.40 were ranked 1st as the major reconstruction process used. Followed by Procedures of requiring construction approvals with 2.28 mean and was ranked 2nd. Next was Decisions on whether to repair, replace or abandon affected houses with 2.21 mean and was ranked 3rd. The Compliance to building code with 2.14 mean which was ranked 4th. Whereas, Lead agency establishment of a working environment in which local, national, and international agencies effectively coordinate reconstruction process with 2.12 mean was ranked 5th and the Lead agency coordination of reconstruction policy decisions with 2.08 mean was ranked 6th. It was revealed that Level of gathering information to understand the impact of the flooding on the environment with 2.07 mean ranked 7th and Building consent application with 2.06 mean was ranked 8th. Then Level of gathering information to understand the housing need of individuals ranked 9th with a mean of 1.95 where the Consultation and planning among all stakeholders during the reconstruction process with 1.82 mean was ranked 10th. Furthermore, the Use of community leaders in charge of reconstruction with 1.75 mean was ranked 11th. The least reconstruction process identified was Use of housing owners in charge of reconstruction with 1.72 mean and was ranked 12th in table 3. This clearly shows that priorities were shifted from the most important stakeholders who are the beneficiaries. The result obtained under this section is similar to the previously discussed results above. It shows that community engagement is far low beyond expectation and also showing complete dominance over the approach of reconstruction by the donors who happen to be the State government. The result obtained under here is in line with previous research that the Nigerian government embraced a contractor driven-approach for the construction of new buildings for flood victims due to the need for quick delivery Bilau et. al. [28]. Holding to the fact that, this method side-lined owners or beneficiaries of housing reconstruction projects, thus denying them reasonable and personal inputs into the housing and other infrastructural provisions meant for them and their family members. The result gained supports from previous researches that top-bottom reconstruction approach is completely controlled or operated by the donor agencies where affected families were only donated houses as witnessed in Sri Lankan government specifications in a new location Ratnayake et. al. [29], Andrew et. al. [26]. Though, there are no precisely definite approaches, suggesting that applied methodologies have to be tailored to the particular post-disaster context is imperative Barakat [14]. However, housing reconstruction is an intricate process and if not approached correctly can weaken State institutions and develop forms of economic and social marginalisation. Richard et. al. [20] recommended that beneficiaries should be involved in Post-disaster housing reconstruction process to enable them to tailor the design and construction of the housing facility to meet their specifications and desires. This opportunity to participate in the housing reconstruction process will ensure satisfaction as well as sustainability of the project. Furthermore, it provides a more complete and structurally integrated solution to PDHR and at the end, produces a more satisfied and empowered homeowner. Thiruppugazh [30] stated that even though the owner-driven model cannot be relied upon in terms of performance in reducing future vulnerability, it is better than the donor-driven model as witnessed in the study area. Abuzayan et. al. [31] promoted the involvement of beneficiaries in the reconstruction efforts so that their needs could be taking into consideration. It is hence not surprising that Vahanvati et. al. [32] advocated the owner-driven model when compared with the donor-driven model. In making a justification for community-driven model, Shafique et. al. [33] established that non-involvement of communities in developing countries contributed to the failure often experienced in post-disaster housing reconstruction and that community participation usually increases the success rate. Based on this, this study joins voice to the established references and call for an all-inclusiveness of the end- users into reconstruction activities as this has to do with their existence and socio-economic, cultural and spiritual livelihoods.

Table 3. Community involvement in reconstruction processes in PDHR.

SN	Variables	Mean	Std. Deviation	Rank
1	Technology-driven housing reconstruction approaches	2.40	.848	1
2	Procedures of requiring construction approvals	2.28	.919	2
3	Decisions on whether to repair, replace or abandon affected houses	2.21	.786	3
4	Compliance to building code	2.14	1.141	4
5	Lead agency establishment of a working environment in which local, national, and international agencies effectively coordinate the reconstruction process	2.12	.787	5
6	Lead agency coordination of reconstruction policy decisions	2.08	.730	6
7	Level of gathering information to understand the impact of the flooding to the environment	2.07	.786	7
8	Building consent application	2.06	.748	8
9	Level of gathering information to understand the housing needs of individuals	1.95	.777	9
10	Consultation and planning among all stakeholders during reconstruction process	1.82	.911	10
11	Use of community leaders in charge of reconstruction	1.75	.863	11
12	Use of housing owners in charge of reconstruction	1.72	.884	12

6. Conclusion

Since renewable energy system is crucial towards green healthcare building and there is a lot of benefits that can be gained through these systems, the renewable energy systems should be utilised properly when designing post-disaster housing reconstruction. Therefore, the application of renewable energy technologies to support green building can reduce the dependency on non-renewable energy and thus, a greener environment can be achieved. There was a misplaced of priority in the housing reconstruction process and resource mobilisation strategies adopted in the study area. The results reflected the misappropriation of priorities through non-involvement of the beneficiaries which, on the long run tends to truncate the success of the project. Beneficiaries' satisfaction which is the prerequisite to the sustainability of post-disaster housing reconstruction, is not in sight as reported in this study. Therefore, this project may be close to failure or can be seen as a colossal failure since the users' requirements were not adequately taken into consideration as required by the World Bank guidelines for housing reconstruction Jha et. al. [34]; the New Sendai Framework for Disaster Risk Reduction 2015-2030, and the "Build Back Better" principles Mannakkara et. al. [24]. Hence, this study suggests the disaster reconstruction and recovery processes that comply with the guidelines mentioned above. Whose aim is

IOP Conf. Series: Materials Science and Engineering **713** (2020) 012044 doi:10.1088/1757-899X/713/1/012044

making the beneficiaries the central focus of the planning, procuring, scheduling, monitoring and allocation of resources mobilisation and reconstruction process. It should be noted that reconstruction is beyond physical housing provision for the affected community but should be seen as an opportunity to incorporate risk reduction measures and increase the resilience of the community to future hazards and climate change effects.

7. References

[1] Recast E P B D 2010 Directive 2010/31/EU of the European Parliament and of the Council of 19 May 2010 on the energy performance of buildings (recast) *Official J. of the Eur. Union* **18** 06-14

[2] Sartori I, Napolitano A and Voss K 2012 Net zero energy buildings: A consistent definition framework *Energy and bld.* **48** 220-232

[3] Chang, Y, Wilkinson S, Brunsdon D, Seville E and Potangaroa R 2011 An integrated approach: Managing resources for post-disaster reconstruction, Dis. **35** 739–765

[4] Nazara S and Resosudarmo B P 2007 *Aceh-Nias reconstruction and rehabilitation: progress and challenges at the end of 2006* (Tokyo: Asian Development Bank Institute) Discussion Papers **70** pp 1-12

[5] Barenstein J D 2010 *Who Governs Reconstruction? Changes and Continuity in Policies, Practices and Outcomes, Rebuilding After Disasters From Emergency to Sustainability* (New York: Routledge) pp 149–176.

[6] Cho S, Gordon P, Moore I, James E, Richardson H W, Shinozuka M and Chang S 2001 Integrating transportation network and regional economic models to estimate the costs of a large urban earthquake *J. of Regional Sci.* **41** 39-65

[7] Singh B and Wilkinson S 2008 Post-disaster resource availability following a wellington earthquake: aggregates, concrete and cement *In I-Rec 2008 Building Resilience: Achieving Effective Post-Disaster Reconstruction, Christchurch* (New Zealand) pp 1-11

[8] Jayasuriya S and McCawley P 2008 Reconstruction after a major disaster: Lessons from the Post-Tsunami Experience in Indonesia, Sri Lanka, and Thailand, ADBI working paper series 4-8

[9] Boen T 2008 Reconstruction of houses in Aceh, three years after the December 26, 2004 Tsunami *Proc. Int. Conf. on Earthquake Engineering and Disaster Mitigation* (Beijing, China) pp 14-15.

[10] Steinberg F 2007 Housing reconstruction and rehabilitation in Aceh and Nias, Indonesia-rebuilding lives *Hab. Int.* **31** 150-166

[11] Manavazhi M R and Adhikari D K 2002 Material and equipment procurement delays in highway projects in Nepal *Int. J. of Prjt. Mgt.* **20** 627-632

[12] Yeo K T and Ning J H 2006 Managing uncertainty in major equipment procurement in engineering projects *Eur. J. of Opt. Resh.* **171** 123-134

[13] Singh B 2007 *Availability of Resources for State Highway Reconstruction : A Wellington Earthquake Scenario* M.SC. Thesis (New Zealand: University Of Auckland) pp 56-62

[14] Barakat S 2003 Housing reconstruction after conflict and disaster, Humanitarian Policy Group *Network Papers* **43** 1-40

[15] O'Brien D, Ahmed I and Hes D 2008 Housing reconstruction in Aceh: relationships between house type and environmental sustainability *In Building Abroad: Procurement of Construction and Reconstruction Projects in the International Context* (Canada: Université de Montréal) pp 361-372.

[16] International Federation of Red Cross 2006 Report of the International Federation of Red Cross (Unpublished work style) Geneva, Unpublished

[17] United Nations International Strategy for Disaster Reduction (UNISDR) 2005 Living with Risk: A global review of disaster reduction initiatives, **1** United Nations Publications

[18] United Nations International Strategy for Disaster Reduction (UNISDR) 2006 Living with Risk: A global review of disaster reduction initiatives, **1** United Nations Publications

[19] Mitchell J K 2004 Re-conceiving recovery *Proc. of the New Zealand Recovery Symposium, Ministry of Civil Defence & Emergency Management Wellington* (New Zealand) pp 47-68

[20] Richard J, Adejo A, James R and Luqman O 2017 Post-disaster housing reconstruction (PDHR) in Ibaji and Lokoja, Kogi State-Nigeria *Malaysian J. of Civil Eng.* **29** 194-215

[21] Chacon T C 2009 *A Survey of School Psychology Faculty Members' Knowledge, Skills, and Attitudes Regarding Distance Education and Distance Education Technologies* (Albany: The State University of New York) pp 13-18

[22] Lloyd-Jones T 2006 *Mind the Gap! Post-Disaster Reconstruction and the Transition From Humanitarian Relief* (London, UK: RICS) pp 30-32

[23] Khasalamwa S 2009 Is 'Build back better'a response to vulnerability? Analysis of the post-Tsunami humanitarian interventions in Sri Lanka *Norwegian J. of Geo.* **63** 73-78

[24] Mannakkara S and Wilkinson S 2013 Post-disaster legislation for building back better *Constr. Law J.* **29** 495-506

[25] Davidson C H, Johnson C, Lizarralde G, Dikmen N and Sliwinski A 2007 Truths and myths about community participation in post-disaster housing projects *Hab. Int.* **31** 100-115

[26] Andrew S A, Arlikatti S, Long L C and Kendra J M 2013 The effect of housing assistance arrangements on household recovery: an empirical test of donor-assisted and owner-driven approaches *J. of Housing and the Built Env.* **28** 17-34

[27] Lawther P M 2010 Community involvement in the post-disaster reconstruction-case study of the British Red Cross Maldives recovery program *Int. J. of Str. Prop. Mgt.* **13** 153-169

[28] Bilau A A, Witt E and Lill I 2016 Housing reconstruction following the 2012 Nigerian floods: was it built back better? *Proc. of the CIB World Building Congress* (Finland) pp 165-176.

[29] Ratnayake R and Rameezdeen R 2008 Post-disaster housing reconstruction: comparative study of donor driven vs owner-driven approach *Proc. of CIB W89 Int. Conf. on Building Education and Research (BEAR)* (Sri Lanka) pp 33-40

[30] Thiruppugazh, V C 2010 *When People are Involved, Beyond Shelter, Architecture and Human Dignity* pp 71-81

[31] Abuzayan K, Whyte A and Bell J 2015 The significant need for end users' involvement in post-conflict conditions In *5th Int. Conf. on Building Resilience, Newcastle, Australia 15-17th July 2015* (Australia: New South Wales) pp 534-1-11

[32] Vahanvati M and Beza B B 2015 Owner-driven reconstruction in India: A case-study of Kosi river floods in Bihar In *5th Int. Conf. on Building Resilience, Newcastle, Australia 15-17th July 2015* (Australia: New South Wales) pp 367-1

[33] Shafique, K and Warren C M J 2015 Significance of community participation in the success of post-natural disaster reconstruction project–evidence from a developing country *5th Int. Conf. on Building Resilience, Newcastle, Australia 15-17th July 2015* (Australia: New South Wales) pp 533-1-12

[34] Jha A K, Barenstein J D, Phelps P M, Pittet D and Sena S 2010 *Safer Homes, Stronger Communities: A Handbook for Reconstruction after Natural Disasters* (Washington DC: The World Bank) pp 1-15

Acknowledgments
The authors would like to thank Ministry of Education (MOE), Pejabat Pendaftar UTHM, Office of Research, Innovation, Commercialisation and RMC UTHM Grant Research (Code H104:TIER 1 and GPPS H042) and FPTP UTHM. Furthermore, our sincere appreciation for the Department of Building Technology, Kogi State Polytechnic, Lokoja-Nigeria to participate in this project.

IOP Conf. Series: Materials Science and Engineering **713** (2020) 012045 doi:10.1088/1757-899X/713/1/012045

Designer's Confidence and Attitude towards Designing-for-Construction-Safety (DfCS) Implementation

N H Abas[1], R Abd Rahman[2], H Mohammad[3], M H Rahmat[1]

[1] Jamilus Research Centre, Faculty of Civil and Environmental Engineering, Universiti Tun Hussein Onn Malaysia, 86400 Batu Pahat, Johor, Malaysia
[2]Iswarabena Sdn Bhd, No. 101 Jalan Tun Sri Lanang, 81900 Kota Tinggi, Johor
[3] Centre for Diploma Studies, Universiti Tun Hussein Onn Malaysia, Hab Pendidikan Tinggi Pagoh, KM1, Jalan Panchor, 86400 Pagoh, Muar, Johor, Malaysia.
Corresponding author: nhaslin@uthm.edu.my

Abstract. The construction industry is known as a hazardous industry, which has been overwhelmed with accidents for a long time. One of the new strategies to reduce the number of accident in Malaysia is through emphasizing the safety responsibility on the parties involved during the design stage. This study investigated the confidence level of designing-for-construction-safety (DfCS) and the perceptions of the hindrance by focusing on the role of designers and its implementation among construction designers. The quantitative approach had been used, in which questionnaire survey forms were distributed via email and self-administered to forty (40) construction designers (i.e. civil and construction engineer, quantity surveyor, architect) in Johor. The findings have revealed that the respondents are confident of their ability to design for construction safety, though they are relatively less willing to implement the concept. This study demonstrates that significant effort is required to promote the benefits of DfCS in improving the safety performance of the construction industry so that it is widely accepted and implemented by the construction stakeholders.

1. Introduction

In many countries in the world, the construction industry is renowned as the most hazardous industry, despite its importance in contributing to the economic growth of the country. In many states, the fatality and incident rates of construction industry are noticeably higher than other sectors. Construction workers are exposed to many dangerous substances from the construction materials, unpleasant dust, unpredicted working environment and other physical hazards [1]. Therefore, construction industry stakeholders need to control safety and health risks at the construction site, and further its safety performance.

In Malaysia, the number of fatalities that occurred in the construction industry is the highest among other sectors, as proven by SOCSO and DOSH reports every year. One of the new strategies to reduce the number of accident in Malaysia is through emphasizing the safety responsibilities on the parties involved during the design stage. Cooke [2] and Gambatese et al. [3] suggested that the lousy safety performance of construction can be improved through preventing accidents and reducing uncertainty before it happens. This statement indicates that, in order to improve safety performance in the construction industry, designing-for-construction-safety or commonly known as DfCS, would be the best option. The influence of the design of construction accidents has been studied in decades. Trethewy et al. [4] argued that designers could directly affect safety and health outcomes through the selection of procurement systems, preparation of contract documentation, construction process sequencing, and contract tenure decisions. It could also indirectly affect safety and health outcomes through the selection of frame types, material specifications and the design itself. Smallwood [5] interviewed general contractors in South Australia and revealed that half of them agreed that "design" is a significant factor that affects health and safety negatively. Majority of these contractors stated that there is a need for architect and design engineers to attend safety education subject at the university or technical college. Meanwhile, a 1991 European study found that about 60% of workplace accidents can be traced back to decisions made before construction work began [6]. More recent research on DfCS is focusing on integrating the concept of DfCS with the use of modern tool such as Building Information Modelling (BIM) [7-10].

In the UK, some rules emphasize the responsibility of designers during the design stage which is the Design and Construction Management (CDM) Regulations 1997 (now known as the CDM Regulations 2007 [11]). This rule requires the construction players, especially designers, to accept national laws to formalize a process to make sure that construction site safety is considered during the design process. However, according to Summerhayes [12], since the introduction of the legislation to construction design professionals in 1997, the designers are still doubtful to comply with the legislation.

In Malaysia, the Master Plan for Occupational Safety and Health in Construction Industry (MPOSHCI) 2005-2010 has introduced this concept as a positive recommendation to improve occupational health and safety performance of the construction industry [13]. A number of OHS-related positive actions related to 'designing construction safety' and incorporating education into the concept of OHS, and providing guidance to clients to conduct safety and health design examinations provided before construction [13-14]. Moreover, the Guidelines on Occupational Safety and Health in Construction Industry (Management) 2017 has been introduced, which assigning specific OHS obligations for construction design professionals [15]. However, it is doubtful that Malaysian construction designers adequately understand how to identify, assess and control OHS risks in their designs. This assumption is based on the 'nature of the job responsibility', in which designers are usually not involved in or responsible for OSH. Moreover, the number of studies that investigate the designer's perspective on the issue is minimal [16]. Although several studies have been conducted to reach the perspectives and insights of the designer's on DfCS concept [16-18], there are still weaknesses in resolving their perspectives on these issues in the Malaysian context. Therefore, this study attempts to contribute to filling this gap by investigating the designers' confidence and attitude towards DfCS concept implementation.

2. DfCS Concept

Traditionally, the parties responsible for preventing injury to workers are contractors and subcontractors. However, the current trend in the construction industry is now focusing on the duty of all construction participants, including owners, designers, contractors, and safety practitioners, to create a safe work environment. In the designing for safety concept, designers assess the identified risks of their designs to the workers on construction sites and eliminate or modify the design to eliminate hazards or minimize the impact of the risks.

The design phase of a new construction project is important since decisions taken in this phase tend to have a significant impact on the construction execution. The degree of uncertainty about the future to be encountered is considered high at this phase. In response to this situation, risk management executed in this phase can play a vital role to avoid uncertainty and risk before the construction phase takes place. Szymberski [19] postulated that by incorporating safety earlier in the project schedule, more considerable influence could be exerted (figure 1). By including construction site safety as a consideration (along with production, quality, project scope, etc.) early in the project's life cycle, one has a more exceptional ability to influence construction site safety positively. Therefore, changes to the design process are needed to enable identification and proper management of decisions with the potential to create a risk to workers on the construction site. Designers are increasingly being pressured to assess the occupational safety and health risk their designs create for the construction worker, and consequently to eliminate or reduce these within their plans.

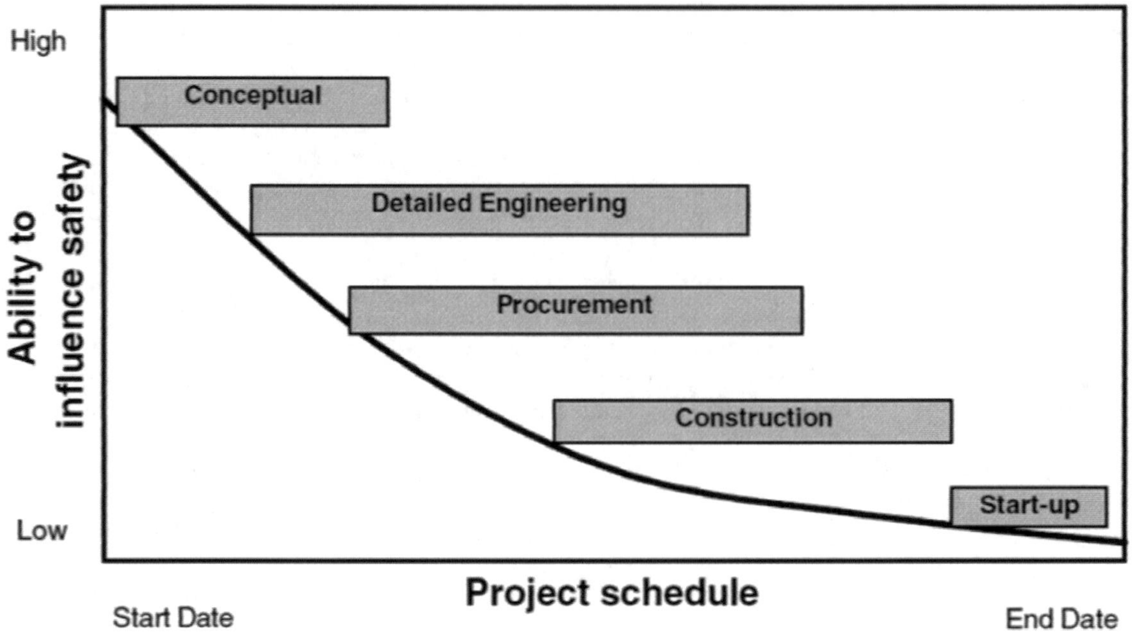

Figure 1. Time/safety influence curve [Szymberski, 19]

2.1 Designer's role in influencing safety through design

There are many design solution suggestions proffered by researchers for better safety performance in construction. Behm [20] gives an example of how the features of the permanent facility can influence the safety of the constructors:

> "When the height of parapet walls is designed to be 42'', the parapet acts as a guardrail and enhances safety. However, when roof perimeters do not contain permanently designed-in fall protection features (i.e. anchorage points and guardrails), worker safety is compromised."

Gambatese et al. [21] suggested the factors that will affect the implementation of designing for safety. The factors include designer's knowledge and acceptance on the concept, designer education and training, designer motivation to implement the concept, ease of implementation of the concept, availability of implementation tools and resources, competing for design/project objectives and design criteria/physical characteristics. The designers may benefit from the implementation of the concept in many ways and that the implementation of this concept may also have positive impacts on construction safety.

3. Methods

3.1 Procedure

This study employed a questionnaire survey method to attain participants' (in particular, the designers) insights on DfCS concepts. The questionnaire form consisted of 3 parts, which are: i) Part 1 – questions concerning the participants' background and awareness on DfCS; ii) Part 2 – designer's confidence in implementing DfCS; and iii) designer's attitude towards DfCS. This study adopted Yacizi and Dulaimi's [16] instrument in developing the questionnaire. For general knowledge questions (Part 1), participants were required to answer the questions in either 'Yes' or 'No' for Part 2 and 3, the questionnaire used the following scales to attain designer's answers on their confidence and attitude towards DfCS: 1 (strongly disagree), 2 (disagree), 3 (neutral), 4 (agree) and 5 (strongly agree).

Before distributing the original questionnaire forms, a pilot study was conducted to validate the questionnaire instrument by sending out the questionnaire to 3 panels of the expert. The experts were among those who had more than 15 years of experience in design activities. All panels agreed with the questionnaire instrument, which denotes that the adopted questionnaire is suitable to be used in the Malaysian environment.

Next, the actual questionnaire form had been distributed to the participants in the form of a self-administered questionnaire. The participated designers were group into architects, civil and structural (C&S) engineers, and quantity surveyors who were currently engaged in design activities for construction projects in Johor. The total number of participants was 40, and the summary of the participants is depicted in Table 1. These surveys were done during working hours, and each respondent took not more than 15 minutes.

Table 1. Number of participants

Type of participant	Number or participant
Architect	13
C&S engineer	16
QS	11
Total	40

3.2 Measures for data analysis

In order to analyze the questionnaire questions for Part 2 and Part 3, this study applied the Average Index method, which provides means to ascertain respondents' validation of a criterion [22]. The average or the mean index of a criterion is evaluated using the following Equation:

$$Average\ or\ Mean\ Index = \sum aiXi/\sum Xi \qquad (1)$$

Where, a = constant, weighing factor for i, {i = 1, 2, 3.... n};

X_i = frequency of respondent

4. Results

4.1 Background and awareness on DfCS of participants

The survey form was distributed to the participants, as depicted in Table 1. The summary of the participants' background is shown in figure 2. As can be seen in the figure, the majority of participants possessed a bachelor's degree and had between 3 to 10 years working experience.

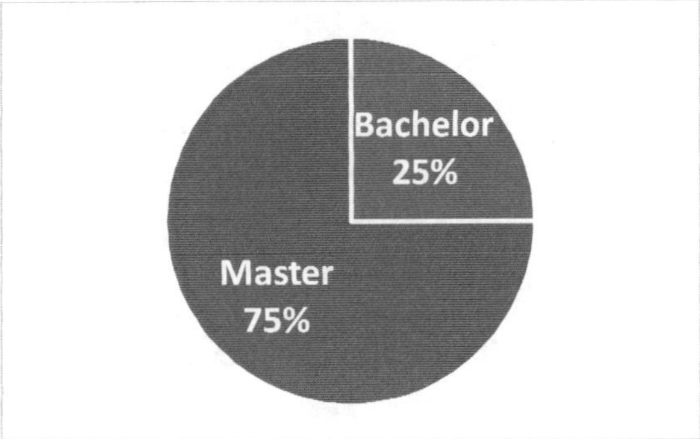

(a)

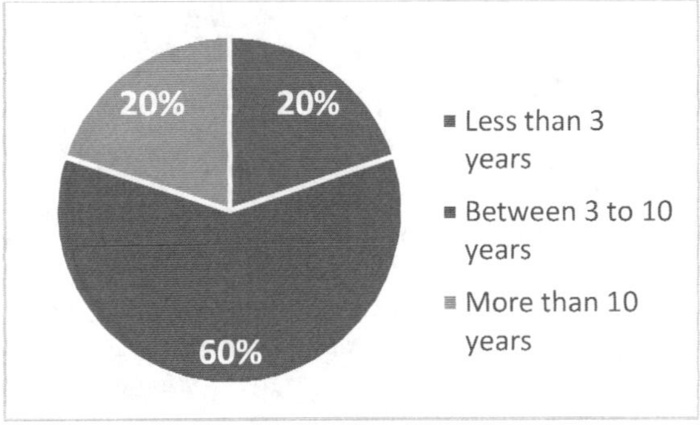

(b)

Figure 2. (a) Level of education, (b) working experience (years)

With regards to awareness on DfCs, majority of total participants are aware of the concept and implement DfCS in their work, as shown in figure 3. However, the selected participants from architects seemed to be less aware of DfCS and did not implement the concept in their work.

IOP Conf. Series: Materials Science and Engineering **713** (2020) 012045 doi:10.1088/1757-899X/713/1/012045

Figure 3. Participants' awareness on DfCS

4.2 Designer's confidence on DfCS concept

In attaining the confidence on DfCS among construction designers, the designers were required to rate their level of confidence based on the 5-point Likert scales given. The results of the designers' answers are shown in Table 2.

Table 2. Average index analysis of respondents' answers on their confidence towards DfCS concept

No.		Architect	C&S engineer	QS	Average for all group
1.	**Influence on end user's safety** *I believe that decisions at the design stage of a project influence end-users' safety and & health*	4.615*	4.438	4.272	4.442
2.	**Influence on workers safety** *I believe that decisions at the design stage of projects can influence workers' safety & health on construction sites*	4.538*	4.563*	4.182	4.427
3.	**Familiarity with the concept** *I am familiar with the concept of designing to enhance construction worker safety & health*	3.538	4.313*	4.000*	3.950
4.	**Capability of identifying hazardous design** *I am capable of identifying potentially hazardous design solutions*	3.769	4.688*	4.636*	4.364
5.	**Awareness on safe design solutions** *I am aware of design solutions that could be used to address construction workers safety & health*	3.923	4.563*	4.273*	4.253
6.	**Safe design development skills** *I am skilled at developing design solutions that can help to avoid hazardous situations on construction sites*	3.923	4.188*	3.909	4.006
7.	**Awareness on OSH legislation** *I fully understand the relevant legislation regarding construction safety in the Malaysia*	3.923	4.313*	3.818	4.018
8.	**Awareness on safe design tools** *I am aware of design tools that could be used to address construction workers safety & health*	3.846	4.600*	4.182	4.422
9.	**Formal education and training on the DfCS concept** *Addressing construction workers safety & health was included in my formal education and training*	3.769	4.750*	4.545*	4.355

**Indicates higher mean value scored than mean*

In general, the participants have high confidence in the DfCS concept. From the data obtained, it could be seen that the top 3 of the designers' confidence in DfCS scored by selected participants are questions number 1, 4, and 9. The lowest scored value is question number 3. It is interesting to note that the C&S engineers possessed higher confidence in implementing DfCS concept, in which they scored the highest value for almost all questions. This is followed by QS. Even though there was no question scored by them as the highest value, most of their scores are higher than the total mean value for all participants. The results are in-line with the awareness of DfCS questions (see Fig. 2) where most selected participants from C&S engineers and QS are aware of the concept and implementing the DfCS in the works. Architects seemed to possess the lowest confidence level in implementing DfCS concept. The majority of the selected architect participants were not aware and not implementing the concept. The results provide evidence that civil and structural engineer were more informed about DfCS concept. C&S engineer is more related to the concept as they do the designing, assessing, and inspecting structures in the construction field, which they know how to prevent the accident from happening. However, there were also several questions in which designers somehow agreed to each other. For example, they believe that the decision at the design stage of a project influences the end user's (questions 1) and the workers' health and safety (question 2). This is in agreement with Dulaimi & Yazici's [16] finding which stated that construction designers are aware of the workers' health and safety.

4.3 Designer's attitude towards DfCS concept

Meanwhile, in obtaining the construction designers' attitude towards DfCS, the designers were required to rate their perceptions on DfCS concept based on ten questions. The results of the designers' answers are shown in Table 3.

Table 3. Average index analysis of respondents' answers on their attitude towards DfCS concept

No.		Architect	C&S engineer	QS	Average for all group
1.	I believe that to design safety & health would limit my creativity as a designer.	3.154	3.125	2.909	3.062
2.	Addressing construction worker safety & health at the design phase will increase my liability exposure.	3.385	3.25	3.091	3.242
3.	I – as a designer- lack motivating forces (legal, contractual, or regulatory) to adopt design for health and safety methods	3.077	3.063	3.091	3.35
4.	I believe that designing for safety & health will increase project cost.	3.038	3.75	3.818	3.535
5.	I believe that designing for safety & health will increase project duration.	3.769	3.875	3.909	3.851
6.	I believe that designing for safety will decrease number of accidents on site.	3.692	4.125	4.091	3.969
7.	I believe that designing for safety will enhance project performance	3.769	3.688	4.812	4.090
8.	The traditional contractual relationship between the designer and contractor impedes me from formally addressing construction worker safety & health in the design phase.	3.462	3.4	3.636	3.499
9.	OSH policies and procedures mean little to me when it comes to get tasks completed.	3.077	1.938	2.455	2.49
10.	As a designer I am not expected to take OSH into consideration when designing a project.	2.846	2	2.364	2.403

From Table 3, it could be seen that in general, the respondents' attitude towards implementing DfCS concept were varied. In some questions, they showed high commitment towards achieving the concept, such as questions number 6 and 7. In these questions, the respondents agreed that the DfCS concept

IOP Conf. Series: Materials Science and Engineering **713** (2020) 012045 doi:10.1088/1757-899X/713/1/012045

could impact project safety performance as a whole. Besides, they also showed positive commitment towards considering DfCS in their designs, as shown in questions number 9 and 10. However, some impediments might hinder the designers from implementing the DfCS concept in their projects, in the aspects of project-specific (increase in the project cost and duration), industry-specific (traditional contractual relations), and country-specific (lack of supportive legal). This is in agreement with Yacizi & Dulami's [16] findings. They indicated that there should be a means to overcome the perceived barriers so that the DfCS concept can be accepted and practiced extensively by all designers. Besides that, the results showed high scores for 'limitation on the creativity; and 'increased liability,' which might hinder the designers from implementing the DfCS concept. Consistently with Yacizi & Dulami's (2015) finding, the respondents showed their disagreement on their roles to implement the concept in current practices.

5. Discussion and Conclusion

This study aims to investigate the confidence and attitude of designers towards the DfCS concept implementation. In general, the results show that the respondents are confident of their ability to design for construction safety. However, they are relatively less willing to implement the concept. It is also worth noting that the respondents' confidence and attitude were varies depending on their professional background. This study provides evidence that an enormous effort is needed to promote the benefits of DfCS in improving the safety performance of the construction industry. This will ensure that it is widely accepted and implemented by the construction stakeholders.

The results of this study may be limited by the number of respondents for each group of professionalism. Thus, further research is needed to increase the number of respondents for each professional group so that the results can be generalized. Moreover, it is also worthwhile to extend the study to involve respondents in other states or regions within Malaysia.

6. References

[1] Snashall D 1990 Safety and health in the construction industry. BMJ, 301(6752), 563-564

[2] Cooke R M 1997 Uncertainty modelling: examples and issues *Safety Science* 26 49–60

[3] Gambatese J A, Behm M and Rajendran S (2008) Design's role in construction accident causality and prevention: Perspectives from an expert panel *Safety Science* **46**(4) 675-691

[4] Trethewy, R. W., Atkinson, M., & Falls, B. (2003). Improved hazard identification forcontractors in the construction industry *Journal of Construction Research* **4**(1) 71–85

[5] Smallwood J and Venter D 2012 The Influence of Project Managers on Construction Health and Safety in South Africa *The Australian Journal of Construction Economics and Building* **2**(1) 57-69

[6] Furst P G 2010 Construction injury prevention through design. Retrieved on November 20, 2018, from http://www.designforconstructionsafety.org/Documents/construction

[7] Khan N, Ali A, Skibniewski M J and Lee D Y 2019 Excavation Safety Modeling Approach Using BIM and VPL *Advances in Civil Engineering* **15** 1-15

[8] Guo H, Yu Y, Weishen Z and Yan L 2016 BIM and Safety Rules Based Automated Identification of Unsafe Design Factors in Construction *Procedia Engineering* **164** 467-472

[9] Zhang S, Sulankivi K, Kiviniemi M and Teizer J 2015 BIM-based Fall Hazard Identification and Prevention in Construction Planning *Safety Science* **72** 31-45

[10] Kim K and Teizer J 2014 Automatic design and planning of scaffolding systems using building information modelling *Advanced Engineering* Informatics **28**(1) 66-80

[11] CDM Regulation 2007 The Construction (Design and Management) Regulations 2007 Retrieved 8 December 2014

[12] Summerhayes S D 2010 *Design Risk Management Contribution to Health and Safety* (New Jersey: Wiley Blackwell)

[13] CIDB 2009 Occupational Safety & Health Master Plan for Malaysia 2015. (Kuala Lumpur: Construction Industry Development Board)
[14] Ministry of Human Resources 2005 Master Plan for Occupational Safety and Health in Construction Industry 2005 – 2010 Retrieved on July 2019 from: https://www.cidb.gov.my/cidbweb/images/pdf/OHSAS2006v2.pdf
[15] Department of Occupational Safety and Health (DOSH) 2017 *Guidelines on Occupational Safety and Health in Construction Industry (Management) 2017* (Malaysia: DOSH)
[16] Yazici E and Dulaimi M 2014 Understanding designing for construction safety: the interaction between confidence and attitude of designers and safety culture *Arch. Eng. and Design Mgmt.* **11** 325-227
[17] Morrow S, Hare B and Cameron Iain 2016 Design engineers' perception of health and safety and its impact in the design process *Eng., Const. and Arch. Mgmt.* **23**(1) 40-59
[18] Creaser W 2008 Prevention through Design (PtD) Safe Design from an Australian Perspective *Journal of Safety Research* **39**(2)131-134
[19] Szymberski R 1997 Construction Project Safety Planning *TAPPI Journal* 80(11) 69–74
[20] Behm M 2004 Establishing the link between construction fatalities and disabling injuries and the design for construction safety concept (Oregon State University: PhD Thesis)
[21] Gambatese J A, Behm M and Hinze J W 2005 Viability of designing for construction worker safety *Journal of Construction Engineering and Management* **131** 1029–1036
[22] Abd. Majid M Z and McCaffer R 1997 Assessment of Work Performance of Maintenance Contractors in Saudi Arabia *Journal of Management in Engineering* **13**(5) 91-91

Acknowledgements

This paper is funded by STG Vot U638 and a part of FRGS Vot 1579 research. The author would like to thank the Universiti Tun Hussein Onn Malaysia (UTHM) and Ministry of Higher Education, Malaysia for the support and funding given towards completing the research.

AUTHOR INDEX

Abas, N H 275, 359, 400
Abas, Norhaslinda 296
Abass, F 65, 351
Abdullah, Siti Radziah 24
Adaji, A A 374, 392
Adnan, S H 40
Ahwang, A 206
Akasah, Z A 143
Alaskar, A 32
Albadri, Wael M. 367
Ali, Noorwirdawati 24, 135
Alisibramulisi, A 40
Alshammari, Faisal 324
Anuar, M N A Wahee 40
Ayop, S S 40
Ayuni, S A 48
Azam, Raden Maizatul Aimi Mohd 152
Azhar, A T S 48
Azim, A 48
Aziman, M 48
Azlan, Ahmad Aiman Bin Ahmad 90
Bakar, A Abu 235
Bakar, Hasniza Abu 152
Boon, Koh Heng 177
Boon, K 169
Bosro, M Z M 97
Budiea, A M A 221
Chan, C-M 113
Chan, Chee-Ming 160
Chik, T N Tuan 235
Chik, T N T 333
Chow, Ming Fai 90, 121
Choy, L J 105
Chuan, D Y Eng 169
Dalmantias, E 228
Daniel, B David 359
Dedi, E 228
Deraman, Rafikullah 296
Ebenehi, I Y 392
Elgadi, A A 65, 351
Embi, M R 343
Ezanee, Shaiifa Ema 284
Fateh, Mohd Ashraf Mohd 284
Ghafar, N Abd 169, 206
Ghafar, NorHayati Abd 177
Ghing, Tong Yean 152
Goh, KC 384
Goh, Lyn Dee 214
Guntor, N A A 18
Guntor, Nickholas Anting Anak 367
Hadipramana, J 198, 235
Hairuddin, M 307
Hamid, A B Abd 343
Hamid, Noor Azlina Abdul 24
Haniza, M M 57
Haris, S 228
Haron, Zaiton Binti 324
Hasan, Nik Nurul Nadia Nik 214
Haslinda, A Nor 307
Hasmori, M F 275
Hasmori, Muhammad Fikri 296
Hassan, N A 105
Hazli, A Roy 392
Hazli, R 374
Higazey, M 32
Hussein, Z H 105
Ibrahim, A 243
Idham, M K 105
Idham, M M 57
Iman, Afif 82
Iman, M. A. 97
Irwan, J M 1
Ismail, L H 65, 351
Ismail, Noorli 152
Ismail, Ruqayyah 214
Ismail, R 243
Izwan, A R Mohd Hilmi 374
Jaini, Z M 198, 206
Jamelloddin, Z 40
Jaya, R P 105
Johan, S F S M 113
Johari, H H Ahmad 235
Jomali, Muhammad Azzim Mohmed 152
Kamarudin, A F 221, 235
Kamarulzaman, N A 40
Kariya, N 275
Kelundapyan, R 359
Khairuddin, Siti Amirah Azra 259
Khalid, F S 1, 18
Khalid, Faisal Sheikh 135
Khalijah, Y Siti 307
Klufallah, Mustafa 296
Kong, S W R 266
Krishnamoorthy, R R 243
Kudus, Sakhiah Abdul 252
Latip, Arisya Abdul 24

Lau, L T ... 266
Lee, M S N ... 384
Majid, M A ... 206
Majid, M ... 40
Manap, N ... 384
Mannan, M A ... 143
Masrom, M A N ... 374, 384
Mohamad, N ... 82, 97
Mohamed, S ... 384
Mohammad, H ... 275, 400
Mokhatar, S N ... 198, 221, 235
Musa, M K ... 235
Mushina, Jalal ... 177
mushina, j ... 169
Mushina, Wissam ... 177
mushina, W ... 169
Mustafa, Anis Nazirah ... 259
Mutalib, N A N A ... 221
Mydin, Azree ... 82
Nagapan, Sasitharan ... 296
Nagapan, S ... 359
Naji, Nahla ... 367
Nawawi, A H ... 317
Nazri, Fadzli Mohamed ... 135
Noh, M S Md ... 221
Nordin, Nurul Syakeera ... 160
Norfarahayu, K ... 307
Norhalim, A F ... 206
Osman, M H ... 40
Othman, Muhammad Amzar ... 135
Othman, N H ... 18
Othman, N ... 1
Panjaitan, T H ... 73
Peter, P. ... 143
Phan, D T ... 266
Putra, N ... 73
Raain, Fasya ... 82
Rahman, Azerai Ali ... 214
Rahman, Norashidah Abd ... 259
Rahman, R Abd ... 400
Rahmat, M H ... 400
Rahmat, N ... 317
Rashid, R Abd ... 127
Razak, A H N Abdul ... 343
Ridzqo, I F ... 73
Roshartini, O ... 392
Rosidan, Muhammad Amin Khalili ... 121
Rossam, Mohd Luqman ... 259
Rozlin, Z ... 392
Saaidin, S H ... 18
Salleh, Norhafizah ... 24
Salleh, N ... 40
Samad, A A A ... 82, 97
Samad, H Abdul ... 127
Sarpin, N ... 384
Segaran, V C ... 359
Seman, M Abu ... 198
Seow, TW ... 374
Sfoog, Ehab Hamad ... 367
Shaghaghi, Taleb Moradi ... 7
Shahidan, Shahiron ... 24, 135
Shahidan, S ... 18, 40
Shamsudin, Shamrul Mar ... 135
Shannag, M J ... 32
Shokravi, Hooman ... 7
Siang, Alvin John Lim Meng ... 367
Sofia, Anis ... 82
Soh, N M Z Nik ... 143
Soh, Nik Mohd Zaini Nik ... 152
Sugiura, Kunitomo ... 252
Suhaini, N A ... 275
Sulzakimin, M ... 374, 392
Susanto, D ... 73
Suzuki, Yasuo ... 252
Syabilee, N L Mohammad ... 392
Syahrul, M R Indera ... 392
Taib, M Z Mohd ... 343
Tambichik, M A ... 97
Teddy, T ... 1
Thamrin, R ... 228
Tong, Y G ... 359
Wahab, I A ... 65, 351
Wong, S Y ... 266
Xian, P C ... 333
Yaacob, H ... 105
Yahya, Khairulzan ... 324
Yassin, N I M ... 40
Yeo, Kevin J.T ... 186
Yeoh, David E.C ... 186
Yeoh, David ... 177
Yi, Sim Sy ... 367
Yunus, Riduan ... 296
Yusoff, N A ... 333
Yusuf, N ... 275
Zakariah, Hanah ... 284
Zakwan, F A A ... 243
Zakwan, Fariz Aswan Ahmad ... 214
Zin, Ahmad Faizul Md ... 296
Zuki, S S Mohd ... 235
Zuki, Sharifah Salwa Mohd ... 135

Institute of Physics
Dirac House, Temple Back
Bristol BS1 6BE UK

ISSN: 1757-8981
ISBN 978-1-7138-2572-2

9 781713 825722

CURRAN ASSOCIATES INC.
proceedings.com

5th International Workshop on Heat/Mass Transfer Advances for Energy Conservation and Pollution Control (IWHT2019)

Journal of Physics: Conference Series Volume 1369

Novosibirsk, Russia
13 - 16 August 2019

IOP
Institute of Physics